Springer Series on

ATOMIC, OPTICAL, AND PLASMA PHYSICS 27

Springer-Verlag Berlin Heidelberg GmbH

Springer Series on
ATOMIC, OPTICAL, AND PLASMA PHYSICS

The Springer Series on Atomic, Optical, and Plasma Physics covers in a comprehensive manner theory and experiment in the entire field of atoms and molecules and their interaction with electromagnetic radiation. Books in the series provide a rich source of new ideas and techniques with wide applications in fields such as chemistry, materials science, astrophysics, surface science, plasma technology, advanced optics, aeronomy, and engineering. Laser physics is a particular connecting theme that has provided much of the continuing impetus for new developments in the field. The purpose of the series is to cover the gap between standard undergraduate textbooks and the research literature with emphasis on the fundamental ideas, methods, techniques, and results in the field.

27 **Quantum Squeezing**
By P.D. Drumond and Z. Ficek

28 **Atom, Molecule, and Cluster Beams I**
Basic Theory, Production and Detection of Thermal Energy Beams
By H. Pauly

29 **Polarization, Alignment and Orientation in Atomic Collisions**
By N. Andersen and K. Bartschat

30 **Physics of Solid-State Laser Physics**
By R.C. Powell
(Published in the former Series on Atomic, Molecular, and Optical Physics)

31 **Plasma Kinetics in Atmospheric Gases**
By M. Capitelli, C.M. Ferreira, B.F. Gordiets, A.I. Osipov

32 **Atom, Molecule, and Cluster Beams II**
Cluster Beams, Fast and Slow Beams, Accessory Equipment and Applications
By H. Pauly

33 **Atom Optics**
By P. Meystre

34 **Laser Physics at Relativistic Intensities**
By A.V. Borovsky, A.L. Galkin, O.B. Shiryaev, T. Auguste

35 **Many-Particle Quantum Dynamics in Atomic and Molecular Fragmentation**
Editors: J. Ullrich and V.P. Shevelko

36 **Atom Tunneling Phenomena in Physics, Chemistry and Biology**
Editor: T. Miyazaki

Series homepage – springeronline.com

Vols. 1–26 of the former Springer Series on Atoms and Plasmas are listed at the end of the book

P. D. Drummond Z. Ficek (Eds.)

Quantum Squeezing

With 52 Figures

Springer

Professor Peter D. Drummond
Dr. Zbigniew Ficek
University of Queensland, Physics Department,
St Lucia 4072, Queensland, Australia
e-mail: drummond@physics.uq.edu.au
ficek@physics.uq.edu.au

ISSN 1615-5653

Library of Congress Cataloging-in-Publication Data applied for
Quantum squeezing/P.D. Drummond, Z. Ficek (eds.).
p. cm. – (Springer series on atomic, optical, and plasma physics, ISSN 1615-5653 ; 27)
Includes biblographical references and index.

DOI 10.1007/978-3-662-09645-1

1. Squeezing light. I. Drummond, P. D. (Peter D.), 1950- II. Ficek, Zbigniew. III. Series
QC446.3.S67Q36 2004 535–dc22 2003060694

springeronline.com

Originally published by Springer-Verlag Berlin Heidelberg New York in 2004
MyCopy version of the original edition 2004

Typesetting: Camera-ready copies by the author
Cover concept by eStudio Calmar Steinen
Cover design: *design & production* GmbH, Heidelberg

Printed on acid-free paper SPIN 10731027 57/3141/YL - 5 4 3 2 1 0
www.springer.com/mycopy

Preface

The concept of squeezing is intimately related to the idea of vacuum fluctuations, once thought to place an absolute limit to the accuracy of measurement. However, vacuum fluctuations are not unchangeable. By recognizing that these quantum fluctuations always occur in two complementary observables, physicists have been able to make an intriguing trade-off. Reduced fluctuations in one variable *can* be realized – at the expense of increased fluctuations in another, according to Heisenberg.

This Heisenberg 'horse-trade' – originally predicted by theorists – was first accomplished experimentally by R. Slusher in 1985. Since then, the various techniques and applications of quantum squeezing have metamorphosed into a central tool in the wider field of quantum information. This book is a summary of the main ideas, methods and applications of quantum squeezing, written by those responsible for some of the chief developments in the field.

The book is divided into three parts, to recognize that there are three areas in this research. These are the fundamental physics of quantum fluctuations, the techniques of generating squeezed radiation, and the potential applications.

Part I of the book, giving the fundamentals, is arranged as follows.

- Chapter 1 introduces the basic ideas about what squeezing of quantum fluctuations is from the quantized free-field perspective. This chapter establishes the definitions and notations used throughout.
- Chapter 2 explains how to quantize radiation in a dielectric, which is the basic technique that is used to make squeezed radiation.
- Chapter 3 explains how to quantize interfaces, where squeezed light is input or output through the dielectric boundaries.

Part II treats methods of generating quantum squeezed radiation.

- Chapter 4 starts with the most commonly used techniques in which squeezed radiation is generated using nonlinear optics. This covers both intra-cavity parametric squeezing and fiber soliton squeezing.
- Chapter 5 describes how lasers with the right kind of pumping may produce squeezed light, typically with squeezing in the intensity.
- Chapter 6 explains how various feedback techniques can be used to also produce non-classical radiation, and describes the distinction between 'in-loop' and 'out-of-loop' or external squeezing.

Part III treats the applications of squeezed radiation.

- In Chap. 7 the ideas of using squeezed and anti-bunched radiation sources for improved communications and measurement are introduced.
- Chapter 8 details applications to spectroscopy of two-level atoms, including the possibility of sub-natural-linewidth spectroscopy.
- Chapter 9 extends this treatment to the spectroscopy of three-level atoms, in which anomalous inversions and pumping rates may be observed.
- Chapter 10 explains how entangled squeezed light can be used to carry out tests of quantum-mechanics and cryptography, based on the original entangled state proposal of Einstein, Podolsky and Rosen.

We emphasize that this is a rapidly changing research field, and the areas of applications, in particular, are the most rapidly changing of them all. While every effort is made to reference the most recent publications, not every development could be covered, for space reasons. However, it is hoped that the fundamental physics issues treated here will be of use to researchers and students in quantum information and related research fields.

Brisbane,
May 2003

Peter D. Drummond
Zbigniew Ficek

Contents

List of Contributors

Peter L. Knight
Blackett Laboratory, Imperial College,
London SW7 2AZ,
United Kingdom
p.knight@imperial.ac.uk

Vladimír Bužek
Institute of Physics, Slovak Academy of Sciences,
Dúbravská cesta 9, 842 28 Bratislava, Slovakia
and
Faculty of Informatics, Masaryk University, Botanická 68a,
602 00 Brno, Czech Republic
fyzibuvl@savba.sk

Mark Hillery
Department of Physics,
Hunter College of the City University of New York,
695 Park Ave, New York,
New York 10021, U.S.A.
mhillery@hunter.cuny.edu

Bernard Yurke
ATT Bell Laboratories, Murray Hill,
New Jersey 07974, U.S.A.
yurke@bell-labs.com

Peter D. Drummond
Department of Physics,
The University of Queensland,
Brisbane, QLD 4072, Australia
drummond@physics.uq.edu.au

Timothy C. Ralph
Department of Physics, The Australian National University,
Canberra, ACT 0200, Australia
and
Department of Physics,
The University of Queensland,
Brisbane, QLD 4072, Australia
ralph@physics.uq.edu.au

Howard M. Wiseman
Department of Physics,
The University of Queensland,
Brisbane, QLD 4072, Australia
and
School of Science, Griffith University,
Nathan, Queensland 4111,
Australia
H.Wiseman@gu.edu.au

Horace P. Yuen
Department of Electrical and Computer Engineering,
Department of Physics and Astronomy,
Northwestern University, Evanston,
IL 60208, U.S.A.
yuen@ece.northwestern.edu

Stuart Swain
School of Mathematics and Physics,
The Queen's University of Belfast,
Belfast BT7 1NN,
United Kingdom
s.swain@qub.ac.uk

Zbigniew Ficek
Department of Physics,
The University of Queensland,
Brisbane, QLD 4072, Australia
ficek@physics.uq.edu.au

Margaret D. Reid
Department of Physics,
The University of Queensland,
Brisbane, QLD 4072, Australia
margaret@physics.uq.edu.au

Part I

Fundamentals

1 Squeezed States: Basic Principles

P. L. Knight and V. Bužek

A squeezed field is one with phase-sensitive quantum fluctuations, which, at certain phase angles are *less* than those of a perfectly coherent field, or of "no" field at all (the vacuum state). Early theoretical work [1–7] (for a historical note on discovery of squeezed states see [8]) in the 1960's and 1970's led in the mid-1980's to the understanding that quantum fluctuations can be reduced below the vacuum-state (shot-noise) limit in various forms of nonlinear optical interactions (see [9–11]). Other authors in this volume will address those realizations and characterize the production and detection of such squeezed states. Our aim in this chapter is to provide the basic theoretical framework which has been utilized to describe squeezed quantum fields.

In this chapter we first describe how a light field may be understood as a collection of harmonic oscillators which is quantized in the standard manner and which fluctuates as a consequence of the basic non-commutability of the canonical field variables (Sect. 1.1). We discuss the mathematical description of quadrature squeezing in terms of the underlying Weyl–Heisenberg algebra. We next turn to how states of quantum fields can be described in terms of phase-space quasi-probability density distributions. In particular, we will utilize the Wigner function to analyze the role of quantum interference in suppression of quantum fluctuations (Sect. 1.2). Our basic quantum building blocks are the Fock number states and coherent states, and we described how each behaves in phase space. Using the insight so derived, we discuss how squeezed states may be described in phase space and characterized by quadrature variances. In Sect. 1.3 we show how interference in phase space may be seen as the underlying cause of squeezing, and we build non-classical states by superposing coherent states.

1.1 Light Fields and Harmonic Oscillators

Let us consider electromagnetic waves in a region of space where the transverse part of the current density is equal to zero ($\boldsymbol{J}_T = 0$). Utilizing the Coulomb gauge $\boldsymbol{\nabla} \cdot \boldsymbol{A}(\boldsymbol{r}, t) = 0$, the vector potential $\boldsymbol{A}(\boldsymbol{r}, t)$ satisfies the wave equation (see for instance [12,13])

$$\nabla^2 \boldsymbol{A}(\boldsymbol{r},t) + \frac{1}{c^2}\frac{\partial^2 \boldsymbol{A}(\boldsymbol{r},t)}{\partial t^2} = 0 . \tag{1.1}$$

To make our discussion more specific, let us imagine the spatial region in which the electromagnetic field exists to be a box (cavity) of side length L. The vector potential then can be expanded in a Fourier series

$$\boldsymbol{A}(\boldsymbol{r},t) = \sum_{\boldsymbol{k}} \left\{ \boldsymbol{A}_{\boldsymbol{k}}(t) e^{i\boldsymbol{k}\cdot\boldsymbol{r}} + \boldsymbol{A}^*_{\boldsymbol{k}}(t) e^{-i\boldsymbol{k}\cdot\boldsymbol{r}} \right\} , \tag{1.2}$$

where components of the wave vector $\boldsymbol{k} = \{k_x, k_y, k_z\}$ take discrete values $k_j = 2\pi\nu_j/L$ with $\nu_j = 0, \pm 1, \pm 2 \ldots$. The Coulomb-gauge condition is satisfied if $\boldsymbol{k}\cdot\boldsymbol{A}_{\boldsymbol{k}}(t) = \boldsymbol{k}\cdot\boldsymbol{A}^*_{\boldsymbol{k}}(t) = 0$, i.e. there are two independent directions of $\boldsymbol{A}_{\boldsymbol{k}}(t)$ for each $\boldsymbol{k}$. The different Fourier components of $\boldsymbol{A}$ are independent and must separately satisfy the field equation

$$\frac{\partial^2}{\partial^2 t}\boldsymbol{A}_{\boldsymbol{k}}(t) + \omega_{\boldsymbol{k}}^2 \boldsymbol{A}_{\boldsymbol{k}}(t) = 0 , \tag{1.3}$$

where $\omega_{\boldsymbol{k}} = c|\boldsymbol{k}|$.

We now introduce an effective position and momentum associated with the given (cavity) electromagnetic mode. To this end, let us evaluate the canonical energy of the cavity normal mode specified by the wave vector $\boldsymbol{k}$. The solution of the simple harmonic equation (1.3) can be taken as $\boldsymbol{A}_{\boldsymbol{k}}(t) = \boldsymbol{A}_{\boldsymbol{k}} \exp(-i\omega_{\boldsymbol{k}} t)$, and the complex vector potential then becomes

$$\boldsymbol{A}_{\boldsymbol{k}}(\boldsymbol{r},t) = \sum_{\boldsymbol{k}} \left\{ \boldsymbol{A}_{\boldsymbol{k}} e^{-i(\omega_{\boldsymbol{k}} t - \boldsymbol{k}\cdot\boldsymbol{r})} + \boldsymbol{A}^*_{\boldsymbol{k}} e^{i(\omega_{\boldsymbol{k}} t - \boldsymbol{k}\cdot\boldsymbol{r})} \right\} . \tag{1.4}$$

The cycle-averaged energy content of a single mode $\boldsymbol{k}$ is

$$\overline{\mathcal{E}_{\boldsymbol{k}}} = \frac{1}{2}\int_{\text{cavity}} \left(\epsilon_0 \overline{\boldsymbol{E}^2_{\boldsymbol{k}}} + \mu_0 \overline{\boldsymbol{H}^2_{\boldsymbol{k}}} \right) d^3V , \tag{1.5}$$

where the bars denote a cycle average. The electric and magnetic fields associated with the mode can be expressed through $\boldsymbol{A}_{\boldsymbol{k}}$ as (here we remind ourselves that in the Coulomb gauge the longitudinal part of the electric field is equal to zero, i.e. $\boldsymbol{E}_L = 0$, so that both $\boldsymbol{E}$ and $\boldsymbol{H}$ have only transverse components):

$$\begin{aligned}
\boldsymbol{E}_{\boldsymbol{k}}(\boldsymbol{r},t) &= -\partial \boldsymbol{A}/\partial t = i\omega_{\boldsymbol{k}} \left\{ \boldsymbol{A}_{\boldsymbol{k}} e^{-i(\omega_{\boldsymbol{k}} t - \boldsymbol{k}\cdot\boldsymbol{r})} - \boldsymbol{A}^*_{\boldsymbol{k}} e^{i(\omega_{\boldsymbol{k}} t - \boldsymbol{k}\cdot\boldsymbol{r})} \right\} , \\
\boldsymbol{H}_{\boldsymbol{k}}(\boldsymbol{r},t) &= \frac{1}{\mu_0}\nabla \times \boldsymbol{A} \\
&= \frac{i}{\mu_0}\boldsymbol{k} \times \left\{ \boldsymbol{A}_{\boldsymbol{k}} e^{-i(\omega_{\boldsymbol{k}} t - \boldsymbol{k}\cdot\boldsymbol{r})} - \boldsymbol{A}^*_{\boldsymbol{k}} e^{i(\omega_{\boldsymbol{k}} t - \boldsymbol{k}\cdot\boldsymbol{r})} \right\} .
\end{aligned} \tag{1.6}$$

From these equations it follows that the magnitudes of $\boldsymbol{E_k}$ and $\boldsymbol{H_k}$ are related as

$$|\boldsymbol{E_k}| = \sqrt{\frac{\epsilon_0}{\mu_0}}|\boldsymbol{H_k}| \ , \tag{1.7}$$

where ϵ_0 and μ_0 are the electric permittivity and magnetic permeability of free space, respectively. Now we can express the cycle-averaged energy content of the single mode $\boldsymbol{k}$ as

$$\overline{\mathcal{E}_{\boldsymbol{k}}} = 2\epsilon_0 V \omega_{\boldsymbol{k}}^2 \boldsymbol{A_k} \boldsymbol{A_k^*} \ , \tag{1.8}$$

where $V = L^3$.

The vector potential $\boldsymbol{A_k}$ associated with a given mode $\boldsymbol{k}$ can be replaced by generalized dimensionless quadratures $x_{\boldsymbol{k}}$ and $y_{\boldsymbol{k}}$ in accordance with the transformation

$$\begin{aligned} \boldsymbol{A_k} &= \frac{1}{2}\sqrt{\frac{\hbar}{2\epsilon_0 V \omega_{\boldsymbol{k}}}}\,(x_{\boldsymbol{k}} + i y_{\boldsymbol{k}})\,\varepsilon_{\boldsymbol{k}} \ , \\ \boldsymbol{A_k^*} &= \frac{1}{2}\sqrt{\frac{\hbar}{2\epsilon_0 V \omega_{\boldsymbol{k}}}}\,(x_{\boldsymbol{k}} - i y_{\boldsymbol{k}})\,\varepsilon_{\boldsymbol{k}} \ . \end{aligned} \tag{1.9}$$

The coordinates $x_{\boldsymbol{k}}$ and $y_{\boldsymbol{k}}$ are scalar quantities, the directional properties of $\boldsymbol{A_k}$ and $\boldsymbol{A_k^*}$ are specified by the introduction of a unit polarization vector $\varepsilon_{\boldsymbol{k}}$ of the mode.

The single-mode energy given by (1.8) can now be expressed as

$$\overline{\mathcal{E}_{\boldsymbol{k}}} = \frac{\hbar\omega_{\boldsymbol{k}}}{4}\left(x_{\boldsymbol{k}}^2 + y_{\boldsymbol{k}}^2\right) \ . \tag{1.10}$$

1.1.1 Quantization of the Electromagnetic Field

The most straightforward approach to the quantization of the electromagnetic field consists in the replacement of classical variables $x_{\boldsymbol{k}}$ and $y_{\boldsymbol{k}}$ by quantum-mechanical operators $\hat{x}_{\boldsymbol{k}}$ and $\hat{y}_{\boldsymbol{k}}$ such that

$$[\hat{x}_{\boldsymbol{k}}, \hat{y}_{\boldsymbol{k}'}] = 2i\delta_{\boldsymbol{k},\boldsymbol{k}'}\hat{1} \ , \tag{1.11}$$

i.e. we associate with each mode $\boldsymbol{k}$ of the electromagnetic field a quantum-mechanical harmonic oscillator. We also introduce two operators $\hat{a}_{\boldsymbol{k}}^\dagger$ and $\hat{a}_{\boldsymbol{k}}$ which create and destroy a quantum of energy $\hbar\omega_{\boldsymbol{k}}$: these quanta are photons of wave vectors $\boldsymbol{k}$, the number of photons in the mode $\boldsymbol{k}$ is determined by the mean value of the photon number operator $\hat{n}_{\boldsymbol{k}} = \hat{a}_{\boldsymbol{k}}^\dagger \hat{a}_{\boldsymbol{k}}$. We note here once again that for each mode $\boldsymbol{k}$ there are two independent directions of the mode polarization $\varepsilon_{\boldsymbol{k}}$.

The classical vector potentials $\boldsymbol{A_k}$ and $\boldsymbol{A_k^*}$ for the field mode $\boldsymbol{k}$ are quantized via the substitution

$$\begin{aligned} \boldsymbol{A_k} &\longrightarrow \sqrt{\frac{\hbar}{2\epsilon_0 V \omega_{\boldsymbol{k}}}}\hat{a}_{\boldsymbol{k}}\varepsilon_{\boldsymbol{k}} \ , \\ \boldsymbol{A_k^*} &\longrightarrow \sqrt{\frac{\hbar}{2\epsilon_0 V \omega_{\boldsymbol{k}}}}\hat{a}_{\boldsymbol{k}}^\dagger\varepsilon_{\boldsymbol{k}} \ . \end{aligned} \tag{1.12}$$

The creation and annihilation operators are related to the quadratures as

$$\hat{a}_{\boldsymbol{k}} = \frac{1}{2}\left(\hat{x}_{\boldsymbol{k}} + i\hat{y}_{\boldsymbol{k}}\right) ,$$
$$\hat{a}^{\dagger}_{\boldsymbol{k}} = \frac{1}{2}\left(\hat{x}_{\boldsymbol{k}} - i\hat{y}_{\boldsymbol{k}}\right) . \tag{1.13}$$

The quantum-mechanical expressions for the electric and magnetic field operators $\hat{\boldsymbol{E}}_{\boldsymbol{k}}$ and $\hat{\boldsymbol{H}}_{\boldsymbol{k}}$ associated with the mode $\boldsymbol{k}$ now read

$$\hat{\boldsymbol{E}}_{\boldsymbol{k}} = i\sqrt{\frac{\hbar\omega_{\boldsymbol{k}}}{2\epsilon_0 V}}\boldsymbol{\varepsilon}_{\boldsymbol{k}}\left[\hat{a}_{\boldsymbol{k}}e^{-i(\omega_{\boldsymbol{k}}t-\boldsymbol{k}\cdot\boldsymbol{r})} - \hat{a}^{\dagger}_{\boldsymbol{k}}e^{i(\omega_{\boldsymbol{k}}t-\boldsymbol{k}\cdot\boldsymbol{r})}\right] ,$$
$$\hat{\boldsymbol{H}}_{\boldsymbol{k}} = i\sqrt{\frac{\hbar c^2}{2\mu_0 V\omega_{\boldsymbol{k}}}}\boldsymbol{k}\times\boldsymbol{\varepsilon}_{\boldsymbol{k}}\left[\hat{a}_{\boldsymbol{k}}e^{-i(\omega_{\boldsymbol{k}}t-\boldsymbol{k}\cdot\boldsymbol{r})} - \hat{a}^{\dagger}_{\boldsymbol{k}}e^{i(\omega_{\boldsymbol{k}}t-\boldsymbol{k}\cdot\boldsymbol{r})}\right] . \tag{1.14}$$

The operators for the total electric and magnetic fields are then $\hat{\boldsymbol{E}} = \hat{\boldsymbol{E}}_T = \sum_{\boldsymbol{k}}\hat{\boldsymbol{E}}_{\boldsymbol{k}}$ and $\hat{\boldsymbol{H}} = \sum_{\boldsymbol{k}}\hat{\boldsymbol{H}}_{\boldsymbol{k}}$, respectively. We note the quantum unit of electric strength is $E_0 = [\hbar\omega_{\boldsymbol{k}}/(2\epsilon_0 V)]^{1/2}$ and when we speak of noise reduction we are discussing modifications at this level.

The Hamiltonian of the free quantized electromagnetic field can be expressed as

$$\hat{H} = \frac{1}{2}\int_{\text{cavity}}\left(\epsilon_0\hat{\boldsymbol{E}}^2_{\boldsymbol{k}} + \mu_0\hat{\boldsymbol{H}}^2_{\boldsymbol{k}}\right)d^3V . \tag{1.15}$$

After the integration is performed we find

$$\hat{H} = \sum_{\boldsymbol{k}}\hbar\omega_{\boldsymbol{k}}\left(\hat{a}^{\dagger}_{\boldsymbol{k}}\hat{a}_{\boldsymbol{k}} + 1/2\right) , \tag{1.16}$$

which means that the quantized electromagnetic field can be expressed as a set of quantum harmonic oscillators.

1.1.2 Uncertainty Relations and Squeezing of Quantum Fluctuations

Let us consider a measure of an uncertainty in the measurement of an observable $\hat{A}$ to be the mean variance ΔA, defined as

$$(\Delta A)^2 = \langle\hat{A}^2\rangle - \langle\hat{A}\rangle^2 . \tag{1.17}$$

We note that for an eigenstate $|\Psi\rangle$ of the observable $\hat{A}$ (i.e. $\hat{A}|\Psi\rangle = a|\Psi\rangle$) the mean variance is equal to zero. For two noncommuting observables $\hat{A}$ and $\hat{B}$ such that

$$[\hat{A},\hat{B}] = i\hat{C} , \tag{1.18}$$

there does not exist a common eigenstate. Consequently, the two observables cannot be determined precisely, in principle. This “incompatibility” of the two observables can be expressed in terms of uncertainty relations. If we utilize the mean variance as a measure of the uncertainty, then for the two observables which fulfill the commutation relation (1.18) we find the uncertainty relation

$$\Delta A \, \Delta B \geq \frac{1}{2}|\langle\hat{C}\rangle| \, . \tag{1.19}$$

In this uncertainty relation, the right-hand side (i.e. the lower bound on the product of the two variances) explicitly depends on the state in which the quantum system is prepared. Those states for which the left and right hand side are mutually equal are called *intelligent* states. If simultaneously, the minimum of both left and right hand sides is attained, then the state under consideration is called a *minimum uncertainty state* (MUS). In either case we can denote $\frac{1}{2}|\langle\hat{C}\rangle| = U^2$. The variances ΔA and ΔB are not necessarily equal. The state $|\Psi\rangle$ is said to be *squeezed* with respect to the observable $\hat{A}$ provided

$$\Delta A < U \, . \tag{1.20}$$

Obviously, this reduction of quantum fluctuations below the limit U can be achieved only at the expense of enhancement of fluctuations in the conjugate observable. The definition of squeezing (1.20) is only meaningful if A and B have the same dimensions, and are related to each other by a symmetry transformation, typically a phase-rotation symmetry.

1.1.3 Weyl–Heisenberg Algebra and Quadrature Squeezing

To make our discussion more specific we describe in more details quadrature squeezing for a singe-mode field.

The two observables $\hat{q}$ and $\hat{p}$ for a mechanical particle satisfy the Heisenberg commutation relation

$$[\hat{q},\hat{p}] = i\hbar\hat{1} \; ; \quad [\hat{q},\hat{1}] = 0 \; ; \quad [\hat{p},\hat{1}] = 0 \, , \tag{1.21}$$

and the uncertainty relation (1.19) for the two observables can be written as

$$(\Delta q)^2(\Delta p)^2 \geq \frac{\hbar^2}{4} \, . \tag{1.22}$$

The group theoretical structure of the commutation (1.21) is described by the so-called Weyl–Heisenberg group (see for instance [14,15]). The operators $\hat{q}$, $\hat{p}$ and $\hat{1}$ generate a Lie algebra usually called the Weyl–Heisenberg Lie algebra.

We now consider how to translate the uncertainty principle into quadratures of a radiation field, defined as

$$\hat{x}_\theta = \hat{a}e^{-i\theta} + \hat{a}^\dagger e^{i\theta} \, . \tag{1.23}$$

Special cases of this are:

$$\begin{aligned} \hat{x} &= \hat{a} + \hat{a}^\dagger = \hat{x}_0 \, , \\ \hat{y} &= \left(\hat{a} - \hat{a}^\dagger\right)/i = \hat{x}_{\pi/2} \, . \end{aligned} \tag{1.24}$$

The annihilation and creation operators satisfy the commutation relations

$$[\hat{a},\hat{a}^\dagger]=\hat{1}\ ;\quad [\hat{a}^\dagger,\hat{1}]=0\ ;\quad [\hat{a},\hat{1}]=0\ , \tag{1.25}$$

and the corresponding commutation relations for the quadrature components are

$$\left[\hat{x}_\theta,\hat{x}_{\theta+\pi/2}\right]=[\hat{x},\hat{y}]=2i\ . \tag{1.26}$$

The uncertainty relation (1.19) for the two quadratures $\hat{x}_\theta$ and $\hat{x}_{\theta+\pi/2}$, or $\hat{x}$ and $\hat{y}$ can be written as

$$\left(\Delta x_\theta\right)^2\left(\Delta x_{\theta+\pi/2}\right)^2\geq 1\ . \tag{1.27}$$

There is a class of states for which the equality in the uncertainty relation (1.27) is satisfied. These are the minimum uncertainty states. In the particular case of the Weyl–Heisenberg algebra these are specific Gaussian states, i.e. states with Gaussian Wigner functions (see below and [16–18]). Among these states, there is a large class of states, called squeezed states, for which

$$\left(\Delta x_\theta\right)^2\leq 1\ . \tag{1.28}$$

Since the free Hamiltonian is invariant under phase rotations, the eigenstates of the free Hamiltonian may be MUS, but they cannot be squeezed.

1.2 Quantum States of Light Fields

1.2.1 Phase-Space Description of Light Fields

In what follows we concentrate our attention mainly on single-mode light fields for a single mode k, and use the notation $\hat{a}=\hat{a}_k$. The parametric space of eigenvalues, i.e. the *phase space* for a single-mode electromagnetic field (harmonic oscillator), is an *infinite* plane of eigenvalues (x,y) of the Hermitean operators $\hat{x}$ and $\hat{y}$. An equivalent phase space is the complex plane of eigenvalues

$$\alpha=\frac{1}{2}\left(x+iy\right) \tag{1.29}$$

of the annihilation operator $\hat{a}$. The two invariant differential elements of the two phase-spaces are related as:

$$\frac{1}{\pi}d^2\alpha=\frac{1}{\pi}d[\mathrm{Re}(\alpha)]\,d[\mathrm{Im}(\alpha)]=\frac{1}{4\pi}dx\,dy\ . \tag{1.30}$$

The phase-space description of a quantum-mechanical oscillator which is in the state described by the density operator $\hat{\rho}$ (in what follows we will consider mainly pure states such that $\hat{\rho}=|\Psi\rangle\langle\Psi|$) is based on the definition of the Wigner function [19,20] $W_{\hat{\rho}}(\alpha)$. Here the subscript $\hat{\rho}$ in the expression

$W_{\hat{\rho}}(\alpha)$ explicitly indicates the state which is described by the given Wigner function .

The Wigner function is related to the characteristic function $C_{\hat{\rho}}^{(W)}(\xi)$ of the Weyl-ordered (i.e. symmetrically-ordered) moments of the annihilation and creation operators of the harmonic oscillator as follows [21]

$$W_{\hat{\rho}}(\alpha) = \frac{1}{\pi}\int C_{\hat{\rho}}^{(W)}(\xi)\, \exp(\alpha\xi^* - \alpha^*\xi)\, d^2\xi \; . \tag{1.31}$$

The characteristic function $C_{\hat{\rho}}^{(W)}(\xi)$ of the system described by the density operator $\hat{\rho}$ is defined as

$$C_{\hat{\rho}}^{(W)}(\xi) \equiv \mathrm{Tr}[\hat{\rho}\hat{D}(\xi)] \; , \tag{1.32}$$

where $\hat{D}(\xi) = \exp[\xi\hat{a}^\dagger - \xi^*\hat{a}]$ is the displacement operator (see below). The characteristic function $C_{\hat{\rho}}^{(W)}(\xi)$ can be used for the evaluation of the Weyl-ordered products of the annihilation and creation operators:

$$\langle\{(\hat{a}^\dagger)^m\hat{a}^n\}\rangle = \left.\frac{\partial^{(m+n)}}{\partial\xi^m\partial(-\xi^*)^n} C_{\hat{\rho}}^{(W)}(\xi)\right|_{\xi=0} \; , \tag{1.33}$$

On the other hand the mean value of the Weyl-ordered product $\langle\{(\hat{a}^\dagger)^m\hat{a}^n\}\rangle$ can be obtained by using the Wigner function directly

$$\mathrm{Tr}\left[\{(\hat{a}^\dagger)^m\hat{a}^n\}\hat{\rho}\right] = \frac{1}{\pi}\int d^2\alpha\, (\alpha^*)^m\alpha^n W_{\hat{\rho}}(\alpha) \; . \tag{1.34}$$

For instance, the Weyl-ordered product $\langle\{\hat{a}^\dagger\hat{a}^2\}\rangle$ can be evaluated as:

$$\langle\{\hat{a}^\dagger\hat{a}^2\}\rangle = \frac{1}{3}\langle\hat{a}^\dagger\hat{a}^2 + \hat{a}\hat{a}^\dagger\hat{a} + \hat{a}^2\hat{a}^\dagger\rangle = \frac{1}{\pi}\int d^2\alpha\, |\alpha|^2\alpha\, W_{\hat{\rho}}(\alpha) \; . \tag{1.35}$$

We note that using the definition given by (1.31) we can associate Wigner functions not only with density operators, but also with any observable $\hat{A}$. We denote the Wigner function associate with $\hat{A}$ as $W_{\hat{A}}(\alpha)$. The mean value $\langle\hat{A}\rangle$ of the observable $\hat{A}$ in the state $\hat{\rho}$ can then be expressed as

$$\langle\hat{A}\rangle = \mathrm{Tr}[\hat{A}\hat{\rho}] = \frac{1}{\pi}\int d^2\alpha W_{\hat{A}}(\alpha) W_{\hat{\rho}}(\alpha) \; . \tag{1.36}$$

Originally the Wigner function was introduced in a form which is quite different from (1.31). Namely, the Wigner function was defined [19] as a particular Fourier transform of the density operator expressed in the basis of the eigenvectors $|q\rangle$ of the position operator $\hat{q}$:

$$W_{\hat{\rho}}(q,p) \equiv \int_{-\infty}^{\infty} d\zeta \langle q - \zeta/2|\hat{\rho}|q + \zeta/2\rangle e^{ip\zeta/\hbar} \; , \tag{1.37}$$

which for a pure state described by a state vector $|\Psi\rangle$ (i.e. $\hat{\rho} = |\Psi\rangle\langle\Psi|$) reads

$$W_{\hat{\rho}}(q,p) \equiv \int_{-\infty}^{\infty} d\zeta \psi(q - \zeta/2)\psi^*(q + \zeta/2)\, e^{ip\zeta/\hbar} \; , \tag{1.38}$$

where $\psi(q) \equiv \langle q|\Psi\rangle$. It can be shown that both definitions (1.31) and (1.37) of the Wigner function are identical (see [20]), providing the position and momentum parameters q and p of a harmonic oscillator of mass m are related to the quadratures x and y as:

$$x = \sqrt{\frac{2m\omega}{\hbar}} q \,, \qquad y = \sqrt{\frac{2}{\hbar m\omega}} p \,, \tag{1.39}$$

i.e.

$$W_{\hat{\rho}}(x,y) = \frac{1}{4\pi} \int C_{\hat{\rho}}^{(W)}(x',y') \exp\left[-\frac{i}{2}(xy' - yx')\right] dx'\, dy' \,, \tag{1.40}$$

where the characteristic function $C_{\hat{\rho}}^{(W)}(x,y)$ is given by the relation

$$C_{\hat{\rho}}^{(W)}(x,y) = \mathrm{Tr}\left[\hat{\rho}\hat{D}(x,y)\right] . \tag{1.41}$$

The displacement operator in terms of the quadrature operators reads

$$\hat{D}(x,y) = \exp\left[\frac{i}{2}(\hat{x}y - \hat{y}x)\right] . \tag{1.42}$$

The Wigner function can be interpreted as the quasi-probability density distribution through which a probability can be expressed to find a quantum-mechanical system (harmonic oscillator) residing at the "point" (x,y) of the phase space.

The quadrature probability distributions $w_{\hat{\rho}}(x)$ and $w_{\hat{\rho}}(y)$ can be determined from the Wigner function $W_{\hat{\rho}}(x,y)$ via marginal integration over the conjugate variable

$$w_{\hat{\rho}}(x) \equiv \frac{1}{4\pi} \int dy\, W_{\hat{\rho}}(x,y) = \langle x|\hat{\rho}|x\rangle \,, \tag{1.43}$$

where $|x\rangle$ is the eigenstate of the quadrature operator $\hat{x}$. The marginal probability distribution $w_{\hat{\rho}}(x)$ is normalized to unity, i.e.

$$\int dx\, w_{\hat{\rho}}(x) = 1 \,. \tag{1.44}$$

From the measurement of the marginals of the "rotated" quadratures it is possible to invert the data to generate a tomographic reconstruction of the relevant Wigner function [22]. In this way there is a direct connection between what may be measured experimentally in (for example) homodyne phase-sensitive field detection and the rather abstract Wigner quasi-probability density distribution.

The relation (1.43) for the probability distribution $w_{\hat{\rho}}(x)$ of the quadrature operator $\hat{x}$ can be generalized to the case of the distribution of the rotated quadrature operator $\hat{x}_\theta$. The rotation (i.e. the linear homogeneous canonical transformation) given by (1.23) can be performed by the unitary operator $\hat{U}(\theta)$:

$$\hat{U}(\theta) = \exp\left(-i\theta \hat{a}^\dagger \hat{a}\right) , \tag{1.45}$$

so that

$$\hat{x}_\theta = \hat{U}^\dagger(\theta)\hat{x}\hat{U}(\theta) ; \qquad \hat{x}_{\theta+\pi/2} = \hat{U}^\dagger(\theta)\hat{y}\hat{U}(\theta) . \tag{1.46}$$

Alternatively, in the vector formalism we can rewrite the transformation (1.46) as

$$\begin{pmatrix} \hat{x}_\theta \\ \hat{x}_{\theta+\pi/2} \end{pmatrix} = \mathbf{F} \begin{pmatrix} \hat{x} \\ \hat{y} \end{pmatrix} ; \qquad \mathbf{F} = \begin{pmatrix} \cos\theta & \sin\theta \\ -\sin\theta & \cos\theta \end{pmatrix} . \tag{1.47}$$

Eigenvalues x_θ and $x_{\theta+\pi/2}$ of the operators $\hat{x}_\theta$ and $\hat{x}_{\theta+\pi/2}$ can be expressed in terms of the eigenvalues x and y of the quadrature operators as

$$\begin{pmatrix} x_\theta \\ x_{\theta+\pi/2} \end{pmatrix} = \mathbf{F} \begin{pmatrix} x \\ y \end{pmatrix} ; \qquad \begin{pmatrix} x \\ y \end{pmatrix} = \mathbf{F}^{-1} \begin{pmatrix} x_\theta \\ x_{\theta+\pi/2} \end{pmatrix} ;$$
$$\mathbf{F}^{-1} = \begin{pmatrix} \cos\theta & -\sin\theta \\ \sin\theta & \cos\theta \end{pmatrix} , \tag{1.48}$$

where the matrix $\mathbf{F}$ is given by (1.47) and $\mathbf{F}^{-1}$ is the corresponding inverse matrix. It has been shown by Ekert and Knight [23] that Wigner functions are transformed under the action of the linear canonical transformation (1.47) as

$$\begin{aligned} W_{\hat{\rho}}(x,y) \to W_{\hat{\rho}}(\mathbf{F}^{-1}(x_\theta, x_{\theta+\pi/2})) &= \\ W_{\hat{\rho}}(x_\theta \cos\theta - x_{\theta+\pi/2}\sin\theta; x_\theta \sin\theta + x_{\theta+\pi/2}\cos\theta) & , \end{aligned} \tag{1.49}$$

which means that the probability distribution $w_{\hat{\rho}}(x_\theta, \theta) = \langle x_\theta|\hat{\rho}|x_\theta\rangle$ can be evaluated as

$$\begin{aligned} w_{\hat{\rho}}(x_\theta,\theta) = \frac{1}{4\pi}\int_{-\infty}^{\infty} dx_{\theta+\pi/2} & \\ \times W_{\hat{\rho}}(x_\theta \cos\theta - x_{\theta+\pi/2}\sin\theta; x_\theta\sin\theta + x_{\theta+\pi/2}\cos\theta) & . \end{aligned} \tag{1.50}$$

As shown by Vogel and Risken [22] (see more details see [24]) the knowledge of $w_{\hat{\rho}}(x_\theta,\theta)$ for all values of θ (such that $[0 < \theta \leq \pi]$) is equivalent to the knowledge of the Wigner function itself. This Wigner function can be obtained from the set of distributions $w_{\hat{\rho}}(x_\theta,\theta)$ via the inverse Radon transformation:

$$\begin{aligned} W_{\hat{\rho}}(x,y) = \frac{1}{4\pi}\int_{-\infty}^{\infty} dx_\theta \int_{-\infty}^{\infty} d\xi\,|\xi| \int_0^\pi d\theta\, w_{\hat{\rho}}(x_\theta,\theta) & \\ \times \exp\left[\frac{i}{2}\xi(x_\theta - x\cos\theta - y\sin\theta)\right] & . \end{aligned} \tag{1.51}$$

Using the method described above Raymer et al. [25] have experimentally reconstructed Wigner functions of optical fields. The optical homodyne tomography is implicitly based on a measurement of all (in principle, infinite number) independent moments of the system operators. Nevertheless, there are states for which the Wigner function can be reconstructed much easier than via the homodyne tomography. These are Gaussian and generalized Gaussian states which are completely characterized by the first two moments of the relevant observables while all higher-order moments can be expressed in terms of the first two moments. On the other hand, if the state under consideration is characterized by an infinite number of independent moments then the homodyne tomography can fail because it does not provide us with a consistent truncation scheme. In this case the MaxEnt principle may help use to reconstruct reliably the Wigner function from incomplete tomographic data (for more details see the recent review [26]).

1.2.2 Fock States

Eigenstates $|n\rangle$ of the photon number operator $\hat{n} = \hat{a}^\dagger\hat{a} = [\frac{1}{4}(\hat{x}^2 + \hat{y}^2) - \frac{1}{2}]$ are called Fock states . These states are orthogonal (i.e. $\langle n|n'\rangle = \delta_{n,n'}$) and complete (i.e. $\sum_n |n\rangle\langle n| = \hat{1}$). The number state $|n\rangle$ can be generated from the vacuum state (i.e. the eigenstate of $\hat{n}$ with the eigenvalue equal to zero) as

$$|n\rangle = \frac{(\hat{a}^\dagger)^n}{\sqrt{n!}}|0\rangle \ . \tag{1.52}$$

The Wigner function of the Fock state $|n\rangle$ in the α-phase space reads

$$W_{|n\rangle}(\alpha) = 2(-1)^n \exp\left(-2|\alpha|^2\right) \mathcal{L}_n\left(4|\alpha|^2\right) \ , \tag{1.53}$$

where $\mathcal{L}_n$ is the Laguerre polynomial of order n [27]. In the (x, y) phase space this Wigner function has the form

$$W_{|n\rangle}(x, y) = 2(-1)^n \exp\left(-\frac{x^2 + y^2)}{2}\right) \mathcal{L}_n\left(x^2 + y^2\right) \ , \tag{1.54}$$

with substantial negativities, indicating the non-classicality of these quasi-probabilities. Although (to our knowledge) the Fock states of a light field have yet to be reconstructed experimentally, the analogous *vibrational* Fock state $n = 1$ of a trapped ion has been [28]. One can find from (1.54) the following expressions for first few moments of the position and momentum operators:

$$\begin{aligned}
\langle\hat{x}\rangle &= \langle\hat{y}\rangle = 0 \ , \\
\langle\hat{x}^2\rangle &= \langle\hat{y}^2\rangle = (2n+1) \ , \\
\langle\hat{x}^4\rangle &= \langle\hat{y}^4\rangle = (6n^2 + 6n + 3) = \frac{3}{2}\frac{\langle\hat{x}^2\rangle^2 + \langle\hat{y}^2\rangle^2}{2} + \frac{3}{2} \ , \\
\langle\hat{x}^2\hat{y}^2\rangle &= \langle\hat{y}^2\hat{x}^2\rangle = (2n^2 + 2n - 1) = \frac{1}{2}\frac{\langle\hat{x}^2\rangle^2 + \langle\hat{y}^2\rangle^2}{2} - \frac{3}{2} \ ,
\end{aligned} \tag{1.55}$$

which clearly illustrates the non-Gaussian character of the Wigner function (1.54). This Wigner function is characterized by an infinite number of

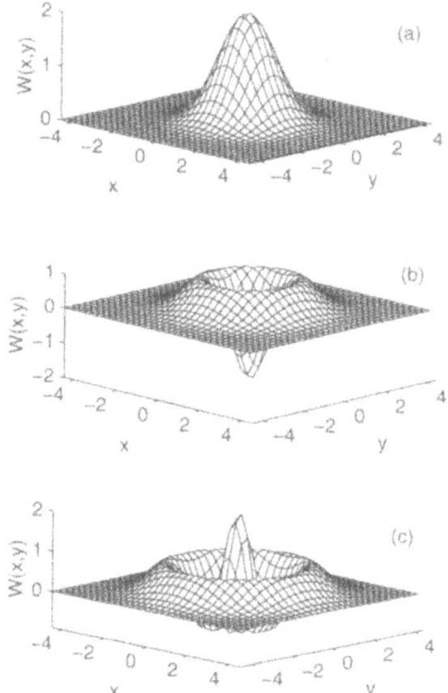

Fig. 1.1. We plot the Wigner function of the first three Fock states, that is the vacuum state $|0\rangle$ (**a**); the Fock state with one photon $|1\rangle$ (**b**), and the Fock state with two photons $|2\rangle$ (**c**). We see that only the Wigner function of the vacuum state is non-negative. The other two Wigner functions take negative values which clearly indicates the non-classical nature of Fock states

independent moments of the position and momentum operators. We plot the Wigner function of the first three Fock states in Fig. 1.1. From (1.55) it follows that the variances of the quadrature operators read $(\Delta x)^2 = (\Delta y)^2 = (2n+1)$.

Moments of the photon number operator $\hat{n}$ in the Fock state $|n\rangle$ are

$$\langle \hat{n}^k \rangle = n^k \ , \tag{1.56}$$

from which it follows higher-order moments of the operator $\hat{n}$ can be expressed in terms of the first-order moment.

1.2.3 Coherent States

Coherent states of a radiation field play a central role in quantum optics [29]. They can be produced by a stabilized laser well above threshold and formally can be described as the consequence of driving of a quantum harmonic oscillator initially prepared in the vacuum state $|0\rangle$ by a classical current of given amplitude and phase. In the interaction picture the corresponding Hamiltonian can be written as (here we assume that the classical current is on resonance with the given quantum oscillator)

$$\hat{H}_I = i\hbar \left(\mathcal{E}\hat{a}^\dagger - \mathcal{E}^*\hat{a} \right) \ . \tag{1.57}$$

The corresponding evolution operator $\hat{U}(t) = \exp(-it\hat{H}_I/\hbar)$ is equal to the displacement operator (1.42) with $\alpha = (x+iy)/2 = t\mathcal{E}$. That is, the coherent state $|\alpha\rangle$ is given by the expression

$$|\alpha\rangle = \hat{D}(\alpha)|0\rangle \ . \tag{1.58}$$

We can decompose the coherent state (1.58) in the Fock basis as the Poissonian distribution

$$|\alpha\rangle = \sum_{n=0}^{\infty} \frac{\alpha^n}{\sqrt{n!}}|n\rangle \ . \tag{1.59}$$

We note that coherent states are eigenstates of the annihilation operator $\hat{a}$, i.e. $|\alpha\rangle$ is not an eigenstate of an observable. The photon number variance of the coherent state is Poissonian, i.e. $(\Delta n)^2 = |\alpha|^2 = \bar{n}$. Fluctuations in the photon number can be described in terms of deviation from the Poissonian variance by the Mandel Q-parameter defined by

$$Q = \frac{\langle(\Delta\hat{n})^2\rangle - \langle\hat{n}\rangle}{\langle\hat{n}\rangle} \ . \tag{1.60}$$

The Wigner function of the coherent state $|\beta\rangle$ in the complex α-phase space is a Gaussian (see Fig. 1.2)

$$W_{|\beta\rangle}(\alpha) = 2\exp\left(-2|\alpha-\beta|^2\right) \ , \tag{1.61}$$

or alternatively, in the (x, y) phase space we have:

$$W_{|\beta\rangle}(x,y) = \frac{2}{\sigma_x\sigma_y}\exp\left[-\frac{(x-\bar{x})^2}{2\sigma_x^2} - \frac{(y-\bar{y})^2}{2\sigma_y^2}\right] \ , \tag{1.62}$$

where $\bar{x} = \beta + \beta^*$, $\bar{y} = -i(\beta - \beta^*)$, and

$$\sigma_x^2 = \sigma_y^2 = 1 \ . \tag{1.63}$$

We note that the only *pure*-state Wigner function which is positive everywhere has this Gaussian form.

The mean photon number in the coherent state is equal to $\bar{n} = |\beta|^2$. The variances for the quadrature operators are

$$\langle\beta|(\Delta\hat{x})^2|\beta\rangle = \sigma_x^2 \ ; \qquad \langle\beta|(\Delta\hat{y})^2|\beta\rangle = \sigma_y^2 \ , \tag{1.64}$$

from which it is seen that the coherent state belongs to the class of minimum uncertainty states for which

$$\langle(\Delta\hat{x})^2\rangle\langle(\Delta\hat{y})^2\rangle = \sigma_x^2\sigma_y^2 = 1 \ . \tag{1.65}$$

The quantum statistical properties of coherent states are completely determined by the mean values of the position and momentum operators and their variances.

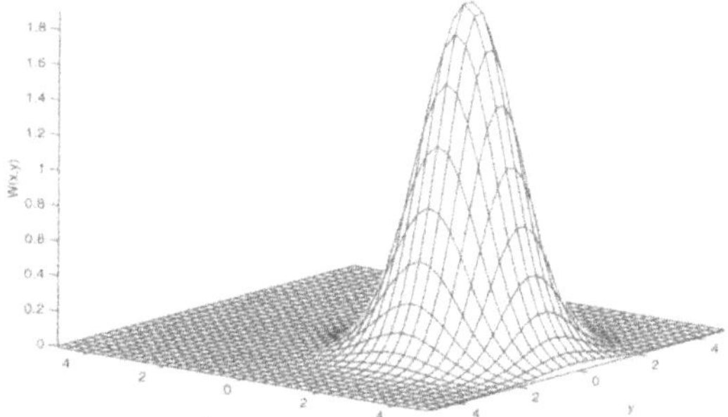

Fig. 1.2. The Wigner function of the coherent state (1.62) with the real amplitude $\beta = 2$

1.2.4 Squeezed States

Squeezed states of radiation are produced in nonlinear processes in which a "classical" electromagnetic field drives a nonlinear medium. In the nonlinear medium, pairs of correlated photons of the same frequency can be generated. In the interaction representation the corresponding effective Hamiltonian for one such nonlinear optical process, that of parametric down-conversion, can be written as [30]

$$\hat{H}_I = \left[\epsilon(\hat{a}^\dagger)^2 + \epsilon^* \hat{a}^2\right] . \tag{1.66}$$

The effective operator describes how a pump field, here approximated as a *classical* field, is down-converted to its sub-harmonics at half the driving frequency. The quantity ϵ includes the amplitude of the driving field as well as the second-order susceptibility for the down-conversion. The operator $\hat{U}(t) = \exp(-i\hat{H}t/\hbar)$ describing the time evolution of the single-mode radiation field under consideration can be written as

$$\hat{S}(\xi) = \exp\left[\xi \frac{(\hat{a}^\dagger)^2}{2} - \xi^* \frac{\hat{a}^2}{2}\right] , \tag{1.67}$$

with $\xi = r\exp(i\phi)$. It is this quantity $\xi = -i\epsilon t/\hbar$ which will determine the size of the squeezing and will depend on interaction time (essentially the length of the nonlinear medium) as well as the susceptibility and the amplitude of the driving field strength. If the electromagnetic mode is initially prepared in the vacuum state, then it evolves under the action of the Hamiltonian (1.66) into the state

$$|\xi\rangle = \hat{S}(\xi)|0\rangle , \tag{1.68}$$

which is usually called the squeezed vacuum state [16]. Taking into account that the operators $\hat{K}_- = \hat{a}^2/2$ and $\hat{K}_+ = (\hat{a}^\dagger)^2/2$ together with the operator $\hat{K}_0 = (\hat{a}^\dagger\hat{a} + 1/2)/2$ constitute a specific Bosonic representation of the

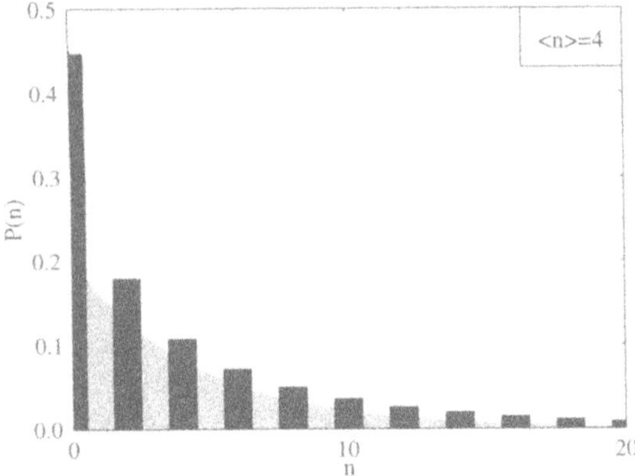

Fig. 1.3. The photon number distribution of the squeezed vacuum state (1.69) with the mean photon number $\bar{n} = 4$ is plotted as bars. The shaded area corresponds to the photon number distribution of a thermal field with the same mean photon number. We see the dramatic oscillations in the photon number distribution of the squeezed vacuum. These oscillations have their origin in specific phase-space interference

$SU(1,1)$ Lie algebra. We can use the $SU(1,1)$ disentangling theorem and obtain for the squeezed vacuum state (1.68) the expression in the Fock basis

$$|\xi\rangle = \frac{1}{\sqrt{\cosh r}} \sum_{n=0}^{\infty} \frac{[(2n)!]^{1/2}}{2^n n!} \left[e^{i\phi} \tanh r\right]^n |2n\rangle \ . \tag{1.69}$$

From here it follows that the squeezed vacuum state is a superposition of only *even* Fock states. The photon number distribution of the state (1.69) reads

$$\begin{aligned} P_{2n} &= \frac{1}{\cosh r} \frac{(2n)!}{(n!)^2 2^n} (\tanh r)^{2n} = \frac{1}{\sqrt{1+\bar{n}}} \frac{(2n)!}{(n!)^2 2^n} \left(\frac{\bar{n}}{1+\bar{n}}\right)^n , \\ P_{2n+1} &= 0 \ , \end{aligned} \tag{1.70}$$

where $\bar{n} = \sinh^2 r$ is the mean photon number in the state (1.68). We plot this oscillating distribution in Fig. 1.3.

The variance in the photon number distribution (1.70) $(\Delta n)^2 = 2\bar{n}(\bar{n}+1)$ is twice that for a thermal state of the same mean photon number. We remind ourselves that the thermal state is described by the density operator

$$\hat{\rho} = \frac{1}{1+\bar{n}} \sum_{n=0}^{\infty} \left(\frac{\bar{n}}{1+\bar{n}}\right)^n |n\rangle\langle n| \ . \tag{1.71}$$

The squeezed vacuum state $|\xi\rangle$ given by (1.67) is an eigenstate with a zero eigenvalue of the annihilation operator $\hat{b}$ (i.e. $\hat{b}|\xi\rangle = 0$) which can be obtained from the operators $\hat{a}$ and $\hat{a}^\dagger$ via the Bogoliubov transformation [16]

$$\begin{aligned} \hat{b} &= \hat{S}(\xi)\hat{a}\hat{S}^\dagger(\xi) = \hat{a}\cosh r - \hat{a}^\dagger e^{i\phi}\sinh r \ , \\ \hat{b}^\dagger &= \hat{S}(\xi)\hat{a}^\dagger\hat{S}^\dagger(\xi) = \hat{a}^\dagger\cosh r - \hat{a}e^{-i\phi}\sinh r \ . \end{aligned} \tag{1.72}$$

This means that the ground state of the transformed oscillator (b) is equal to the squeezed vacuum state of the bare oscillator (a).

Just as for the coherent states, the squeezed vacuum states are not mutually orthogonal. The scalar product of two squeezed states $|\xi\rangle$ and $|\xi'\rangle$ is

$$\langle\xi'|\xi\rangle = \frac{1}{\{\cosh r\cosh r' - \exp[i(\phi-\phi')]\sinh r\sinh r'\}^{1/2}} \ . \tag{1.73}$$

The Wigner function of the squeezed vacuum state (for simplicity we assume here the phase of the squeezing to be equal to zero [1] is of the Gaussian form (see Fig. 1.4)

$$W_{|\xi\rangle}(x,y) = \frac{2}{\sigma_x\sigma_y}\exp\left[-\frac{x^2}{2\sigma_x^2} - \frac{y^2}{2\sigma_y^2}\right] \tag{1.74}$$

from which it follows that the squeezed vacuum is completely characterized by just two variances σ_x and σ_y (and the rotation angle ϕ). Using the Wigner function we find that the mean value of quadrature operator $\hat{x}_\theta$ is zero, while its variance is

$$(\Delta x^\theta)^2 = \mathrm{e}^{2r}\sin^2(\theta-\phi/2) + \mathrm{e}^{-2r}\cos^2(\theta-\phi/2) \ . \tag{1.75}$$

This means that for the chosen phase of squeezing (ϕ) and $r>0$ the quadrature fluctuations in y are reduced at the expense of an increase in the fluctuations in x. Nevertheless, the product of the variances takes the minimum value, i.e. $(\Delta x)^2(\Delta y)^2 = 1$, which demonstrates that the squeezed vacuum is the minimum uncertainty state. We note that infinitely squeezed vacuum states are eigenstates of the rotated quadratures given by (1.23).

The combined action of the squeezing operator $\hat{S}(\xi)$ and the displacement operator $\hat{D}(\alpha)$ on the vacuum state $|0\rangle$ generates a minimum uncertainty squeezed state. There are, however, two equivalent but different definitions and notations of the squeezed states. Yuen [5] introduced the two-photon coherent states (TCS) defined as

$$|\beta,\mu,\nu\rangle = \hat{U}_L\hat{D}(\beta)|0\rangle \ , \tag{1.76}$$

[1] We note that the squeezed vacuum characterized by the complex squeezing parameter $\xi = r\exp(i\phi)$ can be obtained from that squeezed vacuum characterized by $\xi = r$ (i.e. $\phi = 0$) by the action of the rotation operator $\hat{R}(\phi) = \exp\left[i\frac{\phi}{2}\hat{a}^\dagger\hat{a}\right]$, i.e. $|r\exp(i\phi)\rangle = \hat{R}|r\rangle$ [18].

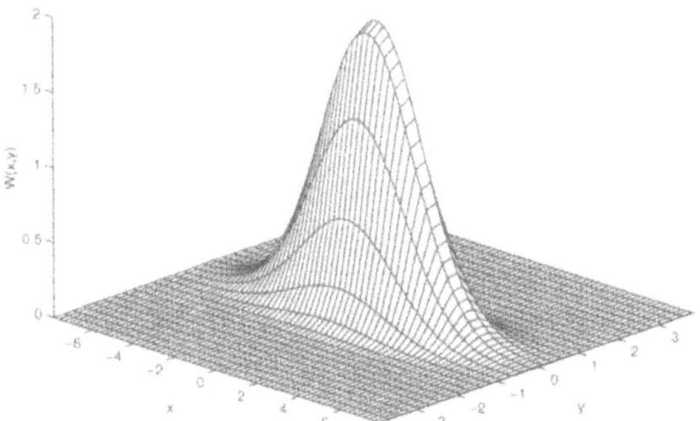

Fig. 1.4. We plot the Wigner function of the squeezed vacuum state given by (1.69). We assume the state is squeezed in the momentum, i.e. in the notation of (1.75) the parameter r is positive, so that $\sigma_p < 1/\sqrt{2}$

where $\hat{U}_L$ is the squeeze operator, equivalent to $\hat{S}(\xi)$ in (1.67). Later, Caves [7] defined a class of Gaussian squeezed displaced states as [2]

$$|\alpha; \xi\rangle = \hat{D}(\alpha)\hat{S}(\xi)|0\rangle \ . \tag{1.77}$$

The Yuen TCS parameters are related to the Caves notation by

$$\begin{aligned} \beta &= \mu\alpha + \nu\alpha^* \ , \\ \mu &= \cosh r \ , \\ \nu &= -\exp(i\phi)\sinh r \ . \end{aligned} \tag{1.78}$$

The Yuen and the Caves notations for squeezed states can be used interchangeably: one or the other my be particularly convenient for a specific calculation.

The Yuen notation is often useful for calculating expectation values of operator moments. This is achieved by inverting the Bogoliubov transformation, to create a 'Heisenberg-like' picture, in which operator moments are all obtained from simple coherent-state expectation values. Thus, let

$$\hat{a} = \hat{S}^\dagger(\xi)\hat{b}\hat{S}(\xi) = \mu^*\hat{b} - \nu\hat{b}^\dagger \ . \tag{1.79}$$

In this way, for example, an operator moment like $\langle \hat{a}^2 \rangle$ can be calculated from the identity

$$\langle \beta, \mu, \nu | \, \hat{a}^2 \, | \beta, \mu, \nu \rangle = \langle \beta, \mu, \nu | \, (\mu^*\hat{b} - \nu\hat{b}^\dagger)^2 \, | \beta, \mu, \nu \rangle \tag{1.80}$$

together with the result that the Yuen coherent states are eigenstates of the (transformed) annihilation operator $\hat{b}$, with eigenvalue β. This result demonstrates that the variances of the quadrature operators are not changed by the

[2] Here we note that the operators $\hat{D}(\alpha)$ and $\hat{S}(\xi)$ do not commute. Consequently the states $\hat{D}(\alpha)\hat{S}(\xi)|0\rangle$ and $\hat{S}(\xi)\hat{D}(\alpha)|0\rangle$ are not equal, but they are related as $\hat{S}(\xi)\hat{D}(\alpha)|0\rangle = \hat{D}(\alpha\cosh r + \alpha^* \exp(i\phi)\sinh r)\hat{S}(\xi)|0\rangle$

displacement. Accordingly, the Wigner function of the displaced squeezed vacuum state can be obtained via a simple displacement of the Wigner function of the squeezed vacuum state (1.74) in the phase space.

It is interesting to note that the mean photon number of the displaced squeezed vacuum can be expressed as $\langle \hat{n} \rangle = \sinh^2 r + |\alpha|^2$, which is the sum of the mean photon number of the cohere state $|\alpha\rangle$ and the squeezed vacuum $|\xi\rangle$. The photon number distribution of the displaced squeezed vacuum has again an oscillatory behavior, but unlike the photon number distribution for the squeezed vacuum, this does not take zero values for the odd Fock states. Moreover, for specific values of the displacement and squeezing parameters, this field can exhibit sub-Poissonian photon statistics (we remind ourselves that the variance of the photon number distribution for the squeezed vacuum is larger than the thermal-state variance). The variance of the photon number distribution of the displaced squeezed state can be expressed as

$$(\Delta n)^2 = |\alpha|^2 \left[e^{-2r} \sin^2(\theta - \phi/2) + e^{2r} \cos^2(\theta - \phi/2) \right] \\ +2 \sinh^2 r \cosh^2 r \ , \tag{1.81}$$

where θ is the argument of the displacement parameter α. If the phase θ of the displacement is chosen so that $\theta - \phi/2 = \pi/2$, then for large values of $|\alpha|$ we obtain from (1.81) the expression

$$(\Delta n)^2 \simeq |\alpha|^2 e^{-2r} \ , \tag{1.82}$$

which shows directly that the displaced squeezed vacuum is described by sub-Poissonian photon number distribution. Moreover (1.82) also indicates that the reduction of the quadrature variances can be identified via the suppression of the photon-number fluctuations (this is the basis for the method of homodyne detection used for measurement of quadrature squeezing [30]).

1.2.5 Two-Mode Squeezed Vacuum

The two-mode squeezed vacuum can be generated in a non-degenerate two-photon down conversion process. Formally the unitary operator associated with this process can be expressed as

$$\hat{S}_{12}(\xi) = \exp\left[\xi \hat{a}_1^\dagger \hat{a}_2^\dagger - \xi^* \hat{a}_1 \hat{a}_2 \right] \ . \tag{1.83}$$

The operators $\hat{K}_- = \hat{a}_1 \hat{a}_2$ and $\hat{K}_+ = \hat{a}_1^\dagger \hat{a}_2^\dagger$ together with the operator $\hat{K}_0 = (\hat{a}_1^\dagger \hat{a}_1 + \hat{a}_2^\dagger \hat{a}_2)/2$ constitutes yet another Bosonic representation of the $SU(1,1)$ Lie algebra with the Bargmann index $k = 1/2$. The two-photon squeezed vacuum is now defined by the action of the operator (1.83) on the vacuum states of the modes a_1 and a_2, i.e.

$$|\xi\rangle_{12} = \hat{S}_{12}(\xi) |0\rangle_1 |0\rangle_2 \ , \tag{1.84}$$

where $\xi = r\exp(i\phi)$. Using the disentangling theorem for the $SU(1,1)$ Lie algebra, we can rewrite the state (1.84) in the Fock basis of the two modes as

$$|\xi\rangle_{12} = \frac{1}{\cosh r}\sum_{n=0}^{\infty} \mathrm{e}^{in\phi}[\tanh r]^n |n\rangle_1 |n\rangle_2 \ . \tag{1.85}$$

From here we see that the two-mode squeezed vacuum is a superposition of only Fock states in which the modes contain the same number of photons, i.e. the two modes are highly correlated. Each mode separately is in the thermal state [13]. To see this, we evaluate the density operator of the mode a_1:

$$\begin{aligned}\hat{\rho}_{a_1} = \mathrm{Tr}_2|\xi\rangle_{12}\langle\xi| &= \frac{1}{\cosh^2 r}\sum_{n=0}^{\infty}[\tanh r]^{2n}|n\rangle_{11}\langle n| \\ &= \frac{1}{1+\bar{n}}\sum_{n=0}^{\infty}\left(\frac{\bar{n}}{1+\bar{n}}\right)^n |n\rangle_{11}\langle n| \ ,\end{aligned} \tag{1.86}$$

where we have used the fact that the mean photon number $\bar{n}$ in each of the modes of the two-mode squeezed vacuum is $\bar{n} = \sinh^2 r$. With this parameterization we can rewrite the expression for the two-mode squeezed vacuum state given by (1.85) as

$$|\xi\rangle_{12} = \frac{1}{\sqrt{1+\bar{n}}}\sum_{n=0}^{\infty}\left(\frac{\bar{n}}{1+\bar{n}}\right)^{n/2} e^{in\phi}|n\rangle_1|n\rangle_2 \ , \tag{1.87}$$

which has a very similar form to the $SU(1,1)$ coherent state except the basis in which the state is expressed.

From the structure of the generators of the two-mode squeezed vacuum it is clear that the two-mode squeezing operator $\hat{\xi}_{12}$ cannot be expressed as the product of the single-mode squeezing operators of the modes a_1 and a_2. On the other hand we can re-express the two-mode squeezing operator in the modes a_1 and a_2 as a product of the single-mode squeezing operators in the new modes a_+ and a_- which are specified by the transformation

$$\hat{a}_+ = \frac{\hat{a}_1 + \hat{a}_2}{\sqrt{2}} \ ; \qquad \hat{a}_- = \frac{\hat{a}_2 - \hat{a}_1}{\sqrt{2}} \ . \tag{1.88}$$

This type of the unitary transformation can be achieved with the help of the 50 : 50 beam splitter. If we rewrite the two-mode squeezing operator $\hat{S}_{12}(\xi)$ in terms of the operators $\hat{a}_+$ and $\hat{a}_-$ we find

$$\hat{S}_{12}(\xi) = \hat{S}_{a_+}(\xi)\hat{S}_{a_-}(-\xi) \ , \tag{1.89}$$

where $\hat{S}_{a_+}(\xi)$ and $\hat{S}_{a_-}(-\xi)$ are single-mode squeezing operators associated with modes c and d, respectively. This formula directly indicates how the two-mode squeezing can be understood: The squeezing of the modes a_+ and a_- is a consequence of the correlations between the original modes a_1 and a_2. This correlation is also responsible for the zero variance of the difference between the photon numbers $\hat{n}_1$ and $\hat{n}_2$, i.e. $[\Delta(n_1 - n_2)]^2 = (\Delta n_1)^2 + (\Delta n_2)^2 - 2(\langle \hat{n}_1 \hat{n}_2 \rangle - \langle \hat{n}_1 \rangle \langle \hat{n}_2 \rangle) = 0$.

The generalization of the squeezed vacuum states to continuum modes is achieved by the action of the multi-mode operator

$$\hat{S}[\xi_k] = \exp\left[\frac{1}{2}\sum_k \left(\xi_k \hat{a}_k^\dagger \hat{a}_{2k_0-k}^\dagger - \text{H.c.}\right)\right] , \tag{1.90}$$

on the multi-mode vacuum $\prod_k |0\rangle_k$, where $k = \omega/c$. The squeezing parameter ξ now reads $\xi_k = r_k \exp(i\phi_k)$ and the bosonic operators $\hat{a}_k$ and $\hat{a}_k^\dagger$ fulfill the commutation relation $[\hat{a}_k, \hat{a}_{k'}^\dagger] = \delta_{k,k'} \hat{1}$. The operator (1.90) is a natural generalization of the two-mode squeezing operator (1.83), and its action on two-mode field states results in correlations between pairs of modes with frequencies symmetrically placed around the carrier frequency $\omega_0 = ck_0$, i.e. the multi-mode state is said to be squeezed at the frequency ω_0. The broadband squeezed vacuum obtained by the action of the multi-mode squeezing operator (1.90) may play a role of a reservoir with unusual phase correlations [32]. This reservoir is characterized by the correlation functions

$$\begin{aligned} \langle \hat{a}_k^\dagger \hat{a}_{k'} \rangle &= \sinh^2 r_k \ \delta_{k,k'} , \\ \langle \hat{a}_k \hat{a}_{k'} \rangle &= e^{i\phi_k} \sinh r_k \cosh r_k \ \delta_{k+k'-2k_0} . \end{aligned} \tag{1.91}$$

These correlations may significantly affect the dynamics of an open system surrounded by the multi-mode squeezed vacuum [33].

1.3 Origin of Squeezing: Quantum Interference in Phase Space

1.3.1 Superpositions of Fock States

To illuminate the basic ideas of quantum interference and in order to link the concept of quantum interference in phase space with the existence of off-diagonal matrix elements of a density operator in the Fock basis we will first study superpositions of Fock states. We note at the outset of our discussion that Fock states are highly non-classical states of light. These states exhibit a high degree of sub-Poissonian photon statistics with the corresponding value of the Mandel Q parameter equal to -1 [see (1.60)]. The Wigner function (1.53) corresponding to the Fock state $|n\rangle$ takes on negative values, which is another indication of the non-classical character of the Fock state. As follows from the analysis presented by Schleich and Wheeler [34], Fock states in the semiclassical approximation can be understood as *non-overlapping* circular (Bohr–Sommerfeld) bands centered at the origin of the phase space. Therefore, using the semiclassical approximation we are not able to describe properly effects which are related to quantum interferences between Fock states.

Let us consider a pure quantum-mechanical superposition of two Fock-states $|n\rangle$ and $|m\rangle$ (for definiteness we assume $n < m$):

$$|\psi\rangle = C_n|n\rangle + C_m|m\rangle \ , \tag{1.92}$$

where the complex probability amplitudes satisfy the normalization condition $|C_n|^2 + |C_m|^2 = 1$, and we denote the relative phase between C_n and C_m as ζ, i.e. $C_n^* C_m = |C_n||C_m|\exp(i\zeta)$.

The Wigner function corresponding to the state (1.92) can be written as a sum of the "mixture" part $W^M(\alpha)$ and the quantum interference part $W^I(\alpha)$:

$$W(\alpha) = W^M(\alpha) + W^I(\alpha) \ . \tag{1.93}$$

The mixture part $W^M(\alpha)$ is equal to the Wigner function of the statistical mixture state described by the density operator $\hat{\rho} = |C_n|^2|n\rangle\langle n| + |C_m|^2|m\rangle\langle m|$, and can be written as:

$$W^M(\alpha) = |C_n|^2 W_{|n\rangle}(\alpha) + |C_m|^2 W_{|m\rangle}(\alpha) \ , \tag{1.94}$$

where $W_{|n\rangle}(\alpha)$ is the Wigner function of the Fock state $|n\rangle$ given by (1.53). The interference part of the Wigner function has the form

$$W^I(\alpha) = C_n^* C_m W_{|n\rangle\langle m|}(\alpha) + C_m^* C_n W_{|m\rangle\langle n|}(\alpha) \ , \tag{1.95}$$

with

$$\begin{aligned} W_{|n\rangle\langle m|}(\alpha) = 2(-1)^n \left(\frac{n!}{m!}\right)^{1/2} (2\alpha^*)^{m-n} \\ \times \exp(-2|\alpha|^2)\mathcal{L}_n^{(m-n)}(4|\alpha|^2) \ , \end{aligned} \tag{1.96}$$

where $\mathcal{L}_n^{(m-n)}(x)$ is the Laguerre polynomial [27].

The interference part $W^I(\alpha)$ can also be rewritten as

$$\begin{aligned} W^I(\alpha) = 4|C_n||C_m|(-1)^n \left(\frac{n!}{m!}\right)^{1/2} (2|\alpha|)^{m-n}\exp(-2|\alpha|^2) \\ \times \mathcal{L}_n^{(m-n)}(4|\alpha|^2)\cos[\zeta - \varphi(m-n)] \ , \end{aligned} \tag{1.97}$$

where φ is the phase of the complex amplitude α. From (1.97) we see that the interference term depends on the phase difference $\Phi = \zeta - \varphi(m-n)$. This phase sensitivity may seem rather surprising because we have superposed two *phase-insensitive* component states $|n\rangle$ and $|m\rangle$ (see for more details [35,36]). Quantum interference given by (1.97) gives rise to the existence of non-vanishing mean values of powers of the annihilation (creation) operator:

$$\langle\psi|\hat{a}^k|\psi\rangle = \frac{1}{\pi}\int d^2\alpha\,\alpha^k W(\alpha) = \sqrt{\frac{m!}{n!}} C_n^* C_m \delta_{n+k,m} \ . \tag{1.98}$$

The non-zero values of $\langle \hat{a}^k \rangle$ represent a *necessary* condition that phase-sensitive non-classical effects such as quadrature squeezing can be observed.

Quadrature squeezing is associated with the reduction of quadrature fluctuations below the vacuum limit, i.e. the variances of the quadrature operators should be less than one. For the superposition state (1.92) we can find these variances in the form:

$$\begin{aligned}
\langle (\Delta\hat{x})^2 \rangle &= 1 + 2\Big[|C_n|^2 n + |C_m|^2 m + \delta_{n+2,m}|C_n||C_m|\sqrt{m(m-1)}\cos\zeta \\
&\quad - 2\delta_{n+1,m}|C_n|^2|C_m|^2 m\cos^2\zeta\Big] , \\
\langle (\Delta\hat{y})^2 \rangle &= 1 + 2\Big[|C_n|^2 n + |C_m|^2 m - \delta_{n+2,m}|C_n||C_m|\sqrt{m(m-1)}\cos\zeta \\
&\quad - 2\delta_{n+1,m}|C_n|^2|C_m|^2 m\sin^2\zeta\Big] .
\end{aligned} \tag{1.99}$$

From the above it follows that the superposition of number states (1.92) is not a minimum uncertainty state. Nevertheless, if $n = m - 1$ or $n = m - 2$, then the squeezing of the quadrature variances can be observed for certain values of the parameters C_n and C_m. Let for simplicity suppose $n = 0$ and $m = 1$ (see [35]) If we take $\cos^2\zeta = 0$, then in this case we find that the variance $\langle (\Delta\hat{x})^2 \rangle \geq 1$ for all values of $|C_1|$, whereas $\langle (\Delta\hat{y})^2 \rangle$ can be squeezed for certain values of $|C_1|$.

On the other hand, for $\cos^2\zeta = 1$ this relationship is reversed and $\langle (\Delta\hat{x})^2 \rangle$ is squeezed. From (1.99) it follows that squeezing in $\langle (\Delta\hat{x})^2 \rangle$ is possible for values of $|C_1| < 1/\sqrt{2}$. Moreover, we can also see that the variance $\langle (\Delta\hat{x})^2 \rangle$ decreases to its minimum value of 3/4 at $|C_1|$=1/2 and then increases to its maximum value of 3 at $|C_1| = 1$. We note that the one-photon superposition state can exhibit squeezing because, although the expectation values of the squares of the quadrature operators are equal, the state has a non-vanishing mean electric field. This electric field provides the required phase sensitivity and is responsible for squeezing. Formally the non-vanishing electric field arises as a consequence of the interference between the states $|0\rangle$ and $|1\rangle$. It should be also said that the superposition state $|\psi\rangle = C_0|0\rangle + C_1|1\rangle$ is a state which in principle can be generated by a photon interaction Hamiltonian which is linear in $\hat{a}$ and $\hat{a}^\dagger$. For example, the Jaynes–Cummings model (see the review [37]) with a cavity field initially prepared in the vacuum state and the atom prepared in the coherent superposition of upper and lower states is good example of a system in which one-photon superposition states can be generated (for more details see [35]).

From the expression for the variance of the quadrature operators (1.99) it follows that the superposition of the vacuum state and the two-photon state, i.e. $|\psi\rangle = C_0|0\rangle + C_2|2\rangle$, can exhibit squeezing as well. In fact the degree of squeezing in this case can be even larger then in the previous case. In particular, if we consider $m = 2$, $n = 0$ and $\cos\zeta = -1$ in (1.99), then we can observe reduced fluctuations in the $\hat{x}$ quadrature for $|C_2| < 1/\sqrt{3}$. The highest degree of squeezing is obtained for $|C_2|^2 = 1/2 - 1/\sqrt{6}$ and is approximately equal to 45% below the standard quantum limit. Superposition

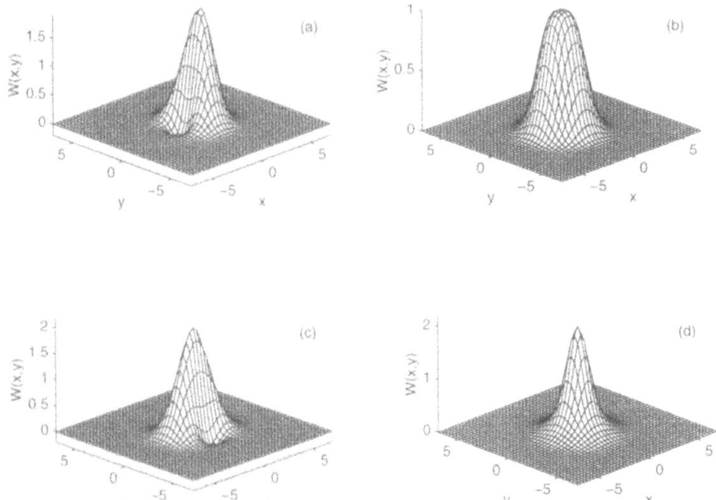

Fig. 1.5. We compare the Wigner function of the one-photon superposition state $|\psi\rangle = C_0|0\rangle + C_1|1\rangle$ with $|C_1| = 1/2$ and $\zeta = 0$ (**a**) with the corresponding statistical mixture of the vacuum state and the one-photon state (**b**). "Squeezing" of the Wigner function of the superposition state is transparent. In figure (**c**) we plot the Wigner function of the superposition state $|\psi\rangle = C_0|0\rangle + C_2|2\rangle$ with $|C_2| = 0.303$ and $\zeta = 0$. In figure (**d**) the Wigner function of the corresponding mixture state is shown. We see that Wigner functions of the statistical mixture state are rotationally invariant, i.e. these states are phase insensitive, while pure superposition states are phase sensitive

of higher numbers of photons does give rise to phase sensitive noise properties. However, the fluctuations in these states are not reduced (squeezed) below the vacuum level. The reason for this is that the noise associated with a number state, other than the vacuum, is too large to be squeezed by the superposition process. To illustrate the effect of the quantum interference we plot in Fig. 1.5 Wigner functions of the one-photon and the two-photon superposition states.

Fock states are highly non-classical states themselves, and therefore one would rather look for quantum interference between coherent states (the most "classical" states in the framework of the quantum theory) as a source of non-classical effects. We will discuss this point later in this Section. Before doing this we would like to discuss briefly how coherent states can be "constructed" through quantum interferences between Fock states (later in this section we will show how Fock states can be represented in terms of coherent states). The coherent state $|\beta\rangle$ can be expressed as a pure superposition of Fock states given by (1.59), where $\beta = |\beta|\exp(i\vartheta)$. The corresponding Wigner function $W_{|\beta\rangle}(\alpha)$ can be expressed in terms of the Wigner functions of Fock states

$W_{|n\rangle}(\alpha)$ given by (1.53) and interference terms $W_{|n\rangle\langle m|}(\alpha)$ given by (1.96):

$$W_{|\beta\rangle}(\alpha) = W^M_{|\beta\rangle}(\alpha) + W^I_{|\beta\rangle}(\alpha) , \qquad (1.100)$$

where

$$W^M_{|\beta\rangle}(\alpha) = \sum_{n=0}^{\infty} |C_n|^2 W_{|n\rangle}(\alpha) \qquad (1.101)$$

and

$$W^I_{|\beta\rangle}(\alpha) = \sum_{n=0}^{\infty} \sum_{m>n}^{\infty} \left[C_n^* C_m W_{|n\rangle\langle m|}(\alpha) + C_m^* C_n W_{|m\rangle\langle n|}(\alpha)\right] . \qquad (1.102)$$

With the use of the explicit expressions for the functions $W_{|n\rangle}(\alpha)$ and $W_{|n\rangle\langle m|}(\alpha)$, we find the mixture part of the Wigner function (1.101):

$$W^M_{|\beta\rangle}(\alpha) = 2\exp\left(-2|\beta|^2 - 2|\alpha|^2\right) J_0(4i|\alpha|\,|\beta|) , \qquad (1.103)$$

where $J_0(z)$ is the Bessel function. If we take into account that (see [27]):

$$J_0(4i|\alpha|\,|\beta|) = \sum_{n=0}^{\infty} \frac{(4|\alpha|^2|\beta|^2)^n}{(n!)^2} , \qquad (1.104)$$

then we can conclude that $W^M_{|\beta\rangle}(\alpha)$ is strictly positive [3]. From (1.103) we see that $W^M_{|\beta\rangle}(\alpha)$ corresponds to the statistical mixture state described by the density operator

$$\hat{\rho} = \sum_{n=0}^{\infty} |C_n|^2 |n\rangle\langle n| = \frac{1}{2\pi}\int_{-\pi}^{\pi} d\theta |\beta\rangle\langle\beta| \qquad (1.105)$$

and it differs significantly from the Wigner function of the coherent state $|\beta\rangle$. Equation (1.105) describes phase-averaged (diffused) coherent state.

For the interference part $W^I_{|\beta\rangle}(\alpha)$ given by (1.102) we find

$$W^I_{|\beta\rangle}(\alpha) = 4\exp(-2|\alpha|^2 - 2|\beta|^2) \sum_{m=0}^{\infty} \frac{\cos(\chi m)}{i^m} J_m(4i|\alpha|\,|\beta|) , \qquad (1.106)$$

where $\chi = \vartheta - \varphi$. As seen from (1.106) the interference term of the Wigner function is again phase sensitive. The phase-sensitivity of the interference part of the Wigner function is reflected also by the fact that when integrated over the whole phase space it gives zero:

$$\frac{1}{\pi}\int d^2\alpha W^I_{|\beta\rangle}(\alpha) = 0 , \qquad (1.107)$$

[3] We should note that this is not the case when just a *finite* number of Fock states contribute to the superposition.

while the mixture part is normalized to unity

$$\frac{1}{\pi}\int d^2\alpha W^M_{|\beta\rangle}(\alpha) = 1 \ . \tag{1.108}$$

Finally, if we use the relation for the Bessel functions

$$J_0(iz) + 2\sum_{m=0} \frac{\cos(\chi m)}{i^m} J_m(iz) = \exp[z\cos\chi] \ , \tag{1.109}$$

then the sum of expressions for $W^I_{|\beta\rangle}$ and $W^M_{|\beta\rangle}$ yields the correct expression for the Wigner of the coherent state [see (1.61)].

In what follows we will discuss how quantum interference between coherent states results in non-classical effects and we will associate quantum coherences between coherent states (i.e. off-diagonal elements of the density matrix in the *coherent*-state basis) with quantum interference in phase space. This correspondence between quantum coherences in the coherent-state basis and quantum interference in the phase space is essentially based on Schrödinger's understanding of coherent states as the quantum counterparts of classical points in phase space. Schrödinger's argument is based on stability and invariance of Gaussian wave packets of an isolated quantum harmonic oscillator.

1.4 Superpositions of Coherent States

We present a general formalism for description of Wigner functions of superpositions of an arbitrary number of coherent states. Nevertheless, to understand the role of quantum interference in phase space we will study in detail the quantum statistical properties of a superposition of just two coherent states. To make our discussion most transparent we will concentrate on two specific cases:

$$|\beta_e\rangle = N_e^{1/2}\left(|\beta\rangle + |-\beta\rangle\right) \ ; \qquad N_e^{-1} = 2\left[1+\exp(-2|\beta|^2)\right] \ , \tag{1.110}$$

and

$$|\beta_o\rangle = N_o^{1/2}\left(|\beta\rangle - |-\beta\rangle\right) \ ; \qquad N_o^{-1} = 2\left[1-\exp(-2|\beta|^2)\right] \ , \tag{1.111}$$

which are called the even and odd coherent states, respectively. These states have been introduced by Dodonov et al. [38] in a formal group-theoretical analysis of various subsystems of coherent states. Both the even and odd coherent states are eigenstates of the square of the annihilation operator $\hat{a}^2$. We note here that superpositions of higher numbers of n of coherent states can be defined as the eigenstates of annihilation operator $\hat{a}^n$ [39,40].

The Wigner function of the even CS (1.110) and odd CS (1.111) can be expressed as (in what follows we assume β to be real):

$$W_{|\beta_e\rangle}(x,y) = N_e\left[W_{|\beta\rangle}(x,y) + W_{|-\beta\rangle}(x,y) + W_{\text{int}}(x,y)\right] \ , \tag{1.112}$$

and

$$W_{|\beta_o\rangle}(x,y) = N_o \left[W_{|\beta\rangle}(x,y) + W_{|-\beta\rangle}(x,y) - W_{\text{int}}(x,y)\right] , \qquad (1.113)$$

respectively, where the Wigner functions $W_{|\pm\beta\rangle}(x,y)$ of coherent states $|\pm\beta\rangle$ are given by (1.62). The interference part of the Wigner functions (1.111) and (1.112) is given by the relation

$$W_{\text{int}}(x,y) = \frac{4}{\sigma_x\sigma_y}\exp\left[-\frac{x^2}{2\sigma_x^2} - \frac{y^2}{2\sigma_y^2}\right]\cos\left(\frac{2\bar{x}y}{\sigma_x\sigma_y}\right) , \qquad (1.114)$$

where $\bar{x} = 2\beta$ (β is real) and the variances σ_x^2 and σ_y^2 are given by (1.64). From (1.112)–(1.113) it follows that the even and odd coherent states differ by the sign of the interference part, which results in completely different quantum-statistical properties of these states. We plot the Wigner functions of the even and odd coherent states in Fig. 1.6 together with the Wigner function of the statistical mixture given by the density operator

$$\hat{\rho} = \frac{1}{2}|\beta\rangle\langle\beta| + \frac{1}{2}|-\beta\rangle\langle-\beta| . \qquad (1.115)$$

From Fig. 1.6 we clearly see that the Wigner function of the statistical mixture is everywhere positive, while the Wigner functions of the even and odd coherent states take negative values which clearly indicates that these two states exhibit non-classical statistical properties.

With the help of the Wigner function (1.112) we evaluate mean values of moments of the quadrature operators $\hat{x}$ and $\hat{y}$. The first moments are equal to zero, i.e. $\langle\hat{x}\rangle = \langle\hat{y}\rangle = 0$, while for higher-order moments we find

$$\begin{aligned} \langle\hat{x}^2\rangle &= \left(1 + 8N_e\beta^2\right) , \\ \langle\hat{y}^2\rangle &= \left(1 - 8N_e\beta^2 e^{-2\beta^2}\right) , \\ \langle\hat{x}^4\rangle &= 3\left[1 + 16N_e\beta^2\left(1 + \tfrac{2}{3}\beta^2\right)\right] , \\ \langle\hat{y}^4\rangle &= 3\left[1 - 16N_e\beta^2 e^{-2\beta^2}\left(1 - \tfrac{2}{3}\beta^2\right)\right] . \end{aligned} \qquad (1.116)$$

From (1.116) it follows that the even coherent state exhibits squeezing in the $\hat{y}$-quadrature [41,42]. This means that the specific interference typical for the even coherent states leads to a reduction of quadrature fluctuations [42]. On the other hand the same interference is responsible for the super-Poissonian character of the photon number distribution of the even coherent state. To see this we evaluate the photon number distribution for the even coherent state.

$$\begin{aligned} P_{2n} &= \frac{2\exp(-\beta^2)}{1+\exp(-2\beta^2)}\frac{\beta^{4n}}{(2n)!} , \\ P_{2n+1} &= 0 . \end{aligned} \qquad (1.117)$$

The oscillations in P_n (1.117) are very similar to those in the case of the squeezed vacuum (see [34]). The mean photon number of the even coherent

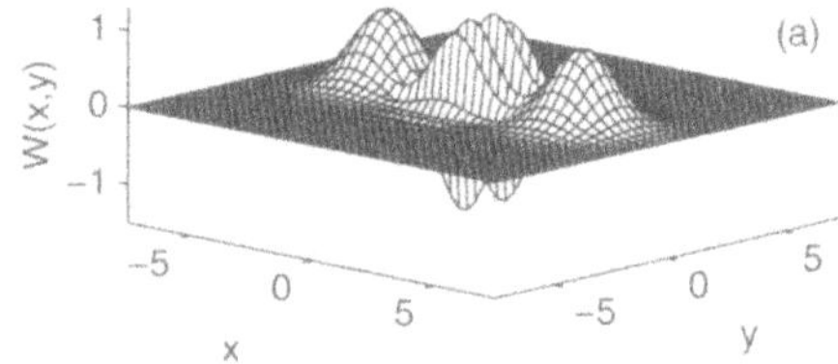

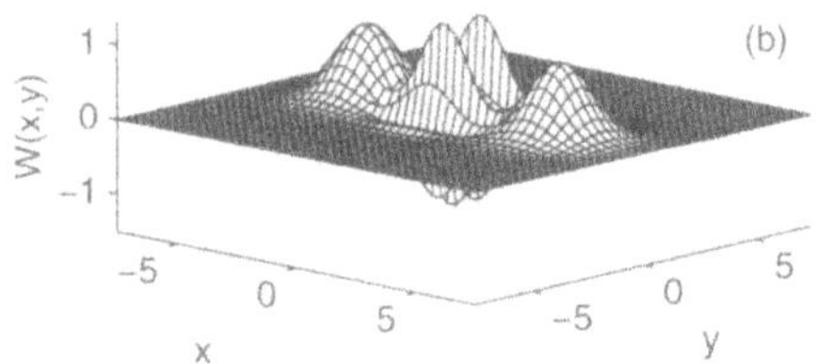

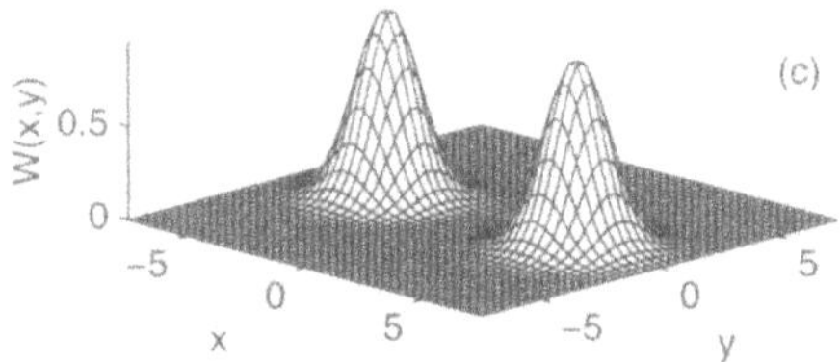

Fig. 1.6. We plot the Wigner function of the even (**a**) and odd (**b**) coherent state as well as the corresponding statistical mixture (**c**). We assume the amplitudes of coherent components $|\beta\rangle$ and $|-\beta\rangle$ to be real and $\beta = 2$

state is $\bar{n} = \beta^2 \tanh \beta^2$ and the Mandel Q parameter

$$Q = \frac{4\beta^2 \exp(-2\beta^2)}{1 - \exp(-4\beta^2)} > 0 \; , \tag{1.118}$$

is positive for any value of β^2 which simply means that the even coherent state has super-Poissonian photon statistics.

On the other hand the odd CS differs from the even CS "just" by a phase factor $\exp(i\pi)$. This difference leads to qualitatively different interference, which result in the fact that the odd coherent states do not exhibit quadrature squeezing, but they have sub-Poissonian photon statistics. The photon number distribution of the odd CS can be written as:

$$\begin{aligned} P_{2n+1} &= \frac{2\exp(-\beta^2)}{1 - \exp(-2\beta^2)} \frac{\beta^{4n+2}}{n!} \; , \\ P_{2n} &= 0 \; . \end{aligned} \tag{1.119}$$

The mean photon number in the odd coherent state is $\bar{n} = \beta^2 \coth \beta^2$ and the Mandel Q parameter of the odd CS is

$$Q = -\frac{4\beta^2 \exp(-2\beta^2)}{1 - \exp(-4\beta^2)} < 0 \,, \tag{1.120}$$

from which it follows that the odd CS has sub-Poissonian photon statistics. It is interesting to note that the odd CS has the maximum degree of sub-Poissonian photon statistics at low intensities $\bar{n}$ of the field, that is for small values of the parameter β^2. In particular, in the limit $\beta^2 \to 0$ (i.e. $\bar{n} \to 1$) we find that $Q \to -1$.

1.5 One-dimensional Continuous Superpositions of Coherent States

We have shown earlier that the even CS exhibits quadrature squeezing. The degree of squeezing can be increased if we add to the superposition (1.110) another pair of coherent states $|\beta'\rangle$ and $|-\beta'\rangle$ (β and β' are supposed to be real)

$$|\Psi\rangle = A^{1/2}[p_\beta(|\beta\rangle + |-\beta\rangle) + p_{\beta'}(|\beta'\rangle + |-\beta'\rangle)] \,, \tag{1.121}$$

where $p_{\beta,\beta'}$ are some numerical parameters and A is the corresponding normalization constant. The state (1.121) for properly chosen values β and β' may exhibit a larger degree of squeezing than the even coherent state (1.110). We can follow this strategy and consider superposition state with more specifically chosen coherent components. In fact, it has been shown [43,44] that a one-dimensional continuous superposition of coherent states of the type:

$$|\xi\rangle = C_F \int_{-\infty}^{\infty} F(\beta, \xi) d\beta |\beta\rangle \,, \tag{1.122}$$

where β is real and

$$C_F^{-2} = \int\int_{-\infty}^{\infty} F(\beta, \xi)\, F(\beta', \xi) \exp[-(\beta - \beta')^2/2] d\beta d\beta' \,, \tag{1.123}$$

with properly chosen weight functions $F(\beta, \xi)$, can exhibit a large degree of squeezing. If $F(\beta, \xi)$ is taken to be the Gaussian function [44]

$$F(\beta, \xi) = \exp\left[-\frac{(1-\xi)}{2\xi}\beta^2\right] \,, \tag{1.124}$$

then the state (4.31) is equal to the squeezed vacuum state, that is:

$$|\xi\rangle = C_F \int_{-\infty}^{\infty} F(\beta, \xi) \hat{D}(\beta)|0\rangle d\beta = \hat{S}(\xi)|0\rangle \,, \tag{1.125}$$

where $\hat{S}(\xi)$ is the squeeze operator and $\hat{S}(\xi)|0\rangle$ is the squeezed vacuum state.

On the other hand the odd CS exhibits a large degree of sub-Poissonian statistics ($Q < 0$). One can construct a superposition of coherent states $||\beta| \exp(-i\varphi)\rangle$ with equal amplitudes $|\beta|$ and suitably chosen distribution of their phases φ, in such a way that the degree of sub-Poissonian photon statistics can be increased (for details see [43,44]). It has been shown by Gardiner [32] that in the continuous limit of coherent states on the circle one can find the following relation

$$|n\rangle = A_n(r) \int_{-\pi}^{\pi} d\varphi e^{-in\varphi} |re^{i\varphi}\rangle , \qquad \beta = re^{i\varphi} , \tag{1.126}$$

where the normalization constant $A_n(r)$ is

$$A_n(r) = \frac{\sqrt{n!} r^{-n}}{2\pi} e^{r^2/2} . \tag{1.127}$$

In other words, a continuous superposition of coherent states on the circle can represent the number state $|n\rangle$, that is, the state with the highest degree of sub-Poissonian photon statistics $Q = -1$.

1.6 Conclusion

In this chapter we have outlined the fundamental properties of squeezed light fields. We have shown how quantum field fluctuations may be manipulated via nonlinear interactions. We have seen that the main feature behind the reduction of quantum fluctuations is the quantum interference between coherent components of the states under consideration. Depending on the character of quantum interference, quadrature squeezing or sub-Poissonian photon statistics can be observed.

Acknowledgements

This work was supported by the UK Engineering and Physical Sciences Research Council, by the IST projects EQUIP (IST-1999-11053) and QUBITS (IST-1999-13021).

References

1. H. Takahasi, Adv. Commun. Syst. **1**, 227 (1965)
2. D.R. Robinson, Commun. Math. Phys. **1**, 159 (1965)
3. D. Stoler, Phys. Rev. D **1**, 3217 (1970); *ibid.* **4**, 1925 (1971)
4. E.Y. Lu, Lett. Nuovo Cimmento **2**, 1241 (1971); *ibid.* **3**, 585 (1972)
5. H.P. Yuen, Phys. Rev. A **13**, 2226 (1976)
6. J.N. Hollenhorst, Phys. Rev. D **19**, 1669 (1979)
7. C.M. Caves, Phys. Rev. D **23**, 1693 (1981)
8. M.M. Nieto, *The discovery of squeezed states – In 1927*, quant-ph/9708012

9. R.E. Slusher, L.W. Hollberg, B. Yurke, J.C. Mertz and J.F. Valley, Phys. Rev. Lett. **55**, 2409 (1985)
10. R.M. Shelby, M.D. Levenson, S.H. Perlmutter, R.G. DeVoe and D.F. Walls, Phys. Rev. Lett. **57**, 691 (1986)
11. L.-A. Wu, H.J. Kimble, J.L. Hall and H. Wu, Phys. Rev. Lett. **57**, 2520 (1986)
12. R. Loudon, *The Quantum Theory of Light* 2nd edition (Clarendon, Oxford 1983)
13. S.M. Barnett and P.M. Radmore, *Methods in Theoretical Quantum Optics* (Clarendon, Oxford 1997)
14. H. Weyl, *The Theory of Groups and Quantum Mechanics* (Dover, New York 1950)
15. A.M. Perelomov, *Generalized Coherent States and Their Applications* (Springer, Berlin Heidelberg New York 1987)
16. R. Loudon and P.L. Knight, J. Mod. Opt. **34**, 709 (1987)
17. K. Zaheer and M.S. Zubairy, in *Advances in Atomic, Molecular, and Optical Physics*, Vol. 28, eds. D. Bates and B. Bederson (Academic Press, New York 1991), p. 143
18. B.L. Schumaker, Phys. Rep. **135**, 317 (1986)
19. E.P. Wigner, Phys. Rev. **40**, 749 (1932); see also E.P. Wigner, in *Perspectives in Quantum Theory*, eds. W. Yourgrau and A. van der Merwe (Dover, New York 1979) p. 25
20. M. Hillery, R.F. O'Connell, M.O. Scully and E.P. Wigner, Phys. Rep. **106**, 121 (1984); V.I. Tatarskij, Sov. Phys. Usp. **26**, 311 (1983)
21. K.E. Cahill and R.J. Glauber, Phys. Rev. **177**, 1857 (1969); *ibid* **177**, 1882 (1969)
22. K. Vogel and H. Risken, Phys. Rev. A **40**, 2847 (1989)
23. A.K. Ekert and P.L. Knight, Phys. Rev. A **43**, 3934 (1991)
24. U. Leonhardt, *Measuring the quantum state of light* (Cambridge University Press, Cambridge 1997)
25. D.T. Smithey, M. Beck, M.G. Raymer and A. Faridani, Phys. Rev. Lett. **70**, 1244 (1993)
26. V. Bužek, G. Drobný, R. Derka, G. Adam and H. Wiedemann, quant-ph/9805086
27. M. Abramowitz and I.A. Stegun, *Handbook of Mathematical Functions* (Dover, New York 1972)
28. D. Leibfried, D.M. Meekhof, B.E. King, C. Monroe, W.M. Itano and D.J. Wineland, Phys. Rev. Lett. **77**, 4281 (1996); D. Leibfried, D.M. Meekhof, C. Monroe, B.E. King, W.M. Itano and D.J. Wineland, J. Mod. Opt. **44**, 2485 (1997); D. Wineland, C. Monroe, D.M. Meekhof, B.E. King, D. Liebfried, W.M. Itano, J.C. Bergquist, D . Berkeland, J.J. Bollinger and J. Miller, Proc. Roy. Soc. A **454**, 411 (1998)
29. R.J. Glauber, Phys. Rev. Lett. **10**, 84 (1963); E.C.G. Sudarshan, Phys. Rev. Lett. **10**, 277 (1963)
30. L. Mandel and E. Wolf, *Optical Coherence and Quantum Optics* (Cambridge University Press, Cambridge 1995)
31. V. Bužek, Phys. Rev. A **39**, 3196 (1989); V. Bužek and G. Adam, Acta Phys. Slov. **45**, 425 (1995)
32. C.W. Gardiner, *Quantum Noise* (Springer, Berlin Heidelberg New York 1991)
33. B.J. Dalton, Z. Ficek and S. Swain, J. Mod. Opt. **46**, 376 (1999)

34. W.P. Schleich, *Interference in phase space, Habilitation thesis.* (Max–Planck–Institute für Quantenoptik, Garching 1988); W.P. Schleich and J.A. Wheeler, Nature **326**, 574 (1987). W.P. Schleich and J.A. Wheeler, in *Proceeding of the First International Conference of Physics in Phase Space*, ed. W.W. Zachary (Springer, Berlin Heidelberg New York 1987), p. 200; W.P. Schleich, D.F. Walls and J.A. Wheeler, Phys. Rev. A **38**, 1177 (1988)
35. K. Wódkiewicz, P.L. Knight, S.J. Buckle and S.M. Barnett, Phys. Rev. A **35**, 2567 (1987)
36. A. Vidiella Baranco, V. Bužek, P.L. Knight and W.K. Lai, in *Quantum Measurements in Optics*, NATO ASI series, eds. P. Tombesi and D.F. Walls (Plenum Press, New York 1992), p. 221
37. B.W. Shore and P.L. Knight, J. Mod. Opt. **40**, 1195 (1993)
38. V.V. Dodonov, I.A. Malkin and V.I. Manko, Nuovo Cimento **24B**, 46 (1974); Physica **72**, 597 (1974); I.A. Malkin and V.I. Manko, *Dynamical Symetries and Coherent States of Quantum Systems*, (Nauka, Moscow 1979)
39. V. Bužek, J. Mod. Opt. **37**, 303 (1990); I. Jex and V. Bužek, J. Mod. Opt. **40**, 771 (1993)
40. E.E. Hach, III and C.C. Gerry, J. Mod. Opt. **39**, 2501 (1992)
41. W.P. Schleich, M. Pernigo and Fam Le Kien, Phys. Rev. A **44**, 2172 (1991)
42. V. Bužek, A. Vidiella–Barranco and P.L. Knight, Phys. Rev. A **45**, 6570 (1992); V. Bužek and P.L. Knight, in *Progress in Optics*, Vol. 34, ed. E. Wolf (North Holland, Amsterdam 1995), p.1
43. D.F. Walls and G.J. Milburn, in *Quantum Optics, Experimental Gravity, and Measurement Theory*, eds. P. Meyster and M.O. Scully (Plenum, New York 1983), p. 209
44. J. Janszky and An.V. Vinogradov, Phys. Rev. Lett. **64**, 2771 (1990); P. Adam, J. Janszky and An.V. Vinogradov, Opt. Commun. **80**, 155 (1990); P. Adam, J. Janszky and An.V. Vinogradov, Phys. Lett. A **160**, 506 (1991)
45. M. Hillery, Opt. Commun. **62**, 135 (1987); Phys. Rev. A **36**, 3796 (1987)

2 Nonlinear Dielectrics

M. Hillery

The interaction of light with a nonlinear dielectric medium is responsible for a number of the non-classical effects which have been studied in quantum optics in recent years. Squeezing and sub-Poissonian photon statistics can be produced in $\chi^{(2)}$ media and quantum phase diffusion occurs in $\chi^{(3)}$ media [1–4]. More recently studies of the propagation of quantized fields in nonlinear media have been undertaken. Squeezing in quantum solitons has been predicted and observed [5,6], as has phase diffusion [6–8]. Collisions of quantum solitons can be used to perform a quantum nondemolition measurement of photon number [9–11].

The first step which is necessary for the description of these phenomena is the quantization of the electromagnetic field in the presence of a nonlinear dielectric medium. This has been done in three ways. The first is the phenomenological approach in which one starts with the classical Hamiltonian and simply substitutes for the classical fields, the usual expressions for field operators in terms of creation and annihilation operators. This is the most commonly employed method, and it is incorrect. Its most glaring defect is that the Heisenberg equations of motion for the field operators which result from it, do not agree with Maxwell's equations [12]. The reason why this procedure fails is not immediately obvious, but can be uncovered if we notice that by retaining the free-field operator expressions in the interacting theory, we have implicitly assumed that the field commutation relations are unaltered by the interaction. As we shall see, this assumption is not justified, because the interaction changes the momentum which is canonical to the vector potential, and this does change the commutation relations. A more methodical way of proceeding is the macroscopic approach, which describes the medium in terms of its linear and nonlinear susceptibilities and then applies the canonical quantization procedure. That is, one finds a Lagrangian which gives the macroscopic Maxwell equations as equations of motion, from the Lagrangian the canonical momenta and Hamiltonian are found, and, finally, the canonical commutation relations are imposed [12]. The third method, the microscopic approach, involves constructing a microscopic model for the medium and retaining the medium degrees of freedom in the theory [13,14].

In this chapter, we would like to discuss the macroscopic and microscopic approaches. They each have advantages and disadvantages. In the macroscopic approach the description of dispersion is more complicated and it is not possible to address questions of operator ordering. On the other hand, the microscopic approach is limited by the model of the medium which has been chosen while the macroscopic theory needs only a set of numbers, the susceptibilities, to characterize the medium. The macroscopic theory has been developed by Carter and Drummond [15] into a useful tool for the study of the propagation of quantized fields in nonlinear media. The microscopic approach is not at this time as well developed, but it is an area of active research.

2.1 Macroscopic Approach

The macroscopic quantum theory of electrodynamics in a linear, homogeneous medium was developed in 1948 is a series of papers by Jauch and Watson [16]. They used their theory to analyze Čerenkov radiation and discussed the problems associated with dispersive media. This theory was subsequently extended to inhomogeneous, non-dispersive media by Carniglia and Mandel [17], who studied quantization in the presence of a single dielectric interface, and Glauber and Lewenstein [18] who considered the general case.

The extension to inhomogeneous media is important because it allows one to examine finite media and thereby scattering. In the laboratory, fields are produced, interact with a medium, and are measured, with both the production and measurement occurring in free space. This means that the experiment can be described as a scattering event in which one has a free field at both the initial and final times. Such a description is impossible within the context of a theory which concerns itself only with homogeneous, infinite media. Such theories are, however, necessary first steps, and we shall consider them here, but ultimately any successful quantum theory of nonlinear optics must be capable of treating scattering problems.

Let us now turn our attention to the nonlinear, macroscopic theory. Maxwell's equations inside a dielectric medium are given in S.I. units, by:

$$\nabla \cdot \mathbf{D} = 0 \, , \qquad \nabla \times \mathbf{E} = -\frac{\partial \mathbf{B}}{\partial t} \, ,$$

$$\nabla \cdot \mathbf{B} = 0 \, , \qquad \nabla \times \mathbf{B} = \mu_0 \frac{\partial \mathbf{D}}{\partial t} \, , \tag{2.1}$$

in the absence of external charges and currents. Here $\mathbf{D} = \varepsilon_0 \mathbf{E} + \mathbf{P}$ is the displacement field and the polarization $\mathbf{P}$ is given by

$$\mathbf{P} = \varepsilon_0 \left[\chi^{(1)} : \mathbf{E} + \chi^{(2)} : \mathbf{EE} + \chi^{(3)} : \mathbf{EEE} + \dots \right] \, . \tag{2.2}$$

The quantities $\chi^{(j)}$ are the $(j+1)$-rank susceptibility tensors. We shall assume that the medium is uniform, lossless, and non-dispersive. We want to find a Lagrangian which has the Maxwell's equations (2.1) as its equations of motion. Before doing so we need to choose a particular field which is to be the basic dynamical variable in the problem. There are two possibilities. The first is the usual vector potential $A = (A_0, \mathbf{A})$ where

$$\mathbf{E} = -\frac{\partial \mathbf{A}}{\partial t} - \nabla A_0 , \qquad \mathbf{B} = \nabla \times \mathbf{A} , \tag{2.3}$$

and the second is the dual potential $\Lambda = (\Lambda_0, \mathbf{\Lambda})$ where

$$\mathbf{B} = \mu_0 \left[\frac{\partial \mathbf{\Lambda}}{\partial t} + \nabla \Lambda_0 \right] , \qquad \mathbf{D} = \nabla \times \mathbf{\Lambda} . \tag{2.4}$$

This potential can only be used if external charges and currents are absent. We shall discuss both approaches [12].

2.1.1 Vector Potential Quantization

If the vector potential A is used as the basic field the Lagrangian density is:

$$\begin{aligned}\mathcal{L}(A, \dot{A}) = \varepsilon_0 [\frac{1}{2}(\mathbf{E}^2 - c^2\mathbf{B}^2) + \frac{1}{2}\chi^{(1)}_{ij} E_i E_j + \frac{1}{3}\chi^{(2)}_{ijk} E_i E_j E_k \\ + \frac{1}{4}\chi^{(3)}_{ijkl} E_i E_j E_k E_l + \ldots] .\end{aligned} \tag{2.5}$$

From this we find that the canonical momentum corresponding to A, which we designate by $\Pi = (\Pi_0, \mathbf{\Pi})$, is

$$\Pi_0 = \frac{\partial \mathcal{L}}{\partial(\partial_0 A_0)} = 0 , \qquad \Pi_i = \frac{\partial \mathcal{L}}{\partial(\partial_0 A_i)} = -D_i . \tag{2.6}$$

Here we note two things. The first is that the canonical momentum is different from that in the noninteracting theory where $\Pi_i = -E_i$. This is a consequence of the fact that the interaction depends on $\dot{\mathbf{A}}$. The second is that the vanishing of Π_0 implies that A_0 is not an independent field. In the case of free fields, if we choose the Coulomb gauge, it is also possible to choose $A_0 = 0$. This follows from the fact that for the free theory, $A_0 = 0$ implies that the time derivative of $\nabla \cdot \mathbf{A}$ is zero, so that if $\nabla \cdot \mathbf{A} = 0$ initially, it will remain zero. Therefore, in this case the Coulomb and temporal ($A_0 = 0$) gauges are consistent. This is, however, no longer true when a nonlinear interaction is present, and then A_0 must be determined by solving the equation

$$\nabla^2 A_0 = -\nabla \cdot \mathbf{E} , \tag{2.7}$$

where the electric field $-\mathbf{E}$ can be expressed in terms of the canonical momentum, $-\mathbf{D}$ (see (2.9)).

Another consequence of the fact that A_0 is not an independent field in the Hamiltonian formulation is that we lose Gauss' law as an equation of motion. However, the other equations of motion imply that $\nabla \cdot \mathbf{D}$ is time independent, and this allows us to impose Gauss' law as an initial condition.

For the Hamiltonian we have [12]

$$H(\mathbf{A},\mathbf{\Pi}) = \varepsilon_0 \int d^3r[\frac{1}{2}(\mathbf{E}^2 + c^2\mathbf{B}^2 + \chi_{ij}^{(1)} E_i E_j) + \frac{2}{3}\chi_{ijk}^{(2)} E_i E_j E_k + \frac{3}{4}\chi_{ijkl}^{(3)} E_i E_j E_k E_l + \ldots] . \tag{2.8}$$

It is useful to express this directly in terms of the canonical momenta, D_i. To this end we define the tensors $\eta^{(i)}$ by

$$E_i = \eta_{ij}^{(1)} D_j + \eta_{ijk}^{(2)} D_j D_k + \ldots . \tag{2.9}$$

These tensors can be expressed in terms of the susceptibility tensors

$$\begin{aligned} \eta^{(1)} &= [\varepsilon_0(1+\chi^{(1)})]^{-1} , \\ \eta_{imn}^{(2)} &= -\varepsilon_0 \eta_{ij}^{(1)} \eta_{km}^{(1)} \eta_{ln}^{(1)} \chi_{jkl}^{(2)} . \end{aligned} \tag{2.10}$$

Making use of (2.8), we find for the Hamiltonian

$$H(\mathbf{A},\mathbf{\Pi}) = \int d^3r[\frac{1}{2\mu_0}\mathbf{B}^2 + \frac{1}{2}\eta_{ij}^{(1)} D_i D_j + \frac{1}{3}\eta_{ijk}^{(2)} D_i D_j D_k + \frac{1}{4}\eta_{ijkl}^{(3)} D_i D_j D_k D_l] . \tag{2.11}$$

The theory is quantized by imposing the equal-time commutation relations:

$$[A_j(\mathbf{r},t), \Pi_k(\mathbf{r}',t)] = i\hbar\delta_{jk}^{(\mathrm{tr})}(\mathbf{r}-\mathbf{r}') . \tag{2.12}$$

Here, as in standard QED, we use the transverse delta function in order to be consistent with both the Coulomb gauge condition, $\nabla\cdot\mathbf{A}=0$, and Gauss' law, $\nabla\cdot\mathbf{D}=0$. As in the case of free QED it is possible to perform a mode expansion for the field and to define creation and annihilation operators. In particular, for the mode with momentum $\mathbf{k}$ and polarization $\hat{\mathbf{e}}_\lambda(\mathbf{k})$ we have the annihilation operator

$$a_{\mathbf{k},\lambda}(t) = \frac{1}{\sqrt{\hbar V}} \int d^3r e^{-i\mathbf{k}\cdot\mathbf{r}} \hat{\mathbf{e}}_\lambda(\mathbf{k}) \cdot \left[\sqrt{\frac{\varepsilon_0\omega_k}{2}}\mathbf{A}(\mathbf{r},t) - \frac{i\mathbf{D}(\mathbf{r},t)}{\sqrt{2\varepsilon_0\omega_k}}\right] , \tag{2.13}$$

where $\omega_k = c|\mathbf{k}|$. Note that because $a_{\mathbf{k},\lambda}$ depends on $\mathbf{D}$, and consequently contains matter degrees of freedom, it is not a pure photon operator. It represents a collective matter-field mode.

2.1.2 Dual Potential Quantization

If one takes the dual potential, Λ, as the basic field the resulting theory is somewhat simpler. First, it is no longer convenient to express the polarization in terms of the electric field, but we instead expand directly from the start for a general inhomogeneous dielectric as:

$$\mathbf{P} = \eta^{(1)}(\mathbf{r}) : \mathbf{D} + \eta^{(2)}(\mathbf{r}) : \mathbf{DD} + \eta^{(3)}(\mathbf{r}) : \mathbf{DDD} + \dots \ . \tag{2.14}$$

The Lagrangian density is now

$$\mathcal{L} = \frac{1}{2\mu_0}\mathbf{B}^2 - \left(\frac{1}{2}\mathbf{D}\cdot\eta^{(1)}(\mathbf{r}) : \mathbf{D} + \frac{1}{3}\mathbf{D}\cdot\eta^{(2)}(\mathbf{r}) : \mathbf{DD} + \dots\right) \ . \tag{2.15}$$

From this one finds for the canonical momenta $\Pi_0 = 0$ and $\Pi_j = -B_j$. A particularly simple case, which will be of importance later, is that of a plane wave field of area A, polarized in the y direction and propagating in the x direction. This one-dimensional field has the Lagrangian density

$$\mathcal{L} = \frac{A}{2\mu_0}\left(\frac{\partial \Lambda}{\partial t}\right)^2 - \frac{\eta(x)A}{2}\left(\frac{\partial \Lambda}{\partial x}\right)^2 \ . \tag{2.16}$$

In this approach the canonical momenta do not depend on the interaction; they are the same in the free and interacting theories. This is what makes the dual potential theory simpler. When the theory is quantized the equal-time commutation relations are the same as in the free theory. As we saw, this was not the case when the usual vector potential was used. Because of this property it is also straightforward to describe inhomogeneous media using the dual potential. This is considerably more complicated if A is taken to be the basic field, because then the inhomogeneity of the medium appears in the fundamental field commutation relations [12]. From now on, for ease of notation, the spatial dependence of the dielectric coefficients will be dropped.

The fact that $\Pi_0 = 0$ again means that Λ_0 is not an independent field. In this case, however, if we impose the Coulomb gauge, $\nabla \cdot \mathbf{\Lambda} = 0$, we can choose $\Pi_0 = 0$ [12]. This represents another simplification over the standard vector potential approach.

The Hamiltonian is now

$$H = \int d^3r[\frac{1}{2\mu_0}\mathbf{B}^2 + \frac{1}{2}\mathbf{D}\cdot\eta^{(1)}(\mathbf{r}) : \mathbf{D} + \frac{1}{3}\mathbf{D}\cdot\eta^{(2)}(\mathbf{r}) : \mathbf{DD} + \frac{1}{4}\mathbf{D}\cdot\eta^{(3)}(\mathbf{r}) : \mathbf{DDD} + \dots] \ . \tag{2.17}$$

This together with the canonical equal-time commutation relations

$$[\dot{\Lambda}_j(\mathbf{r},t), \Lambda_k(\mathbf{r}',t)] = \frac{i\hbar}{\mu_0}\delta^{(\mathrm{tr})}_{jk}(\mathbf{r}-\mathbf{r}') \ , \tag{2.18}$$

define the fully quantized theory.

2.2 Mode Expansion

If we are describing the behavior of the electromagnetic field in a cavity or a waveguide, it is often useful to expand the field operators in terms of the mode functions of the relevant inhomogeneous dielectric system. The mode functions are determined by the linear part of the polarization.

For example, a waveguide can be constructed from a spatially varying dielectric and a cavity by surrounding a region of free space by dielectric sheets or slabs. Therefore, let us suppose that the linear behavior of our system is described by a linear susceptibility $\eta^{(1)}(\mathbf{r})$. Then the mode functions must obey the wave equation

$$\nabla \times [\eta^{(1)}(\mathbf{r})\nabla \times \mathbf{\Lambda}(\mathbf{r},t)] = -\mu_0 \ddot{\mathbf{\Lambda}}(\mathbf{r},t) , \tag{2.19}$$

and the proper boundary conditions which we usually take to be either periodic or vanishing. The modes have well-defined frequencies, ω_n, so that the mode functions are of the form $\mathbf{\Lambda}(\mathbf{r},t) = \mathbf{\Lambda}_n(\mathbf{r})\exp(-i\omega_n t)$, with the functions $\mathbf{\Lambda}_n(\mathbf{r})$ satisfying the equation

$$\nabla \times [\eta^{(1)}(\mathbf{r})\nabla \times \mathbf{\Lambda}_n(\mathbf{r})] = \mu_0 \omega_n^2 \mathbf{\Lambda}_n(\mathbf{r}) . \tag{2.20}$$

Note that because the fields are transverse, we must have $\nabla \cdot \mathbf{\Lambda}_n = 0$. We can show that modes corresponding to different frequencies are orthogonal by taking the scalar product of both sides of this equation with $\mathbf{\Lambda}_{n'}^*$ and integrating over the quantization volume. Applying the vector identity $\nabla \cdot [\mathbf{A} \times \mathbf{A}'] = \mathbf{A}' \cdot [\nabla \times \mathbf{A}] - \mathbf{A} \cdot [\nabla \times \mathbf{A}']$ twice, and assuming vanishing boundary terms, gives us

$$(\omega_n^2 - \omega_{n'}^2) \int d^3r \mathbf{\Lambda}_{n'}^*(\mathbf{r}) \cdot \mathbf{\Lambda}_n(\mathbf{r}) = 0 , \tag{2.21}$$

which immediately implies that if $\omega_n \neq \omega_{n'}$, then the integral must vanish. It should also be possible to choose different modes with the same frequency to be orthogonal, and we shall assume that this has been done. In particular, note that $\mathbf{\Lambda}_n$ and $\mathbf{\Lambda}_n^*$ are both solutions of (2.20). These can always be chosen so that they are orthogonal to each other. Therefore, for the mode functions $\mathbf{\Lambda}_n$ we have

$$\int d^3r \mathbf{\Lambda}_{n'}^*(\mathbf{r}) \cdot \mathbf{\Lambda}_n(\mathbf{r}) = \delta_{n,n'} , \tag{2.22}$$

and we shall assume that the modes form a complete set in the space of transverse functions.

We next define annihilation operators, a_n, by

$$a_n(t) = \sqrt{\frac{\mu_0}{\hbar}} \int d^3r \mathbf{\Lambda}_n^*(\mathbf{r}) \cdot \left[\sqrt{\frac{\omega_n}{2}} \mathbf{\Lambda}(\mathbf{r},t) - \frac{i}{\sqrt{2\omega_n}} \dot{\mathbf{\Lambda}}(\mathbf{r},t)\right] . \tag{2.23}$$

These obey the commutation relations

$$[a_n(t), a_{n'}^\dagger(t)] = \delta_{n,n'} . \tag{2.24}$$

We can also invert (2.23) to express the fields in terms of the creation and annihilation operators. Making use of (2.23) and its adjoint we find that

$$\sum_n \sqrt{\frac{2\hbar}{\mu_0\omega_n}}[a_n(t)\mathbf{\Lambda}_n(\mathbf{r}) + a_n^\dagger(t)\mathbf{\Lambda}_n^*(\mathbf{r})] = 2\mathbf{\Lambda}(\mathbf{r},t)$$
$$+ \int d^3r' \sum_n \frac{i}{\omega_n}[\mathbf{\Lambda}_n^*(\mathbf{r})\mathbf{\Lambda}_n(\mathbf{r}') - \mathbf{\Lambda}_n(\mathbf{r})\mathbf{\Lambda}_n^*(\mathbf{r}')] \cdot \dot{\mathbf{\Lambda}}(\mathbf{r}',t) . \tag{2.25}$$

The second term on the right-hand side vanishes, because both $\mathbf{\Lambda}_n$ and $\mathbf{\Lambda}_n^*$ are mode functions. In particular, for each n there is an n' such that $\mathbf{\Lambda}_{n'} = \mathbf{\Lambda}_n^*$. This implies that

$$\mathbf{\Lambda}_{n'}(\mathbf{r})\mathbf{\Lambda}_{n'}^*(\mathbf{r}') = \mathbf{\Lambda}_n^*(\mathbf{r})\mathbf{\Lambda}_n(\mathbf{r}') , \tag{2.26}$$

so that the first term in the brackets for n is canceled by the second term in the brackets for n'. This finally gives us our mode expansion for the field

$$\mathbf{\Lambda}(\mathbf{r},t) = \sum_n \sqrt{\frac{\hbar}{2\mu_0\omega_n}}[a_n(t)\mathbf{\Lambda}_n(\mathbf{r}) + a_n^\dagger(t)\mathbf{\Lambda}_n^*(\mathbf{r})] . \tag{2.27}$$

The mode expansion of the $\mathbf{D}$ field can be obtained from this equation by taking the curl of both sides. Similar reasoning gives

$$\dot{\mathbf{\Lambda}}(\mathbf{r},t) = i\sum_n \sqrt{\frac{\hbar\omega_n}{2\mu_0}}[a_n(t)\mathbf{\Lambda}_n(\mathbf{r}) - a_n^\dagger(t)\mathbf{\Lambda}_n^*(\mathbf{r})] , \tag{2.28}$$

which immediately yields the mode expansion for the B field. These expressions can be substituted into (2.17), which will give us the Hamiltonian in terms of the mode creation and annihilation operators.

2.3 Dispersion

A realistic description of the propagation of fields in a nonlinear medium must include the effects of linear dispersion (the effects of nonlinear dispersion are small and can be neglected to lowest order). Dispersion, however, is difficult to incorporate into the standard canonical formulation, because it is an effect which is nonlocal in time. It arises from the fact that the polarization of the medium at time t, $\mathbf{P}(t)$, depends not only on the electric field at time t, but also on its values at previous times [19]

$$\mathbf{P}(t) = \varepsilon_0 \int_0^\infty d\tau \chi^{(1)}(\tau) : \mathbf{E}(t-\tau) . \tag{2.29}$$

There are two known approaches to constructing a quantized nonlinear theory which incorporates dispersion. The first, which was pioneered by Drummond [20], has as its basic objects narrow-band fields for which it is possible to derive an approximate theory which is local. In the second, the degrees of freedom of the medium are included in the theory, and the entire theory, fields plus medium, is local. Each approach has its advantages. The second is more fundamental, but requires a model for the medium which, for many systems of interest, will be complicated. The first is more phenomenological, but needs only a set of numbers, the polarizabilities, to describe the medium. Both are useful, and we shall consider each of them.

In the case of linear theories, a number of other methods of quantizing fields in the presence of dispersive media have been developed. One starts with the equations of motion for the field operators, and then introduces frequency dependence into the linear susceptibility and noise currents [21]. A second starts from the results of a microscopic model and generalizes them to be able to treat arbitrary frequency-dependent susceptibilities [22]. A third is able to define a Lagrangian and Hamiltonian for fields in a medium with arbitrary frequency response by introducing auxiliary fields [23]. This theory can then be quantized in the usual way. These methods have not yet been employed to treat nonlinear media.

Let us first look at the approximate macroscopic theory due to Drummond [20]. The basic field in this theory is the dual potential. It is simplest to start by considering a linear, dispersive medium in which case the electric field is related to the dual potential by

$$E_i(\mathbf{r},t) = \int_0^\infty d\tau \eta_{ij}^{(1)}(\mathbf{r},\tau) D_j(\mathbf{r},t-\tau) \ , \tag{2.30}$$

where $\mathbf{D} = \nabla \times \mathbf{\Lambda}$. As is evident from this equation, $\eta^{(1)}$ is in general a tensor, but we shall assume, for the sake of simplicity, that the medium is isotropic which implies that $\eta^{(1)}$ is a scalar. The relation between $\mathbf{E}$ and $\mathbf{D}$ and Maxwell's equations imply that $\mathbf{\Lambda}(\mathbf{r},t)$ satisfies the equation

$$\nabla \times \int_0^\infty d\tau \eta^{(1)}(\mathbf{r},\tau)[\nabla \times \mathbf{\Lambda}(\mathbf{r},t-\tau)] = -\mu_0 \ddot{\mathbf{\Lambda}}(\mathbf{r},t) \ . \tag{2.31}$$

which is clearly nonlocal in time. Now suppose that $\mathbf{\Lambda}^\nu$ is a narrow-band field with frequency components near ω_ν (that is, $\mathbf{\Lambda}^\nu \sim \exp(-i\omega_\nu t)$), and that $\mathbf{\Lambda}$ can be expressed as

$$\mathbf{\Lambda} = \mathbf{\Lambda}^\nu + (\mathbf{\Lambda}^\nu)^* \ . \tag{2.32}$$

The field $\mathbf{\Lambda}^\nu$ will also satisfy (2.31). Let us now define

$$\eta^{(1)}(\mathbf{r},\omega) = \int_0^\infty d\tau e^{i\omega\tau} \eta^{(1)}(\mathbf{r},\tau) \ , \tag{2.33}$$

and, because we are interested in frequencies near ω_ν, expand this quantity up to second order in $\omega - \omega_\nu$

$$\eta^{(1)}(\mathbf{r},\omega) \cong \eta_\nu(\mathbf{r}) + \omega\eta_\nu'(\mathbf{r}) + \frac{1}{2}\omega^2\eta_\nu''(\mathbf{r}) \ , \tag{2.34}$$

where

$$\eta_\nu(\mathbf{r}) \equiv \eta^{(1)}(\mathbf{r},\omega_\nu) - \omega_\nu \frac{d\eta^{(1)}}{d\omega}(\mathbf{r},\omega_\nu) + \frac{1}{2}\omega^2 \frac{d^2\eta^{(1)}}{d\omega^2}(\mathbf{r},\omega_\nu) ,$$
$$\eta'_\nu(\mathbf{r}) \equiv \frac{d\eta^{(1)}}{d\omega}(\mathbf{r},\omega_\nu) - \omega_\nu \frac{d^2\eta^{(1)}}{d\omega^2}(\mathbf{r},\omega_\nu) ,$$
$$\eta''_\nu(\mathbf{r}) \equiv \frac{d^2\eta^{(1)}}{d\omega^2}(\mathbf{r},\omega_\nu) . \tag{2.35}$$

We now consider the wave equation which $\mathbf{\Lambda}^\nu$ satisfies (2.31), and make the following approximation. The quantity $\exp(-i\omega_\nu\tau)\mathbf{\Lambda}^\nu(t-\tau)$ is a slowly varying function of τ, so we expand it in a Taylor series in τ up to second order. Taylor expansions are often used in this way in classical dispersion theory to simplify wave equations, and this technique was first introduced into macroscopic field quantization by Kennedy and Wright [24]. When this expansion is inserted into (2.31) we find

$$-\mu_0\ddot{\mathbf{\Lambda}} = \nabla \times [\eta_\nu \nabla \times \mathbf{\Lambda}^\nu + i\eta'_\nu \nabla \times \dot{\mathbf{\Lambda}}^\nu - \frac{1}{2}\eta''_\nu \nabla \times \ddot{\mathbf{\Lambda}}^\nu] , \tag{2.36}$$

which is a local equation for $\mathbf{\Lambda}^\nu$, which can, in turn, be derived from a local Lagrangian. The Lagrangian density from which it follows is

$$\mathcal{L} = \frac{1}{2}[\mu_0(\dot{\mathbf{\Lambda}}^\nu)^* \cdot \dot{\mathbf{\Lambda}}^\nu - (\nabla \times \mathbf{\Lambda}^\nu)^* \cdot \eta_\nu(\nabla \times \mathbf{\Lambda}^\nu)$$
$$-i(\nabla \times \mathbf{\Lambda}^\nu)^* \cdot \eta'_\nu(\nabla \times \dot{\mathbf{\Lambda}}^\nu) - \frac{1}{2}(\nabla \times \dot{\mathbf{\Lambda}}^\nu)^*\eta''_\nu(\nabla \times \dot{\mathbf{\Lambda}}^\nu)] , \tag{2.37}$$

which gives the canonical momentum

$$\mathbf{\Pi}^\nu = \mu_0(\dot{\mathbf{\Lambda}}^\nu)^* - \frac{1}{2}\nabla \times [\eta''_\nu(\nabla \times \dot{\mathbf{\Lambda}}^\nu)^* + i\eta'_\nu(\nabla \times \mathbf{\Lambda}^\nu)^*] . \tag{2.38}$$

Finally, from the Lagrangian and the canonical momentum we find the Hamiltonian

$$H = \frac{1}{2}\int d^3r[\mu_0(\dot{\mathbf{\Lambda}}^\nu)^* \cdot \dot{\mathbf{\Lambda}}^\nu + (\nabla \times \mathbf{\Lambda}^\nu)^* \cdot \eta_\nu(\nabla \times \mathbf{\Lambda}^\nu)$$
$$-\frac{1}{2}(\nabla \times \dot{\mathbf{\Lambda}}^\nu)^*\eta''_\nu(\nabla \times \dot{\mathbf{\Lambda}}^\nu)] , \tag{2.39}$$

where $\dot{\mathbf{\Lambda}}^\nu$ is to be considered a function of $\mathbf{\Pi}^\nu$. As has been shown by Drummond [20], this Hamiltonian is the energy for a classical field in a dispersive dielectric. He has also emphasized that the Lagrangian for the theory, which is not unique (it can, for example, be scaled by an arbitrary factor and the equations of motion will be unaffected) should be chosen so that it does give a Hamiltonian which is the classical energy.

If the dispersion in the nonlinear interaction is ignored, so that the interaction is considered local, it can be included in the theory by adding the terms

$$-\int d^3r[\frac{1}{3}\mathbf{D} \cdot \eta^{(2)} : \mathbf{DD} + \frac{1}{4}\mathbf{D} \cdot \eta^{(3)} : \mathbf{DDD}] , \tag{2.40}$$

to the Hamiltonian in (2.39). In addition, if fields with several discrete frequencies are present, they can be accommodated by adding additional fields, $\mathbf{\Lambda}^{\nu}$, centered about these frequencies to the theory. This has the effect of adding summations over ν to the Lagrangian density and Hamiltonian in (2.37) and (2.39).

In order to quantize the theory we would like to simply impose the commutation relations

$$[\Lambda_j^{\nu}(\mathbf{r},t), \Pi_{j'}^{\nu}(\mathbf{r}',t)] = i\hbar\delta_{j,j'}^{(\mathrm{tr})}(\mathbf{r}-\mathbf{r}') . \tag{2.41}$$

However, in order for this to be true the fields must have Fourier components of arbitrarily high frequency, while $\mathbf{\Lambda}^{\nu}$ is limited in bandwidth. An alternative, and in this case better, approach is to expand the field in terms of spatial modes and to use the expansion coefficients as coordinates. One then finds the corresponding canonical momentum for each coordinate and then imposes the usual commutation relations between coordinates and momenta.

Before proceeding it is useful to consider a simplified version of the theory. We shall consider a field consisting of plane waves of area A which are polarized in the y direction and propagate in the x direction. In this case the field, $\Lambda^{\nu}(x,t)$ is a scalar and a function of only one spatial coordinate. If, in addition, we assume that the only nonlinearity present is described by $\eta^{(3)}$ and that the medium is homogeneous, we find that the Hamiltonian is

$$\begin{aligned} H = \int dx\{\mu_0|\dot{\Lambda}^{\nu}|^2 + \eta_{\nu}|\partial_x\Lambda^{\nu}|^2 - \frac{1}{2}\eta_{\nu}''|\partial_x\dot{\Lambda}^{\nu}|^2 \\ +\frac{1}{4}\eta^{(3)}[\partial_x(\Lambda^{\nu}+(\Lambda^{\nu})^*)]^4\}A . \end{aligned} \tag{2.42}$$

We begin the quantization of this theory by expanding Λ^{ν} as

$$\Lambda^{\nu}(x,t) = \sqrt{\frac{1}{V}}\sum_k \lambda_k e^{ikx} , \tag{2.43}$$

where $V = LA$, L is the quantization length and the momenta, k, satisfy periodic boundary conditions. In terms of the λ_k the linear Lagrangian density becomes

$$\begin{aligned} \mathcal{L}_0 = \sum_k [\dot{\lambda}_k^*(\mu_0 - \frac{1}{2}k^2\eta_{\nu}'')\dot{\lambda}_k - k^2\lambda_k^*\eta_{\nu}\lambda_k \\ -\frac{i}{2}k^2\eta_{\nu}'(\lambda_k^*\dot{\lambda}_k - \dot{\lambda}_k^*\lambda_k)] . \end{aligned} \tag{2.44}$$

From this, treating λ_k and λ_k^* as independent coordinates, we find that the momentum conjugate to λ_k is (note that this expression will be unchanged by the nonlinear interaction, because this interaction does not depend on $\dot{\lambda}_k$)

$$\pi_k = (\mu_0 - \frac{1}{2}k^2\eta_{\nu}'')\dot{\lambda}_k^* - \frac{i}{2}k^2\eta_{\nu}'\lambda_k^* , \tag{2.45}$$

and that conjugate to λ_k^* is

$$\pi_k^* = (\mu_0 - \frac{1}{2}k^2\eta_\nu'')\dot{\lambda}_k + \frac{i}{2}k^2\eta_\nu'\lambda_k \ . \tag{2.46}$$

The theory is now quantized by imposing the commutation relations

$$[\lambda_k, \pi_{k'}] = i\hbar\delta_{k,k'}, \qquad [\lambda_k^*, \pi_{k'}^*] = i\hbar\delta_{k,k'} \ . \tag{2.47}$$

It is also possible to define two sets of creation and annihilation operators

$$\begin{aligned} a_k &= \frac{1}{\sqrt{2\hbar}}\left[A_k\lambda_k + \left(\frac{i}{A_k^*}\right)\pi_k^*\right] , \\ b_k^\dagger &= \frac{1}{\sqrt{2\hbar}}\left[A_k\lambda_k - \left(\frac{i}{A_k^*}\right)\pi_k^*\right] , \end{aligned} \tag{2.48}$$

where $A_k = [(\mu_0 - k^2\eta_\nu''/2)k^2\eta_\nu + k^4(\eta_\nu')^2]^{1/4}$. Expressing the linear part of the Hamiltonian in terms of the creation and annihilation operators we find

$$H_0 = \hbar\sum_k [\omega_+(k)a_k^\dagger a_k + \omega_-(k)b_k^\dagger b_k] , \tag{2.49}$$

where $\omega_\pm(k)$ are solutions of the equation

$$\mu_0\omega_\pm^2 = k^2(\eta_\nu \pm \omega_\pm\eta_\nu' + \frac{1}{2}\omega_\pm^2\eta_\nu'') \ . \tag{2.50}$$

Note that the expression in parentheses in the above equation is identical to the expansion of $\eta^{(1)}(\omega)$ about ω_ν if the plus sign is used. Because $\eta^{(1)}(\omega) = 1/\varepsilon(\omega)$, where $\varepsilon(\omega)$ is the usual frequency-dependent dielectric function, (2.50) for ω_+ is approximately the same as

$$\omega = \frac{k}{\sqrt{\mu_0\varepsilon(\omega)}} , \tag{2.51}$$

which is the relation between ω and k we would expect for a wave traveling in a linear dielectric medium. This leaves us with the question of how to interpret ω_- and b_k.

An examination of (2.48) shows us that

$$\lambda_k = \frac{\sqrt{\hbar}}{A_k\sqrt{2}}(a_k + b_k^\dagger) , \tag{2.52}$$

while (2.49) implies that $b_k \sim \exp(-i\omega_- t)$. Now ω_- is not too far from ω_ν, which implies that the $b_k^\dagger$ term in λ_k has a time dependence given approximately by $\exp(i\omega_\nu t)$. This places it outside the bandwidth for the field Λ^ν. In order to be consistent we must assume that all of the b_k modes are in the vacuum state and, thereby, drop these operators from the theory. This implies that the Hamiltonian for the full theory (as opposed to just the linear part) is

$$H = \hbar \sum_k \omega_+(k) a_k^\dagger a_k + \frac{1}{4}\eta^{(3)} \int dx (\partial_x (\Lambda^\nu + \Lambda^{\nu\dagger}))^4 \ , \tag{2.53}$$

where

$$\Lambda^\nu = \sqrt{\frac{\hbar}{2V}} \sum_k \frac{1}{A_k} a_k e^{ikx} \ . \tag{2.54}$$

We now have a theory which is capable of describing the propagation of quantum fields in nonlinear, dispersive media. Carter and Drummond have applied this theory to describe fields propagating through a fiber with a $\chi^{(3)}$ nonlinearity [15], and we shall briefly give the theory in the form in which they use it. They first take the limit $L \to \infty$ and replace the operator a_k by $a(k) = a_k \sqrt{L/2\pi}$. These operators have the commutation relations

$$[a(k), a^\dagger(k')] = \hbar\delta(k - k') \ . \tag{2.55}$$

They then define a slowly varying field

$$\Psi(x,t) = \sqrt{\frac{1}{2\pi}} \int dk \ a(k) \exp[i(k - k_\nu) + i\omega_\nu t] \ , \tag{2.56}$$

where k_ν is the wave vector corresponding to ω_ν. This field has equal time commutation relations given by

$$[\Psi(x,t), \Psi^\dagger(x',t)] = \tilde{\delta}(x - x') \ , \tag{2.57}$$

where $\tilde{\delta}$, which because it has a band-limited Fourier transform, is a smoothed version of the Dirac delta function. The Hamiltonian can be expressed in terms of Ψ and is given by

$$\begin{aligned} H = \frac{\hbar}{2} \int_0^L dx [iv \left(\frac{\partial \Psi^\dagger}{\partial x} \Psi - \Psi^\dagger \frac{\partial \Psi}{\partial x} \right) + \omega'' \frac{\partial \Psi^\dagger}{\partial x} \frac{\partial \Psi}{\partial x} \\ - v^2 \chi^E (\Psi^\dagger)^2 \Psi^2] \ , \end{aligned} \tag{2.58}$$

where v is the group velocity, χ^E is proportional to $\chi^{(3)}$ and ω'' is the second derivative of frequency with respect to wave number evaluated at k_ν. The equation of motion for Ψ is essentially a nonlinear Schrödinger equation which has solitons as solutions.

2.4 Microscopic Approach

The fundamental microscopic approach to the theory of quantum fields in a linear medium is contained in a paper by Hopfield [25]. The medium is described by a polarization field, which in the absence of coupling to the electromagnetic field, oscillates freely at a frequency ω_0. This field is then coupled linearly to the electromagnetic field. After diagonalizing the resulting Hamiltonian, one finds a theory of noninteracting excitations, which are known as polaritons. These are mixtures of matter and field excitations, and they obey dispersion relations which are different from that of photons. More recently, the effect of loss reservoirs has been incorporated into this theory by Knöll and Leonhardt [26], who also treated the scattering problem, i.e. a field generated by sources in free space propagates through a finite medium and is subsequently measured, and by Huttner and Barnett [27].

Let us look at Hopfield's theory in more detail. From this point on, to simplify the notation, all units will be chosen with natural (equivalent to Heaviside-Lorentz) dimensions, so that $\hbar = c = \mu_0 = \varepsilon_0 = 1$.

The basic fields are the vector potential, $\mathbf{A}(\mathbf{r}, t)$ and the polarization field, $\mathbf{P}(\mathbf{r}, t)$. The Lagrangian density is given by

$$\begin{aligned}\mathcal{L} = & \frac{1}{2}(\mathbf{E}^2 - \mathbf{B}^2) + \frac{1}{2\omega_0^2 \eta}\left(\frac{\partial \mathbf{P}}{\partial t}\right)^2 - \frac{1}{2\eta}\mathbf{P}^2 \\ & + A_0(\nabla \cdot \mathbf{P})^2 + \mathbf{A} \cdot \frac{\partial \mathbf{P}}{\partial t} ,\end{aligned} \tag{2.59}$$

where η is a constant which describes the coupling of the polarization to the electromagnetic field. This can be seen from the equation of motion for $\mathbf{P}$

$$\frac{1}{\omega_0^2}\frac{\partial^2 \mathbf{P}}{\partial t^2} + \mathbf{P} = \eta \mathbf{E} . \tag{2.60}$$

The Lagrangian density is used to find canonical momenta and a Hamiltonian. If we express the vector potential and the polarization field in terms of creation and annihilation operators

$$\begin{aligned}\mathbf{A} &= \sum_{\mathbf{k},\lambda} \sqrt{\frac{1}{2Vk}}\hat{\mathbf{e}}_\lambda(\mathbf{k})(a_{\mathbf{k},\lambda}e^{i\mathbf{k}\cdot\mathbf{r}} + a_{\mathbf{k},\lambda}^\dagger e^{-i\mathbf{k}\cdot\mathbf{r}}) , \\ \mathbf{P} &= \sum_{\mathbf{k},\lambda} \sqrt{\frac{\omega_0 \eta}{2Vk}}\hat{\mathbf{e}}_\lambda(\mathbf{k})(b_{\mathbf{k},\lambda}e^{i\mathbf{k}\cdot\mathbf{r}} + b_{\mathbf{k},\lambda}^\dagger e^{-i\mathbf{k}\cdot\mathbf{r}}) ,\end{aligned} \tag{2.61}$$

then the Hamiltonian is

$$\begin{aligned}H = & \sum_{\mathbf{k},\lambda}[k a_{\mathbf{k},\lambda}^\dagger a_{\mathbf{k},\lambda} + \omega_0 b_{\mathbf{k},\lambda}^\dagger b_{\mathbf{k},\lambda} + \frac{i\omega_0^2\sqrt{\eta}}{2\sqrt{k\omega_0}}(a_{\mathbf{k},\lambda}^\dagger b_{\mathbf{k},\lambda} - a_{\mathbf{k},\lambda} b_{\mathbf{k},\lambda}^\dagger \\ & + a_{-\mathbf{k},\lambda} a_{\mathbf{k},\lambda} - a_{-\mathbf{k},\lambda}^\dagger b_{\mathbf{k},\lambda}^\dagger) + \frac{\eta\omega_0^2}{k}(a_{\mathbf{k},\lambda} a_{\mathbf{k},\lambda}^\dagger + a_{\mathbf{k},\lambda}^\dagger a_{\mathbf{k},\lambda} \\ & + a_{\mathbf{k},\lambda}^\dagger a_{-\mathbf{k},\lambda}^\dagger + a_{\mathbf{k},\lambda} a_{-\mathbf{k},\lambda})] .\end{aligned} \tag{2.62}$$

In this Hamiltonian the coupling between the polarization and the electromagnetic field is evident from the terms containing both polarization and field operators. What one now wants to do is to express the Hamiltonian in terms of new creation and annihilation operators so that it represents non-interacting excitations. We first note that the different polarizations do not couple to each other and that $\mathbf{k}$ only couples to $-\mathbf{k}$. Therefore, in our diagonalization problem we need only consider one polarization and the two wave vectors $\mathbf{k}$ and $-\mathbf{k}$ at a time. Dropping the polarization index, we define a new annihilation operator

$$\alpha_{\mathbf{k}} = wa_{\mathbf{k}} + xb_{\mathbf{k}} + ya^{\dagger}_{-\mathbf{k}} + zb^{\dagger}_{-\mathbf{k}} , \tag{2.63}$$

where the coefficients w, x, y, and z are determined by the equation

$$[\alpha_{\mathbf{k}}, H] = E_{\mathbf{k}}\alpha_{\mathbf{k}} . \tag{2.64}$$

Equation (2.64) not only guarantees that H will simply be a free field Hamiltonian when expressed in terms of $\alpha_{\mathbf{k}}$ and $\alpha^{\dagger}_{\mathbf{k}}$, but it gives us an equation for $E_{\mathbf{k}}$ as well,

$$\frac{k^2}{E_{\mathbf{k}}} = 1 + \frac{\eta}{1 - E^2_{\mathbf{k}}/\omega_0^2} . \tag{2.65}$$

For each wave vector $\mathbf{k}$ there are two solutions for $E_{\mathbf{k}}$, which means that the dispersion relation has two branches. The first (lower) branch starts at 0 and is linear (photon-like) for k/ω_0 small. As k approaches ω_0, however, this branch bends over and approaches ω_0 asymptotically. The second (upper) branch starts at a value slightly larger that ω_0 and is approximately constant for k/ω_0 small. As k approaches ω_0, the branch bends upward and for large k is photon-like. The Hamiltonian can then be expressed as

$$H = \sum_{\mathbf{k},\lambda} \left[E_+(k)\alpha^{(+)\dagger}_{\mathbf{k},\lambda}\alpha^{(+)}_{\mathbf{k},\lambda} + E_-(k)\alpha^{(-)\dagger}_{\mathbf{k},\lambda}\alpha^{(-)}_{\mathbf{k},\lambda} \right] , \tag{2.66}$$

where the plus sign refers to the upper branch of the dispersion relation and the minus sign to the lower. The excitations created by $\alpha^{(+)\dagger}_{\mathbf{k},\lambda}$ and $\alpha^{(-)\dagger}_{\mathbf{k},\lambda}$ are referred to as polaritons, and as can be seen from (2.63), they represent mixed matter-field modes.

The advent of quantum wells in semiconductors has led to a consideration of polaritons in confined structures [28,29]. Here the polarization is restricted to only a part of the quantization volume with the result that the polariton modes acquire finite lifetimes. The theory developed in this context has been applied to a microscopic treatment of the propagation of light through a finite medium, and it permits one to treat the full scattering problem for linear media [30,31]. This represents one extension of Hopfield's theory.

What we need to do now, however, is extend Hopfield's theory to include nonlinear effects. As a first step we need a model for the nonlinear medium itself [13,14]. We shall consider one consisting of two-level atoms. The atoms will occupy the entire quantization volume, V, and their density ρ will be such that there are a large number of atoms per cubic optical wavelength. Because we shall only consider optical wavelengths we can partition the medium into small boxes. The size of each box is much less than a wavelength, but it, nonetheless, contains a large number of atoms which we shall assume to be the same for each box and shall call n_0. Because the size of a box is small compared to a wavelength each atom in the box sees the same field. Consequently, the atoms in each box can be described as a spin $s = n_0/2$ object which interacts with the field. The Hamiltonian describing the boxes interacting with the electromagnetic field is

$$H = \sum_{l=1}^{N_b} E_0 S_l^{(3)} + \sum_{|\mathbf{k}|<k_u} \omega_k a_{\mathbf{k}}^{\dagger} a_{\mathbf{k}} + \frac{e^2 n_0}{2m} \sum_{l=1}^{N_b} \mathbf{A}^2(\mathbf{r}_l) + H_{\text{int}} , \tag{2.67}$$

where

$$H_{\text{int}} = \sum_{l=1}^{N_b} \sum_{|\mathbf{k}|<k_u} \sqrt{\frac{1}{2\omega_k V}} (a_{\mathbf{k}} e^{i\mathbf{k}\cdot\mathbf{r}_l} + a_{\mathbf{k}}^{\dagger} e^{-i\mathbf{k}\cdot\mathbf{r}_l}) \times i\mu(\mathbf{k})(S_l^{(+)} - S_l^{(-)}) . \tag{2.68}$$

Here we have assumed only one polarization is present, E_0 is the energy difference between the levels, $S_l^{(3)}$, $S_l^{(+)}$, and $S_l^{(-)}$ are the spin operators for the lth box, N_b is the number of boxes, and

$$\mu(\boldsymbol{k}) = eE_0 \langle a|\boldsymbol{x}|b\rangle \cdot \hat{\mathbf{e}}(\boldsymbol{k}) . \tag{2.69}$$

The matrix element $\langle a|\mathbf{x}|b\rangle$ is just the dipole matrix element of the atom and $\hat{\mathbf{e}}(\boldsymbol{k})$ is the field polarization vector. The wave number k_u is a cutoff imposed to guarantee that the wavelength does not become smaller than the box size.

In order to proceed we want to expand the spin operators. This can be done by using the Holstein–Primakoff representation of the spin operators in terms of boson creation and annihilation operators $\zeta^{\dagger}$ and ζ. We have [32]

$$S^{(-)} = (2s - \zeta^{\dagger}\zeta)^{\frac{1}{2}}\zeta , \quad S^{(+)} = \zeta^{\dagger}(2s - \zeta^{\dagger}\zeta)^{\frac{1}{2}} , \quad S^{(3)} = -s + \zeta^{\dagger}\zeta . \tag{2.70}$$

The excitation number for the boson operators, i.e. $\zeta^{\dagger}\zeta$, corresponds to $s_3 + s$, where s_3 is the eigenvalue of $S^{(3)}$. Therefore, the boson vacuum state corresponds to the spin pointing down, i.e. all atoms in their ground states. If we are only considering states whose excitation number is small we can expand the square roots

$$S^{(-)} = \sqrt{2s}(1 - \frac{1}{4s}\zeta^{\dagger}\zeta)\zeta , \qquad S^{(+)} = \sqrt{2s}\zeta^{\dagger}(1 - \frac{1}{4s}\zeta^{\dagger}\zeta) . \tag{2.71}$$

In our model of a nonlinear medium the fraction of atoms in each block which is excited is small because we are off resonance, and, therefore, the use of this expansion is justified.

For each spin operator in the Hamiltonian we substitute the corresponding expressions in terms of the creation and annihilation operators $\zeta_l^\dagger$ and ζ_l (each box has its own set of creation and annihilation operators). In addition we go to a continuous representation where ζ_l is replaced by $\zeta(\mathbf{r})$ where

$$[\zeta(\mathbf{r}), \zeta^\dagger(\mathbf{r}')] = \delta^{(3)}(\mathbf{r} - \mathbf{r}') \; . \tag{2.72}$$

The resulting Hamiltonian is

$$H = H_0 + H_{\mathrm{int}}^{(1)} + H_{\mathrm{int}}^{(2)} \; , \tag{2.73}$$

where

$$\begin{aligned} H_0 &= -\frac{1}{2} N E_0 + E_0 \int_V d^3r \; \zeta^\dagger(\mathbf{r})\zeta(\mathbf{r}) + \sum_{|\mathbf{k}|<k_u} \omega_k a_{\mathbf{k}}^\dagger a_{\mathbf{k}} \\ &\quad + \frac{e^2\rho}{2m} \sum_{|\mathbf{k}|<k_u} \frac{1}{2\omega_k} [a_{\mathbf{k}} a_{-\mathbf{k}} + a_{\mathbf{k}} a_{\mathbf{k}}^\dagger + a_{\mathbf{k}}^\dagger a_{\mathbf{k}} + a_{\mathbf{k}}^\dagger a_{-\mathbf{k}}^\dagger] \; , \\ H_{\mathrm{int}}^{(1)} &= \frac{i}{\sqrt{V}} \sum_{|\mathbf{k}|<k_u} \int_V d^3r \; g_{\mathbf{k}} (a_{\mathbf{k}} e^{i\mathbf{k}\cdot\mathbf{r}} + a_{\mathbf{k}}^\dagger e^{-i\mathbf{k}\cdot\mathbf{r}}) \left[\zeta^\dagger(\mathbf{r}) - \zeta(\mathbf{r})\right] \; , \\ H_{\mathrm{int}}^{(2)} &= -\frac{i}{2\rho\sqrt{V}} \sum_{|\mathbf{k}|<k_u} \int_V d^3r \; g_{\mathbf{k}} (a_{\mathbf{k}} e^{i\mathbf{k}\cdot\mathbf{r}} + a_{\mathbf{k}}^\dagger e^{-i\mathbf{k}\cdot\mathbf{r}}) \\ &\quad \times \left[\zeta^\dagger(\mathbf{r})^2\zeta(\mathbf{r}) - \zeta^\dagger(\mathbf{r})\zeta(\mathbf{r})^2\right] \; . \end{aligned} \tag{2.74}$$

Here $g_{\mathbf{k}} = \mu(\mathbf{k})\sqrt{\rho/2\omega_k}$ and N is the total number of atoms.

The Hamiltonian $H_0 + H_{\mathrm{int}}^{(1)}$ describes the electromagnetic field interacting with a linear medium and is closely related to the Hamiltonian considered by Hopfield [25]. This part of the total Hamiltonian can be diagonalized by transforming to polaritons.

The operator $H_{\mathrm{int}}^{(2)}$ describes the nonlinear interaction between the field and the medium. It is not in a form which is familiar from nonlinear optics. In particular, for a single-mode field interacting with a two-level atom medium we would expect a Hamiltonian more like the one which is used to describe self-phase modulation

$$H_{\mathrm{int}} = \lambda (a^\dagger)^2 a^2 \; . \tag{2.75}$$

Is it possible to extract an interaction of this form from $H_{\mathrm{int}}^{(2)}$?

The answer to this question is yes, and the key to answering it is polaritons. One first diagonalizes $H_0 + H_{\mathrm{int}}^{(1)}$ in terms of new operators $\alpha_{\mathbf{k}}$ and $\eta_{\mathbf{k}}$ which are linear combinations of $a_{\mathbf{k}}$, $\zeta_{\mathbf{k}}$, $a_{-\mathbf{k}}^\dagger$, and $\zeta_{-\mathbf{k}}^\dagger$, and satisfy boson commutation relations. Here

$$\zeta_{\mathbf{k}} = \frac{1}{\sqrt{V}} \int_V d^3r e^{-i\mathbf{k}\cdot\mathbf{r}} \zeta(\mathbf{r}) \; . \tag{2.76}$$

One then has

$$H_0 + H_{\rm int}^{(1)} = \sum_{|\mathbf{k}|<k_u} [E_1(k)\alpha_{\mathbf{k}}^\dagger \alpha_{\mathbf{k}} + E_2(k)\eta_{\mathbf{k}}^\dagger \eta_{\mathbf{k}}] \,, \tag{2.77}$$

where

$$\begin{aligned} E_1(k) &= \frac{1}{\sqrt{2}} \Big\{ [E_0^2 + \omega_k(\omega_k + 2C_0)] \\ &\quad + [[E_0^2 - \omega_k(\omega_k + 2C_0)]^2 + 16E_0\omega_k g_{\mathbf{k}}^2]^{1/2} \Big\}^{1/2} \,, \\ E_2(k) &= \frac{1}{\sqrt{2}} \Big\{ [E_0^2 + \omega_k(\omega_k + 2C_0)] \\ &\quad - [[E_0^2 - \omega_k(\omega_k + 2C_0)]^2 + 16E_0\omega_k g_{\mathbf{k}}^2]^{1/2} \Big\}^{1/2} \,, \end{aligned} \tag{2.78}$$

with $C_0 = (e^2\rho)/(2m\omega_k)$. The next step is to express $H_{\rm int}^{(2)}$ in terms of $\alpha_{\mathbf{k}}$ and $\eta_{\mathbf{k}}$. This results in a great many terms, but a considerable simplification occurs if only a small number of modes are highly excited, and we keep only terms containing these modes. For example, if the $\alpha_{\mathbf{k}_0}$ mode is the only one which is highly excited we find that

$$H_{\rm int}^{(2)} \to \lambda(\mathbf{k}_0)(\alpha_{\mathbf{k}_0}^\dagger)^2 \alpha_{\mathbf{k}_0}^2 \,, \tag{2.79}$$

where $\lambda(\mathbf{k}_0)$ is a constant which is found from the transformation which relates the polariton to the matter and field operators. By going to the polariton basis we have recovered the self-phase modulation interaction. This tells us, in addition, that the operators in (2.75) should not be field operators, but should, in fact, correspond to mixed matter-field modes.

Suppose that instead of a single mode we have a pulse. In particular, let us assume that the pulse consists of $\alpha(\mathbf{k})$ modes where $\mathbf{k}$ is in a small neighborhood, S, of $\mathbf{k}_0$. In that case we have

$$H_{\rm int}^{(2)} \to \lambda(\mathbf{k}_0) \sum_{\mathbf{k}\in S} \sum_{\mathbf{k}_1\in S} \sum_{\mathbf{k}_2\in S} \sum_{\mathbf{k}_3\in S} \delta_{\mathbf{k}+\mathbf{k}_1,\mathbf{k}_2+\mathbf{k}_3} \alpha_{\mathbf{k}}^\dagger \alpha_{\mathbf{k}_1}^\dagger \alpha_{\mathbf{k}_2} \alpha_{\mathbf{k}_3} \,. \tag{2.80}$$

The equations of motion for the operators $\alpha_{\mathbf{k}}$ which emerge from the theory obtained by combining (2.77) and (2.80) are similar in form to those obtained for the mode annihilation operators in the Carter–Drummond theory.

Finally, let us note two features of the microscopic theory. First, dispersion is included in a natural way because the basic modes in the theory correspond to polaritons and not photons. The polariton dispersion relation is different from that of photons. The second concerns operator ordering. The polariton operators do not appear in the Hamiltonian in normal order. In deriving (2.79) and (2.80) we have dropped terms arising from commutators. We expect these terms to be small in the situations considered here, but this will not always be the case. The microscopic theory gives us a way of examining this issue while the macroscopic does not.

2.5 Conclusion

We have examined three different approaches to the quantization of the electromagnetic field in the presence of a nonlinear dielectric medium: the phenomenological, the macroscopic, and the microscopic. The phenomenological approach has found the greatest use, but it is incorrect and does not lead to Maxwell's equations as equations of motion. The macroscopic theory is built on firmer foundations and has been applied to the study of the propagation of quantum fields in fibers. The microscopic approach is the most fundamental, but its development is at an early stage.

The study of the propagation of quantized fields in nonlinear media is an area of increasing experimental and theoretical activity. It is essential that we achieve a better understanding of the underlying theory for the field to proceed.

References

1. D. Stoler, Phys. Rev. D **1**, 3217 (1970)
2. D. Stoler, Phys. Rev. D **4**, 1925 (1971)
3. M. Kozierowski and R. Tanaś, Opt. Commun. **21**, 229 (1977)
4. G.J. Milburn, Phys. Rev. A **33**, 674 (1986)
5. P.D. Drummond and S. Carter, J. Opt. Soc. Am. B **4**, 1565 (1987)
6. M. Rosenbluh and R.M. Shelby, Phys. Rev. Lett. **66**, 153 (1991)
7. S. Carter, P.D. Drummomd, M. Reid, and R. Shelby, Phys. Rev. Lett. **58**, 1841 (1987)
8. H. Haus and Y. Lai, J. Opt. Soc. Am. B **7**, 386 (1990)
9. H. Haus, K. Watanabe and Y. Yamamoto, J. Opt. Soc. Am. B **6**, 1138 (1989)
10. K. Watanabe and Y. Yamamoto, Phys. Rev. A **42**, 1699 (1990)
11. S. Friberg, S. Machida and Y. Yamamoto, Phys. Rev. Lett. **69**, 3165 (1992)
12. M. Hillery and L. Mlodinow, Phys. Rev. A **30**, 1860 (1984)
13. M. Hillery and L. Mlodinow, Phys. Rev. A **31**, 797 (1985)
14. M. Hillery and L. Mlodinow, Phys. Rev. A **55**, 678 (1997)
15. S. Carter and P.D. Drummond, Phys. Rev. Lett. **67**, 3757 (1991); P.D. Drummond, R.M. Shelby, S.R. Friberg and Y. Yamamoto, Nature **365**, 307 (1993); P.D. Drummond, Adv. Chem. Phys. **85**, 379 (1994); P.D. Drummond and J.F. Corney, J. Opt. Soc. Am. B **18**, 139 (2001)
16. J.M. Jauch and K.M. Watson, Phys. Rev. **74**, 950 (1948); **74**, 1485 (1948); **75**, 1249 (1948)
17. C.K. Carniglia and L. Mandel, Phys. Rev. D **3**, 280 (1971)
18. R.J. Glauber and M. Lewenstein, Phys. Rev. A **43**, 467 (1991)
19. L. Landau and E. Lifshitz, *Electrodynamics of Continuous Media* (Pergamon, Oxford 1960)
20. P.D. Drummond, Phys. Rev. A **42**, 6845 (1990)
21. R. Matloob and R. Loudon, Phys. Rev. A **52**, 4823 (1995)
22. T. Gruner and D.-G. Welsch, Phys. Rev. A **51**, 3246 (1995)
23. A. Tip, Phys. Rev. A **57**, 4818 (1997)
24. T.A.B. Kennedy and E.M. Wright, Phys. Rev. A **38**, 212 (1988)

25. J.J. Hopfield, Phys. Rev. **112**, 1555 (1958)
26. L. Knöll and U. Leonhardt, J. Mod. Opt. **39**, 1253 (1992)
27. B. Huttner and S. Barnett, Phys. Rev. A **46**, 4306 (1992)
28. S. Jorda, U. Rössler and D. Broido, Phys. Rev. B **48**, 1669 (1992)
29. V. Savona, Z. Hradil, A. Quattropani and P. Schwendimann, Phys. Rev. B **49**, 8774 (1994)
30. S. Savasta and R. Girlanda, Phys. Rev. A **53**, 2716 (1996)
31. Z. Hradil, Phys. Rev. A **53**, 3687 (1996)
32. T. Holstein and H. Primakoff, Phys. Rev. **58**, 1098 (1940)

3 Input-Output Theory

B. Yurke

In the early 1980's calculations were performed on the degree of squeezing that could be achieved within cavity parametric amplifiers. These calculations indicated that the field inside the cavity could be squeezed by no more than a factor of two under steady state conditions [1]. As a result lore spread that cavities were bad for squeezing. It turns out, however, that because of a remarkable interference effect between the field that is reflected off of the input port mirror, and the field that – after entering the cavity – is squeezed and then re-emitted, the field external to the cavity can exhibit arbitrarily large amounts of squeezing: even though the field inside the cavity only exhibits a factor of two squeezing [2]. How to calculate the quantum statistical properties of the field exiting a cavity given the input field and the internal field had become an important issue that needed to be solved for the advancement of the squeezed state field. Works by Yurke and Denker [3], Yurke [4], and Collett and Gardner [5–7] were instrumental in showing the way. This is, in part, what input-output theory is about.

But input-output theory has much more to offer. The equations of motion of the external fields and the canonical commutation relations that the field operators must satisfy place powerful constraints, in fact, often dictating the quantum behaver of the systems with which they interact. This theme will be developed through much of this chapter. Recent developments in input-output theory include the formulation of better mathematical machinery for dealing with nonlinearities [8], the development of machinery for dealing with random media and cavities of complex shape [9], and the application of input-output theory to fermion systems [10].

For a book whose subject matter currently finds its primary application in optics this chapter, with its emphases on mechanical systems and exactly solvable systems, may seem a bit quirky. However, it is not out of line with how quantum mechanics is usually taught. Quantization of the electromagnetic field is often taught almost as a footnote or afterthought once one has mastered the quantum mechanics of mechanical systems. In most cases it will be obvious how to apply the techniques presented here to the corresponding optics problem. Besides, once the equations are cast in terms of creation and annihilation operators and characteristic time scales, quantities having units such as mass or length can be stripped away and the equations become identical with those used in quantum optics. Also, an emphasis on mechanical systems is, perhaps, not out of line from a historical perspective. Much early development of squeezed state theory concerned its application to gravitational wave detection using mechanical oscillators [11,12]. Quantum mechanical calculations were carried out with one ton mechanical resonators in mind.

To give the reader advance warning of where we are headed, in Sect. 3.1 the canonical quantization of a free massless scalar field (waves propagating on a string under tension) is reviewed. We will get much mileage out of the results presented here. In the next section we show that the quantum behavior of a harmonic oscillator attached to the end of the string is entirely dictated by the quantum mechanics of the signals propagating toward the oscillator along the string. Section 3.3 presents a field theoretic approach toward determining the quantum behavior of a system coupled to a field. This is followed by a section in which through a series of approximations the equations of motion for a nonlinear mechanical oscillator coupled to a string are cast into a form identical with the equations which a quantum optician would write down for cavity with a nonlinear medium. Since simple one-dimensional fields are employed, it is obvious how to identify the incoming and outgoing fields. This serves as a prelude to Sect. 3.5 which describes Collett and Gardner's approach of quantizing systems coupled to baths. Because the identification of the incoming and outgoing fields is not so obvious in this approach the results of the previous section come in handy here. Sections 3.6 and 3.7 provide illustrative examples of how the input and output field commutation relations and causality conditions place constraints on the various types of behavior that a device can exhibit. Section 3.8 provides a discussion of homodyne detection within the context of the Markov approximation employed by Collett and Gardner. Beginning with Sect. 3.9, we return to massless scalar fields, generalizing the results of Sects. 3.1 and 3.3 to the multiple field case. The commutation relations and causality conditions obtained in this section place restrictions on the kind of behavior phase shifters, beam-splitters, cavities, and lossy transmission lines can exhibit, as illustrated in the following sections. Of particular note, it is shown in Sect. 3.12 how the equations describing the Fabry–Perot resonances of a mechanical transmission line resonator can be reduced through suitable approximations to those of a mechanical oscillator. This is offered as justification for the standard practice in quantum optics of representing a cavity resonance as a harmonic oscillator. In Sect. 3.13, as a final example of how the quantum behavior of the input and output fields dictate the quantum behavior of the systems with which they interact, the quantum behavior of a lossy line is obtained by coupling it to the lossless transmission lines of Sect. 3.1. A section of this lossy transmission line is then used in the Sect. 3.14 as a model for an attenuator. The ground covered in this chapter should give one an idea of the scope and utility of input-output theory and should provide one with a new appreciation of the environment which, instead of spoiling quantum behavior, imposes quantum behavior.

3.1 Free Fields

We begin our analysis by reviewing the basic principles of the quantization of a scalar field.

Consider the Lagrangian density

$$\mathcal{L} = \frac{\rho}{2}\left(\frac{\partial Y}{\partial t}\right)^2 - \frac{\sigma}{2}\left(\frac{\partial Y}{\partial x}\right)^2 \tag{3.1}$$

of a string under tension where σ is the tension, ρ is the mass per unit length of the string, and Y is the hight of the string above the x-axis. Y is a real scalar field. Although our model is that for mechanical oscillations on a string, it is mathematically identical with the dual potential model of a one-dimensional electromagnetic waveguide, considered in (2.16); provided one makes the identification of $Y \to \Lambda$, $\rho \to A/\mu_0$ and $\sigma \to A\eta(x)$.

The variation of the action with respect to Y gives the Euler–Lagrange equation

$$\frac{\partial}{\partial t}\left(\frac{\partial \mathcal{L}}{\partial(\partial Y/\partial t)}\right) + \frac{\partial}{\partial x}\left(\frac{\partial \mathcal{L}}{\partial(\partial Y/\partial x)}\right) = 0 \ , \tag{3.2}$$

from which, one obtains the wave equation

$$\rho\frac{\partial^2 Y}{\partial t^2} - \sigma\frac{\partial^2 Y}{\partial x^2} = 0 \ . \tag{3.3}$$

The momentum density canonically conjugate to the field variable Y can be identified from (3.1) via

$$\Pi \equiv \frac{\partial \mathcal{L}}{\partial(\partial Y/\partial t)} = \rho\frac{\partial Y}{\partial t} \ . \tag{3.4}$$

Then the canonical quantization imposes the commutation relations on Y and Π

$$[Y(x,t), \Pi(x',t)] = i\hbar\delta(x-x') \ , \tag{3.5}$$

$$[Y(x,t), Y(x',t)] = [\Pi(x,t), \Pi(x',t)] = 0 \ . \tag{3.6}$$

We can write the field in the Fourier expansion

$$\begin{aligned} Y(x,t) = \frac{1}{2}\sqrt{\frac{\hbar}{\pi\Gamma}}\int_0^\infty \frac{d\omega}{\sqrt{\omega}}[a_L(\omega)e^{-i\omega(t+x/v)} + a_L^\dagger(\omega)e^{i\omega(t+x/v)} \\ +a_R(\omega)e^{-i\omega(t-x/v)} + a_R^\dagger(\omega)e^{i\omega(t-x/v)}] \ , \end{aligned} \tag{3.7}$$

where

$$v = \sqrt{\frac{\sigma}{\rho}} \tag{3.8}$$

is the velocity of propagation and

$$\Gamma = \sqrt{\rho\sigma} \tag{3.9}$$

is the characteristic impedance of the transmission line. The operators $a_L(\omega)$ and $a_R(\omega)$ are annihilation operators for field quanta propagating, respectively, to the left or to the right. The measure $d\omega/\sqrt{\omega}$ looks strange in a Fourier decomposition, but, since the $a_R(\omega)$ and $a_L(\omega)$ can in principle be arbitrary functions of ω, such a measure is perfectly acceptable.

The inverse transforms of (3.7) are

$$a_L(\omega) = \frac{i}{2v}\sqrt{\frac{\Gamma}{\pi\omega}} \int_{-\infty}^{\infty} dx \left[\frac{\partial Y}{\partial t} e^{i\omega(t+x/v)} - Y \frac{\partial}{\partial t} e^{i\omega(t+x/v)} \right] , \tag{3.10}$$

$$a_R(\omega) = \frac{i}{2v}\sqrt{\frac{\Gamma}{\pi\omega}} \int_{-\infty}^{\infty} dx \left[\frac{\partial Y}{\partial t} e^{i\omega(t-x/v)} - Y \frac{\partial}{\partial t} e^{i\omega(t-x/v)} \right] , \tag{3.11}$$

and similar relations are obtained for their Hermitian conjugates. Using the canonical commutation relations (3.5) and (3.6), it is straightforward to show from (3.10) and (3.11) that

$$[a_L(\omega), a_L^\dagger(\omega')] = [a_R(\omega), a_R^\dagger(\omega')] = \delta(\omega - \omega') , \tag{3.12}$$

and

$$\begin{aligned} &[a_L(\omega), a_L(\omega')] = [a_R(\omega), a_R(\omega')] \\ &= [a_L(\omega), a_R(\omega')] = [a_L(\omega), a_R^\dagger(\omega')] = 0 . \end{aligned} \tag{3.13}$$

The measure in (3.7) and the constants in front of the integral were normalized such that simple commutation relations would arise. Note that (3.12) and (3.13) are the standard boson commutation relations.

Consider now the Hamiltonian density, which is given by

$$\mathcal{H} \equiv \Pi \frac{\partial Y}{\partial t} - \mathcal{L} = \frac{\rho}{2}\left(\frac{\partial Y}{\partial t}\right)^2 + \frac{\sigma}{2}\left(\frac{\partial Y}{\partial x}\right)^2 . \tag{3.14}$$

Using (3.7) and the commutation relations (3.12) and (3.13), one obtains

$$H \equiv \int_{-\infty}^{\infty} dx \mathcal{H} = \int_0^{\infty} d\omega [a_L^\dagger(\omega) a_L(\omega) + a_R^\dagger(\omega) a_R(\omega)] , \tag{3.15}$$

where the constant terms arising from the use of the commutation relations have been dropped. Since the total energy should be bounded from below, quantization is completed by postulating the existence of a vacuum state $|0\rangle$ defined by

$$a_L(\omega)|0\rangle = a_R(\omega)|0\rangle = 0 , \tag{3.16}$$

and $\langle 0|0\rangle = 1$.

3.2 Harmonic Oscillator Coupled to a Transmission Line

Let's consider for the moment a classical harmonic oscillator consisting of a mass M and a spring K [13]. Attached to the mass is a string extending to infinity. Let $Y(x,t)$ denote the height of the mass when $x = 0$ and the height of the string when $x > 0$. The equation of motion of the mass is then given by

$$M\frac{d^2Y}{dt^2} + KY = \sigma\left.\frac{\partial Y}{\partial x}\right|_{x=0+} , \tag{3.17}$$

where σ denotes the string tension. As in (3.3) the equation of motion of the string is given by

$$\rho\frac{\partial^2 Y}{\partial t^2} - \sigma\frac{\partial^2 Y}{\partial x^2} = 0 . \tag{3.18}$$

The general solution of the wave equation (3.18) has the form

$$Y = Y_{\text{in}}(t + x/v) + Y_{\text{out}}(t - x/v) , \tag{3.19}$$

where the propagation velocity is given by (3.8), Y_{in} is the field propagating toward the oscillator, and Y_{out} is the field propagating away from the oscillator. Because of causality, the motion of the oscillator is determined only by the incoming field Y_{in}. We now make this explicit. The time derivative of (3.19) gives

$$\frac{\partial Y}{\partial t} = \frac{\partial Y_{\text{in}}}{\partial t} + \frac{\partial Y_{\text{out}}}{\partial t} . \tag{3.20}$$

The derivative of (3.19) with respect to x gives

$$\frac{\partial Y}{\partial x} = \frac{1}{v}\frac{\partial Y_{\text{in}}}{\partial t} - \frac{1}{v}\frac{\partial Y_{\text{out}}}{\partial t} . \tag{3.21}$$

Using (3.19) to eliminate Y_{out} from the above equation, we obtain

$$\frac{\partial Y}{\partial x} = \frac{2}{v}\frac{\partial Y_{\text{in}}}{\partial t} - \frac{1}{v}\frac{\partial Y}{\partial t} . \tag{3.22}$$

Substituting (3.22) into (3.17), we get

$$M\frac{d^2Y}{dt^2} + \Gamma\frac{dY}{dt} + KY = 2\Gamma\frac{dY_{\text{in}}}{dt} , \tag{3.23}$$

where the damping constant Γ is the mechanical impedance (3.10) of the string, and we use dY_{in}/dt to denote $\partial Y_{\text{in}}/\partial t$ evaluated at $x = 0$. The linear damping term $\Gamma dY/dt$ represents the radiation damping experienced by the oscillator as it radiates its energy into the string. The term $2\Gamma dY_{\text{in}}/dt$ is the driving force that the incoming field exerts on the oscillator. For the case when the transmission line represents a heat bath this term is the Langevin term representing the fluctuating force exerted on the oscillator by the bath.

The calculation just performed is classical. We now consider the quantum behavior of the oscillator [14]. Since the incoming field does not know that an oscillator is located at $x = 0$ until it arrives, one can argue that the left propagating part of (3.7) is the correct expression for the quantum mechanical field propagating toward the oscillator. In particular at $x = 0$ one has

$$Y_{\text{in}}(t) = \frac{1}{2}\sqrt{\frac{\hbar}{\pi\Gamma}}\int_0^\infty \frac{d\omega}{\sqrt{\omega}}[a_{\text{in}}(\omega)e^{-i\omega t} + a_{\text{in}}^\dagger(\omega)e^{i\omega t}] \ , \tag{3.24}$$

where we have replaced the subscript L with "in". From (3.12) and (3.13), we find that the creation and annihilation operators satisfy the commutation relations

$$[a_{\text{in}}(\omega), a_{\text{in}}^\dagger(\omega')] = \delta(\omega - \omega') \quad \text{and} \quad [a_{\text{in}}(\omega), a_{\text{in}}(\omega')] = 0 \ . \tag{3.25}$$

Let's substitute the quantum expression (3.24) for Y_{in} into the classical equation of motion (3.17) and see where that gets us. A straightforward exercise in Fourier transforms gives

$$Y(t) = -\sqrt{\frac{\hbar\Gamma}{\pi}}\int_0^\infty d\omega\ \sqrt{\omega}\left[\frac{ia_{\text{in}}(\omega)e^{-i\omega t}}{-\omega^2 M - i\omega\Gamma + K} + \text{H.c.}\right] \ . \tag{3.26}$$

Consider the momentum of the oscillator, defined as

$$P = M\frac{dY}{dt} \ . \tag{3.27}$$

Substituting (3.26) into (3.27) gives an expression for P in terms of the creation and annihilation operators of the incoming field. Using the commutation relations (3.25), one then obtains

$$[Y, P] = i\frac{2\hbar\Gamma M}{\pi}\int_0^\infty d\omega \frac{\omega^2}{(K - \omega^2 M)^2 + \omega^2\Gamma^2} \ . \tag{3.28}$$

Introducing the resonant frequency

$$\omega_0 = \sqrt{\frac{K}{M}} \tag{3.29}$$

and the quality factor

$$Q = \frac{\omega_0 M}{\Gamma} \ , \tag{3.30}$$

one finds that (3.28) can be written in the form

$$[Y, P] = i\frac{2\hbar}{\pi Q}\int_0^\infty d\xi \frac{\xi^2}{(1 - \xi^2)^2 + \xi^2/Q^2} \ . \tag{3.31}$$

A straightforward exercise in contour integration yields

$$[Y, P] = i\hbar \ . \tag{3.32}$$

It is worthwhile to summarize what has just been done. We started with a classical analysis of a harmonic oscillator coupled to a transmission line. In the resulting Langevin equation we substituted the quantum mechanical expression of the signals coming in along the transmission line. Upon solving for the motion of the harmonic oscillator in response to this input, we find that the expressions for the oscillator's position and momentum, which are now operators, satisfy the correct quantum mechanical commutation relations, that is, the oscillator behaves quantum mechanically simply because the signals it is receiving are quantum mechanical.

This point can be explored further by calculating the ground state position probability distribution for the oscillator. Because (3.26) is linear in the creation and annihilation operators, it is immediately evident that

$$\langle 0|Y^{2n+1}|0\rangle = 0 \ , \tag{3.33}$$

where n is an integer, and

$$\langle 0|Y^{2n}|0\rangle = (2n-1)!!(\Delta Y)^{2n} \ , \tag{3.34}$$

with

$$(\Delta Y)^2 = \langle 0|Y^2|0\rangle = \frac{\hbar\Gamma}{\pi}\int_0^\infty \frac{\omega d\omega}{(K - M\omega^2)^2 + \omega^2\Gamma^2} \ . \tag{3.35}$$

The integration in (3.35) can be carried out, and for the case when the oscillator is underdamped, $Q > 1/2$, one obtains [3,15]

$$(\Delta Y)^2 = \frac{\hbar Q}{\sqrt{MK}\sqrt{4Q^2-1}}\left[1 - \frac{1}{\pi}\tan^{-1}\left(\frac{\sqrt{4Q^2-1}}{2Q^2-1}\right)\right] \ . \tag{3.36}$$

Note that (3.33) and (3.34) are the moments for a Gaussian probability distribution

$$P(y) = \frac{1}{\sqrt{2\pi}\Delta Y}\exp\left(-\frac{y^2}{4(\Delta y)^2}\right) \ . \tag{3.37}$$

In the limit when Q goes to infinity, one obtains from (3.36) that

$$(\Delta Y)^2 = \sqrt{\frac{\hbar^2}{4MK}} \ . \tag{3.38}$$

This is the standard result for the variance of the position coordinate of an isolated harmonic oscillator in its ground state. From the above analysis, however, it is evident that the uncertainty in the oscillator's position results from the quantum Langevin noise driving the oscillator and the oscillator's response to this noise does not vanish as the oscillator becomes more and more isolated from the environment, i.e. as Q goes toward infinity. As Q is increased the bandwidth over which the oscillator responds to the Langevin noise shrinks as $1/Q$ but the responsiveness of the oscillator to the Langevin noise within this bandwidth grows as Q. Viewed this way a quantum mechanical oscillator is never isolated from its environment, even in the limit of zero coupling. The time scale with which the oscillator responds to its environment, of course, grows as Q. It is interesting to speculate to what extent Schrödinger quantum mechanics could be viewed as this kind of a weak coupling limit of a system coupled to a bath, a limit in which it is the bath that dictates the quantum behavior of the system.

Before leaving this system a few comments are worth making about the output field Y_{out}. Given the incoming field Y_{in} and the oscillator's position, the output field, according to (3.19), is given by

$$Y_{\text{out}}(t) = Y(t) - Y_{\text{in}}(t) \ . \tag{3.39}$$

From (3.7) the output field is given by

$$Y_{\text{out}}(t) = \frac{1}{2}\sqrt{\frac{\hbar}{\pi\Gamma}} \int_0^\infty \frac{d\omega}{\sqrt{\omega}} [a_{\text{out}}(\omega)e^{-i\omega t} + a_{\text{out}}^\dagger(\omega)e^{i\omega t}] \ . \tag{3.40}$$

Substituting (3.40) and (3.20) into (3.11), we obtain

$$a_{\text{out}}(\omega) = -\frac{K - \omega^2 M + i\omega\Gamma}{K - \omega^2 M - i\omega\Gamma} a_{\text{in}}(\omega) \ . \tag{3.41}$$

The function multiplying a_{in} has unit norm. Thus, one sees that a_{out} is simply a phase shifted version of a_{in}. It is evident that a_{out} satisfies the boson commutation relations

$$[a_{\text{out}}(\omega), a_{\text{out}}^\dagger(\omega')] = \delta(\omega - \omega') \tag{3.42}$$

and

$$[a_{\text{out}}(\omega), a_{\text{out}}(\omega')] = [a_{\text{in}}(\omega), a_{\text{out}}(\omega')] = 0 \ . \tag{3.43}$$

Hence, nearly all the commutation relations (3.12) and (3.13) are satisfied, all except the one for $[a_{\text{in}}(\omega), a_{\text{out}}^\dagger(\omega')]$, which now reads

$$[a_{\text{in}}(\omega), a_{\text{out}}^\dagger(\omega')] = -\frac{K - \omega^2 M - i\omega\Gamma}{K - \omega^2 M + i\omega\Gamma} \delta(\omega - \omega') \ . \tag{3.44}$$

Using the commutation relations (3.42) and (3.43), one obtains

$$[Y_{\text{in}}(t), Y_{\text{out}}(t')] = -\frac{\hbar}{\pi\Gamma} \int_{-\infty}^{\infty} \frac{d\omega}{\omega} \frac{K - \omega^2 M - i\omega\Gamma}{K - \omega^2 M + i\omega\Gamma} e^{i\omega(t-t')} \ . \tag{3.45}$$

This can be integrated to yield

$$[Y_{\text{in}}(t), Y_{\text{out}}(t')] = iu(t'-t)\frac{\hbar}{\Gamma} \times \left\{1 - \frac{2\Gamma e^{-\frac{\Gamma}{2M}(t'-t)}}{M\sqrt{\omega_0^2 - (\Gamma/2M)^2}} \sin\left[\sqrt{\omega_0^2 - (\Gamma/2M)^2}(t'-t)\right]\right\} \tag{3.46}$$

where $u(\tau)$ is the unit step function and is zero for $\tau < 0$. It is evident from this equation that the commutator is zero if t' is less than t, that is, the commutation relation (3.46) expresses a quantum causality condition between the incoming fields and outgoing fields.

3.3 Field Theoretic Approach to Input-Output Theory

To illustrate a field theoretical method for dealing with open quantum systems, for which one is interested in the input and output fields, we consider a particle of mass M moving in a potential V. The particle is directly attached to a string under tension. The string is the transmission line along which the incoming and outgoing fields propagate. The Lagrangian density for this system is given by

$$\mathcal{L} = \delta(x)\left[\frac{M}{2}\left(\frac{\partial Y}{\partial t}\right)^2 - V(Y)\right] + u(x)\left[\frac{\rho}{2}\left(\frac{\partial Y}{\partial t}\right)^2 - \frac{\sigma}{2}\left(\frac{\partial Y}{\partial x}\right)\right]^2 . \tag{3.47}$$

The Dirac delta function $\delta(x)$ serves to localize the particle at $x = 0$. Since the particle is directly attached to the $x = 0$ end of the transmission line, Y is used both for the positional degree of freedom of the particle and the displacement of the string from its equilibrium position. The unit step function $u(x)$ insures that the transmission line exists only along the positive x axis. We now proceed with the canonical quantization procedure. First the Euler–Lagrange equation

$$\frac{\partial}{\partial t}\frac{\partial \mathcal{L}}{\partial(\partial Y/\partial t)} + \frac{\partial}{\partial x}\frac{\partial \mathcal{L}}{\partial(\partial Y/\partial x)} - \frac{\partial V(Y)}{\partial Y} = 0 \tag{3.48}$$

is used to generate the field equation, also referred to as the Heisenberg equation of motion. Substituting (3.47) into (3.48), one obtains

$$\delta(x)\left[M\frac{\partial^2 Y}{\partial t^2} + \frac{\partial V(Y)}{\partial Y}\right] + u(x)\rho\frac{\partial^2 Y}{\partial t^2} - \sigma\frac{\partial}{\partial x}\left(u(x)\frac{\partial Y}{\partial x}\right) = 0 . \tag{3.49}$$

Integrating (3.49) with respect to x over the interval $-\epsilon < x < \epsilon$, and taking the limit as ϵ goes to zero, one obtains

$$M\frac{d^2 Y}{dt^2} + \frac{\partial V(Y)}{\partial Y} - \sigma\left.\frac{\partial Y}{\partial x}\right|_{x=0+} = 0 . \tag{3.50}$$

This is the equation of motion for the particle, and we have used d^2Y/dt^2 to denote $\partial^2 Y/\partial t^2$ evaluated at $x = 0$. When $x > 0$, the equation of motion (3.49) reduces to the field equation for the string

$$\rho\frac{\partial^2 Y}{\partial t^2} - \sigma\frac{\partial^2 Y}{\partial x^2} = 0 . \tag{3.51}$$

One sees that (3.50) is a generalization of (3.17) and that (3.51) is equal to (3.18). Although (3.17) and (3.18) were written as classical equations, one sees from the above discussion that they are the appropriate Heisenberg equations of motion. Hence, it is no surprise that the correct behavior of the harmonic oscillator was obtained when the appropriate quantum mechanical expression was used for the incoming field. Going through the same procedures, as were carried out to obtain (3.23), one gets

$$M\frac{d^2Y}{dt^2} + \Gamma\frac{dY}{dt} + \frac{\partial V(Y)}{\partial Y} = 2\Gamma\frac{dY_{\mathrm{in}}}{dt} . \tag{3.52}$$

Again, the outgoing field is given in terms of the incoming field by

$$Y_{\mathrm{out}}(t) = Y(t) - Y_{\mathrm{in}}(t) , \tag{3.53}$$

and the momentum density Π canonically conjugate to Y is given by

$$\Pi \equiv \frac{\partial \mathcal{L}}{\partial(\partial Y/\partial t)} = [\delta(x)M + u(x)\rho]\frac{\partial Y}{\partial t} . \tag{3.54}$$

The canonical commutation relations are

$$[Y(x,t), \Pi(x',t)] = i\hbar\delta(x - x') \tag{3.55}$$

and

$$[Y(x,t), Y(x,t')] = [\Pi(x,t), \Pi(x',t)] = 0 . \tag{3.56}$$

Equations (3.53), (3.54), and (3.55) are taken to imply the commutation relation

$$[Y, P] = i\hbar \tag{3.57}$$

at $x = 0$, where

$$P = M\frac{dY}{dt} \tag{3.58}$$

and dY/dt is used to denote $\partial Y/\partial t$ evaluated at $x = 0$. This is just the usual position momentum commutation relation. For $x > 0$, (3.54) and (3.55) reduce to those appropriate for a free field, i.e. (3.5) and (3.6). Now that the commutation relations have been established, one can proceed with quantization, but, since the free field has already been quantized in Sect. 3.1, we will use the results of that section by invoking causality to argue that the incoming field is given by

$$Y_{\text{in}}(t) = \frac{1}{2}\sqrt{\frac{\hbar}{\pi\Gamma}} \int_0^\infty \frac{d\omega}{\sqrt{\omega}} [a_{\text{in}}(\omega)e^{-i\omega t} + a_{\text{in}}^\dagger(\omega)e^{i\omega t}] \,, \tag{3.59}$$

where the creation and annihilation operators satisfy the following commutation relations

$$[a_{\text{in}}(\omega), a_{\text{in}}^\dagger(\omega')] = \delta(\omega - \omega') \quad \text{and} \quad [a_{\text{in}}(\omega), a_{\text{in}}(\omega')] = 0 \,. \tag{3.60}$$

Similarly, the outgoing field has the form

$$Y_{\text{out}}(t) = \frac{1}{2}\sqrt{\frac{\hbar}{\pi\Gamma}} \int_0^\infty \frac{d\omega}{\sqrt{\omega}} [a_{\text{out}}(\omega)e^{-i\omega t} + a_{\text{out}}^\dagger(\omega)e^{i\omega t}] \,, \tag{3.61}$$

where the creation and annihilation operators satisfy the following commutation relations

$$[a_{\text{out}}(\omega), a_{\text{out}}^\dagger(\omega')] = \delta(\omega - \omega') \quad \text{and} \quad [a_{\text{out}}(\omega), a_{\text{out}}(\omega')] = 0 \,. \tag{3.62}$$

3.4 Approximations

In this section we will make a slowly varying envelope approximation for the particle and we will make a Markov approximation for the incoming and outgoing fields. For definiteness we will consider the case when the particle's potential V is given by

$$V = \frac{K}{2}Y^2 + \frac{K_N}{4}Y^4 \,, \tag{3.63}$$

where K_N will be regarded as small so that the nonlinearity can be treated perturbatively. Substituting (3.63) into (3.52), we obtain

$$M\frac{d^2Y}{dt^2} + \Gamma\frac{dY}{dt} + KY + K_N Y^3 = 2\Gamma\frac{dY_{\text{in}}}{dt} \,. \tag{3.64}$$

We now write

$$Y = \sqrt{\frac{\hbar}{2\sqrt{KM}}}[a\exp(-i\omega_0 t) + a^\dagger \exp(i\omega_0 t)] \,, \tag{3.65}$$

where ω_0 is the resonant frequency of the linear oscillator (3.29). The amplitude a is to be regarded as slowly varying with time. When the Q of the resonator is high, one expects the resonator to respond to inputs only near the frequency ω_0. With this in mind we approximate $\sqrt{\omega}$, appearing in (3.24), by $\sqrt{\omega_0}$ and write (3.59) in the form

$$Y_{\text{in}}(t) = \frac{1}{2}\sqrt{\frac{\hbar}{\pi\Gamma\omega_0}} \left\{ e^{-i\omega_0 t} \int_{-\omega_0}^\infty d\omega\, b_{\text{in}}(\omega)e^{-i\omega t} \right.$$
$$\left. + e^{i\omega_0 t} \int_{-\omega_0}^\infty d\omega\, b_{\text{in}}^\dagger(\omega)e^{i\omega t} \right\} \,. \tag{3.66}$$

In obtaining (3.66), the change in the variable of integration $\omega \to \omega_0 + \omega$ has been employed so

$$b_{\text{in}}(\omega) = a_{\text{in}}(\omega_0 + \omega) . \quad (3.67)$$

Note that $b_{\text{in}}(\omega)$ has the same commutation relations as $a_{\text{in}}(\omega)$. Since we expect the oscillator to be responsive only at frequencies near ω_0, that is, at frequencies near $\omega = 0$ in the integrals appearing in (3.66), we argue that the lower limit of integration of (3.66) can be extended to $-\infty$. Thus, we can write

$$Y_{\text{in}}(t) = \sqrt{\frac{\hbar}{2\Gamma\omega_0}} \left[e^{-i\omega_0 t} b_{\text{in}}(t) + e^{i\omega_0 t} b_{\text{in}}^\dagger(t) \right] , \quad (3.68)$$

where

$$b_{\text{in}}(t) = \frac{1}{\sqrt{2\pi}} \int_{-\infty}^{\infty} d\omega b_{\text{in}}(\omega) e^{-i\omega t} . \quad (3.69)$$

The oscillator is viewed as responding to the slowly varying part of $b_{\text{in}}(t)$. Hence, (3.68) is viewed as a slowly varying approximation of Y_{in} in which the time evolution of $b_{\text{in}}(t)$ is slow compared to ω_0.

Since $b_{\text{in}}(\omega)$ satisfies the commutation relations

$$[b_{\text{in}}(\omega), b_{\text{in}}^\dagger(\omega')] = \delta(\omega - \omega') \quad \text{and} \quad [b_{\text{in}}(\omega), b_{\text{in}}(\omega')] = 0 , \quad (3.70)$$

it is easy to show that

$$[b_{\text{in}}(t), b_{\text{in}}^\dagger(t')] = \delta(t - t') \quad \text{and} \quad [b_{\text{in}}(t), b_{\text{in}}(t')] = 0 . \quad (3.71)$$

The delta function appearing in (3.70) should be regarded as in some sense broad compared to the oscillation period of the oscillator $T = 2\pi/\omega_0$. We call the approximations resulting in (3.66), (3.68), (3.69), and (3.70) the Markov approximation.

In a similar manner, one has for the outgoing field

$$Y_{\text{out}}(t) = \sqrt{\frac{\hbar}{2\Gamma\omega_0}} [e^{-i\omega_0 t} b_{\text{out}}(t) + e^{i\omega_0 t} b_{\text{out}}^\dagger(t)] , \quad (3.72)$$

with

$$b_{\text{out}}(t) = \frac{1}{\sqrt{2\pi}} \int_{-\infty}^{\infty} d\omega \; b_{\text{out}}(\omega) e^{-i\omega t} , \quad (3.73)$$

where

$$[b_{\text{out}}(\omega), b_{\text{out}}^\dagger(\omega')] = \delta(\omega - \omega') \quad \text{and} \quad [b_{\text{out}}(\omega), b_{\text{out}}(\omega')] = 0 , \quad (3.74)$$

or, in the time domain,

$$[b_{\text{out}}(t), b_{\text{out}}^\dagger(t')] = \delta(t - t') \quad \text{and} \quad [b_{\text{out}}(t), b_{\text{out}}(t')] = 0 . \quad (3.75)$$

Substituting (3.65) and (3.68) into (3.64) and separating the parts oscillating at $\exp(-i\omega_0 t)$ from those oscillating at $\exp(i\omega_0 t)$, we obtain

$$M\frac{d^2}{dt^2}ae^{-i\omega_0 t} + \Gamma\frac{d}{dt}ae^{-i\omega_0 t} + Kae^{-i\omega_0 t} + \frac{3\hbar K_N}{2\sqrt{KM}}a^\dagger aae^{-i\omega_0 t}$$
$$= 2\sqrt{M\Gamma}\frac{d}{dt}b_{\text{in}}(t)e^{-i\omega_0 t} \ . \tag{3.76}$$

In this equation the term $d^2 a\exp(-i\omega_0 t)/dt^2$ will be approximated by

$$\frac{d^2}{dt^2}ae^{-i\omega_0 t} = -\omega_0^2 ae^{-i\omega_0 t} - i2\omega_0\frac{da}{dt}e^{-i\omega_0 t} \ . \tag{3.77}$$

When substituted into (3.76) the first term of this equation will be canceled by the $Ka\exp(-i\omega_0 t)$ because ω_0 is the resonant frequency. Hence, we are neglecting d^2a/dt^2 compared with $\omega_0 da/dt$. Moreover, $da\exp(-i\omega_0 t)/dt$ and $d\exp(i\omega_0 t)b_{\text{in}}(t)/dt$ will be approximated by

$$\frac{d}{dt}a\exp(-i\omega_0 t) = -i\omega_0 a\exp(-i\omega_0 t) \tag{3.78}$$

and

$$\frac{d}{dt}b_{\text{in}}(t)\exp(-i\omega_0 t) = -i\omega_0 b_{\text{in}}(t) \ . \tag{3.79}$$

Equations (3.77), (3.78), and (3.79) constitute the slowly varying envelope approximation. Substituting these equations into (3.76), we obtain

$$\frac{da}{dt} + \frac{\gamma}{2}a + i\beta a^\dagger aa = \sqrt{\gamma}b_{\text{in}}(t) \ , \tag{3.80}$$

where

$$\gamma = \frac{\Gamma}{M} \tag{3.81}$$

is the characteristic damping time for the system and

$$\beta = \frac{3\hbar K_N}{4KM} \tag{3.82}$$

is the strength of the nonlinearity. Substituting (3.68), (3.72), and (3.80) into (3.53) yields

$$b_{\text{out}}(t) = \sqrt{\gamma}\, a - b_{\text{in}}(t) \ . \tag{3.83}$$

Substituting (3.65) into (3.58) and making the slowly varying envelope approximation (3.78) yields an expression for P which, when substituted into (3.57) produces the commutation relation

$$[a, a^\dagger] = 1 \ . \tag{3.84}$$

Although (3.74), (3.75), and (3.83) were derived for a mechanical oscillator, we will show in later sections that they are directly applicable to a cavity mode of the electromagnetic field which is coupled to an external field through a partly transmitting mirror.

The creation and annihilation operators of (3.78) and (3.79) are by construction those of the rotating frame. One can transform back to the non-rotating frame through the transformations

$$c(t) = \exp(-i\omega_0 t)a(t) \ , \tag{3.85}$$

$$d_{\rm in}(t) = \exp(-i\omega_0 t)b_{\rm in}(t) \ , \tag{3.86}$$

and

$$d_{\rm out} = \exp(-i\omega_0 t)b_{\rm out}(t) \ . \tag{3.87}$$

The commutation relations of these operators have the same form as that given in (3.70), (3.71), and (3.84). The equations of motion in the non-rotating frame are:

$$\frac{dc}{dt} + i\omega_0 c + \frac{\gamma}{2}c + i\beta c^\dagger cc = \sqrt{\gamma}d_{\rm in}(t) \tag{3.88}$$

and

$$d_{\rm out}(t) = \sqrt{\gamma}\, c - d_{\rm in}(t) \ . \tag{3.89}$$

Using this equation to eliminate $d_{\rm in}$ from (3.88) gives

$$\frac{dc}{dt} + i\omega_0 c - \frac{\gamma}{2}c + i\beta c^\dagger cc = -\sqrt{\gamma}d_{\rm out}(t). \tag{3.90}$$

Let us consider the exactly solvable case, $\beta = 0$, and derive some commutation relations that will be referred to in the next Section. In this case (3.88) can be integrated to yield

$$c(t) = \sqrt{\gamma}\int_{-\infty}^{t} dt'\ d_{\rm in}(t')\exp[-(i\omega_0 + \gamma/2)(t-t')] \ . \tag{3.91}$$

From this equation one obtains the commutation relations

$$[c(t), d_{\rm in}(t')] = 0 \tag{3.92}$$

and

$$[c(t), d^\dagger_{\rm in}(t')] = \begin{cases} \sqrt{\gamma}\exp[-(i\omega_0 + \gamma/2)(t-t')] & \text{if} \quad t > t' \\ \sqrt{\gamma}/2 \quad \text{if} \quad t = t' \\ 0 \quad \text{if} \quad t < t' \ . \end{cases} \tag{3.93}$$

Moreover, from (3.93) and (3.91), it follows that

$$[\frac{\gamma}{2}c(t) - \sqrt{\gamma}d_{\rm in}(t), c(t)] = 0 \tag{3.94}$$

and

$$[\frac{\gamma}{2}c(t) - \sqrt{\gamma}d_{\text{in}}(t), c^{\dagger}(t)] = 0 . \tag{3.95}$$

For the case when $\beta = 0$, (3.91) yields the following expression for the Fourier transform of c

$$c(\omega) = \frac{\sqrt{\gamma}}{i(\omega_0 - \omega) + \Gamma/2} d_{\text{in}}(\omega) , \tag{3.96}$$

from which we obtain the commutation relation

$$[c(\omega), c(\omega')] = \frac{\Gamma}{(\omega_0 - \omega)^2 + (\Gamma/2)^2}\delta(\omega - \omega') . \tag{3.97}$$

It is evident that even though $c(t)$ satisfies the usual commutation relation for creation and annihilation operators (3.84), the components $c(\omega)$ of its spectral decomposition need not. Taking the inverse transform of (3.96), one obtains

$$c(t) = \frac{1}{\sqrt{2\pi}} \int_{-\infty}^{\infty} d\omega \frac{\sqrt{\gamma}d_{\text{in}}(\omega)\exp(-i\omega t)}{i(\omega_0 - \omega) + \Gamma/2} . \tag{3.98}$$

From (3.65) it follows that the position coordinate of the oscillator is given by

$$Y(t) = \frac{1}{\sqrt{2\pi}}\sqrt{\frac{\hbar}{2\sqrt{KM}}} \int_{-\infty}^{\infty} d\omega \left[\frac{\sqrt{\gamma}d_{\text{in}}(\omega)\exp(-i\omega t)}{i(\omega_0 - \omega) + \Gamma/2} + \text{H.c.}\right] . \tag{3.99}$$

3.5 Alternative Approach to Input-Output Theory

In Sect. 3.3, we considered a general field theoretic approach to input-output theory. Because transmission lines are used for the environmental modes it is obvious how to identify the incoming and outgoing fields. In Sect. 3.4 a series of approximations were made to reduce (3.52), (3.53), (3.59), and (3.61) to a form that is commonly encountered in quantum optics when a cavity mode of the electromagnetic field is coupled to an external field through a partly transparent mirror. When one is willing to make such approximations, there is a more direct route to the equations which will be presented here. The account of this method closely follows that presented by Gardiner and Collett [5]. See also [6,7].

We begin by considering a system linearly coupled to N heat baths. The Hamiltonian is given by

$$H = H_{\text{sys}} + H_B + H_{\text{int}} , \tag{3.100}$$

where H_{sys} is the Hamiltonian of the finite system,

$$H_B = \hbar \sum_n \int_{-\infty}^{\infty} d\omega\ \omega b_n^{\dagger}(\omega) b_n(\omega) \tag{3.101}$$

is the sum over the heat bath Hamiltonians, and

$$H_{\text{int}} = i\hbar \sum_n \sqrt{\frac{\kappa_n}{2\pi}} \int_{-\infty}^{\infty} d\omega [b_n^\dagger(\omega) c_n - c_n^\dagger b_n(\omega)] \tag{3.102}$$

is the interaction Hamiltonian between the system and the baths. This Hamiltonian is quadratic and gives rize to linear coupling between the system varaibles and the bath modes. Since more than one bath could be coupled to the same system variable, the c_n appearing in (3.102) need not be distinct system operators. The $b_n(\omega)$ are the boson annihilation operators of the baths and satisfy the equal time commutation relations

$$[b_n(\omega), b_{n'}^\dagger(\omega')] = \delta_{nn'}\delta(\omega - \omega') \tag{3.103}$$

and

$$[b_n(\omega), b_{n'}(\omega')] = [b_n(\omega), b_{n'}^\dagger(\omega')] = 0 \ . \tag{3.104}$$

Hamiltonians (3.101) and (3.102) look unrealistic. One would have expected the range of ω integration to extend from 0 to infinity and one would in general expect the coupling of the bath modes to the system operators c_n to be frequency dependent. By writing the bath and interaction Hamiltonians in the form of (3.101) and (3.102), the Markov approximation has been built in right from the beginning, as shall be seen shortly. As was argued in connection with (3.66), since cavity resonances and atomic line widths tend to be sharp, one does not expect to introduce much of an error by using a frequency independent coupling to the bath and by extending the range of ω integration.

The Heisenberg equations of motion for an operator $\mathcal{O}$ are generated by

$$i\hbar \frac{d\mathcal{O}}{dt} = [\mathcal{O}, H] \ . \tag{3.105}$$

Using this formula, one obtains the following Heisenberg equations of motion for the mode operators of the baths

$$\frac{d}{dt} b_n(\omega) = -i\omega b_n(\omega) + \sqrt{\frac{\kappa_n}{2\pi}} c_n \ . \tag{3.106}$$

For a system operator a one obtains the following Heisenberg equation of motion

$$\begin{aligned} \frac{da}{dt} &= -\frac{i}{\hbar}[a, H_{\text{sys}}] \\ &\quad + \sum_n \sqrt{\frac{\kappa_n}{2\pi}} \int_{-\infty}^{\infty} d\omega \ \{b_n^\dagger(\omega)[a, c_n] - [a, c_n^\dagger] b_n(\omega)\} \ . \end{aligned} \tag{3.107}$$

Integrating (3.106), we obtain

$$b_n(\omega) = e^{-i\omega(t-t_0)} b_n^0(\omega) + \sqrt{\frac{\kappa_n}{2\pi}} \int_{t_0}^{t} e^{-i\omega(t-t')} c_n(t') dt' \ , \tag{3.108}$$

where $b_n^0(\omega)$ is the annihilation operator for the mode at frequency ω of the nth bath at time $t = t_0$. Consequently, the b_n^0 satisfies the same commutation relations as the $b_n(\omega)$. Substituting this equation into (3.107) yields

$$\frac{da}{dt} = -\frac{i}{\hbar}[c, H_{\rm sys}] + \sum_n \frac{\kappa_n}{2\pi}\int_{-\infty}^{\infty} d\omega \int_{t_0}^{t} dt' \{e^{i\omega(t-t')} c_n^\dagger(t')[a,c] - \text{H.c.}\}$$
$$+\sqrt{\frac{\kappa_n}{2\pi}}\int_{-\infty}^{\infty} d\omega \{e^{i\omega(t-t_0)} b_n^{0\dagger}(\omega)[a, c_n] - \text{H.c.}\} . \qquad (3.109)$$

The second term on the right-hand side of this equation can be evaluated using

$$\int_{-\infty}^{\infty} d\omega\ \exp[-i\omega(t-t')] = 2\pi\delta(t-t') \qquad (3.110)$$

and

$$\int_{t_0}^{t} c(t')\delta(t-t')dt' = \frac{1}{2}\text{sgn}(t-t_0)c(t) , \qquad (3.111)$$

where $\text{sgn}(t-t_0)$ denotes the sign of $t-t_0$. Provided that $t > t_0$, then (3.109) becomes

$$\frac{da}{dt} = -\frac{i}{\hbar}[c, H_{\rm sys}]$$
$$+\sum_n \left\{\frac{\kappa_n}{2} c_n^\dagger [a, c_n] + \sqrt{\frac{\kappa_n}{2\pi}}\int_{-\infty}^{\infty} d\omega e^{i\omega(t-t_0)} b_n^{0\dagger}(\omega)[a, c_n]\right\}$$
$$- \sum_n [a, c_n^\dagger]\left\{\frac{\kappa_n}{2} c_n + \sqrt{\frac{\kappa_n}{2\pi}}\int_{-\infty}^{\infty} d\omega e^{-i\omega(t-t_0)} b_n^{0\dagger}(\omega)\right\} . \qquad (3.112)$$

It is instructive to consider the case when there is only one bath and one system mode $c_1 = c$ governed by the Hamiltonian

$$H_{\rm sys} = \hbar\omega_0 c^\dagger c + \hbar\frac{\beta}{2} c^\dagger c^\dagger c c . \qquad (3.113)$$

Letting $a = c$ and suppressing the bath mode subscript, (3.112) yields

$$\frac{dc}{dt} = -i\omega_0 c - i\beta c^\dagger c c - \frac{\kappa}{2} c - \sqrt{\frac{\kappa}{2\pi}}\int_{-\infty}^{\infty} d\omega e^{-i\omega(t-t')} b_n^0(\omega) . \qquad (3.114)$$

By comparison with (3.88) one identifies the last term on the right-hand side of (3.114) as the driving term due to the incoming bath field. Hence, we define

$$b_n^{\rm in}(t) = -\frac{1}{\sqrt{2\pi}}\int_{-\infty}^{\infty} d\omega\ \exp[-i\omega(t-t_0)] b_n^0(\omega) \qquad (3.115)$$

as the incoming field from the nth bath and then (3.112) becomes

$$\frac{da}{dt} = -\frac{i}{\hbar}[a, H_{\rm sys}] + \sum_n \left\{ \left[\frac{\kappa}{2}c^\dagger - \sqrt{\kappa}b^\dagger_{\rm in}(t)\right][a,c] - [a,c^\dagger]\left[\frac{\kappa}{2}c - \sqrt{\kappa}b_{\rm in}(t)\right]\right\} . \quad (3.116)$$

Using the bath mode commutation relations (3.103) and (3.104), one obtains from (3.115) the commutation relations

$$[b_n^{\rm in}(t), b_{n'}^{\rm in\dagger}(t')] = \delta_{nn'}\delta(t-t') \quad \text{and} \quad [b_n^{\rm in}(t), b_{n'}^{\rm in}(t')] = 0 . \quad (3.117)$$

These commutation relations demonstrate that the Markov approximation has been built into the Hamiltonian (3.100). From (3.108), (3.111), and (3.115) one obtains

$$\frac{1}{\sqrt{2\pi}}\int_{-\infty}^{\infty} d\omega\; b_n(\omega) = \frac{\sqrt{\kappa_n}}{2}c_n(t) - b_n^{\rm in}(t) . \quad (3.118)$$

Because the equal time commutation relation of $b_n(\omega)$ with any system operator is zero, it is clear that $\frac{1}{2}\kappa c_n(t) - \sqrt{\kappa_n}b_n^{\rm in}(t)$ and $\frac{1}{2}\kappa_n c_n^\dagger(t) - \sqrt{\kappa_n}b_n^{\rm in\dagger}(t)$ commute with any system operator a. This is a generalization of (3.94) and (3.95), which were obtained for the special case of a linear system.

To identify the output fields, we integrate (3.108) backward in time, starting from initial conditions at t_1:

$$b_n(\omega) = e^{-i\omega(t-t_1)}b_n^1(\omega) - \sqrt{\frac{\kappa_n}{2\pi}}\int_t^{t_1} e^{-i\omega(t-t')}c_n(t')dt' . \quad (3.119)$$

Carrying out the same procedure as before, and using (3.113) to aid in the identification of the outgoing bath fields, one obtains

$$b_n^{\rm out}(t) = \frac{1}{\sqrt{2\pi}}\int_{-\infty}^{\infty} d\omega e^{-i\omega(t-t_1)}b_n^1(\omega) \quad (3.120)$$

and

$$\frac{da}{dt} = -\frac{i}{\hbar}[a, H_{\rm sys}] - \sum_n \left\{ \left[\frac{\kappa_n}{2}c_n^\dagger - \sqrt{\kappa_n}b_n^{\rm out\dagger}(t)\right][a, c_n] - [a, c_n^\dagger]\left[\frac{\kappa_n}{2}c_n - \sqrt{\kappa_n}b_n^{\rm out}(t)\right]\right\} . \quad (3.121)$$

It is worth noting that one must have $t_1 > t$ for (3.119) and (3.120) to hold, just as one must have $t > t_0$ for (3.108) and (3.109) to hold. Using the bath mode commutation relations (3.103) and (3.104), one obtains from (3.120) the following commutation relations for the outgoing bath fields

$$[b_n^{\rm out}(t), b_{n'}^{\rm out\dagger}(t')] = \delta_{nn'}\delta(t-t') \quad \text{and} \quad [b_n^{\rm out}(t), b_{n'}^{\rm out}(t')] = 0 . \quad (3.122)$$

In analogy with (3.118), one finds

$$\frac{1}{\sqrt{2\pi}}\int_{-\infty}^{\infty} b(\omega)d\omega = b_n^{\text{out}}(t) - \frac{\sqrt{\kappa_n}}{2}c_n(t) . \tag{3.123}$$

Equations (3.119) and (3.123) yield the following generalization of (3.89):

$$b_n^{\text{out}}(t) = \sqrt{\kappa_n}c_n(t) - b_n^{\text{in}}(t) . \tag{3.124}$$

Equation (3.121) can in principle be solved to obtain the system operators in terms of their past values and those of the $b_n^{\text{in}}(t)$. Consequently, one has

$$[a(t), b_n^{\text{in}}(t')] = 0, \qquad t' > t . \tag{3.125}$$

Likewise, one obtains

$$[a(t), b_n^{\text{out}}(t')] = 0, \qquad t' < t . \tag{3.126}$$

These two equations maintain quantum causality. Defining the unit step function by

$$u(t) = \begin{cases} 1, & t > 0 \\ 1/2 & t = 0 \\ 0 & t < 0 \end{cases} \tag{3.127}$$

and substituting (3.124) into the commutation relations (3.125) and (3.126), we obtain

$$[a(t), b_n^{\text{in}}(t')] = u(t - t')\sqrt{\kappa}[a(t), c_n(t')] . \tag{3.128}$$

Similarly,

$$[a(t), b_n^{\text{out}}(t')] = u(t' - t)\sqrt{\kappa}[a(t), c_n(t')] . \tag{3.129}$$

Hence, the commutators for system operators with the incoming or outgoing field operators can be calculated from the system operators alone. As illustrated by examples in Sects. 3.6 and 3.7, one is often interested in obtaining restrictions on the of scattering matrix between the incoming and outgoing fields. From (3.124) one obtains the following expression for c_n

$$c_n = \frac{1}{\sqrt{\kappa_n}}[b_n^{\text{out}}(t) + b_n^{\text{in}}(t)] . \tag{3.130}$$

Substituting this or its Hermitian conjugate into (3.125), one obtains the following causality restrictions on the commutators between the incoming and outgoing fields:

$$[b_n^{\text{in}}(t), b_n^{\text{out}}(t')] = [b_n^{\text{in}}(t), b_n^{\text{out}\dagger}(t')] = 0 , \qquad t > t' . \tag{3.131}$$

3.6 Scattering Matrices Within the Markov Approximation

Caves has exploited the restrictions that commutation relations place on the relations between incoming and outgoing fields in order to derive quantum limits on linear amplifier noise [16]. Here we exhibit some input output relationships that satisfy the commutation relations established in Sect. 3.5 for the incoming and outgoing fields. In Sects. 3.9 through 3.11 a treatment will be given that does not rely on the Markov approximation.

It is easy to check that the phase shifter scattering relation

$$b_{\rm out}(t) = e^{i\phi} b_{\rm in}(t) \ , \tag{3.132}$$

where ϕ is a constant phase, and the beam-splitter scattering relation

$$\begin{pmatrix} b_1^{\rm out}(t) \\ b_2^{\rm out}(t) \end{pmatrix} = \begin{pmatrix} S_{11} & S_{12} \\ S_{21} & S_{22} \end{pmatrix} \begin{pmatrix} b_1^{\rm in}(t) \\ b_2^{\rm in}(t) \end{pmatrix} \ , \tag{3.133}$$

where

$$\begin{pmatrix} S_{11} & S_{12} \\ S_{21} & S_{22} \end{pmatrix} \tag{3.134}$$

is any constant unitary matrix, satisfies the commutation relations (3.117), (3.122), and (3.131). Hence, these are allowed transformations within the Markov approximation.

As an example of a scatterer that produces frequency conversion, consider the mode transformation performed by a noiseless squeezer

$$b_{\rm out}(\omega_0+\omega) = \mu(\omega_0+\omega) b_{\rm in}(\omega_0+\omega) + \nu(\omega_0-\omega) b_{\rm in}^\dagger(\omega_0-\omega) \ , \tag{3.135}$$

where $b_{in}(\omega)$ and $b_{out}(\omega)$ are given by

$$b_{\rm out}(\omega) = \frac{1}{\sqrt{2\pi}} \int_{-\infty}^{\infty} dt e^{i\omega t} b_{\rm out}(t) \tag{3.136}$$

and

$$b_{\rm in}(\omega) = \frac{1}{\sqrt{2\pi}} \int_{-\infty}^{\infty} dt e^{i\omega t} b_{\rm in}(t) \ . \tag{3.137}$$

The commutation relations (3.117) and (3.122), Fourier transformed, yield

$$[b_{\rm in}(\omega), b_{\rm in}^\dagger(\omega')] = [b_{\rm out}(\omega), b_{\rm out}^\dagger(\omega')] = \delta(\omega-\omega') \ , \tag{3.138}$$

$$[b_{\rm in}(\omega), b_{\rm in}(\omega')] = [b_{\rm out}(\omega), b_{\rm out}(\omega')] = 0 \ . \tag{3.139}$$

These commutation relations lead to the restrictions

$$1 = |\mu(\omega_0+\omega)|^2 - |\nu(\omega_0-\omega)|^2 \tag{3.140}$$

and

$$\mu(\omega_0 + \omega)\nu(\omega_0 + \omega) = \mu(\omega_0 - \omega)\nu(\omega_0 - \omega) \ . \tag{3.141}$$

From (3.140) and (3.141), one can obtain

$$|\nu(\omega_0 + \omega)| = |\nu(\omega_0 - \omega)| \ , \tag{3.142}$$

$$|\mu(\omega_0 + \omega)| = |\mu(\omega_0 - \omega)| \ . \tag{3.143}$$

Equations (3.135) through (3.139) also yield

$$[b_{\text{in}}(t), b_{\text{out}}(t')] = \frac{e^{-2i\omega_0 t'}}{2\pi} \int_{-\infty}^{\infty} d\omega e^{-i\omega(t-t')}\nu(\omega) \tag{3.144}$$

and

$$[b_{\text{in}}(t), b_{\text{out}}^\dagger(t')] = \frac{1}{2\pi} \int_{-\infty}^{\infty} d\omega e^{-i\omega(t-t')}\mu^*(\omega) \ . \tag{3.145}$$

The causality conditions (3.131) thus lead to the following additional restrictions on $\mu(\omega)$ and $\nu(\omega)$

$$\int_{-\infty}^{\infty} e^{i\omega t}\mu(\omega)d\omega = \int_{-\infty}^{\infty} e^{i\omega t}\nu^*(\omega)d\omega = 0 \text{ for } t > 0 \ . \tag{3.146}$$

An example of a pair of functions satisfying (3.140), (3.141), and (3.146) will be obtained in the next section.

3.7 Reflection Parametric Amplifier

Here we analyze a degenerate parametric amplifier [5] whose system Hamiltonian is given by

$$H_{\text{sys}} = \hbar\omega_0 + i\frac{\hbar\epsilon\kappa}{4}[e^{-2i\omega_0 t}a^\dagger a^\dagger - e^{2i\omega_0 t}aa] \tag{3.147}$$

and whose coupling to the bath is given by

$$H_{\text{int}} = i\hbar\sqrt{\frac{\kappa}{2\pi}} \int_{-\infty}^{\infty} d\omega[b^\dagger(\omega)a - a^\dagger b(\omega)] \ . \tag{3.148}$$

For this system, (3.116) reduces to

$$\frac{da}{dt} + i\omega_0 a + \frac{\kappa}{2}a - \frac{\epsilon\kappa}{2}e^{-2i\omega_0 t}a^\dagger = \sqrt{\kappa}b_{\text{in}}(t) \ . \tag{3.149}$$

Taking the Fourier transform of this equation, one obtains

$$b_{\text{out}}(\omega_0 + \omega) = \mu(\omega_0 + \omega)b_{\text{in}}(\omega_0 + \omega) + \nu(\omega_0 - \omega)b_{\text{in}}^\dagger(\omega_0 - \omega) \ , \tag{3.150}$$

where

$$\mu(\omega_0 + \omega) = -\left[1 - \frac{\kappa(-i\omega + \kappa/2)}{(-i\omega + \kappa/2)^2 - \epsilon^2\kappa^2/4}\right] \tag{3.151}$$

and

$$\nu(\omega_0 - \omega) = \frac{\epsilon\kappa^2/2}{(-i\omega + \kappa/2)^2 - \epsilon^2\kappa^2/4} \ . \tag{3.152}$$

With a bit of algebra one can demonstrate that (3.151) and (3.152) satisfies (3.140) through (3.143). Also, from (3.151) and (3.152) one can show that the poles of $\mu(\omega)$ and $\nu^*(\omega)$ occur in the lower half plane at

$$\omega = \omega_0 - i\left(\frac{\kappa}{2} \pm \frac{\epsilon\kappa}{2}\right) \tag{3.153}$$

for the physical regime

$$\epsilon < 1 \ . \tag{3.154}$$

From this it follows that the causality conditions (3.146) are met. That these functions meet the causality requirement is no surprise since they were obtained as the solution to causal equations of motion.

3.8 Homodyne Detection

Suppose that $b_{\text{in}}(t)$ is the field delivered to a photodetector. Within the Markov approximation the operator for the photon current I is given by

$$I(t) = b_{\text{in}}^{\dagger}(t) b_{\text{in}}(t) \ . \tag{3.155}$$

Consider the unnormalized state

$$|t_1, t_2, \ldots, t_N\rangle = b_{\text{in}}^{\dagger}(t_1) b_{\text{in}}^{\dagger}(t_1) \ldots b_{\text{in}}^{\dagger}(t_1)|0\rangle \ . \tag{3.156}$$

Operating on this state with $I(t)$ and using the commutation relations (3.117), we get

$$I(t)|t_1, t_2, \ldots, t_N\rangle = \sum_{n=1}^{N} \delta(t - t_n)|t_1, t_2, \ldots, t_N\rangle \ . \tag{3.157}$$

Thus the state is an eigenstate of the photon current operator. The photocurrent i produced by the photodetector in response to this state is a series of delta function pulses

$$i = \sum_{n=1}^{N} \delta(t - t_n) \tag{3.158}$$

occurring at the arrival times $t_1, t_2, \ldots, t_N$.

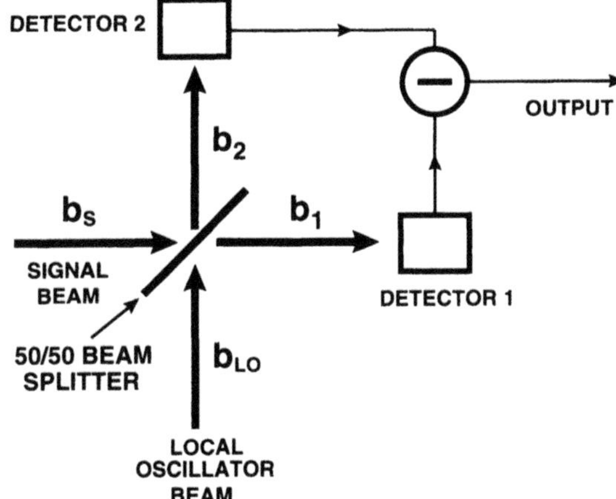

Fig. 3.1. Balanced-homodyne detector. Output current is proportional to the difference between the photocurrents developed in detectors 1 and 2

Although photodetectors count photons, through optical mixing they can also be used to measure wave properties of the electromagnetic field. An example of an optical mixer is a balanced homodyne detector [17,18]. Such a detector, illustrated in Fig. 3.1, consists of a 50/50 beam splitter and two photodetectors. The signal field $b_S(t)$ is fed into one port of the beam-splitter and a local oscillator field $b_{LO}(t)$ is fed into the other port. The two light fields are combined by the beam splitter and delivered to the two photodetectors. The difference in the photocurrents produced by the two photodetectors is monitored. Let the field delivered to one photodetector be denoted by $b_1(t)$ and that delivered to the other photodetector be denoted $b_2(t)$.

Let the beam-splitter mode transformation be given by the following unitary transformation

$$\begin{pmatrix} b_1(t) \\ b_2(t) \end{pmatrix} = \frac{1}{\sqrt{2}} \begin{pmatrix} 1 & 1 \\ -1 & 1 \end{pmatrix} \begin{pmatrix} b_S(t) \\ b_{LO}(t) \end{pmatrix} . \tag{3.159}$$

The operator for the differenced photocurrent reported by the homodyne detector is

$$I_D(t) = b_1^\dagger(t) b_1(t) - b_2^\dagger(t) b_2(t) . \tag{3.160}$$

Since the photocurrents delivered by the photodetectors are classical, the summed photocurrent is available for simultaneous measurement with the differenced photocurrent. It is given by

$$I_S(t) = b_1^\dagger(t) b_1(t) + b_2^\dagger(t) b_2(t) . \tag{3.161}$$

Using the transformation (3.159) to re-express the differenced and summed photocurrents in terms of the signal and local oscillator field operators, one obtains

$$I_D(t) = b_{LO}^\dagger(t) b_S(t) + b_S^\dagger(t) b_{LO}(t) \ , \tag{3.162}$$

$$I_S(t) = b_{LO}^\dagger(t) b_{LO}(t) + b_S^\dagger(t) b_S(t) \ . \tag{3.163}$$

Since $I_D(t)$ and $I_S(t)$ are simultaneously measurable, states must exist that are simultaneous eigenstates for both of these operators. Such a state representing a positive going spike for the differenced photocurrent at time t_1 is

$$|t_1+\rangle = \frac{1}{\sqrt{2}}[b_{LO}^\dagger(t_1) + b_S^\dagger(t_1)]|0\rangle \ . \tag{3.164}$$

For this state

$$I_D(t)|t_1+\rangle = \delta(t - t_1)|t_1+\rangle \ , \tag{3.165}$$

$$I_S(t)|t_1+\rangle = \delta(t - t_1)|t_1+\rangle \ . \tag{3.166}$$

An eigenstate representing a negative going spike in the differenced photocurrent at time t_1 is

$$|t_1-\rangle = \frac{1}{\sqrt{2}}[b_{LO}^\dagger(t_1) - b_S^\dagger(t_1)]|0\rangle \ . \tag{3.167}$$

For this state

$$I_D(t)|t_1-\rangle = -\delta(t - t_1)|t_1-\rangle \ , \tag{3.168}$$

$$I_S(t)|t_1-\rangle = \delta(t - t_1)|t_1-\rangle \ . \tag{3.169}$$

The equivalence of (3.161) with (3.163) is an expression of photon number conservation, thus the photocurrent spikes exhibited by the detectors can be attributed to photons in the signal and local oscillator field. However, as illustrated by (3.164) and (3.167), which detector a given photon materializes at, that is, whether the difference current spike is positive or negative, is determine by the interference between the field amplitudes $b_{LO}(t)$ and $b_S(t)$. This is also evident from the expression for the difference photocurrent operator (3.162). This expression does not involve the signal or local oscillator photon current operators $b_S^\dagger(t)b_S(t)$ or $b_{LO}^\dagger(t)b_{LO}(t)$. The differenced output of a homodyne detector reveals field properties of the signal by how it interferes with the local oscillator field which serves as a reference field.

Coherent light with a photon flux much higher than the photon flux of the signal is generally employed for the local oscillator. To this end we introduce the coherent state defined by

$$b_{LO}(t)|\sqrt{I_{LO}}e^{-i\psi}\rangle = \sqrt{I_{LO}}e^{-i\psi}|\sqrt{I_{LO}}e^{-i\psi}\rangle \tag{3.170}$$

and

$$\langle\sqrt{I_{LO}}e^{-i\psi}|\sqrt{I_{LO}}e^{-i\psi}\rangle = 1 \ , \tag{3.171}$$

where I_{LO} is the local oscillator photon current and ψ is the instantaneous phase of the local oscillator which usually is of the form $\psi = -\omega_{LO}t - \phi$, where ω_{LO} is the local oscillator frequency and ϕ is a phase that can be adjusted via a phase shifter in the path of the local oscillator beam. We note however that I_{LO} and ψ could be arbitrary functions of time, in particular, the expressions are applicable even for the case when the local oscillator consists of a train of pulses with chirp. We now introduce the local oscillator operator

$$d_{LO}(t) = b_{LO}(t) - \sqrt{I_{LO}}e^{-i\psi} . \tag{3.172}$$

The local oscillator state vector acts like the vacuum state for this operator, that is,

$$d_{LO}(t)|\sqrt{I_{LO}}e^{-i\psi}\rangle = 0 . \tag{3.173}$$

By introducing $d_{LO}(t)$ we have separated the local oscillator field into a coherent mean field and quantum fluctuations. Using (3.172) to eliminate the b_{LO} from (3.162), we obtain

$$\begin{aligned} I_D(t) = &\sqrt{I_{LO}}[e^{i\psi}b_S(t) + e^{-i\psi}b_S^\dagger(t)] \\ &+[e^{i\psi}d_{LO}^\dagger(t)b_S(t) + e^{-i\psi}b_S^\dagger(t)d_{LO}(t)] . \end{aligned} \tag{3.174}$$

If the local oscillator intensity is sufficiently large, the terms in the second set of square brackets can be neglected and the operator corresponding to the observable reported by the homodyne detector becomes

$$I_D(t) = \sqrt{I_{LO}}[e^{i\psi}b_S(t) + e^{-i\psi}b_S^\dagger(t)] . \tag{3.175}$$

Using the free field commutation relations (3.117), one finds

$$[I_D(t), I_D(t')] = 0 . \tag{3.176}$$

From this one sees that a measurement of I_D at time t does not disturb a successive measurement of I_D at time t'. From the commutation relation it also follows that a product such as $I_D(t)I_D(t')$ is Hermitian, and, hence, measurable.

We now specialize to the case when

$$\psi = \omega_0 t + \phi , \tag{3.177}$$

and I_{LO} is a constant. Equation (3.175) then becomes

$$I_D(t) = \sqrt{I_{LO}}[e^{i\phi}b_S(t)e^{i\omega_0 t} + e^{-i\phi}b_S^\dagger(t)e^{-i\omega_0 t}] . \tag{3.178}$$

The Fourier transform of (3.178) is

$$I_D(\omega) = \sqrt{I_{LO}}[e^{i\phi}b_S(\omega_0 + \omega) + e^{-i\phi}b_S^\dagger(\omega_0 - \omega)] . \tag{3.179}$$

Hence, the output of the homodyne detector at frequency ω is caused by the components of the signal field at frequency $\omega_0 + \omega$ and $\omega_0 - \omega$. These components are often referred to as signal and image sidebands. The down conversion of these signals to the frequency ω results from the beating of the local oscillator field at frequency ω_0 with the signal fields at the photodetectors. For reference, $I_D(\omega)$ satisfies the commutation relation

$$[I_D(\omega), I_D^\dagger(\omega')] = 0 . \tag{3.180}$$

For squeezed state experiments one is often interested in measuring the power spectrum of I_D. To this end it is instructive to contemplate how one might build an instrument to measure the power spectrum. To measure the amount of signal power at frequency ω, we can first multiply the classical output $I_D(t)$ delivered by the homodyne detector by $\cos\omega t$ and $\sin\omega t$ and then run this through a low-pass filter. That is, from the $I_D(t)$ one can generate the classical signals

$$\int_{-T/2}^{T/2} I_D(t)\cos\omega t \, dt \tag{3.181}$$

and

$$\int_{-T/2}^{T/2} I_D(t)\sin\omega t \, dt . \tag{3.182}$$

Then we can square these signals, add them together, and divide the resulting quantity by $1/T$ so that the noise floor will remain fixed as the filter bandwidth is reduced, i.e. as T is made larger. Finally, we can repeat this procedure multiple times on I_D for successive time intervals of length T and then average the results. According to this procedure, the power spectrum is defined by

$$\begin{aligned} P(\omega) = \lim_{T\to\infty} \frac{1}{T} \Bigg\langle \left[\int_{-T/2}^{T/2} I_D(t)\cos\omega t \, dt\right]^2 \\ + \left[\int_{-T/2}^{T/2} I_D(t)\sin\omega t \, dt\right]^2 \Bigg\rangle . \end{aligned} \tag{3.183}$$

This expression can be reduced to

$$P(\omega) = \lim_{T\to\infty} \frac{1}{T} \int_{-T/2}^{T/2} dt \int_{-T/2}^{T/2} dt' \langle I_D(t) I_D(t')\rangle \cos\omega(t-t') . \tag{3.184}$$

Substituting the Fourier transform

$$I_D(t) = \frac{1}{\sqrt{2\pi}} \int_{-\infty}^{\infty} d\omega I_D(\omega) e^{-i\omega t} \tag{3.185}$$

into (3.184), one obtains after some manipulation

$$P(\omega) = \lim_{T\to\infty} \frac{T}{2\pi} \int_{-\infty}^{\infty} d\omega' \int_{-\infty}^{\infty} d\omega'' \frac{\sin[(\omega' - \omega)T/2]}{(\omega' - \omega)T/2} \times \frac{\sin[(\omega' - \omega)T/2]}{(\omega' - \omega)T/2} \langle I_D^\dagger(\omega') I_D(\omega'')\rangle \ . \tag{3.186}$$

Suppose that $\langle I_D^\dagger(\omega') I_D(\omega'')\rangle$ has the form

$$\langle I_D^\dagger(\omega) I_D(\omega')\rangle = S(\omega)\delta(\omega - \omega') \ . \tag{3.187}$$

Then, substituting (3.187) into (3.186), we obtain

$$P(\omega) = S(\omega) \ . \tag{3.188}$$

Hence, if $\langle I_D^\dagger(\omega') I_D(\omega'')\rangle$ has the form (3.187), we can, upon evaluating the expectation value, read off the power spectrum $P(\omega)$ from

$$P(\omega)\delta(\omega - \omega') = \langle I_D^\dagger(\omega) I_D(\omega')\rangle \ . \tag{3.189}$$

As a simple example, consider the case when the signal field consists of vacuum fluctuations, that is, the state vector is the vacuum state $|0\rangle_b$ defined by

$$b_S(\omega)|0\rangle_b = 0 \ . \tag{3.190}$$

Using (3.179), one finds

$${}_b\langle 0|I_D^\dagger(\omega) I_D(\omega')|0\rangle_b = I_{LO}\delta(\omega - \omega') \ . \tag{3.191}$$

Hence,

$$P(\omega) = I_{LO} \ . \tag{3.192}$$

This is local oscillator shot noise.

As a second example, consider the case where the signal field has passed through a squeezer performing the mode transformation of the form given in (3.135) as

$$b_S(\omega_0 + \omega) = \mu(\omega_0 + \omega) a_S(\omega_0 + \omega) + \nu(\omega_0 - \omega) a_S^\dagger(\omega_0 - \omega) \ , \tag{3.193}$$

and let the state vector be the vacuum state $|0\rangle_a$ for the input to the squeezer, that is,

$$a_S(\omega)|0\rangle_a = 0 \ . \tag{3.194}$$

Using (3.179) and (3.189), one then finds

$$P(\omega) = I_{LO} \left\{ |\mu(\omega_0 - \omega)|^2 + |\nu(\omega_0 - \omega)|^2 + 2\mathrm{Re}\left[e^{i2\phi}\mu(\omega_0 - \omega)\nu(\omega_0 - \omega)\right] \right\} \ . \tag{3.195}$$

As the phase ϕ is varied $P(\omega)$ varies between the minimum and maximum values of

$$P_{\min}(\omega) = I_{LO}\{|\mu(\omega_0 - \omega)| - |\nu(\omega_0 - \omega)|\}^2 , \tag{3.196}$$

$$P_{\max}(\omega) = I_{LO}\{|\mu(\omega_0 - \omega)| + |\nu(\omega_0 - \omega)|\}^2 . \tag{3.197}$$

Using the relation (3.140), one finds

$$P_{\min}(\omega) P_{\max}(\omega) = I_{LO}^2 . \tag{3.198}$$

Consider the case when $|\mu(\omega_0 - \omega)|$ is large compared with one. Then, from (3.140) one obtains, to order $1/|\mu(\omega_0 - \omega)|^2$,

$$|\nu(\omega_0 - \omega)| = |\mu(\omega_0 - \omega)| - \frac{1}{2|\mu(\omega_0 - \omega)|} . \tag{3.199}$$

The minimum value of the power spectrum at ω is then given by:

$$P(\omega) = \frac{I_0}{4|\mu(\omega_0 - \omega)|} . \tag{3.200}$$

Hence, by making $|\mu(\omega_0 - \omega)|$ very large the noise floor of a homodyne detector can be made much smaller than the shot noise floor given by (3.192). This noise reduction is called squeezing. Squeezed state generation involves the establishment of correlations between signals at frequencies $\omega_0 + \omega$ and $\omega_0 - \omega$, Homodyne detection is sensitive to these correlations through mixing with a local oscillator at frequency ω_0. Schumaker and Caves [19,20] have developed an efficient formalism for dealing with such two mode systems. For a treatment of photon counting and homodyne detection that does not make the Markov approximation see [21].

3.9 Multiple Ports

We now return to the approach of Sect. 3.3 for the case when there are multiple input fields. The Lagrangian density of the system is

$$\mathcal{L} = \delta(x) L_{\text{sys}} + u(x) \sum_{n=1}^{m} \left\{ \frac{\rho_n}{2} \left(\frac{\partial Y_n}{\partial t} \right)^2 - \frac{\sigma_n}{2} \left(\frac{\partial Y_n}{\partial x} \right)^2 \right\} . \tag{3.201}$$

For $x > 0$ the canonical momentum for the nth field is given by

$$\Pi_n(x,t) = \frac{\partial \mathcal{L}}{\partial(\partial Y_n / \partial t)} = \rho_n \frac{\partial Y_n}{\partial t} . \tag{3.202}$$

The canonical commutation relations become

$$[Y_n(x,t), \Pi_{n'}(x',t)] = i\hbar \delta_{nn'} \delta(x - x') \tag{3.203}$$

and

$$[Y_n(x,t), Y_{n'}(x',t)] = [\Pi_n(x,t), \Pi_{n'}(x',t)] = 0 \ . \tag{3.204}$$

To simplify the notation we will use a_n to denote the incoming field annihilation operators and b_n to denote the outgoing field annihilation operators. The nth field is thus given by

$$Y_n(x,t) = Y_n^{\text{in}}(x,t) + Y_n^{\text{out}}(x,t) \ , \tag{3.205}$$

where

$$Y_n^{\text{in}}(x,t) = \frac{1}{2}\sqrt{\frac{\hbar}{\pi \Gamma_n}} \int_0^\infty \frac{d\omega}{\sqrt{\omega}} \left[a_n(\omega) e^{-i\omega(t+x/v_n)} + \text{H.c.} \right] \tag{3.206}$$

and

$$Y_n^{\text{out}}(x,t) = \frac{1}{2}\sqrt{\frac{\hbar}{\pi \Gamma_n}} \int_0^\infty \frac{d\omega}{\sqrt{\omega}} \left[b_n(\omega) e^{-i\omega(t-x/v_n)} + \text{H.c.} \right] \ . \tag{3.207}$$

If the transmission lines had extended to minus infinity, then a straightforward generalization of (3.10) and (3.11) could be used to obtain generalized commutation relations of the form (3.12) and (3.13). However, because of scattering by the system from one mode to another, the complete set of such commutation relations does not have to be satisfied. Since an incoming field does not know that a system capable of scattering exists at $x = 0$ until it arrives, the incoming fields, at least, should behave as if they were propagating on independent transmission lines extending to infinity in both directions. From this we conclude that the following commutation relations are satisfied

$$[a_n(\omega), a_{n'}^\dagger(\omega')] = \delta_{nn'}\delta(\omega - \omega') \quad \text{and} \quad [a_n(\omega), a_{n'}(\omega')] = 0 \ . \tag{3.208}$$

If one separates the incoming and outgoing fields via a circulator, (which separates the fields propagating toward and away from the scattering system), then by restricting oneself to observations on a single transmission line, there is no way of inferring the location of the scattering system. Thus, the outgoing field should behave as if it were residing on a transmission line of infinite extent in both directions. Hence, we conclude that

$$[b_n(\omega), b_n^\dagger(\omega')] = \delta(\omega - \omega') \quad \text{and} \quad [b_n(\omega), b_n(\omega')] = 0 \ . \tag{3.209}$$

Since the outward propagating fields reside on separate transmission lines, a disturbance of the outgoing field in one transmission line will not propagate into the outgoing fields of the other transmission lines. Hence, (3.209) generalizes to

$$[b_n(\omega), b_{n'}^\dagger(\omega')] = \delta_{nn'}\delta(\omega - \omega') \quad \text{and} \quad [b_n(\omega), b_{n'}(\omega')] = 0 \ . \tag{3.210}$$

Because of the scattering by the system at $x = 0$, a disturbance in one of the incoming fields may propagate into any one of the outgoing fields. Hence, commutation relations of the form $[a_n(\omega), b_{n'}(\omega')]$ and $[a_n(\omega), b^\dagger_{n'}(\omega')]$ are not necessarily zero, that is, they need not satisfy the commutation relations between independent scalar fields. However, because the disturbances are propagated causally, one has

$$[Y_n^{\text{out}}(x,t), Y_{n'}^{\text{in}}(x',t')] = 0 \;, \quad \text{for} \quad t - t' < \frac{x}{v} + \frac{x'}{v'} \;. \tag{3.211}$$

Equations (3.206) and (3.207) and the commutation relations (3.208) and (3.210) yield

$$[Y_n^{\text{in}}(x,t), Y_{n'}^{\text{in}}(x',t)] + [Y_n^{\text{out}}(x,t), Y_{n'}^{\text{out}}(x',t)] = 0 \;, \tag{3.212}$$
$$[\Pi_n^{\text{in}}(x,t), \Pi_{n'}^{\text{in}}(x',t)] + [\Pi_n^{\text{out}}(x,t), \Pi_{n'}^{\text{out}}(x',t)] = 0 \;, \tag{3.213}$$
$$\begin{aligned}&[Y_n^{\text{in}}(x,t), \Pi_{n'}^{\text{in}}(x',t)] + [Y_n^{\text{out}}(x,t), \Pi_{n'}^{\text{out}}(x',t)] \\ &= i\hbar\delta_{n,n'}\delta(\omega-\omega') \;,\end{aligned} \tag{3.214}$$

From (3.212)–(3.214) and the causality relation (3.211), it is straightforward to show that the canonical commutation relations (3.203) and (3.204) are satisfied.

When $x = x' = 0$, the causality condition (3.211) can be written in the simpler form

$$[Y_n^{\text{out}}(t), Y_{n'}^{\text{in}}(t')] = 0 \quad \text{for} \quad t - t' < 0 \;. \tag{3.215}$$

The commutation relations established in this section place restrictions on the kind of input-output relations a device can display. We will make use of these restrictions in the next sections.

3.10 Phase Shifters

As an example of the application of the above theory to the one port case, suppose the scatterer performs a frequency dependent phase shift of the incoming signal of the form

$$b(\omega) = \exp[i(\phi_0 + \tau_0\omega)]a(\omega) \;. \tag{3.216}$$

This transformation satisfies the commutation relations (3.210), provided the input commutation relations (3.208) are satisfied. One also has

$$[b(\omega), a(\omega')] = e^{i\phi(\omega)}\delta(\omega-\omega') \;. \tag{3.217}$$

Using (3.206) and (3.207), one obtains

$$\begin{aligned}[Y^{\text{out}}(t), Y^{\text{in}}(t')] = &-i\frac{\hbar\cos\phi_0}{2\Gamma}u(t-t'-\tau_0) \\ &+ i\frac{\hbar\sin\phi_0}{\pi\Gamma}\int_0^\infty \frac{d\omega}{\omega}\cos[\omega(t-t'-\tau_0)] \;.\end{aligned} \tag{3.218}$$

The first term on the right-hand side of this equation satisfies the causality condition (3.215), provided that

$$\tau_0 > 0 \; . \tag{3.219}$$

However, the functional in the second term of (3.218) is not causal. Hence, one must have

$$\sin \phi_0 = 0 \; , \tag{3.220}$$

that is,

$$\phi_0 = \pi n \; , \tag{3.221}$$

where n is an integer. Substituting (3.216) together with (3.221) into (3.207) yields

$$Y_{\mathrm{out}}(t) = e^{i\pi n} Y_{\mathrm{in}}(t - \tau) \; , \tag{3.222}$$

which is manifestly causal when (3.221) is satisfied. A phase shifter satisfying (3.216), (3.219), and (3.221) can indeed be built. It consists of a section of transmission line of length L such that the round-trip travel time τ is given by

$$\tau = \frac{2L}{v} \; . \tag{3.223}$$

The n even case is that for which the far end of the transmission line is terminated so that the reflected signal is not inverted. The n odd case is that for which the far end to the transmission line is terminated so that the reflected signal is inverted. We note that phase shifters in quantum optics are often modeled by

$$b = e^{i\phi} a \; , \tag{3.224}$$

where ϕ is a frequency independent constant. It was shown via (3.220) that such a transformation is causal only under the special cases $\exp(i\phi) = \pm 1$. Apart from these special cases, (3.224) can accurately represent the performance of a phase shifter only over some finite frequency interval.

3.11 Beam Splitters

We now consider the case of two transmission lines for which the scatterer performs the following transformation:

$$\begin{pmatrix} b_1(\omega) \\ b_2(\omega) \end{pmatrix} = \begin{pmatrix} S_{11}(\omega) & S_{12}(\omega) \\ S_{21}(\omega) & S_{22}(\omega) \end{pmatrix} \begin{pmatrix} a_1(\omega) \\ a_2(\omega) \end{pmatrix} \; . \tag{3.225}$$

The commutation relations (3.208) and (3.210) require that the scattering matrix

$$\mathbf{S} = \begin{pmatrix} S_{11}(\omega) & S_{12}(\omega) \\ S_{21}(\omega) & S_{22}(\omega) \end{pmatrix} \tag{3.226}$$

be unitary, i.e.

$$\mathbf{S}^{\dagger} = \mathbf{S}^{-1} \ . \tag{3.227}$$

For the case when the $S_{nn'}$ are real and independent of ω one obtains

$$[Y_n^{\text{out}}(x,t), Y_{n'}^{\text{in}}(x',t')] = -i\frac{\hbar S_{nn'}}{2\sqrt{\Gamma_n \Gamma_n'}} u[t - t' - (x/v_n + x'/v_n')] \ . \tag{3.228}$$

It is clear that (3.228) satisfies the causality condition (3.211). An example of such a causal input-output transformation is the beam splitter transformation

$$\mathbf{S} = \begin{pmatrix} -\sqrt{R} & \sqrt{T} \\ \sqrt{T} & \sqrt{R} \end{pmatrix} , \tag{3.229}$$

where

$$T + R = 1 \tag{3.230}$$

and T and R are the transmission and reflection coefficients, respectively. This type of scattering is produced by an abrupt change of index of refraction as light propagates from one dielectric medium to another.

This type of scattering is also produced by a sharp discontinuity in the impedance of a transmission line, as will now be shown. Consider two transmission lines governed by wave equations of the form (3.51). One transmission line lies along the positive x-axis. Its field will be denoted by Y_1 and its mass density and tension are denoted by ρ_1 and σ_1 respectively. The other transmission line lies along the negative x-axis. Its field is denoted by Y_2 and its mass density and tension are denoted by ρ_2 and σ_2, respectively. One thus has

$$Y_1(x,t) = Y_1^{\text{in}}(t + x/v_1) + Y_1^{\text{in}}(t - x/v_1) \ , \tag{3.231}$$

$$Y_2(x,t) = Y_2^{\text{in}}(t - x/v_2) + Y_2^{\text{in}}(t + x/v_2) \ . \tag{3.232}$$

The transmission lines are attached end to end at the interface $x = 0$. Hence, continuity at $x = 0$ yields

$$Y_1(t) = Y_2(t) \ , \tag{3.233}$$

where we have suppressed the position coordinate, since it is zero. The balance of forces at the interface requires

$$\sigma_1 \left.\frac{\partial Y_1}{\partial x}\right|_{x=0+} = \sigma_2 \left.\frac{\partial Y_2}{\partial x}\right|_{x=0+} \ . \tag{3.234}$$

The boundary conditions (3.233) and (3.234) yield, respectively,

$$Y_1^{\mathrm{out}}(t) + Y_1^{\mathrm{in}}(t) = Y_2^{\mathrm{out}} + Y_2^{\mathrm{in}}, \tag{3.235}$$

$$\Gamma_1[Y_1^{\mathrm{out}}(t) - Y_1^{\mathrm{in}}(t)] = -\Gamma_2[Y_2^{\mathrm{out}} - Y_2^{\mathrm{in}}] . \tag{3.236}$$

Making use of (3.235) and (3.236) and sticking with the convention that a_n and b_n denote the creation operators for the incoming and outgoing fields, respectively, these equations yield

$$\frac{1}{\sqrt{\Gamma_1}}[b_1(\omega) + a_1(\omega)] = \frac{1}{\sqrt{\Gamma_2}}[b_2(\omega) + a_2(\omega)] , \tag{3.237}$$

$$\sqrt{\Gamma_1}[b_1(\omega) - a_1(\omega)] = -\sqrt{\Gamma_2}[b_2(\omega) - a_2(\omega)] . \tag{3.238}$$

Solving for the output fields in terms of the input fields yields

$$b_1(\omega) = \frac{\Gamma_1 - \Gamma_2}{\Gamma_1 + \Gamma_2} a_1(\omega) + \frac{2\sqrt{\Gamma_1 \Gamma_2}}{\Gamma_1 + \Gamma_2} a_2(\omega) , \tag{3.239}$$

and

$$b_2(\omega) = \frac{2\sqrt{\Gamma_1 \Gamma_2}}{\Gamma_1 + \Gamma_2} a_1(\omega) - \frac{\Gamma_1 - \Gamma_2}{\Gamma_1 + \Gamma_2} a_2(\omega) . \tag{3.240}$$

Equations (3.239) and (3.240) yield the scattering matrix (3.229) when the mechanical impedance of transmission line 2 is larger than that of transmission line 1, i.e. $\Gamma_2 > \Gamma_1$. In this case

$$\sqrt{R} = \frac{\Gamma_2 - \Gamma_1}{\Gamma_2 + \Gamma_1} \tag{3.241}$$

and

$$\sqrt{T} = \frac{2\sqrt{\Gamma_2 \Gamma_1}}{\Gamma_2 + \Gamma_1} . \tag{3.242}$$

3.12 Resonators

In a Fabry–Perot resonator the field is reflected back and forth between two mirrors. Because of this, the arguments made in Sect. 3.1 that led to the commutation relations (3.12) and (3.13) cannot be carried out for the fields inside the resonator. Thus, although the commutation relations (3.5) and (3.6) must hold for the fields inside the cavity, the commutation relations (3.12) and (3.13) need not hold in this region. That (3.12) and (3.13) do not hold is demonstrated in the analysis of a transmission line resonator presented here. A second point of this presentation is to justify the treatment of the field inside a cavity as a single harmonic oscillator characterized by equations of the form given in Sect. 3.2. To this end, it is shown how (3.26) can be arrived at by making suitable approximations to the equations for the fields inside the cavity resonator.

Consider a cavity consisting of a transmission line of length L, lying along the x-axis between $-L < x < 0$. Let its mechanical impedance be denoted Γ_2. At $x = 0$ a transmission line of impedance Γ_1 is attached. We can thus use the results of the last section. In particular, at $x = 0$, (3.233) and (3.234) hold. At $x = -L$ the boundary condition

$$Y(-L,t) = 0 \tag{3.243}$$

is taken to hold. A signal of frequency ω making a round trip in this cavity accumulates a phase shift $\omega\tau_0$, where τ_0 is the round-trip travel time:

$$\tau_0 = \frac{2L}{v_2} \ . \tag{3.244}$$

Hence, one has

$$a_2(\omega) = -e^{i\omega\tau_0} b_2(\omega) \ . \tag{3.245}$$

Solving (3.240) for the operators in terms of the incoming field creation operator yields

$$b_2(\omega) = \frac{\sqrt{T}}{1+\sqrt{R}e^{i\omega\tau_0}} a_1(\omega) \ , \tag{3.246}$$

$$a_2(\omega) = -\frac{\sqrt{T}e^{i\omega\tau_0}}{1+\sqrt{R}e^{i\omega\tau_0}} a_1(\omega) \ , \tag{3.247}$$

and

$$b_2(\omega) = -e^{i\omega\tau_0}\frac{1+\sqrt{R}e^{-i\omega\tau_0}}{1+\sqrt{R}e^{i\omega\tau_0}} a_1(\omega) \ . \tag{3.248}$$

One finds that

$$[a_2(\omega), a_2^\dagger(\omega)] = [a_2(\omega), a_2^\dagger(\omega)] = \frac{T}{1+R-2\sqrt{R}\cos\omega\tau_0}\delta(\omega-\omega') \ . \tag{3.249}$$

It is thus evident that the commutation relations (3.12) are not satisfied within the cavity. Substituting (3.246) and (3.247) into (3.7), remembering that the a's and b's have to be interchanged since, consistent with the convention of the last section, for $-L < x < 0$, the a's are associated with right propagating fields and the b's are associated with the left propagating fields, one obtains the following expression for the field inside the resonator

$$Y_2(x,t) = \sqrt{\frac{\hbar}{\pi\Gamma_2}} \int_0^\infty \frac{d\omega}{\sqrt{\omega}} \times \left\{ \frac{i\sqrt{T}e^{i\omega\tau_0/2}\sin[\omega(L+x)/v_2]}{1+\sqrt{R}e^{i\omega\tau_0}} a_1(\omega)e^{-i\omega t} + \text{H.c.} \right\} \ . \tag{3.250}$$

From any of (3.246) through (3.248) one sees that the resonances occur when

$$\exp(i\omega\tau_0) = -1 \; , \tag{3.251}$$

that is, the resonant frequencies are

$$\omega_n = \frac{(2n+1)\pi}{\tau_0} \; , \tag{3.252}$$

where n is a non-negative integer. When R approaches 1 these resonances become quite sharp. The spatial dependence $\sin[\omega(L+x)/v_2]$ at the lowest resonant frequency indicates that for this frequency the resonator is a quarter-wave resonator with the amplitude clamped at zero at $x = -L$ and the maximum excursion occurring at $x = 0$. Motivated by the shape of the mode function at resonance, we introduce the function

$$f_n(x) = (-1)^n \sin\left[\frac{(2n-1)\pi}{2L}(L+x)\right] = \cos\left[\frac{(2n+1)\pi}{2L}x\right] \tag{3.253}$$

and use it to project out the nth mode from $Y_2(x,t)$, that is,

$$y_n(t) = \frac{2}{L}\int_{-L}^{0} f_n(x) Y_2(x,t) dx \; . \tag{3.254}$$

We also introduce the momentum

$$p_n = \frac{\rho_2 L}{2}\frac{\partial y_n(t)}{\partial t} \; , \tag{3.255}$$

which can be written as

$$p_n = \int_{-L}^{0} f_n(x) \Pi_2(x,t) dx \; . \tag{3.256}$$

The functions $f_n(x)$ satisfy the orthogonality relation

$$\int_{-L}^{0} f_n(x) f_{n'}(x) dx = \frac{L}{2}\delta_{nn'} \; . \tag{3.257}$$

Making use of this orthogonality condition and the canonical commutation relations (3.5), one obtains

$$[y_n, p_{n'}] = i\hbar\delta_{nn'} \; . \tag{3.258}$$

Hence, this method of projection at least picks out a portion of the field Y_2 that satisfies the appropriate position and momentum commutation relations for a particle of mass

$$M = \rho_2 L/2 \; . \tag{3.259}$$

If the resonances are narrow, i.e. $\sqrt{R}$ is close to unity, then, upon substitution of (3.250) into (3.254) one expects contributions from frequencies only near ω_n. In this case one can make the approximation

$$\sqrt{R} = \sqrt{1-T} = 1 - T/2 \; . \tag{3.260}$$

The resonant denominator can be approximated as

$$1 + \sqrt{R}\exp(i\omega\tau_0) = \tau_0[i(\omega_0 - \omega) + \kappa/2] \; , \tag{3.261}$$

where the damping constant κ is given by

$$\kappa = T/\tau_0 \; . \tag{3.262}$$

Defining the spring constant K_n through the relation

$$\omega_n = \sqrt{\frac{K_n}{M}} \tag{3.263}$$

and keeping terms to lowest order in $\delta\omega = \omega - \omega_n$, one obtains

$$y_n(t) = \frac{1}{\sqrt{2\pi}}\sqrt{\frac{\hbar}{2\sqrt{MK_n}}}\int_{-\infty}^{\infty} d\omega \left[\frac{\sqrt{\kappa}a_1(\omega)e^{-i\omega t}}{i(\omega_n - \omega) + \kappa/2} + \text{H.c.}\right] \; , \tag{3.264}$$

where again we have extended the range of integration on the grounds that contributions from frequency components in the wings of the resonance will be negligible. This has the same form as (3.99). Hence, a procedure has been demonstrated by which the field inside a resonator can be decomposed into a discrete set of independent damped harmonic oscillators.

3.13 Lossy Transmission Lines

As another example of how the input-output theory can be used to deduce the quantum behavior of a system, in this section we quantize a lossy transmission line by coupling it to a lossless transmission line for which the commutation relations for the operators are known. This system will be used to give a quantum mechanical description to an attenuator. The transmission line analyzed here is electrical and is governed by macroscopic equations

$$-\frac{\partial V}{\partial x} = L\frac{\partial I}{\partial t} + RI \tag{3.265}$$

and

$$-\frac{\partial I}{\partial x} = C\frac{\partial V}{\partial t} + GV \; , \tag{3.266}$$

where I and V are the current and voltage, L is the inductance per unit length, C is the capacitance per unit length, R is the resistance per unit length along the transmission line, and G is the conductance per unit length across the transmission line. To keep the analysis simple we will assume that the transmission line is a distortionless line. The condition that must be met for distortionless transmission is

$$\frac{R}{L} = \frac{G}{C} . \tag{3.267}$$

Introducing the propagation velocity

$$v = \frac{1}{\sqrt{LC}} \tag{3.268}$$

and the attenuation constant

$$\alpha = \sqrt{RG} , \tag{3.269}$$

we find that (3.265) can be solved to yield the wave equation

$$\frac{\partial^2 Q}{\partial t^2} - v^2 \frac{\partial^2 Q}{\partial x^2} + 2\alpha v \frac{\partial Q}{\partial t} + \alpha^2 v^2 Q = 0 , \tag{3.270}$$

where Q is the charge along the transmission line and is given in terms of the current by

$$I = \frac{\partial Q}{\partial t} . \tag{3.271}$$

Let Q have the time dependence

$$Q = Q(x, \omega) \exp(-i\omega t) . \tag{3.272}$$

Then the wave equation (3.270) can be written in the form

$$\left[\frac{\partial}{\partial x} + i\frac{\omega}{v} - \alpha\right]\left[\frac{\partial}{\partial x} - i\frac{\omega}{v} + \alpha\right] Q(x, \omega) = 0 . \tag{3.273}$$

Hence, the charge $Q(x, \omega)$ can be decomposed into left-traveling, $Q_L(x, \omega)$, and right-traveling, $Q_R(x, \omega)$, parts as

$$Q(x, \omega) = Q_L(x, \omega) + Q_R(x, \omega), \tag{3.274}$$

where $Q_L(x, \omega)$ and $Q_R(x, \omega)$ satisfy the equations

$$\left[\frac{\partial}{\partial x} + i\frac{\omega}{v} - \alpha\right] Q_L(x, \omega) = 0 , \tag{3.275}$$

and

$$\left[\frac{\partial}{\partial x} - i\frac{\omega}{v} + \alpha\right] Q_R(x, \omega) = 0 . \tag{3.276}$$

Equations (3.275) and (3.276) have the solutions

$$Q_L(x, \omega) = Q_L(\omega) \exp[(\alpha - i\omega/v)x] , \tag{3.277}$$
$$Q_R(x, \omega) = Q_R(\omega) \exp[(-\alpha + i\omega/v)x] . \tag{3.278}$$

From (3.271) one obtains for the current

$$I(x,\omega) = -i\omega \{Q_R(\omega)\exp[(\alpha - i\omega/v)x] \\ +Q_L(\omega)\exp[(\alpha - i\omega/v)x]\} \ . \tag{3.279}$$

Equation (3.265) can now be solved to yield

$$V(x,\omega) = -i\omega Z[Q_R(\omega)e^{-\alpha x}e^{i\omega x/v} - Q_L(\omega)e^{\alpha x}e^{-i\omega x/v}] \ , \tag{3.280}$$

where the transmission line impedance is given by

$$Z = \sqrt{\frac{L}{C}} \ . \tag{3.281}$$

We now consider the case when our transmission line extends along the x-axis from minus infinity to zero. At $x = 0$ the transmission line is connected to a lossless transmission line, $R = G = 0$ but having the same L and C as the lossy transmission line. Since v and Z depend only on L and C, it is evident that (3.279) and (3.280) hold for $x > 0$ but with $\alpha = 0$. At the boundary the current and voltage are required to be continuous, that is,

$$I(0+,\omega) = I(0-,\omega) \ , \tag{3.282}$$

$$V(0+,\omega) = V(0-,\omega) \ . \tag{3.283}$$

Introducing the superscripts + and − to distinguish quantities evaluated on the right or left side of the boundary, respectively, one obtains

$$I(0-,\omega) = -i\omega[Q_R^-(\omega) + Q_L^-(\omega)] \ , \tag{3.284}$$

$$I(0+,\omega) = -i\omega[Q_R^+(\omega) + Q_L^+(\omega)] \ , \tag{3.285}$$

$$V(0-,\omega) = -i\omega Z[Q_R^-(\omega) - Q_L^-(\omega)] \ , \tag{3.286}$$

$$V(0+,\omega) = -i\omega Z[Q_R^+(\omega) - Q_L^+(\omega)] \ . \tag{3.287}$$

These equations can be solved to yield

$$Q_R^-(\omega) = Q_R^+(\omega) \ , \tag{3.288}$$

$$Q_L^-(\omega) = Q_L^+(\omega) \ , \tag{3.289}$$

that is, there are no reflections at the boundary between the lossless line and the lossy line. The results obtained in Sect. 3.1 for lossless mechanical transmission lines can be translated to those for a lossless electrical transmission line through the identification

$$Y(x,t) \to Q(x,t) \ , \tag{3.290}$$

$$\rho \to L \ , \tag{3.291}$$

$$\sigma \to 1/C \ . \tag{3.292}$$

In particular, under this transformation (3.7) becomes

$$Q(x,t) = \frac{1}{2}\sqrt{\frac{\hbar}{\pi Z}}\int_0^\infty \frac{d\omega}{\sqrt{\omega}} \times \left[a_L(\omega)e^{-i\omega(t+x/v)} + a_R(\omega)e^{-i\omega(t-x/v)} + \text{H.c.}\right] . \tag{3.293}$$

By (3.288) and (3.289) we have established that there are no reflections at the boundary between the lossless and lossy transmission lines, that is, each frequency component of the charge remains the same immediately across the boundary. Consequently, the commutation relations for each frequency component of the charge remains unchanged across the boundary. Hence, one concludes that the quantized expression for the field in the lossy transmission line must have the form

$$Q(x,t) = \frac{1}{2}\sqrt{\frac{\hbar}{\pi Z}}\int_0^\infty \frac{d\omega}{\sqrt{\omega}} \left\{\left[a_L(x,\omega) + a_R(x,\omega)\right] e^{-i\omega t} + \text{H.c.}\right\} , \tag{3.294}$$

where the creation and annihilation operators satisfy the commutation relations

$$[a_L(x,\omega), a_L^\dagger(x,\omega')] = [a_R(x,\omega), a_R^\dagger(x,\omega')] = \delta(\omega - \omega') \tag{3.295}$$

and

$$\begin{aligned}[a_L(x,\omega), a_L(x,\omega')] &= [a_R(x,\omega), a_R(x,\omega')] = [a_L(x,\omega), a_R(x,\omega')] \\ &= [a_L(x,\omega), a_R^\dagger(x,\omega')] = 0 .\end{aligned} \tag{3.296}$$

Requiring this field to satisfy (3.273) will yield unphysical results, the a_L and a_R will exhibit exponential decay as a function of x. What is missing from (3.294) are appropriate Langevin noise sources associated with the loss. One expects there to be voltage and current noise sources distributed continuously along the transmission line. Hence, keeping (3.294) in mind we require the annihilation operators to satisfy the following equations

$$\left(\frac{\partial}{\partial x} + i\frac{\omega}{v} - \alpha\right) a_L(x,\omega) = N\Gamma_L(\omega, x) \tag{3.297}$$

and

$$\left(\frac{\partial}{\partial x} - i\frac{\omega}{v} + \alpha\right) a_R(x,\omega) = N\Gamma_R(\omega, x) , \tag{3.298}$$

where N is a normalization constant and Γ_L and Γ_R are noise operators. One expects noise sources separated in position to be uncorrelated, hence, keeping the commutation relations (3.295) through (3.296) in mind, we require that the noise operators satisfy the following commutation relations

$$\begin{aligned}[\Gamma_L(x,\omega), \Gamma_L^\dagger(x',\omega')] &= [\Gamma_R(x,\omega), \Gamma_R^\dagger(x',\omega')] \\ &= \delta(\omega - \omega')\delta(x - x')\end{aligned} \tag{3.299}$$

and

$$[\Gamma_L(x,\omega),\Gamma_L(x',\omega')] = [\Gamma_R(x,\omega),\Gamma_R(x',\omega')]$$
$$= [\Gamma_L(x,\omega),\Gamma_R(x',\omega')] = [\Gamma_L(x,\omega),\Gamma_R^\dagger(x',\omega')] = 0 \,. \quad (3.300)$$

Considering, for now, the case when the lossless transmission line extends to infinity in both directions along the x-axis, (3.297) and (3.298) have the solutions

$$a_L(x,\omega) = N\int_{-\infty}^{\infty} G_L(\omega, x-x')\Gamma_L(x',\omega)dx' \quad (3.301)$$

and

$$a_R(x,\omega) = N\int_{-\infty}^{\infty} G_R(\omega, x-x')\Gamma_R(x',\omega)dx' \,, \quad (3.302)$$

where the propagators are given by

$$G_L(\omega,x) = -u(-x)\exp[(\alpha - i\omega/v)x] \,, \quad (3.303)$$
$$G_R(\omega,x) = u(x)\exp[(-\alpha + i\omega/v)x] \,. \quad (3.304)$$

Using (3.301) or (3.302) and the commutation relations (3.295) to evaluate $[a_L(x,\omega),a_L^\dagger(x',\omega')]$ or $[a_R(x,\omega),a_R^\dagger(x',\omega')]$, and comparing the results with (3.299), one finds that the normalization constant is given by

$$N = \sqrt{2\alpha} \quad (3.305)$$

and one obtains the following unequal position generalization of (3.295) and (3.296) as

$$[a_L(x,\omega),a_L^\dagger(x',\omega')] = \delta(\omega-\omega')e^{-\alpha|x-x'|}e^{-i\omega(x-x')/v} \,, \quad (3.306)$$
$$[a_R(x,\omega),a_R^\dagger(x',\omega')] = \delta(\omega-\omega')e^{-\alpha|x-x'|}e^{i\omega(x-x')/v} \,, \quad (3.307)$$

and

$$[a_L(x,\omega),a_L(x',\omega')] = [a_R(x,\omega),a_R(x',\omega')] = [a_L(x,\omega),a_R(x',\omega')]$$
$$= [a_L(x,\omega),a_R^\dagger(x',\omega')] = 0 \,. \quad (3.308)$$

Using these commutation relations, it is straightforward to show from (3.294) that the charge along the lossy transmission line satisfies the canonical commutation relations

$$[Q(x,t),\Pi(x',t)] = i\hbar\delta(x-x') \quad (3.309)$$

and

$$[Q(x,t),Q(x',t)] = [\Pi(x,t),\Pi(x',t)] = 0 \,, \quad (3.310)$$

where

$$\Pi(x,t) = L\frac{\partial Q(x,t)}{\partial t} \ . \tag{3.311}$$

Consider the frequency component of the charge Q oscillating at frequency $\exp(-i\omega t)$. In suitable units this is given by

$$a(\omega, x) = a_L(\omega) + a_R(\omega) \ . \tag{3.312}$$

Substituting (3.301) and (3.302) into (3.312), and multiplying both sides of the resulting equation by the operator

$$\left[\frac{\partial}{\partial x} + i\frac{\omega}{v} - \alpha\right]\left[\frac{\partial}{\partial x} - i\frac{\omega}{v} + \alpha\right] , \tag{3.313}$$

we obtain the differential equation

$$\begin{aligned}\left[\frac{\partial^2}{\partial x^2} + \left(\frac{\omega}{v} + i\alpha\right)^2\right] \quad a(\omega, x) = \sqrt{2\alpha}\, \{&\frac{\partial}{\partial x}[\Gamma_L(x,\omega) + \Gamma_R(x,\omega)] \\ &- i\left(\frac{\omega}{v} + i\omega\right)[\Gamma_L(x,\omega) - \Gamma_R(x,\omega)]\} \ .\end{aligned} \tag{3.314}$$

We now compare this result with a more conventional approach to Langevin equations for a lossy transmission line. From the harmonic oscillator example in Sect. 3.2, we have learned to associate a Langevin noise term $2\Gamma dY_{\rm in}/dt$ with the viscous damping term $\Gamma dY/dt$. Similarly, one can associate a fluctuating voltage term of the form $2RI_{\rm in}$ with the resistive voltage drop RI and a fluctuating current term $2GV_{\rm in}$ with the resistive current GV. Hence, when the appropriate Langevin noise terms are added, the transmission line equations (3.265) and (3.266) take the form

$$-\frac{\partial V}{\partial x} = L\frac{\partial I}{\partial t} + RI - 2RI_{\rm in}^R \tag{3.315}$$

and

$$-\frac{\partial I}{\partial x} = C\frac{\partial V}{\partial t} + GV - 2GV_{\rm in}^G , \tag{3.316}$$

where the superscripts R and G have been included on the noise terms to remind ourselves that these have different origins. One comes from resistance along the length of the transmission line, the other arises from conductance across the transmission line. These transmission line equations can be solved to yield

$$\begin{aligned}&\frac{\partial^2 I}{\partial t^2} - \frac{1}{LC}\frac{\partial^2 I}{\partial x^2} + \left(\frac{R}{L} + \frac{G}{C}\right)\frac{\partial I}{\partial t} + \frac{GR}{LC}I = \\ &2\frac{R}{L}\frac{\partial I_{\rm in}^R}{\partial t} + 2\frac{RG}{LC}I_{\rm in}^R - 2\frac{G}{LC}\frac{\partial V_{\rm in}^G}{\partial x} \ .\end{aligned} \tag{3.317}$$

Again, restricting ourselves to the distortionless case (3.267), using (3.268), (3.269), (3.271), and defining the effective current

$$I_{\text{in}}^G = \frac{V_{\text{in}}^G}{Z} , \tag{3.318}$$

the wave equation (3.317) leads to the following equation for the charge

$$\frac{\partial^2 Q}{\partial t^2} - v^2 \frac{\partial^2 Q}{\partial x^2} + 2\alpha v \frac{\partial Q}{\partial t} + \alpha^2 v^2 Q = 2\alpha v \frac{\partial Q_{\text{in}}^R}{\partial t} + 2\alpha^2 v^2 Q_{\text{in}}^R - 2\alpha v^2 \frac{\partial Q_{\text{in}}^G}{\partial x} . \tag{3.319}$$

For the frequency components oscillating as $\exp(-i\omega t)$ this equation becomes

$$\left[\frac{\partial^2}{\partial x^2} + \left(\frac{\omega}{v} + i\omega\right)^2\right] a(x,\omega) = 2i\alpha \left(\frac{\omega}{v} + i\omega\right) a_{\text{in}}^R(x,\omega) + 2\alpha \frac{\partial a_{\text{in}}^G(x,\omega)}{\partial x} . \tag{3.320}$$

Comparing (3.320) with (3.314), we find the following expressions for the Langevin noise operators in terms of the Γ_R and Γ_L:

$$a_{\text{in}}^G(x,\omega) = \frac{1}{\sqrt{2\alpha}}[\Gamma_R(x,\omega) + \Gamma_L(x,\omega)] \tag{3.321}$$

and

$$a_{\text{in}}^R(x,\omega) = \frac{1}{\sqrt{2\alpha}}[\Gamma_R(x,\omega) - \Gamma_L(x,\omega)] . \tag{3.322}$$

3.14 Attenuators

An attenuator can be implemented as a section of a lossy transmission line. Using a section of the transmission line, discussed in Sect. 3.3, running from $x = 0$ to $x = L$ to implement an attenuator, one has

$$a_R(L,\omega) = e^{i\omega(L/v)-\alpha L} a_R(0,\omega) + \sqrt{2\alpha} \int_0^L G_R(\omega, L-x)\Gamma_R(x,\omega)dx , \tag{3.323}$$

where the propagator is given by (3.304). Introducing the noise operator

$$a_N(\omega) = \sqrt{\frac{2\alpha}{1-e^{-\alpha L}}} \int_0^L G_R(\omega, L-x)\Gamma_R(x,\omega)dx , \tag{3.324}$$

(3.323) can be put into the form

$$a_R(L,\omega) = e^{-\alpha L} e^{i\omega L/v} a_R(0,\omega) + \sqrt{1-e^{-2\alpha L}} a_N(\omega) . \tag{3.325}$$

Because the annihilation operator $a_R(0,\omega)$ for the input and the annihilation operator $a_R(L,\omega)$ for the output satisfy the commutation relations

$$[a_R(x,\omega), a_R^\dagger(x,\omega')] = \delta(\omega-\omega') \ , \tag{3.326}$$

$a_N(\omega)$ must satisfy the commutation relation

$$[a_N(\omega), a_N^\dagger(\omega')] = \delta(\omega-\omega') \ . \tag{3.327}$$

That it does so can be demonstrated by an explicit calculation using (3.324) and the commutation relation (3.299).

Denoting the output creation operator $a_R(L,\omega)$ by $b(\omega)$ and the input creation operator $a_R(0,\omega)$ by $a(\omega)$ and taking the limit L goes to zero while holding αL constant and defining η by

$$\eta^{1/2} = \exp(-\alpha L) \ , \tag{3.328}$$

(3.325) becomes

$$b(\omega) = \eta^{1/2} a(\omega) + \sqrt{1-\eta}\, a_N(\omega) \ . \tag{3.329}$$

By taking this limit we have eliminated the propagation effects contained in the phase factor $\exp(i\omega L/v)$ appearing in (3.325). We take (3.329) to be the mode transformation performed by an ideal attenuator. This transformation can be completed into a unitary transformation by introducing the mode $b_N(\omega)$, such that

$$\begin{pmatrix} b(\omega) \\ b_N(\omega) \end{pmatrix} = \begin{pmatrix} \eta^{1/2} & \sqrt{1-\eta} \\ -\sqrt{1-\eta} & \eta^{1/2} \end{pmatrix} \begin{pmatrix} a(\omega) \\ a_N(\omega) \end{pmatrix} \ . \tag{3.330}$$

Such a transformation can be realized by a beam splitter in which vacuum enters the a_N port and unwanted power is dumped into the b_N port. In the analysis of the quantum performance of optical systems it is often convenient to use beam splitters to model attenuators and losses. See, for example [21]. Although I have dealt here with linear losses, nonlinear loss can also be treated through nonlinear coupling of the bath modes to the system variables. Along this line, Collett and Levien [22] have analyzed a two-photon-loss model for intracavity second-harmonic generation.

References

1. G.J. Milburn and D.F. Walls, Optics Commun. **39**, 401 (1981)
2. B. Yurke, Phys. Rev. A **29**, 408 (1984)
3. B. Yurke and J.S. Denker, Phys. Rev. A **29**, 1419 (1984)
4. B. Yurke, Phys. Rev. A **32**, 300 (1985)
5. M.J. Collett and C.W. Gardiner, Phys. Rev. A **30**, 1386 (1984).
6. C.W. Gardiner and M.J. Collett, Phys. Rev. A **31**, 3761 (1985)
7. C.W. Gardiner and P. Zoller, *Quantum Noise* (Springer, Berlin Heidelberg New York 2000)

8. J.M. Courty and S. Reynaud, Phys. Rev. A **46**, 2766 (1992)
9. C. Viviescas and G. Hackenbroich, Phys. Rev. A **67**, 013805 (2003)
10. C.P. Search, S. Pötting, W. Zhang and P. Meystre, Phys. Rev. A **66**, 043616 (2002)
11. J.N. Hollenhorst, Phys. Rev. D **19**, 1669 (1979)
12. C.M. Caves, K.S. Thorne, R.W.P. Drever, V.D. Sandberg and M. Zimmermann, Rev. Mod. Phys. **52**, 341 (1980)
13. B. Yurke, Am. J. Phys. **52**, 1099 (1984)
14. B. Yurke, Am. J. Phys. **54**, 1133 (1986)
15. A.O. Caldeira and A.J. Leggett, Ann. Phys. **149**, 374 (1983)
16. C.M. Caves, Phys. Rev. D **26**, 1817 (1982)
17. H.P. Yuen and V.W.S. Chen, Opt. Lett. **8**, 177 (1983)
18. J.H. Shapiro, IEEE J. Quantum Electron. **QE-21**, 237 (1985)
19. B.L. Schumaker and C.M. Caves, Phys. Rev. A **31**, 3093 (1985)
20. C.M. Caves and B.L. Schumaker, Phys. Rev. A **31**, 3068 (1985)
21. B. Yurke, Phys. Rev. A **32**, 311 (1985)
22. M.J. Collett and R.B. Levien, Phys. Rev. A **43**, 5068 (1991)

Part II

Generation of Quantum Squeezing

4 Squeezing with Nonlinear Optics

P. D. Drummond

The earliest experiments on nonclassical light sources were able to produce photon-antibunching, or sub-Poissonian light, from single-atom resonance fluorescence [1]. However, the most useful experimental techniques for generating squeezing [2,3] have relied on nonlinear optical [4] techniques. There are a variety of possible methods [5], depending on the type of nonlinearity, and number of interacting modes involved. The choice of optical nonlinearity reduces to either three-wave or four-wave mixing, which correspond to quadratic or cubic nonlinear response functions respectively. Each of these has advantages and disadvantages. Three wave mixing requires a pump field at twice the frequency of the field-mode being squeezed. While this is a complication, it is also an advantage in some respects; any sidebands induced on the pump due to phase noise have a large frequency offset from the squeezed fields. On the other hand, four-wave mixing does not require any frequency-doubling step, which has advantages in terms of simplicity. This was also the first method [2], used to obtain squeezing with quadrature noise below the vacuum level.

Most of the early experiments attempted to obtain squeezing initially in one or two cavity modes, with the squeezed fields being transferred to propagating external modes for measurement purposes, through a beam-splitter or mirror [6–8]. This procedure is typically rather narrow-band. In current practice, this technique is most commonly used for three-wave (parametric) squeezing. Later, it was realized that propagating modes in a waveguide could be squeezed directly, giving a relatively simple, broad-band implementation of quadrature squeezing [9,10]. Due to the ready availability of high-quality single-mode silica fiber, this is most often carried out via four-wave mixing with short pulses or solitons [11–13]. Finally, it is possible to obtain squeezing directly through propagation in a bulk crystal, although at signal levels too low to measure quadrature noise-reduction. For this reason, these bulk experiments are usually referred to as spontaneous down-conversion or correlated photon experiments.

To understand the physics of these different techniques, and how they relate to each other, we start with an elementary model using reversible single mode theory in the undepleted pump approximation. These types of model describe idealized, lossless, single-mode interferometers, and are unrealistic for most practical purposes. However, they do show in principle how a nonlinear optical interaction can lead to quantum states which satisfy the fundamental definition of squeezed states. A central problem is that one must then 'extract' the squeezed state – or transfer the internal squeezing to a propagating external mode. This is achieved either using an output coupler, described by quantum input-output theory, or even more simply by the direct action of nonlinearities on the propagating modes themselves. An equally important problem is the question of optimization, since one faces the paradox that the undepleted pump approximation can only be valid for weakly-squeezed fields, while it is usually necessary to increase the squeezing levels as much as possible, for maximum noise-reduction [14] or other quantum information applications [15].

4.1 Transient Squeezing

Firstly, we consider transient intra-cavity three-wave mixing, or transient parametric amplification. While this method is not a very practical one, nor was it the first used to achieve squeezing, it is closely related to the theoretical definition of an idealized squeezed state. The model considered here is the degenerate parametric oscillator.

The system is an idealized interferometer, which is resonant at two frequencies, ω_1 and $\omega_2 \approx 2\omega_1$. It is initialized with a coherent field at the larger of the two frequencies. Down conversion of the pump photons to resonant sub-harmonic mode photons occurs due to a parametric (second-order) nonlinearity present inside the cavity, which is assumed non-dispersive for simplicity. From (2.17), using dual potential quantization, the relevant nonlinear interaction Hamiltonian can be written [16,17] in terms of the nonlinear inverse permittivity tensor $\underline{\eta}^{(2)}$ as

$$\hat{H}_I = \frac{1}{3}\int \hat{\mathbf{D}}(\mathbf{r}) \cdot \underline{\eta}^{(2)}(\mathbf{r}) : \hat{\mathbf{D}}^2(\mathbf{r}) d^3\mathbf{r} \,. \tag{4.1}$$

Starting with the two-mode case, the quantum displacement field may be expanded as

$$\hat{\mathbf{D}}(\mathbf{r}) = \sum_{n=1}^{2} \left(\hat{a}_n^\dagger \mathbf{D}_n^*(\mathbf{r}) + \hat{a}_n \mathbf{D}_n(\mathbf{r})\right) \,, \tag{4.2}$$

where the displacement field mode function in a volume V is defined by

$$\mathbf{D}_n(\mathbf{r}) = \sqrt{\frac{\hbar}{2\omega_n \mu_0 V}}\, \nabla \times \Lambda_n(\mathbf{r}) \,. \tag{4.3}$$

As in (2.21), the functions Λ_n are orthonormal solutions to the wave-equation for the dual potential with normalization:

$$\int_V \Lambda_n^*(\mathbf{r})\Lambda_{n'}(\mathbf{r})d^3\mathbf{r} = \delta_{nn'} \ . \tag{4.4}$$

This mode expansion allows the quantum Hamiltonian to be greatly simplified. All non-resonant terms of form $\hat{a}_n^3$ (and similar terms) are rapidly oscillating at optical frequencies, and can be neglected to an excellent approximation, as long as typical interaction time-scales are much longer than an optical cycle of 10^{-15} Hz. This leads to a parametric Hamiltonian in the form

$$\hat{H} = \sum_{n=1,2} \hbar\omega_n \hat{a}_n^\dagger \hat{a}_n + i\hbar\frac{\chi}{2}\left(\hat{a}_1^{\dagger 2}\hat{a}_2 - \hat{a}_1^2\hat{a}_2^\dagger\right) \ . \tag{4.5}$$

Here the mode-function phase definitions have been chosen so that χ is real, where

$$\chi = \frac{2}{i\hbar}\int\int \mathbf{D}_2(\mathbf{r})\cdot\underline{\eta}^{(2)}(\mathbf{r}) : \mathbf{D}_1^{*2}(\mathbf{r})d^3\mathbf{r} \ . \tag{4.6}$$

Next, consider an interaction picture in which the Heisenberg equations of motion for the operators are given by the free-field terms, causing a phase-rotation at frequency ω_n; that is, the free field Hamiltonian is

$$\hat{H}_0 = \sum_{n=1,2} \hbar\omega_n \hat{a}_n^\dagger(t)\hat{a}_n(t) \ . \tag{4.7}$$

For ease of notation, we can denote the time-dependent operators in the interaction picture by $\hat{a}_n(t) = \hat{a}_n \exp(-i\omega_n t)$, where $\hat{a}_n = \hat{a}_n(0)$ are the time-independent initial conditions (or Schroedinger picture operators).

Suppose, in addition, that the quantum state of the interferometer is prepared with the pump field $\hat{a}_2$ in a coherent state of large amplitude α_p, and the subharmonic mode prepared in a vacuum state. Under these circumstances, the pump field is nearly classical, and can approximately be replaced by its mean value – the c-number amplitude α_p.

For the resonant case (with $\omega_2 = 2\omega_1$), this gives an approximate time-independent effective interaction Hamiltonian

$$\hat{H}_{\text{eff}} = i\frac{\hbar\chi\alpha_p}{2}\left(\hat{a}_1^{\dagger 2} - \hat{a}_1^2\right) \ . \tag{4.8}$$

The resulting interaction-picture time-dependence of the quantum state of the subharmonic mode, resonant at frequency ω_1, is then

$$\begin{aligned} |\Psi(t)\rangle &= e^{-i\hat{H}_{\text{eff}}t}\,|0\rangle \\ &= e^{\left(\xi\hat{a}_1^{\dagger 2} - \xi^*\hat{a}_1^2\right)/2}\,|0\rangle \ , \end{aligned} \tag{4.9}$$

where

$$\xi = \hbar \chi \alpha_p t \ . \tag{4.10}$$

In this idealized model, the state generated by the transient nonlinear optical interactions inside the cavity, is precisely a squeezed state in the usual definitions, as can be seen from (1.67).

4.2 Driven Parametric Oscillator

There are clearly many idealizations in the idealized single-mode theory, including the neglect of the pump dynamics. More interesting is the case of a squeezing field that can propagate outside of a cavity. This requires the theory of a driven, open system.

Thus, the simplest class of practical quantum devices that produce non-classical or 'squeezed' states of the radiation field are parametric oscillators with external coherent driving lasers. While an ideal squeezer as described above involves a unitary transformation of a vacuum state, any useful device must be a non-unitary, open system. The reason for this is the same reason that a laser is an open system. It is necessary to have an output coupler in order to produce the squeezed beam, and this behaves exactly like a loss for the cavity mode.

Steady-state optical parametric oscillators of this type are one of the most well characterized devices in nonlinear optics, with many useful applications, especially as frequency converters and wide-band amplifiers. Novel discoveries made with them in the quantum domain include demonstrations of large amounts of squeezing [6] and significant quantum intensity correlations [7], together with a quadrature correlation measurement that provided the first experimental demonstration of the original EPR paradox. Practical applications include their use as highly efficient and tunable frequency converters. To understand this, we will now treat the quantum fluctuations in the parametric amplifier, including the nonlinear region near threshold. This allows us to analytically treat the limits to quantum squeezing and noise reduction imposed by nonlinearities. An intriguing property of parametric oscillators is that they provide an example of a critical point phase transition which is far from thermal equilibrium, and exhibits strong quantum effects.

Since quantum noise effects scale inversely with the number of quanta, N, it is usually possible to linearize the equations of motion for a system about its steady states in the limit of $N \gg 1$. Unfortunately this approach fails at the bifurcation point for a second order phase transition like that seen in the parametric oscillator. This is because only cubic and higher order terms are left at threshold. Since a cubic decay term is smaller than a linear term for a given small displacement, the fluctuations about the threshold steady state (the critical fluctuations) are much larger than those elsewhere. Fortunately, the "small noise" asymptotic theory can be readily and systematically extended to the critical region by careful attention to the scaling properties of the physical variables – and without requiring a linearization approximation.

While results for this non-equilibrium system cannot be obtained using traditional canonical ensemble methods, it is possible to use operator representations to solve the time-evolution analytically in the time-domain, thus obtaining the steady-state correlation functions. The calculations can be compared to numerical quantum stochastic simulations of the fully nonlinear coupled mode equations. The critical squeezing variance in the output field scales as $N^{-1/2}$, while the zero-frequency squeezing spectrum gives an $N^{-2/3}$ scaling law, except for the case of a high-Q pump, where the squeezing limit can be further improved to an N^{-1} scaling law. In all cases, there is the novel feature that the squeezed fluctuations are reduced rather than enhanced near the critical point. By comparison, in thermal equilibrium, all fluctuations tend to *increase* at the critical point.

From the input-output theory of reservoirs, given in (3.100), the Heisenberg picture Hamiltonian that describes this open system is

$$\hat{H} = \hat{H}_S + \hat{H}_B + \hat{H}_{SB} \ , \tag{4.11}$$

where

$$\begin{aligned} \hat{H}_S = & \sum_{n=1,2} \hbar\omega_n \hat{a}_n^\dagger(t)\hat{a}_n(t) \\ & + i\hbar\frac{\chi}{2}\left(\hat{a}_1^{\dagger 2}(t)\hat{a}_2(t) - \hat{a}_1^2(t)\hat{a}_2^\dagger(t)\right) \\ & + i\hbar\sum_{n=1,2}\left(E_n e^{-i\omega_{Lj}t}\hat{a}_n^\dagger(t) - E^* e^{i\omega_{Lj}t}\hat{a}_n(t)\right) \ . \end{aligned} \tag{4.12}$$

and the reservoir Hamiltonians are given by

$$\begin{aligned} \hat{H}_B &= \hbar\sum_{n=1,2}\int_{-\infty}^{\infty} d\omega\, \omega\hat{b}_n^\dagger(\omega)\hat{b}_n(\omega) \\ \hat{H}_{SB} &= i\hbar\sum_{n=1,2}\left(\hat{a}_n(t)\hat{\Gamma}_n^\dagger(t) - \hat{a}_n^\dagger(t)\hat{\Gamma}_n(t)\right) \ . \end{aligned} \tag{4.13}$$

Here the mode operators $\hat{a}_n(t)$ are defined as previously, while E_n represents the j-th external driving field at frequency ω_{Lj}, and the driving laser(s) are assumed to be frequency locked with $\omega_{L2} = 2\omega_{L1} = 2\omega_0$. The coupling χ is the internal nonlinearity due to a $\chi^{(2)}$ nonlinear medium internal to the cavity, and Γ_n are reservoir operators that describe the losses of the internal mode of frequency ω_n, which occur at rate κ_n. These in turn can be split into *input* and *output* reservoirs, following (3.118-3.124), so that

$$\begin{aligned} \hat{\Gamma}_n(t) &= \frac{\sqrt{\kappa_n}}{2}\left(\hat{b}_n^{\rm out}(t) - \hat{b}_n^{\rm in}(t)\right) \\ &= \frac{\sqrt{\kappa_n}}{2}\sqrt{\frac{1}{2\pi}}\int_{-\infty}^{\infty} d\omega\, e^{-i\omega t}\left(\hat{b}_n^{\rm out}(\omega) - \hat{b}_n^{\rm in}(\omega)\right) \ . \end{aligned} \tag{4.14}$$

Solving the Heisenberg equations directly to eliminate the output reservoirs, leads to the quantum Langevin equations which have damping terms, and quantum noise terms proportional to the input fields. These can be solved only in linear cases in general.

Next, we wish to consider an interaction picture, obtained with the definition that

$$\hat{H}_0 = \sum_{n=1,2} \hbar\omega_{Ln}\hat{a}_n^\dagger(t)\hat{a}_n(t) \ . \tag{4.15}$$

In other words, the operators will evolve according to the laser frequency, while the states evolve according to the rest of the system Hamiltonian, including part of the cavity energy term if the laser is not on resonance. This is slightly different to the previous interaction picture if the lasers are not on resonance with the cavity modes.

Having said this, one then makes a further simplification. Since (in this picture) the operators all have a known time-evolution, it is possible to cancel all of the explicit time dependent terms, giving an interaction picture with rotating-frame operators $\hat{a}_n$ that are defined as:

$$\hat{a}_n = \hat{a}_n(t)\exp(i\omega_{Ln}t) \ . \tag{4.16}$$

Combined with the Markovian approximation, the time-evolution of the density matrix in this picture leads to an equivalent master equation [18,19] in the Markovian approximation of

$$\frac{\partial\hat{\rho}}{\partial t} = \frac{1}{i\hbar}\left[\hat{H}_{\rm int},\hat{\rho}\right] + \sum_{n=1,2}\frac{1}{2}\kappa_n(2\hat{a}_n\hat{\rho}\hat{a}_n^\dagger - \hat{a}_n^\dagger\hat{a}_n\hat{\rho} - \hat{\rho}\hat{a}_n^\dagger\hat{a}_n) \ , \tag{4.17}$$

where the interaction Hamiltonian is given, in this picture, by

$$\begin{aligned}\hat{H}_{\rm int}/\hbar = & \sum_{n=1,2} \Delta_n \left[\hat{a}_n^\dagger\hat{a}_n,\hat{\rho}\right] \\ & + i\frac{\chi}{2}\left(\hat{a}_1^{\dagger 2}\hat{a}_2 - \hat{a}_1^2\hat{a}_2^\dagger\right) \\ & + i\sum_{n=1,2}\left(\mathcal{E}_n\hat{a}_n^\dagger - \mathcal{E}_n^*\hat{a}_n\right) \ . \end{aligned} \tag{4.18}$$

Thus, $\Delta_n = \omega_n - \omega_{Lj}$ represents the detuning between the cavity mode resonance and the j-th laser (angular or carrier) frequency. In practical terms, this must be much less than the cavity mode-spacing for the above approximations to be applicable. A much more serious issue is the possibility that pairs of neighboring cavity modes may be coupled as well, and the down-conversion process could, indeed, be resonant to a whole 'comb' of modes. It is possible to arrange cavity dispersion parameters so that this doesn't occur – especially when the mode-spacings are large, as they would be in a small cavity. We will assume, for the sake of simplicity, that no other modes are excited.

In the limit in which only the second-harmonic field is driven, and all interactions are on-resonance, the interaction Hamiltonian reduces to the standard one [18] for a non-degenerate, single-mode parametric amplifier or oscillator

$$\hat{H}_{\text{int}}/\hbar = i\mathcal{E}\left(\hat{a}_2^\dagger - \hat{a}_2\right) + i\frac{\chi}{2}\left(\hat{a}_2\hat{a}_1^{\dagger 2} - \hat{a}_2^\dagger\hat{a}_1^2\right) . \tag{4.19}$$

Here $\mathcal{E} = \mathcal{E}_2$ is called the pump field, at frequency $\omega_p = \omega_{L2}$. This is proportional to the coherent input or driving field at the second-harmonic frequency, assumed to be at exact resonance with the cavity mode. The term χ is a coupling parameter for the $\chi^{(2)}$ nonlinearity of the medium. For simplicity, we have chosen the field mode-functions so that $\mathcal{E}$ and χ are real.

4.3 Observable Moments and Spectra

The crucial quadrature variables of the system have the definitions, in terms of the rotating-frame operators, of

$$\begin{aligned}\hat{x}_n &= \left(\hat{a}_n + \hat{a}_n^\dagger\right) , \\ \hat{y}_n &= \frac{1}{i}\left(\hat{a}_n - \hat{a}_n^\dagger\right) .\end{aligned} \tag{4.20}$$

The choice of rotating-frame operators as quadrature variables is essential. It is only in this rotating frame that one can obtain quadratures whose variance is reduced below the shot-noise level. These must therefore be measured using a local oscillator at precisely half the pump frequency ω_{2L}. In the laboratory, one must typically start with the local oscillator source at frequency $\omega_0 = \omega_{L1}$, then frequency-double part of this to obtain a pump field, while still retaining enough of the original laser intensity to use as a local oscillator for detection of the squeezed quadrature.

Of especial interest is $\hat{y}_1 = (\hat{a}_1 - \hat{a}_1^\dagger)/i$, since $\hat{y}_1$ turns out to be the low-noise, squeezed quadrature. Here we note that the instantaneous correlation functions of the intra-cavity field operators are called the moments. Typically, they are not easily measurable, when compared to output moments or spectra, but they are useful in that they provide a check on the accuracy of the calculation of measurable spectra. We consider the moments of $\hat{y}_1$ for definiteness, where

$$\langle : \hat{y}_1\hat{y}_1 : \rangle \equiv \mathrm{Tr}\left[\rho_{ss} : \hat{y}_1\hat{y}_1 :\right] . \tag{4.21}$$

The squeezing in terms of the intra-cavity quadrature variances corresponds to an instantaneous measurement of the field moments. If such a measurements were possible, it would include contributions from all frequencies. The normally-ordered moments of the signal quadratures are related to the symmetrically ordered moments by

$$\begin{aligned}\langle \hat{x}_1^2 \rangle &= \langle : \hat{x}_1^2 : \rangle + 1 , \\ \langle \hat{y}_1^2 \rangle &= \langle : \hat{y}_1^2 : \rangle + 1 .\end{aligned} \tag{4.22}$$

However, it is the field outside the cavity which is usually measured. The technique for treating external field squeezing spectra was introduced originally by Yurke [21] and by Collett and Gardiner [19]. Following the notation in (3.124), the output (measured) flux operators $\hat{b}_n^{\mathrm{out}}(t)$ – for mode-matched external modes of a high-Q interferometer – are related to the internal mode operators by

$$\hat{b}_n^{\mathrm{out}}(t) = \sqrt{\kappa_n}\hat{a}_n(t) - \hat{b}_n^{\mathrm{in}}(t) \ . \tag{4.23}$$

One can usefully introduce external output quadrature fields, measurable with homodyne detection and a local oscillator, according to

$$\hat{X}_n^\theta(t) = \left(e^{i(\omega_0 t-\theta)}\hat{b}_n^{\mathrm{out}}(t) + e^{i(\theta-\omega_0 t)}\hat{b}_n^{\mathrm{out}\dagger}(t)\right) \ . \tag{4.24}$$

These are local quadrature field operators for the outward propagating field components, as measured with an appropriate local oscillator with frequency ω_0. They have commutators of:

$$[\hat{X}_n^0(\omega), \hat{X}_n^{\pi/2}(\omega')] = 2i\delta(\omega-\omega') \ . \tag{4.25}$$

The frequency argument denotes a Fourier transform, where

$$\hat{X}_n^\theta(\omega) = \int \frac{\exp(i\omega t)}{\sqrt{2\pi}} \hat{X}_n^\theta(t)\, dt \ . \tag{4.26}$$

The usual quantity of interest is the variance in the quadrature at a given frequency, defined by

$$\left\langle \hat{X}_n^\theta(-\omega), \hat{X}_n^\theta(\omega) \right\rangle \ , \tag{4.27}$$

where the comma notation means that

$$\left\langle : \hat{A}, \hat{B} : \right\rangle = \left\langle : [\hat{A} - \langle\hat{A}\rangle][\hat{B} - \langle\hat{B}\rangle] : \right\rangle \ . \tag{4.28}$$

However, in practical experiments there are always losses from sources other than output mirrors, together with detector inefficiencies, which act like beam-splitters (see (3.229)), and hence reduce the signal further. This causes an increased vacuum noise arising from the 'additional' input ports – which has no effect on a normally ordered correlation. One can introduce κ_n^0 to refer to the decay rate due to the detected signal, which is always less than the total interferometer decay rate! Taking this into account, the output measured spectral variance $V_n^\theta(\omega)$ of a general quadrature, normalized relative to a total detection time T, is given by

$$V_n^\theta(\omega) = 1 + \frac{2\pi\kappa_n^0}{T\kappa_n} \left\langle : \hat{X}_n^\theta(-\omega), \hat{X}_n^\theta(\omega) : \right\rangle \ , \tag{4.29}$$

where the :: indicate normal ordering.

Using the input-output relation (3.124) or (4.23), the squeezing variance can also be written as

$$V_n^\theta(\omega) = 1 + \frac{2\pi\kappa_n^o}{T}\left\langle :\hat{x}_n^\theta(-\omega), \hat{x}_n^\theta(\omega): \right\rangle . \tag{4.30}$$

Here, the input fields must be in a vacuum or coherent state, while the internal operator terms must be both time and normal ordered in order to eliminate cross-correlation terms that would otherwise result [19].

When using master equation methods, it is the intra-cavity dynamics that are calculated. Hence, it useful to introduce a normally-ordered spectrum of the intra-cavity quantum fluctuations normalized by the intra-cavity decay rate, so that

$$S_n^\theta(\omega) = \frac{\pi\kappa_n}{T}\left\langle :\hat{x}_n^\theta(-\omega), \hat{x}_n^\theta(\omega): \right\rangle . \tag{4.31}$$

With this definition, the output measured spectral result simplifies to

$$V_n^\theta(\omega) = \frac{2\kappa_n^o}{\kappa_n} S_n^\theta(\omega) + 1 . \tag{4.32}$$

The optimal value of the internal squeezing $S_n^\theta(\omega)$ is bounded below by -0.5, which corresponds to zero fluctuations in the external squeezed quadrature, for a perfect interferometer with no losses except for the output coupler itself.

4.4 Heisenberg and Classical Equations

A commonly used approach is to treat the operator equations directly in the Heisenberg picture. Suppose there is just a second-harmonic driving field at frequency $\omega_{L2} = 2\omega_0$, then, if we omit damping

$$\begin{aligned} \frac{d\hat{a}_1(t)}{dt} &= -i\omega_1\hat{a}_1(t) + \chi\hat{a}_1^*(t)\hat{a}_2(t) , \\ \frac{d\hat{a}_2(t)}{dt} &= \mathcal{E}e^{-i\omega_p t} - i\omega_2\hat{a}_2(t) - \frac{1}{2}\chi\hat{a}_1^2(t) . \end{aligned} \tag{4.33}$$

If one includes damping in these equations, one obtains the so-called quantum Langevin equations – but these must also involve quantum noise terms as in (3.149), to give

$$\begin{aligned} \frac{d\hat{a}_1(t)}{dt} &= -\left(\frac{1}{2}\kappa_1 + i\omega_1\right)\hat{a}_1(t) + \chi\hat{a}_1^*(t)\hat{a}_2(t) + \sqrt{\kappa_1}\hat{b}_1^{\rm in}(t) , \\ \frac{d\hat{a}_2(t)}{dt} &= \mathcal{E}e^{-i\omega_p t} - \left(\frac{1}{2}\kappa_2 + i\omega_2\right)\hat{a}_2(t) - \frac{1}{2}\chi\hat{a}_1^2(t) + \sqrt{\kappa_2}\hat{b}_2^{\rm in}(t) . \end{aligned} \tag{4.34}$$

These quantum Langevin equations, as they stand, are operator equations which are not easily solved except in linear cases. One obtains the following equations for the expectation value of the rotating-frame picture operators $\hat{a}_n$. These were defined to be time-independent in the interaction picture, though they must evolve in the Heisenberg picture due to interactions. We distinguish the two types of operator by just dropping the time-argument when writing down annihilation and creation operators in the rotating frame. Thus

$$\begin{aligned}\frac{d\langle\hat{a}_1\rangle}{dt} &= -\tilde{\kappa}_1\langle\hat{a}_1\rangle + \chi\langle\hat{a}_1^*\hat{a}_2\rangle \ , \\ \frac{d\langle\hat{a}_2\rangle}{dt} &= \mathcal{E} - \tilde{\kappa}_2\langle\hat{a}_2\rangle - \frac{1}{2}\chi\langle\hat{a}_1^2\rangle \ .\end{aligned} \tag{4.35}$$

Here the complex decay rate of the amplitude, $\tilde{\kappa}_n$ is defined through

$$\tilde{\kappa}_n = i(\omega_n - j\omega_L) + \kappa_n/2 = i\Delta_n + \kappa_n/2 \ . \tag{4.36}$$

Similar results are obtained including absorption, if the interaction picture master equation is used as a starting point. The advantage of these equations is that they eliminate the time-dependence of the driving field term. However, it is important to recall that the full operator time-dependence of the expectation values in this 'laser' interaction picture requires use of the mapping

$$\langle\hat{a}_n(t)\rangle = \langle\hat{a}_n\rangle\exp(ij\omega_L t) \ . \tag{4.37}$$

We use the notation that a fully explicit time-dependence of the operator indicates that all time-dependences are included, even the carrier frequency terms themselves. Otherwise, the equations are in a rotating frame, and only include the time-dependence due to interactions, damping, and detuning.

In the classical limit, all noise and commutation terms are neglected, and all operator expectation values are assumed to factorize. If classical expectation values are defined through the relation

$$\langle\hat{a}_n(t)\rangle = \langle\hat{a}_n\rangle\exp(ij\omega_L t) = \alpha_n\exp(ij\omega_L t) \ , \tag{4.38}$$

then the usual results of classical nonlinear optics are regained, giving

$$\begin{aligned}\frac{d\alpha_1}{dt} &= -\tilde{\kappa}_1\alpha_1 + \chi\alpha_1^*\alpha_2 \ , \\ \frac{d\alpha_2}{dt} &= \mathcal{E} - \tilde{\kappa}_2\alpha_2 - \frac{1}{2}\chi\alpha_1^2 \ .\end{aligned} \tag{4.39}$$

The steady state is approached with a time-scale of κ_1^{-1} or κ_2^{-1}, depending on which is the slower time-scale. This gives steady-state equations to be solved of

$$\begin{aligned}\alpha_1 &= \chi\alpha_1^*\alpha_2/\tilde{\kappa}_1 \ , \\ \alpha_2 &= \left(\mathcal{E} - \frac{1}{2}\chi\alpha_1^2\right)/\tilde{\kappa}_2 \ .\end{aligned} \tag{4.40}$$

In general, there is always one trivial solution, which is called the below-threshold solution

$$\alpha_1 = 0 ,$$
$$\alpha_2 = \mathcal{E}/\tilde{\kappa}_2 . \tag{4.41}$$

In addition, there can be a second solution in which $\alpha_1 \neq 0$. In this solution, clearly

$$\alpha_1 = \chi \alpha_1^* \left(\mathcal{E} - \frac{1}{2} \chi \alpha_1^2 \right) / (\tilde{\kappa}_1 \tilde{\kappa}_2) . \tag{4.42}$$

With no loss of generality, we suppose that the input $\mathcal{E}$ and coupling χ are real – or, in other words, this condition is used to define the phase reference for the other fields. Next, suppose that $\alpha_1 = |\alpha_1| \exp(i\theta)$, and $\tilde{\kappa}_1 \tilde{\kappa}_2 = |\tilde{\kappa}_1 \tilde{\kappa}_2| \exp(i\phi)$. This allows us to solve for the phase and amplitude, via

$$|\tilde{\kappa}_1 \tilde{\kappa}_2| \exp(i\phi) = \chi \left(\mathcal{E} \exp(-2i\theta) - \frac{|\alpha_1|^2 \chi}{2} \right) . \tag{4.43}$$

Taking the modulus of both sides, it is clear that

$$|\tilde{\kappa}_1 \tilde{\kappa}_2|^2 / \chi^2 = \mathcal{E}^2 \sin^2(2\theta) + \left(\mathcal{E} \cos(2\theta) - \frac{|\alpha_1|^2 \chi}{2} \right)^2 . \tag{4.44}$$

This is generally rather complex, except in the simplest case where the cavity is doubly resonant (or symmetrically off-resonant) with the driving field, so that $\phi = 0$, and

$$\alpha_1 = \pm \sqrt{\frac{2}{\chi} \left(\mathcal{E} - |\tilde{\kappa}_1 \tilde{\kappa}_2| / \chi \right)} ,$$
$$\alpha_2 = \mathcal{E}_c / \tilde{\kappa}_2 . \tag{4.45}$$

Here $\mathcal{E}_c = |\tilde{\kappa}_1 \tilde{\kappa}_2| / \chi$ is the critical or threshold point, and clearly it is necessary that $\mathcal{E}$ must be larger than this to obtain a solution of this type, which is called the above-threshold solution. At this point, the below-threshold solutions themselves become unstable. The above-threshold solutions are stable themselves up to a second threshold, where an oscillatory behavior occurs. Above this point there are no stable steady-state solutions, although limit-cycle behavior (self-pulsing) occurs.

It is possible to take the above equations in the Heisenberg picture, assume that the pump is completely classical, and then neglect the pump depletion term.

This approach is valid sufficiently far below the critical point, as we will see in the next section, using a more careful asymptotic expansion approach. The immediate result of this operator linearization approximation is that it gives linear operator equations which can be solved directly in terms of the Langevin noise terms. To see this, suppose we take the previous operator equations just for the signal field alone, and assume a resonant interferometer with a classical intra-cavity pump. Then, in the rotating frame

$$\frac{d\hat{a}_1}{dt} = -\frac{\kappa_1}{2}\left(\hat{a}_1 + \varepsilon\hat{a}_1^*\right) + \sqrt{\kappa_1}\hat{b}_1^{\text{in}}(t) \ , \tag{4.46}$$

where $\varepsilon = 4\chi\mathcal{E}/(\kappa_1\kappa_2) < 1$, and the reservoir operators as finite temperature free boson fields, have the correlations

$$\begin{aligned}\langle\hat{b}_1(t)\hat{b}_1^\dagger(t')\rangle &= [n_{\text{th}}(\omega_1)+1]\,\delta(t-t') \ ,\\ \langle\hat{b}_1^\dagger(t)\hat{b}_1(t')\rangle &= n_{\text{th}}(\omega_1)\,\delta(t-t') \ .\end{aligned} \tag{4.47}$$

Here $n_{\text{th}}(\omega_1)$ is the Bose occupation number of the thermal photons at the resonant signal frequency ω_1. At optical frequencies, and room temperature external reservoirs, this can usually be neglected: $n_{\text{th}}(\omega_1) \ll 1$ since $k_BT \ll \hbar\omega_1$. Of course, there may be noise sources due to imposed phase noise on the laser [22] or mirror instability in the interferometer. All of these additional noise sources must be reduced to a sufficiently low level.

The resulting quadrature equations are given by

$$\begin{aligned}\frac{d\hat{x}_1}{dt} &= -\kappa_1\,(1-\varepsilon)\,\hat{x}_1/2 + \sqrt{\kappa_1}\hat{X}_1^{\text{in}}(t) \ ,\\ \frac{d\hat{y}_1}{dt} &= -\kappa_1\,(1+\varepsilon)\,\hat{y}_1/2 + \sqrt{\kappa_1}\hat{Y}_1^{\text{in}}(t) \ .\end{aligned} \tag{4.48}$$

In the zero temperature limit, the correlations of the quadrature noise operators, $\hat{X}_n^{\text{in}} = \hat{b}_n^{\text{in}} + \hat{b}_n^{\text{in}\dagger}$ and $\hat{Y}_n^{\text{in}} = -i(\hat{b}_n^{\text{in}} - \hat{b}_n^{\text{in}\dagger})$, are

$$\langle\hat{X}_n^{\text{in}}(t)\hat{X}_{n'}^{\text{in}}(t')\rangle = \langle\hat{Y}_n^{\text{in}}(t)\hat{Y}_{n'}^{\text{in}}(t')\rangle = 2\delta_{nn'}\delta(t-t') \ . \tag{4.49}$$

The steady-state (long-time) solution to these equations is directly

$$\begin{aligned}\hat{x}_1 &= \sqrt{\kappa_1}\int_{-\infty}^{t} e^{-\kappa_1(1-\varepsilon)(t-t')/2}\hat{X}_1^{\text{in}}(t')dt' \ ,\\ \hat{y}_1 &= \sqrt{\kappa_1}\int_{-\infty}^{t} e^{-\kappa_1(1+\varepsilon)(t-t')/2}\hat{Y}_1^{\text{in}}(t')dt' \ .\end{aligned} \tag{4.50}$$

The resulting *intra-cavity* operator moments or equal-time correlations are

$$\begin{aligned}\langle\hat{x}_1^2\rangle &= \frac{1}{1-\varepsilon} \ ,\\ \langle\hat{y}_1^2\rangle &= \frac{1}{1+\varepsilon} \ .\end{aligned} \tag{4.51}$$

The uncertainty product is

$$\Delta x^2\Delta y^2 = \frac{1}{1-\varepsilon^2} \ . \tag{4.52}$$

This is always greater than the Heisenberg uncertainty limit, since the greatest reduction in the moment of the squeezed (y) quadrature is 50%, while the unsqueezed quadrature can have extremely large fluctuations at the critical point. The linearization method can also be used to obtain the external squeezing spectrum, which is both more useful and has a lower noise floor. This is most readily worked out in frequency space, where the corresponding results are

$$\begin{aligned}\hat{x}_1(\omega) &= \frac{\sqrt{\kappa_1}}{\lambda_- - i\omega}\hat{X}_1^{\text{in}}(\omega) \ , \\ \hat{y}_1(\omega) &= \frac{\sqrt{\kappa_1}}{\lambda_+ - i\omega}\hat{Y}_1^{\text{in}}(\omega) \ , \end{aligned} \tag{4.53}$$

where the renormalized amplitude decay rates $\lambda_\pm$ are introduced, with: $\lambda_+ = (1+\varepsilon)\kappa/2$, $\lambda_- = (1-\varepsilon)\kappa/2$.

Using the input-output relations again, in the form

$$\hat{X}_n^{\theta\text{out}}(\omega) = \sqrt{\kappa_n}\hat{x}_n^{\theta}(\omega) - \hat{X}_n^{\theta\text{in}}(\omega) \ , \tag{4.54}$$

leads to the output quadratures in frequency space

$$\begin{aligned}\hat{X}_1^{\text{out}}(\omega) &= -\frac{\lambda_+ - i\omega}{\lambda_- - i\omega}\hat{X}_1^{\text{in}}(\omega) \ , \\ \hat{Y}_1^{\text{out}}(\omega) &= -\frac{\lambda_- - i\omega}{\lambda_+ - i\omega}\hat{Y}_1^{\text{in}}(\omega) \ . \end{aligned} \tag{4.55}$$

Noting that the input reservoir operators, being in the vacuum state, have frequency-space correlations given by

$$\langle \hat{X}_n^{\text{in}}(\omega)\hat{X}_{n'}^{\text{in}}(\omega')\rangle = \langle \hat{Y}_n^{\text{in}}(\omega)\hat{Y}_{n'}^{\text{in}}(\omega')\rangle = \delta_{nn'}\delta(\omega+\omega') \ , \tag{4.56}$$

the correlations of the quadrature output fields are therefore

$$\begin{aligned}\langle \hat{X}_1^{\text{out}}(\omega)\hat{X}_1^{\text{out}}(\omega')\rangle &= \frac{\lambda_+^2 + \omega^2}{\lambda_-^2 + \omega^2}\delta(\omega+\omega') \ , \\ \langle \hat{Y}_1^{\text{out}}(\omega)\hat{Y}_1^{\text{out}}(\omega')\rangle &= \frac{\lambda_-^2 + \omega^2}{\lambda_+^2 + \omega^2}\delta(\omega+\omega') \ . \end{aligned} \tag{4.57}$$

Here $\delta(0) = T/(2\pi)$ is the delta function at zero frequency, for an integration time T. Using the quadrature variance input-output relations of (4.30), the output quadrature variances are

$$\begin{aligned}V_1^{\pi/2}(\omega) &= 1 - \left(\frac{\kappa_1^0}{\kappa_1}\right)\frac{\kappa_1^2\varepsilon}{\omega^2 + \lambda_+^2} \ . \\ V_1^{0}(\omega) &= 1 + \left(\frac{\kappa_1^0}{\kappa_1}\right)\frac{\kappa_1^2\varepsilon}{\omega^2 + \lambda_-^2} \ . \end{aligned} \tag{4.58}$$

These results show that fluctuations can (apparently) be reduced to zero in the squeezed quadrature at the critical point where $\varepsilon = 1$, provided the interferometer has the idealized property that $\kappa_1^0 = \kappa_1$; and with the price that the conjugate fluctuations become infinite. This implies an infinite energy in the output field, which is clearly not possible, indicating a breakdown of the simple operator linearization method.

These results can also be written in terms of the expectation values of the external field flux operators, instead of quadratures

$$\left\langle \hat{b}^{\dagger}\left(\omega_0+\omega\right)\hat{b}\left(\omega_0+\omega'\right)\right\rangle = N\left(\omega\right)\delta\left(\omega-\omega'\right) \ , \tag{4.59}$$

$$\left\langle \hat{b}\left(\omega_0+\omega\right)\hat{b}\left(\omega_0+\omega'\right)\right\rangle = M\left(\omega\right)e^{-2i\omega_0 t}\delta\left(\omega+\omega'\right) \ , \tag{4.60}$$

where, for an ideal degenerate parametric oscillator

$$\begin{aligned} N\left(\omega\right) &= \frac{\kappa^2\varepsilon}{4}\left(\frac{1}{\omega^2+\lambda_-^2}-\frac{1}{\omega^2+\lambda_+^2}\right) \ , \\ M\left(\omega\right) &= \frac{\kappa^2\varepsilon}{4}\left(\frac{1}{\omega^2+\lambda_-^2}+\frac{1}{\omega^2+\lambda_+^2}\right) \ . \end{aligned} \tag{4.61}$$

4.5 Fokker–Planck and Stochastic Equations

The operator linearization approach has the serious drawback that it is difficult to determine when the method breaks down, as it must when the threshold point is approached. We can already see that it is at threshold where the squeezing is optimized, so this is an important limitation. Accordingly, we shall utilize a more sophisticated method here, which allows us to deduce what happens both well below and near the critical point.

We will use an operator representation to solve for the quantum fluctuations. In this method, the density matrix is expanded in a quasi-probability representation called the positive P-representation [23] $P(\boldsymbol{\alpha},\boldsymbol{\beta})$, which generalizes the traditional Glauber–Sudarshan P-representation [24] to allow for non-classical quantum fluctuations found in squeezed fields. This is defined as follows

$$\hat{\rho} = \int\int P(\boldsymbol{\alpha},\boldsymbol{\beta})\frac{|\boldsymbol{\alpha}\rangle\langle\boldsymbol{\beta}^*|}{\langle\boldsymbol{\beta}^*|\,\boldsymbol{\alpha}\rangle}d^4\boldsymbol{\alpha}d^4\boldsymbol{\beta} \ . \tag{4.62}$$

Here the complex vector $\boldsymbol{\alpha} = \boldsymbol{\alpha}_x + i\boldsymbol{\alpha}_y$ corresponds to one classical complex mode amplitude for each of the modes present, while $\boldsymbol{\beta}^* = \boldsymbol{\beta}_x - i\boldsymbol{\beta}_y$ corresponds to another mode amplitude, with the physical state being a quantum superposition of the two. This is the most general coherent state expansion possible. The state $|\boldsymbol{\alpha}\rangle$ is a multi-mode coherent state, defined as usual as

$$|\boldsymbol{\alpha}\rangle = \exp\left(\sum_n \alpha_n\hat{a}_n^{\dagger} - |\boldsymbol{\alpha}|^2/2\right)|0\rangle \ , \tag{4.63}$$

where $|0\rangle$ is the ground state. It is often useful to replace $\boldsymbol{\beta}$ with $\boldsymbol{\alpha}^+$ to indicate notationally that it is the left eigenvalue of a creation operator $\hat{a}_n^\dagger$, just as α_n is the right eigenvalue of an annihilation operator $\hat{a}_n$.

The Glauber–Sudarshan $P(\alpha)$-function exists and is most fruitfully employed for describing quantum statistical properties of thermal light fields and coherent fields. However, for fields with non-classical photon statistics $P(\alpha)$ does not exist as a well behaved positive function or is highly singular. The function $P(\boldsymbol{\alpha}, \boldsymbol{\beta})$ can always be chosen as a positive (generalized) probability function. Following standard procedures, the assumption of vanishing boundary terms allows the master equation to be re-written as a Fokker–Planck equation in $P(\boldsymbol{\alpha}, \boldsymbol{\beta})$ and hence as a stochastic equation [25] with real noise. The assumption of vanishing boundary terms is critical to this procedure. This occurs when the ratio of nonlinearity to damping is small [26]. The stochastic positive-P representation is therefore only valid in this limit, in which case the boundary terms are exponentially suppressed. The required ratio of nonlinearity to damping is extremely well-satisfied in all current experiments, where the ratio is typically 10^{-6} or less.

An asymptotic approach can be used to calculate the size of the quantum fluctuations both well below threshold, and in the near threshold region where nonlinear effects are significant. In this region, there are large critical fluctuations which control the dynamics of the system at threshold because of their size and slowness. Stochastic methods provide an extremely straightforward technique for both analytic approximation and numerical simulations for this problem.

The first step is to use operators identities to transform the master equation or quantum Langevin equation, into an equivalent Fokker–Planck formulation. For brevity, the equation is best written down using complex derivatives, with the understanding that the operator equivalences can also be used to obtain the positive-definite diffusion form when required. The full equation, including both types of driving field and detunings, is

$$\begin{aligned}\frac{\partial P(\boldsymbol{\alpha},\boldsymbol{\beta})}{\partial t} = \Bigg\{ & \frac{\partial}{\partial \alpha_1}\left[-\mathcal{E}_1 + \tilde{\kappa}_1\alpha_1 - \chi\alpha_2\beta\right] \\ & + \frac{\partial}{\partial \beta_1}\left[-\mathcal{E}_1^* + \tilde{\kappa}_1^*\beta_1 - \chi\beta_2\alpha\right] \\ & + \frac{1}{2}\frac{\partial^2}{\partial \alpha_1^2}\chi\alpha_2 + \frac{1}{2}\frac{\partial^2}{\partial \beta_1^2}\chi\beta_2 \\ & + \frac{\partial}{\partial \alpha_2}\left[-\mathcal{E}_2 + \tilde{\kappa}_2\alpha_2 + \chi\alpha_1^2\right] \\ & + \frac{\partial}{\partial \beta_2}\left[-\mathcal{E}_2^* + \tilde{\kappa}_2^*\beta_2 + \chi\beta_1^2\right]\Bigg\} P(\boldsymbol{\alpha},\boldsymbol{\beta}) \ . \end{aligned} \tag{4.64}$$

The positive-P Ito stochastic differential equations for the degenerate parametric oscillator were derived by Drummond et al. [18]. The pump mode is described by the complex variables α_2 and β_2. The signal mode, which is at half the pump frequency, is described by α_1 and β_1. Both modes are damped, and these losses are represented by the decay rates κ_2 and κ_1. The non linear coupling χ determines the rate at which a pump mode photon would be converted into a pair of signal mode photons, and vice versa.

These equations are only equivalent to the original quantum operator equations when the phase-space boundary terms are negligible. However, in stochastic numerical simulations which are equivalent to the Fokker–Planck approach, the trajectories are all bounded in a region near the classically stable point. This provides strong evidence that boundary terms at infinity are indeed negligible for the parameter values appropriate to experiments.

Given this assumption, the following Ito stochastic equations are obtained for any driving field $\mathcal{E} = \mathcal{E}_2$; that is, either below or above threshold, where we now simplify to the important case of resonance with a single real pump driving field

$$\begin{aligned}
d\alpha_1 &= (-\tilde{\kappa}_1\alpha_1 + \chi\beta_1\alpha_2)\,dt + \sqrt{\chi\alpha_2}dW_1(t) \ , \\
d\beta_1 &= (-\tilde{\kappa}_1^*\beta_1 + \chi\alpha_1\beta_2)\,dt + \sqrt{\chi\beta_2}dW_2(t) \ , \\
d\alpha_2 &= \left(-\tilde{\kappa}_2\alpha_2 + \mathcal{E} - \frac{1}{2}\chi\alpha_1^2\right) dt \ , \\
d\beta_2 &= \left(-\tilde{\kappa}_2^*\beta_2 + \mathcal{E} - \frac{1}{2}\chi\beta_1^2\right) dt \ .
\end{aligned} \tag{4.65}$$

Here the terms κ_k represent the amplitude damping rates. The stochastic correlations are given by

$$\begin{aligned}
\langle dW_n(t)\rangle &= 0 \ , \\
\langle dW_n(t)dW_l(t)\rangle &= \delta_{nl}dt \ .
\end{aligned} \tag{4.66}$$

This means that $dW_n(t)$ represent two real Gaussian and uncorrelated stochastic processes. Our derivation is formally based on the Itô stochastic calculus. However, in this case, either Itô or Stratonovich stochastic calculus gives identical results [25].

The quadrature operators have corresponding stochastic variables, namely

$$\begin{aligned}
x_n &= (\alpha_n + \beta_n) \sim \left(\hat{a}_n + \hat{a}_n^\dagger\right) \ , \\
y_n &= \frac{1}{i}(\alpha_n - \beta_n) \ \sim \frac{1}{i}\left(\hat{a}_n - \hat{a}_n^\dagger\right) \ .
\end{aligned} \tag{4.67}$$

The normally-ordered variances of the signal quadratures are simply

$$\begin{aligned}
\left\langle x_1^2\right\rangle &= \left\langle \alpha_1^2 + 2\beta_1\alpha_1 + {\beta_1}^2\right\rangle \ , \\
\left\langle y_1^2\right\rangle &= \left\langle -\alpha_1^2 + 2\beta_1\alpha_1 - \beta_1^2\right\rangle \ ,
\end{aligned} \tag{4.68}$$

provided $\langle x_1\rangle = 0$ and $\langle y_1\rangle = 0$ in the steady state. Note that $\langle x_1\rangle = 0$ is only true above threshold if the system is equally likely to be in either of the steady states. This means that $\left\langle x_1^2\right\rangle$ is the variance of x_1, which corresponds to the normally ordered variance of the operator quadrature $\hat{x}_1 = \hat{a}_1 + \hat{a}_1^\dagger$. Similarly, $\langle y_1^2 \rangle$ is the variance of y_1, which corresponds to the normally ordered variance of the orthogonal operator quadrature $\hat{y}_1 = -i(\hat{a}_1 - \hat{a}_1^\dagger)$.

More generally

$$x_n^\theta = \left(\alpha_n e^{-i\theta} + \beta_n e^{i\theta}\right) . \tag{4.69}$$

Since the P-representation is normally ordered, it automatically provides spectral correlations that correspond to the relevant normally ordered, time-ordered operator correlations of the measured output fields. We therefore define Fourier components of the normalized quadratures as

$$\tilde{x}_n^\theta(\omega) = \int \frac{\exp(i\omega t)}{\sqrt{2\pi}} x_n^\theta(t)\, d\omega . \tag{4.70}$$

This leads to the following result for the internal normally-ordered squeezing spectrum

$$\langle \tilde{x}_n^\theta(\omega_1)\, \tilde{x}_n^\theta(\omega_2) \rangle = \frac{\kappa_n}{2} \delta(\omega_1 + \omega_2) S_n^\theta(\omega_1) . \tag{4.71}$$

4.6 Below-Threshold Perturbation Theory

It is usual to introduce a scaling parameter $g^2 = 1/\lambda = 1/(2N\kappa_r)$ at this stage - where $N = (\kappa_1/\chi)^2$ is the threshold pump photon number, and $\kappa_r = \kappa_2/\kappa_1$ is the ratio of decay rates between the pump and signal modes. The g parameter determines the ratio of nonlinear to linear rates of change, and typically $g \ll 1$. It can alternatively be written as

$$g = \sqrt{\frac{2\chi^2}{\kappa_1 \kappa_2 / 2}} . \tag{4.72}$$

We also introduce a scaled time $\tau = \kappa_1 t/2$. In terms of the quadrature variables, the relevant equations become

$$\begin{aligned}
dx_1 &= \left[-x_1 + g\left(x_1 x_p + y_1 y_p\right)\right] d\tau \\
&\quad + \sqrt{\frac{g}{2}} \left[\sqrt{x_p + i y_p}\, dw_1(\tau) + \sqrt{x_p - i y_p}\, dw_2(\tau)\right] , \\
dy_1 &= \left[-y_1 + \frac{g}{2}\left(x_1 y_p + x_p y_1\right)\right] d\tau \\
&\quad + \sqrt{\frac{-g}{2}} \left[\sqrt{x_p + i y_p}\, dw_1(\tau) - \sqrt{x_p - i y_p}\, dw_2(\tau)\right] , \\
dx_p &= -\kappa_r \left[x_p + \frac{g}{2}\left(x_1^2 - y_1^2\right) - 2\varepsilon\right] d\tau , \\
dy_p &= -\kappa_r (y_p + g x_1 y_1) d\tau .
\end{aligned} \tag{4.73}$$

Here we have introduced a dimensionless driving field as $\varepsilon = 4\chi\mathcal{E}/(\kappa_1\kappa_2)$, as before. The stochastic correlations of $dw_n(\tau)$ are exactly those as for $dW_n(t)$, although expressed in terms of τ variables rather than t variables, and to simplify notation, the pump quadratures are rescaled as $x(y)_p = \sqrt{2\kappa_r} x(y)_2$.

Mean-field theories and related linearization methods have the obvious drawback that it is rather difficult to determine the accuracy of a decorrelation assumption. In order to solve these coupled equations more systematically, a formal perturbation expansion in powers of g is now introduced, where

$$\begin{aligned} x_1 &= \sum_{n=0}^{\infty} g^{n-1} x_1^{(n)} , \\ x_2 &= \frac{1}{\sqrt{2\kappa_r}} \sum_{n=0}^{\infty} g^{n-1} x_2^{(n)} , \\ y_1 &= \sum_{n=0}^{\infty} g^{n-1} y_1^{(n)} , \\ y_2 &= \frac{1}{\sqrt{2\kappa_r}} \sum_{n=0}^{\infty} g^{n-1} y_2^{(n)} . \end{aligned} \tag{4.74}$$

4.6.1 Matched Power Equations

The stochastic equations are now solved by the technique of matching powers of g in the corresponding time-evolution equations. This technique can be analyzed diagrammatically, and so can be termed the 'stochastic diagram' method [27]. The zero-th order solution is

$$\begin{aligned} dx_1^{(0)} &= \left[-x_1^{(0)} + \frac{1}{2} \left(x_1^{(0)} x_2^{(0)} + y_1^{(0)} y_2^{(0)} \right) \right] d\tau , \\ dy_1^{(0)} &= \left[-y_1^{(0)} + \frac{1}{2} \left(x_1^{(0)} x_2^{(0)} - x_2^{(0)} y_1^{(0)} \right) \right] d\tau , \\ dx_2^{(0)} &= -\kappa_r \left[x_2^{(0)} + \frac{1}{2} \left(x_1^{(0)2} - y_1^{(0)2} \right) - 2\varepsilon \right] d\tau , \\ dy_2^{(0)} &= -\kappa_r \left(y_2^{(0)} + x_1^{(0)} y_1^{(0)} \right) d\tau . \end{aligned} \tag{4.75}$$

These equations are the classical nonlinear equations for the cavity, expressed in terms of the quadrature amplitudes of dimensionless scaled fields. The steady-state solution below threshold is well-known, and is given by

$$x_1^{(0)} = y_1^{(0)} = y_2^{(0)} = 0; \qquad x_2^{(0)} = 2\varepsilon . \tag{4.76}$$

With no loss of generality, we can set all odd orders of $x_2^{(n)}$, $y_2^{(n)}$, and all even orders of $x_1^{(n)}$, $y_1^{(n)}$ to zero. To first order, the equations are given by

$$\begin{aligned} dx_1^{(1)} &= - (1-\varepsilon) x_1^{(1)} d\tau + \sqrt{2\varepsilon} dw_x(\tau) , \\ dy_1^{(1)} &= - (1+\varepsilon) y_1^{(1)} d\tau - i\sqrt{2\varepsilon} dw_y(\tau) , \end{aligned} \tag{4.77}$$

where $dw_{x(y)}(\tau) = (dw_1(\tau) \pm dw_2(\tau))/\sqrt{2}$. These equations are exactly equivalent to the operator equations ones often used to calculate squeezing. They are non-classical, but correspond to a very simple form of linear, non-classical fluctuation which has a Gaussian quasi-probability distribution. In other words, if no higher-order terms existed, the result would be a Gaussian state in the sub-harmonic, together with an ideal coherent state in the pump. The moments are given by

$$\begin{aligned}\langle \hat{y}_1^2 \rangle &= 1 + \left\langle y_1^{(1)} y_1^{(1)} \right\rangle = \frac{1}{1+\varepsilon} \,, \\ \langle \hat{x}_1^2 \rangle &= 1 + \left\langle x_1^{(1)} x_1^{(1)} \right\rangle = \frac{1}{1-\varepsilon} \,. \end{aligned} \tag{4.78}$$

This is exactly the same result as was obtained with the linearized Heisenberg equations, as expected.

The squeezing spectrum is the directly observable external signature of a squeezed field. This can be calculated directly from the Fourier transform of the stochastic equations. It is most convenient to carry out the calculations using scaled frequency variables $\Omega = 2\omega/\kappa_1$. We can represent the white noise that drives the stochastic equations by its Fourier transform $\xi_{x,y}(\Omega)$, where the spectral moments of the stochastic processes are

$$\begin{aligned}\langle \xi_a(\Omega) \rangle &= 0 \,, \\ \langle \xi_a(\Omega)\, \xi_b(\Omega') \rangle &= \delta_{ab}\delta(\Omega + \Omega') \,. \end{aligned} \tag{4.79}$$

It is also useful to introduce a standard convolution notation, where

$$[A \star B](\Omega) = \int \frac{d\Omega'}{\sqrt{2\pi}} A(\Omega')B(\Omega - \Omega') \,. \tag{4.80}$$

The stochastic equations may now be rewritten in the frequency domain as:

- First order:

$$\begin{aligned}\tilde{x}_1^{(1)}(\Omega) &= \frac{\sqrt{2\varepsilon}\xi_x(\Omega)}{(i\Omega + 1 - \varepsilon)} \,, \\ \tilde{y}_1^{(1)}(\Omega) &= \frac{-i\sqrt{2\varepsilon}\xi_y(\Omega)}{(i\Omega + 1 + \varepsilon)} \,. \end{aligned} \tag{4.81}$$

- Second order:

$$\begin{aligned}\tilde{x}_2^{(2)}(\Omega) &= -\frac{\kappa_r \left[\tilde{x}_1^{(1)} \star \tilde{x}_1^{(1)} - \tilde{y}_1^{(1)} \star \tilde{y}_1^{(1)}\right](\Omega)}{2(i\Omega + \kappa_r)} \,, \\ \tilde{y}_2^{(2)}(\Omega) &= -\frac{\kappa_r \left[\tilde{x}_1^{(1)} \star \tilde{y}_1^{(1)}\right](\Omega)}{(i\Omega + \kappa_r)} \,. \end{aligned} \tag{4.82}$$

- Third order:

$$\tilde{x}_1^{(3)}(\Omega) = \frac{\left[\tilde{x}_2^{(2)} \star \left(\tilde{x}_1^{(1)} + \xi_x/\sqrt{2\varepsilon}\right) + \tilde{y}_2^{(2)} \star \left(\tilde{y}_1^{(1)} + i\xi_y/\sqrt{2\varepsilon}\right)\right](\Omega)}{2\left[i\Omega + 1 - \varepsilon\right]},$$
$$\tilde{y}_1^{(3)}(\Omega) = \frac{\left[\tilde{y}_2^{(2)} \star \left(\tilde{x}_1^{(1)} + \xi_x/\sqrt{2\varepsilon}\right) - \tilde{x}_2^{(2)} \star \left(\tilde{y}_1^{(1)} + i\xi_y/\sqrt{2\varepsilon}\right)\right](\Omega)}{2\left[i\Omega + 1 + \varepsilon\right]}. \tag{4.83}$$

4.6.2 External Squeezing Correlations

We now calculate the spectrum of the squeezed field y_1, which is given by $\langle \tilde{y}_1(\Omega_1)\tilde{y}_1(\Omega_2)\rangle$. Thus, we obtain

$$\langle \tilde{y}_1(\Omega_1)\tilde{y}_1(\Omega_2)\rangle = \left\langle \tilde{y}_1^{(1)}(\Omega_1)\tilde{y}_1^{(1)}(\Omega_2)\right\rangle + \\ +g^2\left\langle \tilde{y}_1^{(1)}(\Omega_2)\tilde{y}_1^{(3)}(\Omega_1) + [\Omega_1 \leftrightarrow \Omega_2]\right\rangle + \cdots \tag{4.84}$$

The contribution from the first order perturbation theory is the usual linearized squeezing result, given in this case by

$$\left\langle \tilde{y}_1^{(1)}(\Omega_1)\tilde{y}_1^{(1)}(\Omega_2)\right\rangle = -\frac{2\varepsilon\delta(\Omega_1+\Omega_2)}{\left[\Omega_1^2 + (1+\varepsilon)^2\right]}. \tag{4.85}$$

Similarly, the complementary (unsqueezed) spectrum is

$$\left\langle \tilde{x}_1^{(1)}(\Omega_1)\tilde{x}_1^{(1)}(\Omega_2)\right\rangle = \frac{2\varepsilon\delta(\Omega_1+\Omega_2)}{\left[\Omega_1^2 + (1-\varepsilon)^2\right]}. \tag{4.86}$$

This gives rise to a well-known result: the linearized spectrum of external squeezing in an OPO, for the de-amplified quadrature

$$V_1^{\pi/2}(\omega) = 1 + \frac{2\pi\kappa_{\text{out}}}{T}\langle :\hat{y}_1(-\omega), \hat{y}_1(\omega):\rangle \\ = 1 - \left(\frac{\kappa_1^0}{\kappa_1}\right)\frac{4\varepsilon}{\Omega_1^2 + (1+\varepsilon)^2}. \tag{4.87}$$

The amplified, or unsqueezed fluctuations, after linearizing, are given by

$$V_1^{0}(\omega) = 1 + \frac{2\pi\kappa_{\text{out}}}{T}\langle :\hat{x}_1(-\omega), \hat{x}_1(\omega):\rangle \\ = 1 + \left(\frac{\kappa_1^0}{\kappa_1}\right)\frac{4\varepsilon}{\Omega_1^2 + (1-\varepsilon)^2}. \tag{4.88}$$

In the limit of a perfect 'one-sided' cavity, where all the losses occur in the output coupler so that $\kappa_1^o = \kappa_1$, these have the interesting behavior that they are complementary to each other

$$V_1^{0}(\omega)V_1^{\pi/2}(\omega) = \frac{\Omega_1^2 + (1-\varepsilon)^2}{\Omega_1^2 + (1+\varepsilon)^2} \times \frac{\Omega_1^2 + (1+\varepsilon)^2}{\Omega_1^2 + (1-\varepsilon)^2}. \tag{4.89}$$

The Heisenberg uncertainty relation is therefore satisfied exactly, since

$$[\Delta X(\omega)\Delta Y(\omega)]^2 = 1 \ . \tag{4.90}$$

Thus, for a perfect, lossless interferometer (i.e. no losses apart from the output coupler), and in the limit of weak coupling (i.e. negligible pump depletion), the output field satisfies the requirements of a perfect squeezed state.

We can obtain the next order contribution to the squeezing, by calculating $\langle \tilde{y}_1^{(3)}(\Omega_1)\,\tilde{y}_1^{(1)}(\Omega_2)\rangle$.

Using these results, we find that the spectrum of the squeezed quadrature, to this order, is given by

$$\langle \tilde{y}_1(\Omega_1)\,\tilde{y}_1(\Omega_2)\rangle = \delta(\Omega_1+\Omega_2)S(\Omega_1) \ , \tag{4.91}$$

which – for a perfect cavity, with no losses except the output coupler – corresponds to a squeezing spectrum of

$$\begin{aligned} V(\Omega) = {} & \frac{\Omega^2+(1-\varepsilon)^2}{\Omega^2+(1+\varepsilon)^2} + \frac{4g^2\varepsilon^2\kappa_r}{[\Omega^2+(1+\varepsilon)^2]^2} \\ & \times \left\{ \frac{(\Omega^2+1-\varepsilon^2)}{2\varepsilon\kappa_r(1-\varepsilon^2)} + \frac{(1-\varepsilon+\kappa_r)(1+\varepsilon)-\Omega^2}{(1-\varepsilon)[\Omega^2+(1-\varepsilon+\kappa_r)^2]} \right. \\ & \left. - \frac{(1+\varepsilon+\kappa_r)(1+\varepsilon)-\Omega^2}{(1+\varepsilon)[\Omega^2+(1+\varepsilon+\kappa_r)^2]} \right\} \ . \end{aligned} \tag{4.92}$$

This equation gives the complete linear and nonlinear squeezing spectrum, including all the nonlinear correction terms that contribute to order g^2 or $1/N$. This is graphed in Fig. 4.1, while Fig. 4.2 gives the nonlinear contribution to the spectrum, with its characteristic divergence as $\varepsilon \to 1$.

4.6.3 Optimal Squeezing

It is interesting to evaluate the squeezing or low-noise correlations in the limit of zero frequency, which is generally the frequency of maximum squeezing. For the ideal interferometer case, we obtain

$$V(0) = \frac{(1-\varepsilon)^2}{(1+\varepsilon)^2} + \frac{2\varepsilon g^2}{(1+\varepsilon)^4}\left\{1 + \frac{4\kappa_r\varepsilon^2(\kappa_r+2)}{(1-\varepsilon)[(1+\kappa_r)^2-\varepsilon^2]}\right\} \ . \tag{4.93}$$

This is graphed in Fig. 4.3.

To understand these results, and their dependence on the relative decay rates, we consider an approximation valid near threshold, where $\varepsilon \approx 1$. We can set $\varepsilon = 1+\delta$, and expand in powers of δ:

- For the case that $\kappa_r \gg g^{2/3}$, this gives the following result, ignoring small terms of order $g^2\delta$

$$V(0) = \frac{1}{4}\left(\delta^2 + \frac{g^2}{2} - \frac{2g^2}{\delta}\right) \ . \tag{4.94}$$

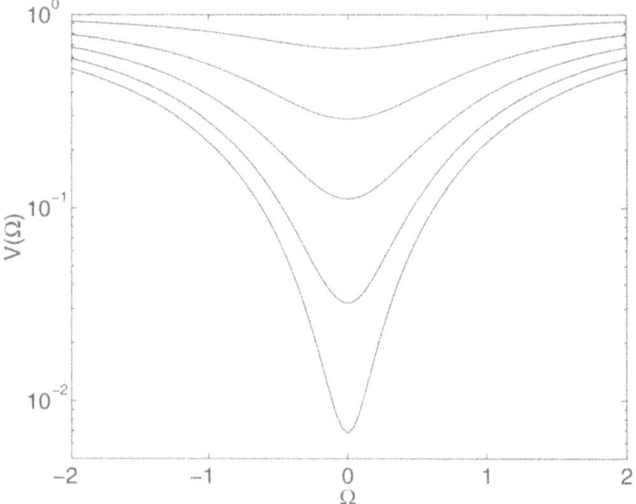

Fig. 4.1. Total OPO squeezing spectrum with $g^2 = 0.001$, $\kappa_r = 0.5$. The ε values plotted are $\varepsilon = 0.1, 0.3, 0.5, 0.7, 0.9$; larger values of ε give the most squeezing (lowest spectral variance)

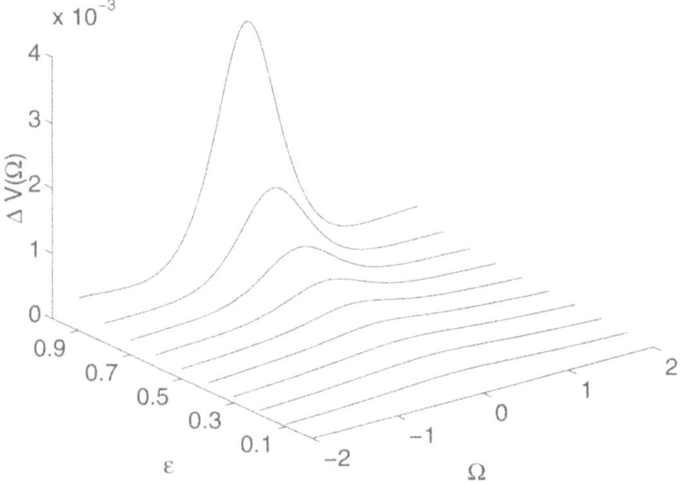

Fig. 4.2. Nonlinear OPO squeezing spectrum with $g^2 = 0.001$, $\kappa_r = 0.5$. The maximum ε value plotted is $\varepsilon = 0.95$

Minimizing this result with respect to δ, we find that

$$V_{\mathrm{opt}}(0) = \frac{3}{4} g^{4/3} \ . \tag{4.95}$$

This result confirms an approximate calculation of Plimak and Walls [30], although the self-consistent method used by these authors makes it

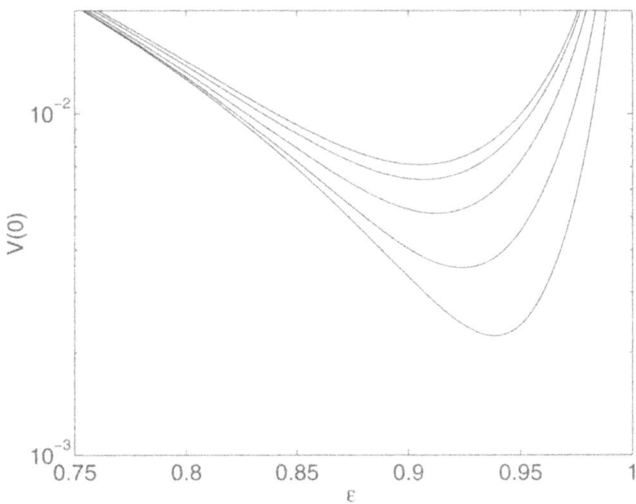

Fig. 4.3. Zero-frequency squeezing spectrum versus driving field, with $g^2 = 0.001$. Values of $\kappa_r = 0.001, 0.01, 0.1, 1.0, 10.0$ are used for the different lines plotted, with the lowest values of κ_r giving the best results for squeezing

difficult to obtain the relevant driving field. In fact, this minimum level of internal fluctuations occurs at a driving field just below threshold, with

$$\varepsilon_{\text{opt}} = 1 - g^{2/3} \,. \tag{4.96}$$

- For the case that $\kappa_r \ll g^{2/3}$, which is an extremely high-Q pump mode, a different result is obtained. Here one finds that

$$V(0) = \frac{1}{4}\left(\delta^2 + \frac{g^2}{2} - \frac{2g^2\kappa_r}{\delta^2}\right) \,. \tag{4.97}$$

Minimizing this result with respect to δ, we find the minimum level of internal fluctuations occurs at a driving field of

$$\varepsilon_{\text{opt}} = 1 - g^{1/2}(2\kappa_r)^{1/4} \,, \tag{4.98}$$

with an optimum squeezing of

$$V_{\text{opt}}(0) = g\sqrt{\kappa_r/2} \,. \tag{4.99}$$

This result gives a much lower level of fluctuations – an improved squeezing bound. This operating regime also has the property that the optimum frequency of noise reduction moves away from zero frequency as the driving field is increased above the optimum value, toward threshold. At slightly higher driving fields than the optimum squeezing point, a bifurcation to a spectrum with two minima occurs. This is shown in detail in Fig. 4.4.

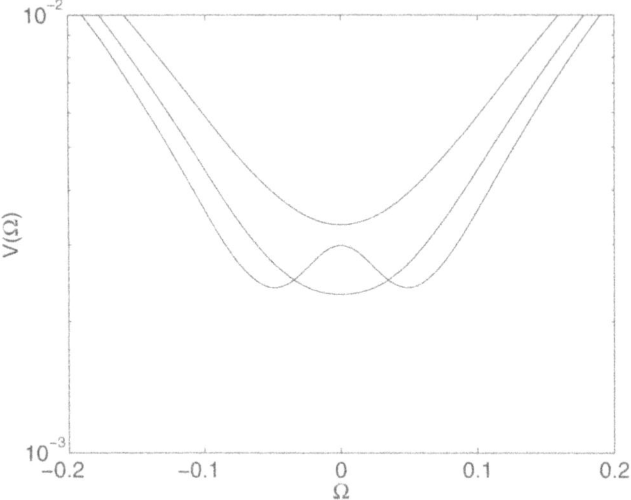

Fig. 4.4. Optimal zero-frequency squeezing spectrum versus frequency using the positive-P method, with $g^2 = 0.001$, $\kappa_r = 0.01$. Driving fields of $\varepsilon = 0.9$, 0.93, 0.96 are used for the different lines plotted, with the higher driving fields giving the best results for squeezing, except at zero frequency

The physics of this is clearly that the onset of critical fluctuations starts to spoil the spectral noise-reduction even before the critical point is reached at $\varepsilon = 1$. This can be seen from the way that the third order term includes contributions from the critical fluctuations in x_1. However, critical fluctuations are less effective at spoiling the optimum squeezing in the limit of a high-Q pump.

4.6.4 Experiments

Although optimum noise reduction is obtained very close to threshold, in current experiments it is more common to choose to operate below the critical region of maximum squeezing. This reduces the squeezing, but allows production of a stable, well-defined squeezed vacuum source. Operation below threshold is advantageous from the viewpoint of eliminating technical noise, which tends to result in 'leakage' of the large fluctuations in the unsqueezed quadrature, into the squeezed quadrature. At the same time, detector inefficiency and optical losses limits the maximum noise reduction possible, well before the ideal threshold point is reached. Because of these practical issues, the best experiments to date [33,34] have observed maximum noise reductions of $\simeq 6$ dB (i.e. $V(0) \simeq 0.25$), with a current limit of around 7 dB [35] . This basic quantum resource can be easily modified to give a squeezed bright beam, or (with two squeezers), an EPR-entangled beam, as discussed in later chapters. The theory [36,37] and experiments [7,8,38] described here can also be extended to treat non-degenerate parametric oscillation, in which the squeezing occurs directly in an EPR-entangled correlation between quadratures of different beams.

In general, the minimum-uncertainty properties of an OPO operated below threshold make it an extremely useful tool for a variety of squeezing and quantum information experiments.

4.7 Waveguides and Fibers

Squeezing interactions can occur inside interferometers, wave-guides, or in bulk crystals. In parametric systems, two low-frequency photons can be combined to give one higher-frequency photon, or vice-versa. The two processes are termed second-harmonic generation (up-conversion), or sub-harmonic generation (down-conversion). However, squeezing is also possible via four-wave mixing, which occurs inside optical fibers as part of the normal dielectric response function.

In this section, we consider the optical fiber and the traveling-wave parametric amplifier, which can also be modeled as a nonlinear, dispersive dielectric waveguide. In this case, we suppose there is propagation in the x-direction in single transverse modes for both the fundamental (signal) and second harmonic (pump), with a medium oriented so that type-I phase matching for the $\chi^{(2)}$ process is dominant. This restriction is imposed since a single polarization mode for both fields is included but this is for simplicity only. It is easy to generalize to non-degenerate type II phase-matching for collinear propagation, which involves three modes.

4.7.1 Fiber Optics Hamiltonian

The optical fiber treated will be a single-transverse mode fiber with dispersion and nonlinearity. Since boundary effects are usually negligible in experiments, it is useful to first take the infinite volume limit, which effectively replaces a summation over wave-vectors with the corresponding integral. The effect of a transverse mode structure will also be included, to show how the simplified theory is applicable in real three-dimensional fibers. From (2.52), the nonlinear Hamiltonian in this case is

$$\hat{H} = \int dk\hbar\omega(k)\hat{a}^\dagger(k)\hat{a}(k) + \frac{1}{4}\eta^{(3)} : \int \hat{\mathbf{D}}^4(\mathbf{r})d^3\mathbf{r} : \; . \tag{4.100}$$

Here $\omega(k)$ is the angular frequency of modes with wave-vector k, describing the *linear* polariton excitations in the fiber, including dispersion. Also, $\hat{a}(k)$ is a corresponding annihilation operator defined so that, as before

$$\left[\hat{a}(k'), \hat{a}^\dagger(k)\right] = \delta(k - k') \; . \tag{4.101}$$

The coefficient $\eta^{(3)}$ is the nonlinear coefficient arising when the electronic polarization field is expanded as a function of the electric displacement. Compared to the commonly used Bloembergen coefficient

$$\eta^{(3)} = -\varepsilon_0 \chi^{(3)}/\varepsilon^4 \ . \tag{4.102}$$

(The units are S.I. units, following current standard usage). In terms of modes of the waveguide, and neglecting modal dispersion, the electric displacement field operator $\hat{\mathbf{D}}(\mathbf{r})$ is

$$\hat{\mathbf{D}}(\mathbf{r}) = i \int dkk \left(\frac{\hbar \varepsilon(k) v(k)}{4\pi k} \right)^{\frac{1}{2}} \hat{a}(k) \mathbf{u}(\mathbf{r}) e^{ikx} + \text{H.c.} \ , \tag{4.103}$$

where

$$\int |\mathbf{u}(\mathbf{r})|^2 d^2\mathbf{r} = 1 \ . \tag{4.104}$$

Here $v(k)$ is the group velocity, and $\varepsilon(k)$ the dielectric permittivity. The mode function $\mathbf{u}(\mathbf{r})$ is included here in its usual three-dimensional form, to show how the simplified one-dimensional quantum theory relates to vector mode theory.

In the infinite volume limit, the annihilation and creation operators can be recovered using

$$\hat{a}(t,k) = \frac{1}{\sqrt{2\pi}} \int dx \ \hat{\Psi}(t,x) e^{-i(k-k_1)x - i\omega_1 t} dx \ . \tag{4.105}$$

The Hamiltonian can now be rewritten approximately, as

$$\begin{aligned} \hat{H} = \hbar \int dx \int dx' \omega(x,x') \hat{\Psi}^\dagger(t,x) \hat{\Psi}(t,x') \\ - \frac{\hbar}{2} \chi_e \int dx \hat{\Psi}^{\dagger 2}(t,x) \hat{\Psi}^2(t,x) dx \ . \end{aligned} \tag{4.106}$$

Here, we have introduced the quantity

$$\omega(x,x') = \int \frac{dk}{2\pi} \omega(k) e^{i(k-k_1)(x-x')} \ . \tag{4.107}$$

On Taylor expanding the mode frequency around $k = k_1$, this can be approximated to quadratic order in $(k - k_1)$, by

$$\begin{aligned} \omega(x,x') = \omega_1 \delta(x - x') \\ + \int \frac{dk}{4\pi} \left[i\omega_1'(\partial_{x'} - \partial_x) + \omega_1''(\partial_x \partial_{x'}) + \cdots \right] e^{ik(x-x')} \ . \end{aligned} \tag{4.108}$$

In addition, a nonlinear coupling term was also introduced, together with the assumption that the frequency dependence in the nonlinear coupling can be neglected to a good approximation for relatively narrow band-widths. Thus, we must define

$$\chi_e = -\frac{3}{4}\hbar\eta^{(3)}\varepsilon_1^2 k_1^2 v^2 \int |\mathbf{u}(\mathbf{r})|^4 d^2\mathrm{r} \ . \tag{4.109}$$

Alternative forms that are sometimes used for the nonlinear coupling are

$$\chi_e \equiv \left[\frac{3\hbar\chi^{(3)} w_1^2 v(k_1)^2}{4\varepsilon(k_1)c^2}\right] \int |\mathbf{u}(\mathbf{r})|^4 d^2\mathrm{r} \equiv \left[\frac{\hbar n_2 \omega_1^2 v^2}{Ac}\right] \ . \tag{4.110}$$

Here $A = [\int |\mathbf{u}(\mathbf{r})|^4 d^2\mathrm{r}]^{-1}$ is the effective modal cross-section, and n_2 is the refractive index change per unit field intensity. The free evolution part of the total Hamiltonian, which is removed here, just describes the carrier-frequency rotation at frequency ω_1. We now proceed as in the case of a purely dispersive fiber. On partial integration of the derivative terms and Fourier transforming, the interaction Hamiltonian describing the evolution of $\hat{\Psi}$ in the slowly-varying envelope and rotating-wave approximations is

$$\begin{aligned}\hat{\mathcal{H}}_I = \frac{\hbar}{2}\int \Bigg\{ & iv\left[\frac{\partial}{\partial x}\hat{\Psi}^\dagger \cdot \hat{\Psi} - \hat{\Psi}^\dagger \frac{\partial}{\partial x}\hat{\Psi}\right] \\ & + w'' \frac{\partial}{\partial x}\hat{\Psi}^\dagger \cdot \frac{\partial}{\partial x}\hat{\Psi} - \chi_e \hat{\Psi}^{\dagger 2}\hat{\Psi}^2 \Bigg\} dx \ . \end{aligned} \tag{4.111}$$

After taking the free evolution into account, we find the following Heisenberg equation of motion for the field operator propagating in the $+x$ direction

$$\left[v\frac{\partial}{\partial x} + \frac{\partial}{\partial t}\right]\hat{\Psi}(t,x) = \left[\frac{iw''}{2}\frac{\partial^2}{\partial x^2} + i\chi_e \hat{\Psi}^\dagger\hat{\Psi}\right]\hat{\Psi}(t,x) \ , \tag{4.112}$$

where, as before, $v = v(k_1) = \partial w/\partial k|_{n=k_1}$, $w'' = \partial^2 w/\partial k^2|_{n=k_1}$, and $w(k)$ is expanded quadratically in a narrow band around $\omega_1 = \omega(k_1)$, which is the carrier frequency.

4.7.2 Nonlinear Schrödinger Equation

In a co-moving reference frame defined by $x_v = x - vt$, this reduces to the quantum nonlinear Schrödinger equation

$$i\frac{\partial}{\partial t}\hat{\Psi}(t,x_v) = \left[-\frac{\omega''}{2}\frac{\partial^2}{\partial x_v^2} - \chi_e\hat{\Psi}^\dagger\hat{\Psi}\right]\hat{\Psi}(t,x_v) \ . \tag{4.113}$$

This equation has a very simple physical meaning. As previously, the particles of the theory – have acquired a nonzero effective mass

$$m = \frac{\hbar}{\omega''} \ . \tag{4.114}$$

The nonlinear term χ_e describes an interaction potential which couples the particles together. This interaction potential is attractive when χ_e is positive, as it is in most Kerr media, and the value of the potential is

$$V(x_v - x'_v) = -\hbar\chi_e\delta(x - x') \ . \tag{4.115}$$

In calculations it is preferable to scale the equations into a dimensionless form, which results in

$$\frac{\partial}{\partial\tau}\hat{\psi}(\tau,\zeta_v) = \left[\pm\frac{i}{2}\frac{\partial^2}{\partial\zeta_v^2} + i\hat{\psi}^\dagger\hat{\psi}\right]\hat{\psi}(\tau,\zeta_v) \ . \tag{4.116}$$

Here we have obviously made the assumption that $\chi_e > 0$, which is the usual case in dielectrics with a positive nonlinear index of refraction. The $\pm$ sign corresponds to the sign of ω'', and so is positive for anomalous dispersion. The photon number scaling parameter, $\overline{n}$ is defined to remove all nonlinear coefficients from the equation. This is obtained with the choice

$$\overline{n} = \frac{|k''|v^2}{\chi_e t_0} \ . \tag{4.117}$$

There is another description of this physical system that has an approximate validity, and is more convenient for some purposes. The polariton flux must be invariant at the dielectric boundaries because energy conservation demands it, even though the changing group velocity means that particle density must change. This leads to a description in terms of the flux operators, with an (approximate) equation for narrow-band fields, where $\hat{\Phi} = v\hat{\psi}$ is a quantum flux operator

$$\left(v\frac{\partial}{\partial x} + \frac{\partial}{\partial t}\right)\hat{\Phi}(t,x) = \left[\frac{i\omega''}{2}\frac{\partial^2}{\partial x^2} + \frac{i\chi_e}{v}\hat{\Phi}^\dagger\hat{\Phi}\right]\hat{\Phi}(t,x) \ . \tag{4.118}$$

Because flux is most naturally regarded as evolving in space, it is also common in laser physics to make a somewhat different choice of coordinates. In this reference frame, the space variable is left unchanged, and the time variable is modified so that $t_v = t - x/v$. This leads to an approximate form of the quantum nonlinear Schrödinger equation in dimensionless variables, in which the time variable has changed places with the space variable

$$\frac{\partial}{\partial\zeta}\hat{\phi}(\tau_v,\zeta) \simeq \left[\pm\frac{i}{2}\frac{\partial^2}{\partial\tau_v^2} + i\hat{\phi}^\dagger\hat{\phi}\right]\hat{\phi}(\tau_v,\zeta) \ . \tag{4.119}$$

This version of the nonlinear Schrödinger equation, (4.119), is often used as an alternative to the more precise (4.116).

For some experiments, there is a constant pump field, and fluctuations develop through four-wave mixing on this background. This can be easily described in a linearized, small fluctuation regime by the definition of:

$$\hat{\phi}(\tau_v,\zeta) = \delta\hat{\phi}(\tau_v,\zeta) + \phi_0 = \delta\hat{\phi}(\tau_v,\zeta) + \langle\hat{\phi}\rangle \ . \tag{4.120}$$

On linearizing the operator equations, one obtains:

$$\frac{\partial}{\partial \tau}\delta\hat{\phi}(\tau,\zeta_v) = \left[\pm\frac{i}{2}\frac{\partial^2}{\partial \zeta_v^2} + 2i|\phi_0|^2\right]\delta\hat{\phi} + i\phi_0^2\delta\hat{\phi}^\dagger \ . \tag{4.121}$$

However, in practical applications, it is typically the case that there are additional phase fluctuations due to thermal fluctuations generated from acoustic (Brillouin) and Raman phonons. These cases have been analyzed elsewhere, and add additional noise terms to the propagation equations.

4.7.3 Parametric Operator Equations

For a parametric waveguide, the Hamiltonian used here is the same as appears in the earlier work of Raymer et al. [10], using $\eta^{(2)} = -\varepsilon_0\chi^{(2)}/\varepsilon^3$ so that

$$\begin{aligned}\hat{H} = &\int_{n\simeq k_1} dk\hbar\omega(k)\hat{a}_k^\dagger\hat{a}_n + \int_{k\simeq k_2} dk\hbar\omega(k)\hat{a}_k^\dagger\hat{a}_k \\ &+\frac{\eta^{(2)}}{3}\int d^3x\left[\hat{D}_1(\mathbf{r}) + \hat{D}_2(\mathbf{r})\right]^3 \ .\end{aligned} \tag{4.122}$$

The electric displacement envelopes $\hat{D}_n(\mathbf{r})$ in the nonlinear term are expanded in terms of the boson field operators

$$\hat{D}_n(\mathbf{r}) = i\int_{k\simeq k_n} dk\left(\frac{\varepsilon_n\hbar v_n k_n}{4\pi}\right)^{1/2}\hat{a}_k u^{(j)}(\mathbf{r})e^{ikx} + \text{H.c.} \ , \tag{4.123}$$

where the frequency dependence of the parameters has been kept only for the phase-shift term $\exp(ikx)$. The electric permittivity, group velocity and wave-frequency at wave-numbers k_1 and $k_2 = 2k_1$ are given here by ε_1, v_1, ω_1 and ε_2, v_2, ω_2 respectively.

Including dispersion, the derivation of the Heisenberg equations of motion for the photon fields [31], defined as

$$\hat{\Psi}_n(x) = \frac{1}{\sqrt{2\pi}}\int_{k\simeq k_n} dk\exp(ikx)\hat{a}_k \tag{4.124}$$

is obtained, with a free-field Hamiltonian of

$$\hat{H}_0 = \hbar\sum_{n=1,2}\int dx[j\omega_1\hat{\Psi}_n^\dagger(x)\hat{\Psi}_n(x)] \ , \tag{4.125}$$

and an interaction Hamiltonian of

$$\begin{aligned}\hat{H}_I = &\frac{\hbar}{2}\sum_{n=1,2}\int dx\left\{iv_n\left[\frac{\partial}{\partial x}\hat{\Psi}_n^\dagger\cdot\hat{\Psi}_n - \hat{\Psi}_n^\dagger\frac{\partial}{\partial x}\hat{\Psi}_n\right] + w_n''\frac{\partial}{\partial x}\hat{\Psi}_n^\dagger\frac{\partial}{\partial x}\hat{\Psi}_n\right\} \\ &+\hbar\int dx\left[\Delta\hat{\Psi}_1^\dagger(x)\hat{\Psi}_1(x) + \left(\frac{i\chi_P}{2}\hat{\Psi}_2^\dagger(x)\hat{\Psi}_1^2(x) + \text{H.c.}\right)\right] \ ,\end{aligned} \tag{4.126}$$

where $\Delta = \omega_1 - \omega_2/2$ is a phase mismatch term, and

$$\chi_p = \eta^{(2)} k_1^{3/2} v_1 \varepsilon_1 \sqrt{\hbar v_2 \varepsilon_2} \int d^2\mathbf{r} \left(u^{(1)}(\mathbf{r})\right)^2 u^{(2)*}(\mathbf{r}) . \tag{4.127}$$

This leads to the following propagation equations

$$\begin{aligned} \left[v_1 \frac{\partial}{\partial x} + \frac{\partial}{\partial t}\right] \hat{\Psi}_1 &= \frac{i\omega_1''}{2}\left(\frac{\partial^2}{\partial x^2} - i\Delta\right)\hat{\Psi}_1 - \chi_p^* \hat{\Psi}_2 \hat{\Psi}_1^\dagger , \\ \left[v_2 \frac{\partial}{\partial x} + \frac{\partial}{\partial t}\right] \hat{\Psi}_2 &= \frac{i\omega_2''}{2}\frac{\partial^2}{\partial x^2}\hat{\Psi}_2 + \frac{\chi_p}{2}\hat{\Psi}_1^2 . \end{aligned} \tag{4.128}$$

This is most simply treated by just considering the limiting case of an undepleted pump, in which case one can write the simplified operator equation

$$\left[v_1 \frac{\partial}{\partial x} + \frac{\partial}{\partial t}\right] \hat{\Psi}_1 = \frac{i\omega_1''}{2}\left(\frac{\partial^2}{\partial x^2} - i\Delta\right)\hat{\Psi}_1 - G_p^*(t,x)\hat{\Psi}_1^\dagger . \tag{4.129}$$

This can be treated as though the fields were generated by an effective interaction Hamiltonian defined in terms of $G_p(t,x) = \chi_p^* \Psi_2(t,x)$, i.e.

$$H_{\text{eff}} = \hbar \int dx \left(\frac{iG_p(t,x)}{2}\hat{\Psi}_1^2(x) + \text{H.c.}\right) . \tag{4.130}$$

Such an interaction Hamiltonian is also called a squeezing Hamiltonian, as it is formally identical to the generator of multi-mode squeezing

The general operator equations given above, can be simplified in the moving frame. Following a similar technique to that for an optical fiber, one then obtains the following equation of motion for the subharmonic field:

$$\frac{\partial}{\partial \zeta}\hat{\phi}(\tau_v,\zeta) \simeq \left[\pm\frac{i}{2}\frac{\partial^2}{\partial \tau_v^2} - i\delta + g\right]\hat{\phi}(\tau_v,\zeta) . \tag{4.131}$$

4.7.4 Squeezed Propagation

Squeezing is obtained either for four-wave mixing or parametric coupling. In practice, one often wishes to include the pump time-dependence, and it is often most convenient to treat soliton fields in practical applications, so as to avoid some of the deleterious effects of dispersion and thermal noise. For simplicity, we will focus here on the case of a c.w. pump field, which gives similar results for either parametric or four-wave mixing.

The linearized Heisenberg equations of motion for either the parametric waveguide or the previous case of an optical fiber with a nonlinear refractive index are formally identical. In either case the equations in the moving frame, in dimensionless form are

$$\frac{\partial}{\partial \zeta}\hat{\phi}(\tau_v,\zeta) \simeq \left[\pm\frac{i}{2}\frac{\partial^2}{\partial \tau_v^2} - i\delta + g\right]\hat{\phi}(\tau_v,\zeta) , \tag{4.132}$$

together with the Hermitian conjugate equation.

To gain some insight into the underlying physics of these squeezed propagating photon beams, we first consider an idealized and analytically solvable model corresponding to a uniform dielectric that fills the entire space, with periodic boundary conditions at $-\tau_m/2$ and $\tau_m/2$. The nonlinear coupling is assumed to be constant during the whole evolution. For simplicity we suppose the gain g is real; other cases are then obtained by a simple rotation of the solution in phase-space, corresponding to a rotation in the squeezed and unsqueezed quadratures.

Solutions to the equations of motion are easily found in frequency space, where we expand $\hat{\phi}(\tau_v, \zeta)$ in terms of single-mode annihilation operators: $\hat{\phi}(\tau_v, \zeta) = \sum_\Omega \hat{a}_\Omega(\zeta) \exp(-i\Omega\tau_v)/\sqrt{\tau_m}$, where Ω is a dimensionless frequency. The single-mode operators $\hat{a}_\Omega$ are assumed to (approximately) satisfy the usual commutation relations of

$$[\hat{a}_\Omega(\zeta), \hat{a}^\dagger_{\Omega'}(\zeta)] \approx \delta_{\Omega\Omega'} \ . \tag{4.133}$$

While this is not exact – since interacting fields only have well-defined equal-time commutators – the procedure can be justified in terms of more rigorous representation theory techniques which are outlined in the next section.

The corresponding equations of motion are given by

$$\begin{aligned} \frac{\partial \hat{a}_\Omega}{\partial \zeta} &= -i\left(\pm \Omega^2/2 + \delta\right) \hat{a}_\Omega + g\hat{a}^\dagger_{-\Omega} \ , \\ \frac{\partial \hat{a}^\dagger_{-\Omega}}{\partial \zeta} &= i\left(\pm \Omega^2/2 + \delta\right) \hat{a}^\dagger_{-\Omega} + g\hat{a}_\Omega \ , \end{aligned} \tag{4.134}$$

and have the following solution

$$\begin{aligned} \hat{a}_\Omega(\zeta) &= \mu_\Omega(\zeta)\hat{a}_\Omega(0) + \nu_\Omega(\zeta)\hat{a}^\dagger_{-\Omega}(0) \ , \\ \hat{a}^\dagger_{-\Omega}(\zeta) &= \mu^*_\Omega(\zeta)\hat{a}^\dagger_{-\Omega}(0) + \nu_\Omega(\zeta)\hat{a}_\Omega(0) + \ . \end{aligned} \tag{4.135}$$

Here $|\mu_\Omega|^2 - |\nu_\Omega|^2 = 1$, and the values of these squeezing parameters are given by:

$$\begin{aligned} \mu_\Omega(\zeta) &= \cosh\left(g_\Omega\zeta\right) - i\lambda_\Omega \sinh\left(g_\Omega\zeta\right)/g_\Omega \ , \\ \nu_\Omega(\zeta) &= g\sinh\left(g_\Omega\zeta\right)/g_\Omega \ , \end{aligned} \tag{4.136}$$

with $\lambda_\Omega \equiv \Omega^2/2 + \delta$, and $g_\Omega = (g^2 - \lambda^2_\Omega)^{1/2}$. These solutions to the classical counterpart of the operator equations are well known in optics, while in quantum optics the operator equations in the context of squeezing on nonlinear propagating fields were studied in [39].

The knowledge of the initial state of the photon field, together with the commutation relations, allows us to calculate operator moments at spatial location ζ. The parameter g_Ω is the gain coefficient; if real, it causes a growing correlated output for the frequency component Ω, while if imaginary it leads to oscillations.

4.7.5 Quadrature Variances

Experimentally, it is possible to perform either a squeezing quadrature measurement, or else a measurement of correlated photon number. Suppose we start by considering a squeezed quadrature measurement. In this case, the squeezing is essentially a two-mode squeezing, and the appropriate quadrature operators are the complex quadrature operators:

$$\hat{x}^{\theta}_{\Omega} = \hat{a}_{\Omega} e^{-i\theta} + \hat{a}^{\dagger}_{-\Omega} e^{i\theta} . \tag{4.137}$$

The squeezing spectrum is then obtained as a variance in these complex quadratures, given by:

$$\begin{aligned} \langle \hat{x}^{\theta\dagger}_{\Omega} \hat{x}^{\theta}_{\Omega} \rangle = & \langle \hat{a}^{\dagger}_{\Omega} \hat{a}_{\Omega} + \hat{a}_{-\Omega} \hat{a}^{\dagger}_{-\Omega} \rangle \\ & + \langle \hat{a}^{\dagger}_{\Omega} \hat{a}^{\dagger}_{-\Omega} e^{2i\theta} + \hat{a}_{\Omega} \hat{a}_{-\Omega} e^{-2i\theta} \rangle . \end{aligned} \tag{4.138}$$

Next, taking expectation values in terms of the initial operator vacuum state means that only terms involving the initial creation operators on the right, or annihilation operators on the left, will give non-vanishing results. Thus:

$$\langle \hat{x}^{\theta\dagger}_{\Omega} \hat{x}^{\theta}_{\Omega} \rangle = 1 + 2\nu^2_{\Omega} + \nu_{\Omega} \left(e^{2i\theta} \mu^*_{\Omega} + e^{2i\theta} \mu_{\Omega} \right) . \tag{4.139}$$

The phase angle θ that minimizes the variance – or maximizes the variance – is therefore the one for which both terms on the RHS above are real and negative, or real and positive. This variation can be experimentally obtained by varying a local oscillator phase at the location of the homodyne detector, after propagating through the nonlinear waveguide.

Defining $r = g_{\Omega}\zeta$, this gives a minimum or maximum variance of:

$$\begin{aligned} \langle \hat{x}^{\theta\dagger}_{\Omega} \hat{x}^{\theta}_{\Omega} \rangle_{\max} &= 1 + 2\nu_{\Omega} \left(\nu_{\Omega} + |\mu_{\Omega}| \right) , \\ \langle \hat{x}^{\theta\dagger}_{\Omega} \hat{x}^{\theta}_{\Omega} \rangle_{\min} &= 1 + 2\nu_{\Omega} \left(\nu_{\Omega} - |\mu_{\Omega}| \right) . \end{aligned} \tag{4.140}$$

It is easily verified that this is a minimum uncertainty product state, given the rather idealized assumptions made here. Substituting in the solutions of the operator equations gives the following result:

$$\begin{aligned} \langle \hat{x}^{\theta\dagger}_{\Omega} \hat{x}^{\theta}_{\Omega} \rangle_{\max} &= 1 + 2 \left(\frac{g}{g_{\Omega}} \right)^2 \sinh r \left(\sinh r + \sqrt{\cosh^2 r - \lambda^2_{\Omega}/g^2} \right) , \\ \langle \hat{x}^{\theta\dagger}_{\Omega} \hat{x}^{\theta}_{\Omega} \rangle_{\min} &= 1 + 2 \left(\frac{g}{g_{\Omega}} \right)^2 \sinh r \left(\sinh r - \sqrt{\cosh^2 r - \lambda^2_{\Omega}/g^2} \right) . \end{aligned} \tag{4.141}$$

The optimum squeezing is obtained when $\lambda_{\Omega} = 0$, in which case $g = g_{\Omega}$, and the results reduce to a familiar exponential growth in one variance, together with exponential squeezing in the other variance:

$$\begin{aligned} \langle \hat{x}^{\theta\dagger}_{\Omega} \hat{x}^{\theta}_{\Omega} \rangle_{\max} &= e^{2g\zeta} , \\ \langle \hat{x}^{\theta\dagger}_{\Omega} \hat{x}^{\theta}_{\Omega} \rangle_{\min} &= e^{-2g\zeta} . \end{aligned} \tag{4.142}$$

The need to use complex quadratures is due to the fact that the interaction produces an entanglement or EPR correlation between real quadratures of opposite (relative) frequencies. This is described in greater detail in later chapters. However, this type of experiment is often somewhat difficult to carry out, due to waveguide imperfections and phase noise effects, that tend to destroy phase information in these types of c.w. waveguide experiments.

4.7.6 Photon Number Correlations

To describe a less phase-sensitive type of experiment, where number correlations are measured, one may introduce number operators $\hat{N}_-(\zeta)$ and $\hat{N}_+(\zeta)$ containing only negative or positive wave-vectors relative to the central carrier frequency. That is:

$$\hat{N}_{-(+)}(\zeta) = \sum_{\Omega<0(\Omega>0)} \hat{a}^\dagger_\Omega(\zeta)\hat{a}_\Omega(\zeta) \ . \tag{4.143}$$

The normalized variance $V(\zeta)$ of the particle number difference $[\hat{N}_-(\zeta) - \hat{N}_+(\zeta)]$ is then given by:

$$V(\zeta) = \langle[\Delta(\hat{N}_- - \hat{N}_+)]^2\rangle \Big/ (\langle\hat{N}_-\rangle + \langle\hat{N}_+\rangle) \ , \tag{4.144}$$

where $\Delta\hat{X} \equiv \hat{X} - \langle\hat{X}\rangle$. Rewriting this in normally ordered operator product form, and using the symmetry properties of these solutions one can show that $\langle\hat{N}_-\rangle = \langle\hat{N}_+\rangle$, $\langle: (\hat{N}_-)^2 :\rangle = \langle: (\hat{N}_+)^2 :\rangle$, and $\langle\hat{N}_-\hat{N}_+\rangle = \langle\hat{N}_+\hat{N}_-\rangle$. The variance then becomes:

$$V(\zeta) = 1 + [\langle: (\hat{N}_+)^2 :\rangle - \langle\hat{N}_-\hat{N}_+\rangle] \Big/ \langle\hat{N}_+\rangle \ . \tag{4.145}$$

Here the first term on the right hand side represents the classical lower bound that can be obtained with coherent fields, while $V(\zeta) < 1$ implies non-classical correlation between $\hat{N}_-$ and $\hat{N}_+$.

Calculating $\langle: (\hat{N}_+)^2 :\rangle$ and $\langle\hat{N}_-\hat{N}_+\rangle$ gives

$$V(\zeta) = 1 + \frac{\sum_{\Omega>0} \nu^2_\Omega \left(\nu^2_\Omega - |\mu_\Omega|^2\right)}{\sum_{\Omega>0} \nu^2_\Omega} = 0 \ , \tag{4.146}$$

where we have used the fact that $\nu^2_\Omega - |\mu_\Omega|^2 = -1$. This result implies perfect (100%) squeezing in the photon number difference.

4.8 Solitons

In practical experiments, cw squeezing is seldom employed in waveguides. This is because the higher intensities of pulsed lasers allow much shorter interaction distances, thereby eliminating some of the deleterious effects of refractive index fluctuations, which causes phase noise. This effect was neglected, for simplicity, in the previous section.

The physical origin of these effects is in coupling to phonons. This is one property of a solid medium that is difficult to remove – except by cooling to cryogenic temperatures. The presence of Raman interactions plays a major role in perturbing the fundamental soliton behavior of the nonlinear Schrödinger equation in optical fibers.

The complete derivation of the quantum theory for optical fibers is given in the literature [40]. This paper presents a detailed derivation of the quantum Hamiltonian, and includes quantum noise effects due to nonlinearities, gain, loss, Raman reservoirs and Brillouin scattering.

Phase-space techniques then allow the quantum Heisenberg equations of motion to be mapped onto stochastic partial differential equations, as in the previous sections.

We therefore start with the phase-space equation for the case of a single polarization mode, obtained using a truncated Wigner representation, which is accurate in the limit of large photon number. One can then use either soliton perturbation theory or numerical integration of the phase-space equation to calculate effects on soliton propagation of all known quantum noise sources.

The Raman noise due to thermal phonon reservoirs is strongly dependent on both temperature and pulse intensity. At room temperature, this means that Raman pulse-position jitter and phase noise become steadily more important as the pulse intensity is increased, which occurs when a shorter soliton pulse is required for a given fiber dispersion.

4.8.1 Raman–Schrödinger Model

We begin with the Raman-modified stochastic nonlinear Schrödinger equation [40], obtained using the Wigner representation, using flux-type variables in the moving frame:

$$
\begin{aligned}
\frac{\partial}{\partial \zeta}\phi(\tau,\zeta) = & \frac{(\alpha^G-\alpha^A)}{2}\phi(\tau,\zeta)+\Gamma(\tau,\zeta)+ \\
& +i\left\{\pm\frac{1}{2}\frac{\partial^2\phi}{\partial\tau^2}+\int_0^\infty d\tau' h(\tau-\tau')\phi^*(\tau',\zeta)\phi(\tau',\zeta)\right. \\
& \left. +\, i\Gamma^R(\tau,\zeta)\right\}\phi(\tau,\zeta)\ . \qquad (4.147)
\end{aligned}
$$

Here $\phi = \Psi\sqrt{vt_0/\overline{n}}$ is a dimensionless photon flux amplitude, while $\tau = (t-x/v)/t_0$ and $\zeta = x/x_0$, where t_0 is a typical pulse duration used for scaling purposes and $x_0 = t_0^2/|k''|$ is a characteristic dispersion length. The group velocity v and the dispersion relation k'' are calculated at the carrier frequency ω_0.

Apart from a cut-off dependent vacuum noise, the photon flux is $\mathcal{J} = |\phi|^2\overline{n}/t_0$, where $\overline{n} = |k''|Ac/(n_2\hbar\omega_0^2 t_0) = v^2 t_0/\chi x_0$ is the typical number of photons in a soliton pulse of width t_0, again for scaling purposes. In this definition, the fiber is assumed to have a modal cross-sectional area $\mathcal{A}$ and a change in refractive index per unit intensity of n_2. The positive sign in front of the second derivative term applies for anomalous dispersion ($k'' < 0$), and the negative sign applies for normal dispersion ($k'' > 0$). The terms α^G, α^A) and h are gain/loss and Raman scattering response functions respectively, while Γ and Γ^R are stochastic terms, discussed below.

Similar, but more accurate, equations occur with the positive-P representation, although in this case, the phase-space dimension is doubled. It should be noted that these equations do not involve any assumptions about equal-space commutators, and are derived using standard equal-time commutators. For this reason, they provide a more complete and rigorous derivation of results obtained using linearized operator equations. They can also be easily extended to include nonlinear behavior, just as in the single-mode case.

4.8.2 Initial Conditions and Quantum Evolution

Equation (4.147) is a complex-number equation that is able to accurately represent quantum operator evolution through the inclusion of various noise sources. In the absence of any noise sources, this equation reduces to the classical non-linear Schrödinger equation, with Raman corrections. This deterministic limit corresponds to taking $\overline{n} \to \infty$. As well as the noise sources explicitly appearing in (4.147), there must be noise in the initial conditions to properly represent a quantum state in the Wigner representation. Regardless of the initial quantum state chosen, there must be at least minimal level of initial fluctuations in ϕ to satisfy Heisenberg's uncertainty principle. We choose to begin with a multi-mode coherent state, which contains this minimal level of initial quantum noise and which is an accurate model of mode-locked laser output. This is also the simplest model for the output of mode-locked lasers, and we note that, in general, there could be extra technical noise. For coherent inputs, the Wigner vacuum fluctuations are Gaussian, and are correlated as

$$\langle \Delta\phi(\tau, 0)\Delta\phi^*(\tau', 0)\rangle = \frac{1}{2\overline{n}}\delta(\tau - \tau') \ . \tag{4.148}$$

Physical quantities can be calculated from this phase-space simulation by averaging products of ϕ and ϕ^* over many stochastic trajectories. In this Wigner representation, these stochastic averages correspond to the ensemble averages of symmetrically-ordered products of quantum operators, such as those representing homodyne measurements and other measurements of phase.

4.8.3 Wigner Noise

Both fiber loss and the presence of a gain medium each contribute quantum noise to the equations in the symmetrically-ordered Wigner representation.

The complex gain/absorption noise enters the nonlinear-Schrödinger (NLS) equation through an additive stochastic term Γ, whose correlations are

$$\langle \Gamma(\Omega,\zeta)\Gamma^*(\Omega',\zeta')\rangle = \frac{(\alpha^G+\alpha^A)}{2\overline{n}}\delta(\zeta-\zeta')\delta(\Omega+\Omega') \; , \tag{4.149}$$

where $\Gamma(\Omega,\zeta)$ is the Fourier transform of the noise source

$$\Gamma(\Omega,\zeta) = \frac{1}{\sqrt{2\pi}}\int_{-\infty}^{\infty} d\tau \Gamma(\tau,\zeta)\exp(i\Omega\tau) \; . \tag{4.150}$$

The dimensionless intensity gain and loss are given by α^G and α^A, respectively.

Similarly, the real Raman noise, which appears as a multiplicative stochastic variable Γ^R, has correlations

$$\begin{aligned}\langle \Gamma^R(\Omega,\zeta)\Gamma^R(\Omega',\zeta')\rangle &= \frac{1}{\overline{n}}\left[n_{\mathrm{th}}(\Omega)+\frac{1}{2}\right]\alpha^R(\Omega) \\ &\quad \times\delta(\zeta-\zeta')\delta(\Omega+\Omega') \; ,\end{aligned} \tag{4.151}$$

where the thermal Bose distribution is given by

$$n_{\mathrm{th}}(\Omega) = \left[\exp\left(\hbar|\Omega|/k_B T t_0\right)-1\right]^{-1} \tag{4.152}$$

and where $\alpha^R(\Omega)$ is the Raman gain. Thus the Raman noise is strongly temperature-dependent, but it also contains a spontaneous component which provides vacuum fluctuations even at $T=0$.

In the case of a cw input, one can solve these equations in a similar way to the linearized operator equations, except we now expand $\phi(\tau_v,\zeta)$ in terms of single-mode Wigner amplitudes:

$$\Delta\phi(\tau_v,\zeta) = \frac{1}{\sqrt{2\pi}}\int d\Omega \Delta\phi(\Omega,\zeta)\exp(i\Omega\tau_v) \; . \tag{4.153}$$

This requires no assumptions about commutation relations.

The corresponding equations of motion are similar to those obtained previously, provided there is no Raman/Brillouin response, and the gain/loss terms are also absent:

$$\begin{aligned}\frac{\partial\Delta\phi(\Omega,\zeta)}{\partial\zeta} &= -i\left(\pm\Omega^2/2+\delta\right)\Delta\phi(\Omega,\zeta)+g\Delta\phi^*(-\Omega,\zeta) \; , \\ \frac{\Delta\phi^*(-\Omega,\zeta)}{\partial\zeta} &= i\left(\pm\Omega^2/2+\delta\right)\Delta\phi^*(-\Omega,\zeta)+g\Delta\phi(\Omega,\zeta) \; .\end{aligned} \tag{4.154}$$

These have the following solution

$$\Delta\phi(\Omega,\zeta) = \mu_\Omega(\zeta)\Delta\phi(\Omega,0)+\nu_\Omega(\zeta)\Delta\phi^*(\Omega,0) \; . \tag{4.155}$$

Here, as previously, $|\mu_\Omega|^2 - |\nu_\Omega|^2 = 1$, and the values of these squeezing parameters are exactly as in the operator equations. Thus, in the Wigner linearized equations, the change in quantum noise is reflected by a corresponding change in the vacuum noise initial conditions.

As the $\overline{n}$ dependence of all the noise correlations show, the classical limit of these quantum calculations is the deterministic nonlinear Schrödinger equation. The problem of jitter in soliton communications is an example of how intrinsic quantum features can have a direct macroscopic consequence, even in a way that impinges on current developments of applied technology. There are, of course, classical contributions to jitter, such as noise arising from technical sources. However, it is the jitter contributions from essentially quantum processes, namely spontaneous emission in fiber amplifiers, that is the current limiting factor in soliton based communications systems. Other jitter calculations rely on a classical formulation with an empirical addition of amplifier noise, and important predictions of the Gordon–Haus effect have been obtained. Nevertheless, this quantum treatment presented here of all known noise sources is necessary to determine the limiting effects of other intrinsic noise sources, which become important for shorter pulses and longer dispersion lengths.

In the absence of the noise sources, the phase-space equations have stationary solutions in the form of bright (+) or dark (–) solitons. Solitons are solitary waves in which the effects of dispersion are balanced by nonlinear effects, to produce a stationary pulse that is robust in the presence of perturbations. We note here that the Raman response function is non-instantaneous, which causes a red-shift in the soliton frequency. In general, a detailed, quantitative understanding of propagation experiments requires the solution of these coupled equations for photons and phonons with quantum noise included. This can be undertaken either through operator linearization methods – which is restricted to small quantum fluctuations – or else through a direct numerical solution of the relevant phase-space equations [11]. These detailed calculations are beyond the scope of the present volume. However, in many cases the existence of phonon noise sources that add thermal phase fluctuations to the coherent propagating field, provide strong limitations on squeezing.

4.8.4 Experiments

Propagation experiments have been very successful in producing large degrees of squeezing over broad band-widths. In contrast to the intra-cavity situation, four-wave mixing or $\chi^{(3)}$ techniques in fibers – using solitons – have proved more successful to date than the cw parametric waveguide approach. In fact, $\chi^{(3)}$ techniques have actually produced substantially more squeezing [42,43] together with entanglement [44], than parametric waveguide experiments have. This is just the reverse of the intra-cavity situation. The main reason for this appears to be simply in material properties – fibers have lower losses and more careful production facilities than parametric waveguides, which require considerable effort to obtain refractive-index (phase) matching of the two frequencies.

However, this is not a fundamental problem. In general terms it seems possible that the fact that additional wide-band phase noise exists with $\chi^{(3)}$ techniques in fibers, due to phase-modulation of the carrier, will provide a substantial limitation to this technique: unless cryogenic fiber sources are used. By comparison, parametric waveguide squeezing is relatively immune to carrier phase-noise, as this takes place near the carrier frequency, far removed from the squeezed wavelengths of interest. In the future, therefore, it would not be unexpected if interest in parametric waveguide techniques was renewed, owing to the obvious potential for very substantial noise reduction over wide bandwidths, together with a minimum level of extraneous phase-noise from carrier interactions with phonons.

4.9 Conclusion

The quantum fluctuations in squeezed fields generated from nonlinear optics can be readily obtained using a nonlinear stochastic theory, and applying either asymptotic approximations or numerical techniques. Corresponding results for the Feynman diagram method [32] would require a summation over infinite sets of diagrams, in order to fully include the reservoirs. The advantage of the representation theory method is due to the fact that the coherent state basis is a more natural basis set for an open system, since it allows the damping reservoirs to be treated non-perturbatively.

Optimal squeezing in the output spectra corresponding to these moments can be estimated for the interferometer (OPO) case. The best squeezing in the zero frequency part of the squeezing spectrum scales like $N^{-2/3}$ just below threshold, unless an extremely low-loss pump is employed, in which case the squeezing improves to give an N^{-1} scaling. In other words, at the true critical threshold – where the linear squeezing is apparently optimized – the nonlinear corrections are too large to give the lowest overall zero-frequency squeezing. Instead, one should operate below the critical point to optimize the spectral squeezing.

At the critical point ($\varepsilon = 1$), the scaling behavior is quite different to the behavior just below threshold, and must be calculated by using a distinct asymptotic perturbation theory, valid at the threshold itself. The critical fluctuations occur in the unsqueezed quadrature, as might be expected, but are not infinite – in contradiction to the predictions of a linearized theory.

The representation theory methods can also be used for waveguide situations, where squeezing is improved as the coupled fields co-propagate down a wave-guide. In these cases, it is vital to notice that there are always many modes involved, as there is no frequency-selective interferometer to limit couplings to just one or two significant driven modes. The many-mode nature of this problem also means that a consideration of frequency-dispersion becomes extremely important, as this is the factor that limits the range of frequencies over which there is parametric gain, and hence vacuum squeezing. This factor is well-recognized in the theory of both solitons in fibers, and of optical parametric amplification (OPA). Of course the Heisenberg uncertainty principle shows us that squeezing and amplification are necessarily found together.

The theoretical methods given here are readily extended to other cases, including non-degenerate parametric oscillators, squeezing generated from four-wave mixing in fiber solitons [41,42], and many other examples. In summary, squeezed radiation generation using nonlinear optics techniques is now a well understood and highly effective means of producing radiation with quantum noise levels in one quadrature suppressed well below the usual vacuum noise levels.

The main limiting factors that remain include control over the driving laser phase and intensity noise, absorption in the dielectric material and associated devices, and phase fluctuations caused both by low-frequency technical noise in components, and by intrinsic material properties of phonon interactions and refractive index fluctuations. In future it appears not impossible to reach squeezing levels of 10 dB or more below shot noise, over bandwidths of up to 1 Thz. At this level of noise reduction, the main problem will become the perennial issue of improving detector technology to allow such low noise levels to be detected.

References

1. H.J. Carmichael and D.F. Walls, J. Phys. B **9**, 1199 (1976); H.J. Kimble, M. Dagenais and L. Mandel, Phys. Rev. Lett. **39**, 691 (1977)
2. R.E. Slusher, L.W. Hollberg, B. Yurke, J.C. Mertz and J.F. Valley, Phys. Rev. Lett. **55**, 2409 (1985)
3. D.F. Walls, Nature **306,** 141 (1983)
4. N. Bloembergen, *Nonlinear Optics* (Benjamin, New York 1965)
5. Squeezed states and non-classical light have been topics of several special journal issues: H.J. Kimble and D.F. Walls, eds., J. Opt. Soc. Am. B **4,** 1450 (1987); R. Loudon and P.L. Knight, eds., J. Mod. Optics **34,** 709 (1987); E. Giacobino and C. Fabre, eds., Appl. Phys. B **55,** 189 (1992)
6. L.A. Wu, H.J. Kimble, J. Hall and H. Wu, Phys. Rev. Lett. **57**, 2520 (1986).
7. A. Heidmann, R.J. Horowicz, S. Reynaud, E. Giacobino, C. Fabre and G. Camy, Phys. Rev. Lett. **59**, 2555 (1987)
8. Z.Y. Ou, S.F. Pereira and H.J. Kimble, Phys. Rev. Lett. **68**, 3663 (1992)
9. R.E. Slusher, P. Grangier, A. LaPorta, B. Yurke and M.J. Potasek, Phys. Rev. Lett. **59**, 2566 (1987)
10. C.M. Caves and D. D. Crouch, J. Opt. Soc. Am. B **4**, 1535 (1987); M.G. Raymer, P.D. Drummond and S.J. Carter, Optics Lett. **16**, 1189 (1991)
11. S.J. Carter, P.D. Drummond, M.D. Reid and R.M. Shelby, Phys. Rev. Lett. **58**, 1841 (1987); P.D. Drummond and S.J. Carter, J. Opt. Soc. Am. B **4**, 1565 (1987); R.M. Shelby, P.D. Drummond and S.J. Carter, Phys. Rev. A **42**, 2966 (1990)
12. Y. Lai and H.A. Haus, Phys. Rev. A **40**, 844 (1989); H.A. Haus and Y. Lai, J. Opt. Soc. Am. B **7**, 386 (1990)

13. M. Rosenbluh and R.M. Shelby, Phys. Rev. Lett. **66**, 153 (1991); P.D. Drummond, R.M. Shelby, S.R. Friberg and Y. Yamamoto, Nature **365**, 307 (1993)
14. C.M. Caves, K.S. Thorne, R.W.P. Drever, V.D. Sandberg and M. Zimmermann, Rev. Mod. Phys. **52**, 341 (1980)
15. V.B. Braginsky and F.Ya. Khalili, *Quantum Measurement*, (Cambridge University Press, Cambridge, 1992)
16. M. Hillery and L.D. Mlodinow, Phys. Rev. A **30**, 1860 (1984); Phys. Rev. A **55**, 678 (1997)
17. P.D. Drummond, Phys. Rev. A **42**, 6845 (1990)
18. P.D. Drummond, K.J. McNeil and D.F. Walls, Optica Acta **27**, 321 (1980); P.D. Drummond, K.J. McNeil and D.F. Walls, Optica Acta **28**, 211 (1981)
19. C.W. Gardiner and M.J. Collett, Phys. Rev. A **31**, 3761 (1985); M.J. Collett and D.F. Walls, Phys. Rev. A **32**, 2887 (1985)
20. C.W. Gardiner: *Handbook of Stochastic Methods* (Springer, Berlin Heidelberg New York 1983); C.W. Gardiner, *Quantum Noise* (Springer, Berlin Heidelberg New York 1992)
21. B. Yurke, Phys. Rev. A **32**, 300 (1985)
22. P.D. Drummond and M.D. Reid, Phys. Rev. A **37**, 1806 (1988)
23. S. Chaturvedi, P.D. Drummond and D.F. Walls, J. Phys. A **10**, L187 (1977); P.D. Drummond and C.W. Gardiner, J. Phys. A **13**, 2353 (1980)
24. R.J. Glauber, Phys. Rev. **130**, 2529 (1963); E.C.G. Sudarshan, Phys. Rev. Lett. **10**, 277 (1963)
25. L. Arnold, *Stochastic Differential Equations: Theory and Applications*, (John Wiley and Sons, New York 1974)
26. A. Gilchrist, C.W. Gardiner and P.D. Drummond, Phys. Rev. A **55**, 3014 (1997)
27. S. Chaturvedi and P.D. Drummond, Eur. Phys. J. B **8**, 251 (1999)
28. P. Kinsler and P.D. Drummond, Phys. Rev. A **52**, 783 (1995)
29. P.D. Drummond, Phys. Rev. A **33**, 4462 (1986)
30. L.I. Plimak and D.F. Walls, Phys. Rev. A **50**, 2627 (1994)
31. T.A.B. Kennedy and E.M. Wright, Phys. Rev. A **38**, 212 (1988)
32. C.J. Mertens, T.A.B. Kennedy and S. Swain, Phys. Rev. Lett. **71**, 2014 (1993); C.J. Mertens, T.A.B. Kennedy and S. Swain, Phys. Rev. A **48**, 2374 (1993); C.J. Mertens, J.M. Hasty, H.H. Roark III, D. Nowakowski, T.A.B. Kennedy and S. Swain, Phys. Rev. A **52**, 742 (1995); C.J. Mertens and T.A.B. Kennedy, Phys. Rev. A **53**, 3497 (1996)
33. E. S. Polzik, J. Carri, and H.J. Kimble, Appl. Phys. B. **55**, 279 (1992)
34. K. Schneider, M. Lang, J. Mlynek and S. Schiller, Optics Express **2**, 59 (1998)
35. P.K. Lam, T.C. Ralph, B.C. Buchler, D.E. McClelland, H.A. Bachor and J. Gao, J. Opt. B **1**, 469 (1999)
36. M. J. Collett and R. Loudon, J. Opt. Soc. Am. B **4**, 1525 (1987)
37. P.D. Drummond and M.D. Reid, Phys. Rev A **41**, 3930 (1990)
38. C. Schori, J.L. Sorensen, E.S. Polzik, quant-ph/0205015 (2002)
39. M.J. Werner, M.G. Raymer, M. Beck and P.D. Drummond, Phys. Rev. A **52**, 4202 (1995); M.J. Werner and P.D. Drummond, Phys. Rev. A **56**, 1508 (1997)
40. P.D. Drummond and J.F. Corney, J. Opt. Soc. Am. B **18**, 139 (2001)
41. M. Werner, Phys. Rev. A **54**, R2567 (1996); M.J. Werner and S.R. Friberg, Phys. Rev. Lett. **79**, 4143 (1997)
42. S. Spalter, M. Burk and G. Leuchs, Europhys. Lett. **38**, 335 (1997)

43. M. Fiorentino, J.E. Sharping, P. Kumar, D. Levandovsky and M. Vasilyev Phys. Rev. A **64**, 031801 (2001)
44. Ch. Silberhorn, P.K. Lam, O. Weiss, N. Korolkova and G. Leuchs, Phys. Rev. Lett. **86**, 4267 (2001)

5 Squeezing from Lasers

T. C. Ralph

The quantum noise properties of the laser were studied intensively after its invention in the early sixties. The conclusion of these studies was that the laser output approaches quite closely the noise properties of a coherent state, provided the laser is operated well above threshold and only short time intervals (high frequencies) are considered. That is the spectral variance of the photon number and phase fluctuations at high enough Fourier frequencies will be at the Poissonian or quantum noise limit (QNL) [1] (also referred to as the shot noise level). These predictions were subsequently confirmed experimentally [2]. At lower frequencies excess noise, both intrinsic and technical is predicted. None-the-less, under ideal conditions, the intensity noise of the laser tends to the QNL, for all frequencies, as the laser is pumped very far above threshold. On the other hand the phase of the laser is unconstrained and drifts as a function of time, so-called laser phase diffusion [3]. As a result even under ideal conditions the phase fluctuations show excess noise at low frequencies.

The prediction that the intensity noise would be at the QNL for all frequencies was based on four assumptions, namely; (i) the pumping process has Poissonian intensity fluctuations, (ii) the lifetime of the lower lasing level is much shorter than the cavity lifetime, (iii) only the ground level and upper lasing level contain a significant population and (iv) collisional or lattice induced dephasing effects are sufficiently rapid such that coherent atomic effects can be neglected. In the early eighties it was realized that if the pumping mechanism for the laser could be made sub-Poissonian then amplitude squeezing could be obtained at low frequency [4]. Yamamoto showed a few years later that sub-Poissonian pumping could be realized in semi-conductor lasers by regularizing the pump current [5]. An experimental demonstration of amplitude squeezing from semi-conductor lasers was soon achieved [6]. Although still requiring rather specialized lasers, many groups now routinely obtain squeezing from semi-conductor lasers.

There are a number of disadvantages to semi-conductor lasers, including their poor beam quality and short coherence length. As a result interest in other possible squeezing mechanisms continued. In the early nineties theoretical investigations which relaxed the other three assumptions found that intrinsic dynamical effects could also lead to amplitude squeezing in the laser output [7–12].

Laser squeezing is of a quite different character to that produced by more "traditional squeezers". For example in parametric amplification the de-amplification of phase fluctuations is accompanied by an equal and opposite amplification of amplitude fluctuations which preserves the uncertainty relationship between the them [13]. However, in laser squeezing the reduction in amplitude fluctuations is not linked to any increase in phase fluctuations. This is allowed because, as already noted, the laser output exhibits excess phase fluctuations. Thus in laser squeezing we take advantage of the fact that the laser output is not in a minimum uncertainty state to suppress amplitude noise. This means though that laser squeezing is always amplitude or intensity squeezing[1].

In this chapter we will investigate the various mechanisms via which lasers can produce squeezing. In the first section we will set up a quantum mechanical theory for a four-level laser and use it to obtain the operator equations of motion. We proceed to obtain the semi-classical steady state. Then by linearization around this steady state and adiabatic elimination of the atomic coherences we solve for the quadrature fluctuation spectra. The adiabatic elimination is justified by assumption (iv). Assumptions (i) through (iii) are not made. In the second section we investigate the various noise sources in the laser model and their importance. We then investigate the mechanisms via which squeezing can be produced by lasers. In the final section we relax assumption (iv) and investigate squeezing from coherent atomic effects.

5.1 The Laser Model

The quantum theory of the laser was developed by the groups of H. Haken, W.E. Lamb and M. Lax during the sixties. In the Lamb [14] approach the photon statistics of the internal field of the laser was computed. Pumping was modeled by the injection of inverted atoms into the laser cavity. In the initial treatments a Poissonian distributed sequence of injected atoms was assumed, but this was later generalized to include sub-Poissonian pumping. Haken [15] and Lax [16] developed techniques for converting operator master equations into c-number Fokker–Planck or equivalent Langevin equations. Linearization of these equations enables internal spectral correlations to be calculated. With the development of the input/output boundary conditions [17] output noise spectra could be calculated from the internal correlations.

More recently linearized input/output approaches have been used in which the operator equations of motion are linearized and solved directly, bypassing the master equation [5,18,19]. We will follow this approach here using the quantum Langevin equation to obtain operator equations of motion directly

[1] Frequency pulling effects, not considered here, can under certain circumstances produce small rotations away from pure amplitude squeezing.

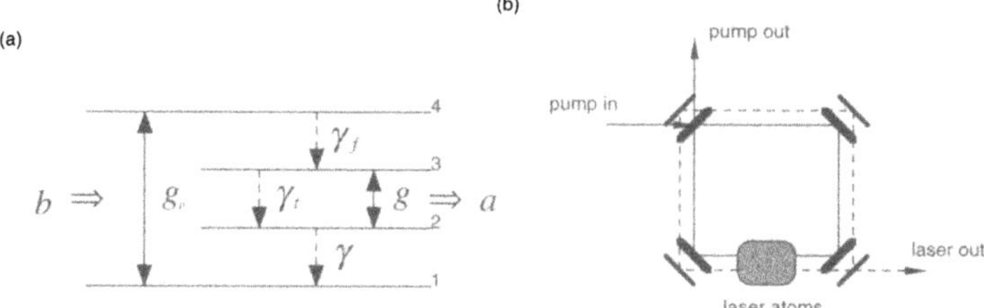

Fig. 5.1. Schematic representation of the laser model showing **(a)** the energy level scheme of the active atoms and **(b)** the cavity arrangement

from the Hamiltonian. From these, linearized equations of motion for the internal quadrature fluctuations will be obtained which, using the boundary conditions, can then be solved in Fourier space for the amplitude and phase noise spectra. The main advantage of this approach is that the physical sources of the various contributions to the spectra can be identified.

Our model is designed to capture the essential features of an optically pumped, continuous wave (CW) atomic laser. Our operator equations include the full dynamics of a four-level atomic energy system coupled to a single lasing mode and a single pump mode. The dipole interactions between the atoms and the two modes are as follows (see Fig. 5.1); the pump mode is resonant with the transition between level $|1\rangle$, the ground state, and level $|4\rangle$, the upper pump level; the laser mode is resonant with the transition between level $|3\rangle$, the upper lasing level, and level $|2\rangle$, the lower lasing level. Both modes are resonant cavity modes. The effects of dissipation and output coupling are modeled by coupling vacuum reservoirs to the various atomic and optical system operators. A colored reservoir is coupled into the pump mode to model the spectral properties of the pump laser.

We make the following simplifying assumptions in our model. We assume a high finesse ring cavity for our laser resonator. For lasers such as diode lasers and fiber lasers this is, in general, not a good assumption. However studies of diode lasers have shown that provided intra-cavity losses are not too large and the laser has a single output coupler, the physics remains essentially the same [20–22]. We neglect spatial variations of the laser and pump field. This has also been shown to have only minor effects [23]. We neglect amplitude/phase correlations due to the linewidth enhancement factor [24]. This is an excellent approximation for solid state lasers but is violated for semi-conductor lasers. Although amplitude/phase coupling leads to excess phase noise it does not effect the amplitude noise with which we are most interested here [25]. Finally we assume single mode operation. Multi-mode operation can have significant effects [26] however various experimental techniques can be used to enforce single mode operation [27,28].

5.1.1 The Hamiltonian

We consider N_A four-level atoms (see Fig. 5.1) interacting with two optical ring cavity modes. The first mode, represented by the annihilation and creation operators $\hat{a}$ and $\hat{a}^\dagger$ respectively, is the lasing mode, with self energy

$$\hat{H}_a = \hbar\omega_a \hat{a}^\dagger \hat{a} \ . \tag{5.1}$$

It interacts with the active atoms via the resonant Jaynes–Cummings Hamiltonian

$$\hat{H}_{\text{rev1}} = i\hbar g(\hat{a}^\dagger \hat{\sigma}_{23} \ - \hat{a}\, \hat{\sigma}_{23}^{+} \ ) \ , \tag{5.2}$$

where carets indicate operators, g is the dipole coupling strength between the atoms and the cavity and $\hat{\sigma}_{ij}$ and $\hat{\sigma}_{ij}^{+}$ are the collective hermitean conjugate atomic lowering and raising operators between the ith and jth levels. The field phase factors have been absorbed into the definition of the atomic operators. The second mode, represented by the annihilation and creation operators $\hat{b}$ and $\hat{b}^\dagger$ respectively, is the pump mode with self energy

$$\hat{H}_b = \hbar\omega_b \hat{b}^\dagger \hat{b} \ . \tag{5.3}$$

It interacts with the active atoms via the resonant Jaynes–Cummings Hamiltonian

$$\hat{H}_{\text{rev2}} = i\hbar g_p(\hat{b}^\dagger \hat{\sigma}_{14} - \hat{b}\, \hat{\sigma}_{14}^{+} \ ) \ , \tag{5.4}$$

where g_p is the dipole coupling strength between the atoms and the cavity.

We couple the atoms and cavities to reservoirs to describe the irreversible transitions of the system. The reservoir self energies are

$$\begin{aligned}\hat{H}_{\text{res}} = \hbar \int_{-\infty}^{\infty} d\omega\ \omega \Big\{ & \hat{A}^\dagger(\omega)\hat{A}(\omega) + \hat{A}_l^\dagger(\omega)\hat{A}_l(\omega) + \hat{B}^\dagger(\omega)\hat{B}(\omega) \\ & + \hat{C}_f^\dagger(\omega)\hat{C}_f(\omega) + \hat{C}_t^\dagger(\omega)\hat{C}_t(\omega) + \hat{C}^\dagger(\omega)\hat{C}(\omega) \\ & + \hat{C}_P^\dagger(\omega)\hat{C}_P(\omega) + \hat{C}_Q^\dagger(\omega)\hat{C}_Q(\omega) \Big\} \ ,\end{aligned} \tag{5.5}$$

where reservoir operators are indicated by capitals. Under the Markov approximation that the coupling constants between the system and the reservoirs are frequency independent the interaction Hamiltonian between the system and the reservoirs is

$$\begin{aligned}\hat{H}_{\text{irrev}} = ih \int_{-\infty}^{\infty} d\omega \Bigg\{ & \sqrt{\frac{\kappa_m}{2\pi}} \left(\hat{A}(\omega)\, \hat{a}^\dagger - \hat{A}^\dagger(\omega)\, \hat{a}\right) \\ & + \sqrt{\frac{\kappa_l}{2\pi}} \left(\hat{A}_l(\omega)\, \hat{a}^\dagger - \hat{A}_l^\dagger(\omega)\, \hat{a}\right) \\ & + \sqrt{\frac{\kappa_b}{2\pi}} \left(\hat{B}(\omega)\, \hat{b}^\dagger - \hat{B}^\dagger(\omega)\, \hat{b}\right)\end{aligned}$$

$$+\sqrt{\frac{\gamma_f}{2\pi}}\left(\hat{C}_f(\omega)\hat{\sigma}_{34}^{+}-\hat{C}_f^{\dagger}(\omega)\hat{\sigma}_{34}\right)$$
$$+\sqrt{\frac{\gamma_t}{2\pi}}\left(\hat{C}_t(\omega)\hat{\sigma}_{23}^{+}-\hat{C}_t^{\dagger}(\omega)\hat{\sigma}_{23}\right)$$
$$+\sqrt{\frac{\gamma}{2\pi}}\left(\hat{C}(\omega)\hat{\sigma}_{12}^{+}-\hat{C}^{\dagger}(\omega)\hat{\sigma}_{12}\right)$$
$$+\sqrt{\frac{\gamma_P}{4\pi}}\left(\hat{C}_P(\omega)-\hat{C}_P^{\dagger}(\omega)\right)(\hat{\sigma}_3-\hat{\sigma}_2)$$
$$+\sqrt{\frac{\gamma_Q}{4\pi}}\left(\hat{C}_Q(\omega)-\hat{C}_Q^{\dagger}(\omega)\right)(\hat{\sigma}_4-\hat{\sigma}_1)\Bigg\}\ , \tag{5.6}$$

where $\hat{\sigma}_i$ is the collective population operator for the ith level. Included in our laser model are atomic spontaneous emission from level $|4\rangle$ to level $|3\rangle$, from level $|3\rangle$ to level $|2\rangle$, and from level $|2\rangle$ to level $|1\rangle$, at rates γ_f, γ_t and γ, respectively. The rate of collisional or lattice induced phase decay of the lasing coherence and pump coherence are γ_P and γ_Q respectively. The laser cavity damping rate due to the output mirror is κ_m. The laser cavity damping rate due to other losses is κ_l. The total laser cavity damping rate is κ. The pump mode damping and pumping occurs through an input output mirror with a rate κ_b. The total Hamiltonian for the laser plus environment is

$$H = H_a + H_{\mathrm{rev1}} + H_b + H_{\mathrm{rev2}} + H_{\mathrm{res}} + H_{\mathrm{irrev}}\ . \tag{5.7}$$

5.1.2 The Quantum Langevin Equation

Following standard procedure [15,17] we define the input fields

$$\delta\hat{A}(t) = -\frac{1}{\sqrt{2\pi}}\int d\omega\ e^{-i\omega(t-t_0)}\hat{A}_0(\omega)\ ,$$
$$\delta\hat{A}_l(t) = -\frac{1}{\sqrt{2\pi}}\int d\omega\ e^{-i\omega(t-t_0)}\hat{A}_{l0}(\omega)\ ,$$
$$\hat{B}(t) = -\frac{1}{\sqrt{2\pi}}\int d\omega\ e^{-i\omega(t-t_0)}\hat{B}_0(\omega)\ ,$$
$$\delta\hat{C}_f(t) = -\frac{1}{\sqrt{2\pi}}\int d\omega\ e^{-i\omega(t-t_0)}\hat{C}_{f0}(\omega)\ ,$$
$$\delta\hat{C}_t(t) = -\frac{1}{\sqrt{2\pi}}\int d\omega\ e^{-i\omega(t-t_0)}\hat{C}_{t0}(\omega)\ ,$$
$$\delta\hat{C}(t) = -\frac{1}{\sqrt{2\pi}}\int d\omega\ e^{-i\omega(t-t_0)}\hat{C}_0(\omega)\ ,$$
$$\delta\hat{C}_P(t) = -\frac{1}{\sqrt{2\pi}}\int d\omega\ e^{-i\omega(t-t_0)}\hat{C}_{P0}(\omega)\ ,$$
$$\delta\hat{C}_Q(t) = -\frac{1}{\sqrt{2\pi}}\int d\omega\ e^{-i\omega(t-t_0)}\hat{C}_{Q0}(\omega)\ , \tag{5.8}$$

where $\hat{A}_0, \ldots, \hat{C}_{Q0}$ are the values of $\hat{A}, \ldots, \hat{C}_Q$ at some arbitrary initial time, t_0. Zero point fields are indicated by δ's (e.g. $\langle \delta C \rangle = 0$). In the interaction picture (or equivalently in a frame rotating at the optical frequency) it can be shown that the dynamics of any system operator ($\hat{O}$) are given by the quantum Langevin equation

$$\dot{\hat{O}} = -\frac{i}{\hbar} \sum_{i=1,2} \left[\hat{O}, H_{\mathrm{rev}\ i}\right] - \sum_{j=1..7} \left\{ \left[\hat{O}, \hat{a}_j^\dagger\right] \left(\frac{\gamma_j}{2}\hat{a}_j - \sqrt{\gamma_j}\hat{A}_j\right) - \left(\frac{\gamma_j}{2}\hat{a}_j^\dagger - \sqrt{\gamma_j}\hat{A}_j^\dagger\right) \left[\hat{O}, \hat{a}_j\right] \right\} , \tag{5.9}$$

where we have used the notation for the system operators

$$(\hat{a}_1, \hat{a}_2, \hat{a}_3, \hat{a}_4, \hat{a}_5, \hat{a}_6, \hat{a}_7, \hat{a}_8) \equiv \left(\hat{a}, \hat{a}, \hat{b}, \hat{\sigma}_{34}, \hat{\sigma}_{23}, \hat{\sigma}_{12}, \hat{\sigma}_3 - \hat{\sigma}_2, \hat{\sigma}_4 - \hat{\sigma}_1\right) , \tag{5.10}$$

for the input fields

$$\left(\hat{A}_1, \hat{A}_2, \hat{A}_3, \hat{A}_4, \hat{A}_5, \hat{A}_6, \hat{A}_7, \hat{A}_8\right) \equiv \left(\delta\hat{A}, \delta\hat{A}_l, \hat{B}, \delta\hat{C}_f, \delta\hat{C}_t, \delta\hat{C}, \delta\hat{C}_P, \delta\hat{C}_Q\right) , \tag{5.11}$$

and for the coupling constants

$$(\gamma_1, \gamma_2, \gamma_3, \gamma_4, \gamma_5, \gamma_6, \gamma_7, \gamma_8) \equiv (\kappa_m, \kappa_l, \kappa_b, \gamma_f, \gamma_t, \gamma, \gamma_P, \gamma_Q) . \tag{5.12}$$

Using (5.9) we can write down the following operator equations of motion for the laser

$$\dot{\hat{a}} = g\hat{\sigma}_{23} - \frac{1}{2}\kappa\hat{a} + \sqrt{\kappa_m}\delta\hat{A}_m + \sqrt{\kappa_l}\delta\hat{A}_l , \tag{5.13}$$

$$\dot{\hat{b}} = g_p\hat{\sigma}_{14} - \frac{1}{2}\kappa_b\hat{b} + \sqrt{\kappa_b}\hat{B} , \tag{5.14}$$

$$\dot{\hat{\sigma}}_{14} = g_p(\hat{\sigma}_4 - \hat{\sigma}_1)\hat{b} - \left(\frac{\gamma_f}{2} + \gamma_Q\right)\hat{\sigma}_{14} + \sqrt{\gamma_f}\hat{\sigma}_{13}\delta\hat{C}_f - \sqrt{\gamma}\hat{\sigma}_{24}\delta\hat{C} - \sqrt{\gamma_Q}\left(\delta\hat{C}_Q - \delta\hat{C}_Q^\dagger\right)\hat{\sigma}_{14} , \tag{5.15}$$

$$\dot{\hat{\sigma}}_{23} = g(\hat{\sigma}_3 - \hat{\sigma}_2)\hat{a} - \frac{1}{2}(2\gamma_P + \gamma_t + \gamma)\hat{\sigma}_{23} - \sqrt{\gamma_t}(\hat{\sigma}_3 - \hat{\sigma}_2)\delta\hat{C}_t + \sqrt{\gamma_f}\delta\hat{C}_f^\dagger\hat{\sigma}_{24} + \sqrt{\gamma}\delta\hat{C}^\dagger\hat{\sigma}_{13} - \sqrt{\gamma_P}\left(\delta\hat{C}_P - \delta\hat{C}_P^\dagger\right)\hat{\sigma}_{23} , \tag{5.16}$$

$$\dot{\hat{\sigma}}_1 = g_p\left(\hat{\sigma}_{14}\hat{b}^\dagger + \hat{\sigma}_{14}^+\hat{b}\right) + \gamma\hat{\sigma}_2 - \sqrt{\gamma}\left(\delta\hat{C}^\dagger\hat{\sigma}_{12} + \hat{\sigma}_{12}^+\delta\hat{C}\right) , \tag{5.17}$$

$$\dot{\hat{\sigma}}_2 = g(\hat{\sigma}_{23}\hat{a}^\dagger + \hat{\sigma}_{23}^+\hat{a}) + \gamma_t\hat{\sigma}_3 - \gamma\hat{\sigma}_2 + \sqrt{\gamma}\left(\delta\hat{C}^\dagger\hat{\sigma}_{12} + \hat{\sigma}_{12}^+\delta\hat{C}\right) - \sqrt{\gamma_t}\left(\delta\hat{C}_t^\dagger\hat{\sigma}_{23} + \hat{\sigma}_{23}^+\delta\hat{C}_t\right) , \tag{5.18}$$

$$\dot{\hat{\sigma}}_3 = -g\left(\hat{\sigma}_{23}\hat{a}^\dagger + \hat{\sigma}_{23}^+\hat{a}\right) - \gamma_t\hat{\sigma}_3 + \gamma_f\hat{\sigma}_4 - \sqrt{\gamma_f}\left(\delta\hat{C}_f^\dagger\hat{\sigma}_{34} + \hat{\sigma}_{34}^+\delta\hat{C}_f\right) + \sqrt{\gamma_t}\left(\delta\hat{C}_t^\dagger\hat{\sigma}_{23} + \hat{\sigma}_{23}^+\delta\hat{C}_t\right) , \quad (5.19)$$

$$\dot{\hat{\sigma}}_4 = -g_p\left(\hat{\sigma}_{14}\hat{b}^\dagger + \hat{\sigma}_{14}^+\hat{b}\right) - \gamma_f\hat{\sigma}_4 + \sqrt{\gamma_f}\left(\hat{\sigma}_{34}\delta\hat{C}_f^\dagger\hat{\sigma}_{34} + \hat{\sigma}_{34}^+\delta\hat{C}_f\right) . \quad (5.20)$$

5.1.3 Laser Rate Equations

We now make a number of assumptions consistent with most laser systems. In Sect. 5.3 we will relax some of these assumptions.

(i) the pump cavity decays very rapidly (κ_b very large). This is consistent with the normal experimental situation in which there is no pump cavity.

(ii) the upper pump level decays very rapidly (γ_f very large). This is a desirable condition for efficient pumping and is normally satisfied.

(iii) the phase decay of the laser and pump coherences are very rapid (γ_P, γ_Q very large). This is also a good assumption for most lasers.

These assumptions mean that the pump mode, the upper pump level, and the atomic coherences all evolve on much shorter time scales than the other variables. Hence we can make the approximation of adiabatically eliminating the equations for $\hat{b}$, $\hat{\sigma}_4$, $\hat{\sigma}_{14}$ and $\hat{\sigma}_{23}$ by setting their time derivatives to zero. From (5.13)–(5.20), we obtain

$$\dot{\hat{a}} = \frac{G}{2}\left(\hat{\sigma}_3 - \hat{\sigma}_2\right)\hat{a} - \frac{1}{2}\kappa\hat{a} + \sqrt{\kappa_m}\delta\hat{A}_m + \sqrt{\kappa_l}\delta\hat{A}_l - \sqrt{G}\delta\hat{A}_p , \quad (5.21)$$

$$\begin{aligned}\dot{\hat{\sigma}}_1 = &-\frac{2G_p\kappa_b\hat{\sigma}_1}{\left(\frac{1}{2}\kappa_b + 2G_p\hat{\sigma}_1\right)^2}\left(\hat{B}^\dagger\hat{B} + \hat{B}\hat{B}^\dagger\right) + \gamma\hat{\sigma}_2 - \sqrt{\gamma}\delta\hat{A} \\ &+\frac{2G_p\kappa_b}{\left(\frac{1}{2}\kappa_b + 2G_p\hat{\sigma}_1\right)^2}\left(\delta\hat{A}_q^\dagger\delta\hat{A}_q + \delta\hat{A}_q\delta\hat{A}_q^\dagger\right) \\ &+\sqrt{4G_p\kappa_b}\frac{\left(\frac{1}{2}\kappa_b - 2G_p\hat{\sigma}_1\right)}{\left(\frac{1}{2}\kappa_b + 2G_p\hat{\sigma}_1\right)^2}\left(\delta\hat{A}_q^\dagger\hat{B} + \delta\hat{A}_q\hat{B}^\dagger\right) , \end{aligned} \quad (5.22)$$

$$\begin{aligned}\dot{\hat{\sigma}}_2 = &\frac{G}{2}\left(\hat{\sigma}_3 - \hat{\sigma}_2\right)\left(\hat{a}\hat{a}^\dagger + \hat{a}^\dagger\hat{a}\right) - \sqrt{G}\left(\delta\hat{A}_p\hat{a}^\dagger + \delta\hat{A}_p^\dagger\hat{a}\right) \\ &+\gamma_t\hat{\sigma}_3 - \gamma\hat{\sigma}_2 + \sqrt{\gamma}\delta\hat{A} - \sqrt{\gamma_t}\delta\hat{A}_t , \end{aligned} \quad (5.23)$$

$$\begin{aligned}\dot{\hat{\sigma}}_3 = &\frac{G}{2}\left(\hat{\sigma}_3 - \hat{\sigma}_2\right)\left(\hat{a}\hat{a}^\dagger + \hat{a}^\dagger\hat{a}\right) - \sqrt{G}\left(\delta\hat{A}_p\hat{a}^\dagger + \delta\hat{A}_p^\dagger\hat{a}\right) \\ &+\frac{2G_p\kappa_b\hat{\sigma}_1}{\left(\frac{1}{2}\kappa_b + 2G_p\hat{\sigma}_1\right)^2}\left(\hat{B}^\dagger\hat{B} + \hat{B}\hat{B}^\dagger\right) - \gamma_t\hat{\sigma}_3 + \sqrt{\gamma_t}\delta\hat{A}_t \\ &-\frac{2G_p\kappa_b}{\left(\frac{1}{2}\kappa_b + 2G_p\hat{\sigma}_1\right)^2}\left(\delta\hat{A}_q^\dagger\delta\hat{A}_q + \delta\hat{A}_q\delta\hat{A}_q^\dagger\right) \\ &-\sqrt{4G_p\kappa_b}\frac{\left(\frac{1}{2}\kappa_b - 2G_p\hat{\sigma}_1\right)}{\left(\frac{1}{2}\kappa_b + 2G_p\hat{\sigma}_1\right)^2}\left(\delta\hat{A}_q^\dagger\hat{B} + \delta\hat{A}_q\hat{B}^\dagger\right) , \end{aligned} \quad (5.24)$$

where

$$G = \frac{2g^2}{\gamma_P} \tag{5.25}$$

and

$$G_p = \frac{g_p^2}{2\gamma_Q} \tag{5.26}$$

are the stimulated emission/absorption rates per photon for the laser and pump modes respectively. Also

$$\delta\hat{\Lambda} = \delta\hat{C}^\dagger\hat{\sigma}_{12} + \hat{\sigma}_{12}^{+}\delta\hat{C} \ , \tag{5.27}$$

$$\delta\hat{\Lambda}_t = \delta\hat{C}^\dagger\hat{\sigma}_{23} + \hat{\sigma}_{23}^{+}\delta\hat{C} \ , \tag{5.28}$$

$$\delta\hat{\Lambda}_p = \left(\delta\hat{C}_P - \delta\hat{C}_P^\dagger\right)\sigma_{23} \ , \tag{5.29}$$

$$\delta\hat{\Lambda}_q = \left(\delta\hat{C}_Q - \delta\hat{C}_Q^\dagger\right)\sigma_{14} \ , \tag{5.30}$$

are noise terms originating from the various vacuum reservoirs. They all have zero first order expectation values ($\langle\delta\hat{\Lambda}_i\rangle = 0$). Using the operator relationships $\hat{\sigma}_{ij}\hat{\sigma}_{ij}^{+} = \hat{\sigma}_i$, $\hat{\sigma}_{ij}^{+}\hat{\sigma}_{ij} = \hat{\sigma}_j$ and $\hat{\sigma}_{ij}^{+}\hat{\sigma}_{ij}^{+} = \hat{\sigma}_{ij}\hat{\sigma}_{ij} = 0$, the commutator relations between system and input fields, and the properties of the vacuum fields, the following second order expectation values can be calculated (see Sect. 5.5)

$$\langle\delta\hat{\Lambda}(t)\delta\hat{\Lambda}(t')\rangle = \langle\hat{\sigma}_2\rangle\delta(t-t') \ , \tag{5.31}$$

$$\langle\delta\hat{\Lambda}_t(t)\delta\hat{\Lambda}_t(t')\rangle = \langle\hat{\sigma}_3\rangle\delta(t-t') \ , \tag{5.32}$$

$$\langle\delta\hat{\Lambda}_Q(t)\delta\hat{\Lambda}_Q(t')\rangle = \langle\hat{\sigma}_1 + \hat{\sigma}_4\rangle\delta(t-t') \ , \tag{5.33}$$

$$\langle\delta\hat{\Lambda}_P(t)\delta\hat{\Lambda}_P(t')\rangle = \langle\hat{\sigma}_3 + \hat{\sigma}_2\rangle\delta(t-t') \ , \tag{5.34}$$

$$\langle\delta\hat{\Lambda}_{\overline{P}}^{-}(t)\delta\hat{\Lambda}_{\overline{P}}^{-}(t')\rangle = \langle\hat{\sigma}_3 + \hat{\sigma}_2\rangle\delta(t-t') \ , \tag{5.35}$$

where $\delta\hat{\Lambda}_Q = \delta\hat{\Lambda}_q^\dagger + \delta\hat{\Lambda}_q$, $\delta\hat{\Lambda}_P = \delta\hat{\Lambda}_p^\dagger + \delta\hat{\Lambda}_p$ and $\delta\hat{\Lambda}_{\overline{P}}^{-} = i(\delta\hat{\Lambda}_p^\dagger - \delta\hat{\Lambda}_p)$.

By taking expectation values of (5.21)–(5.24) and using the semi-classical approximation that expectation values can be factorized we obtain the following laser rate equations

$$\dot{\alpha} = \frac{G}{2}\left(J_3 - J_2\right)\alpha - \frac{1}{2}\kappa\alpha \ , \tag{5.36}$$

$$\dot{J}_1 = -\Gamma J_1 + \gamma J_2 \ , \tag{5.37}$$

$$\dot{J}_2 = G\left(J_3 - J_2\right)\alpha\alpha^\star + \gamma_t J_3 - \gamma J_2 \ , \tag{5.38}$$

$$\dot{J}_3 = -G\left(J_3 - J_2\right)\alpha\alpha^\star - \gamma_t J_3 + \Gamma J_1 \ , \tag{5.39}$$

where $\alpha = \langle\hat{a}\rangle$ and $J_i = \langle\sigma_i\rangle$ are the expectation values of the laser mode and atomic populations and

$$\Gamma = \frac{4G_p\kappa_b}{\left(\frac{1}{2}\kappa_b + 2G_pJ_1\right)^2}\langle\hat{B}^\dagger\hat{B}\rangle \tag{5.40}$$

is the pump rate. Normally one assumes either; the ground state is not significantly depleted by the pumping ($J_1 \approx constant$) or the pump beam itself is not significantly depleted ($J_1 \ll \kappa_b/(2G_p)$). In both cases the pump rate then becomes proportional to the pump beam intensity.

The semi-classical steady state solution can be obtained by setting the time derivatives to zero in (5.36)–(5.39). The full steady state solutions are given in Sect. 5.6. We wish first to consider conditions under which laser squeezing is optimized. Hence we consider laser operation well above threshold such that stimulated emission has become the most rapid rate in the problem, i.e. $G|\alpha|^2 \gg \Gamma, \gamma, \gamma_t, \kappa$. Also we assume that losses are negligible such that $\Gamma, \gamma \gg \gamma_t$ and $\kappa_m \cong \kappa$. Under these approximations the steady state value of the internal photon number is given by

$$n = |\alpha_0|^2 = \frac{\Gamma \gamma N}{\kappa (2\Gamma + \gamma)} , \tag{5.41}$$

where the subscript "0" indicates steady-state values. The phase of the laser field cannot be determined from (5.36)–(5.39) as α_0 cancels and hence the equations can only be solved for $|\alpha_0|^2$. Physically this indicates that the laser phase is meta-stable and will tend to drift in time. This effect is known as phase diffusion. The output photon flux is just $n_{\text{out}} = \kappa_m n$. The relationship between the pump rate and the pump field flux is given by

$$n_b = B^2 = \left(\frac{\gamma}{2\Gamma + \gamma} + \frac{\kappa_b}{4G_p} \right)^2 \frac{\Gamma G_p}{\kappa_b} . \tag{5.42}$$

The non-linearity of this relationship arises both from depletion of the pump field and depletion of the ground state. In extreme cases this non-linearity can result in bi-stable laser operation [29].

5.1.4 Linearized Fluctuation Equations

Suppose we write the annihilation operator describing a particular light field in the form

$$\hat{A}(t) = (|A_0| + \delta\hat{A}(t))e^{i\epsilon(t)} , \tag{5.43}$$

where $|A_0|$ is the absolute value of the semi-classical, steady state coherent amplitude of the field (a c-number), whilst $\delta\hat{A}(t)$ is an operator carrying all the temporal and quantum information. The phase $\epsilon(t)$ is an arbitrary function of time. If this field is directly detected we will obtain a photocurrent proportional to

$$\hat{A}(t)^\dagger \hat{A}(t) = |A_0|^2 + |A_0|(\delta\hat{A}(t)^\dagger + \delta\hat{A}(t)) + \delta\hat{A}(t)^\dagger \delta\hat{A}(t) . \tag{5.44}$$

Notice first that direct detection is phase insensitive, thus the photocurrent is not a function of $\epsilon(t)$. Now suppose that the field amplitude is macroscopic and the temporal behavior of the operators can be considered small perturbations to this steady state. Under such conditions the contribution to the dynamics from the last term in (5.44) can be neglected and we can write the linearized expression

$$\hat{A}(t)^{\dagger}\hat{A}(t) = |A_0|^2 + |A_0|\delta\hat{X}(t) \ , \tag{5.45}$$

where all the dynamics are carried by the amplitude quadrature fluctuations $\delta\hat{X}(t) = \delta\hat{A}(t)^{\dagger} + \delta\hat{A}(t)$.

Thus we are motivated to proceed by writing our solutions in the form

$$\begin{aligned} \hat{a}\,(t) &= (|\alpha_0| + \delta\hat{a}\,(t))e^{i\phi(t)} \ , \quad \hat{\sigma}_i\,(t) = J_{i0} + \delta\hat{\sigma}_i\,(t) \ , \\ \hat{B}\,(t) &= (|B_0| + \delta\hat{B}\,(t))e^{i\phi_b(t)} \ , \end{aligned} \tag{5.46}$$

where $|\alpha_0|$ and J_{i0} are the stable semi-classical steady state solutions to (5.36)–(5.39) for the absolute value of the laser field amplitude and population of level i respectively. $|B_0|$ is the absolute value of the coherent amplitude of the pump mode. The quantum fluctuations, $\delta\hat{a}$, $\delta\hat{\sigma}_i$ and $\delta\hat{B}$, are considered small, zero-mean perturbations to the steady state. This is normally an excellent assumption for macroscopic, continuous wave lasers. As we saw above the phase of the field is of no consequence for direct detection of intensity fluctuations. Thus we can set it constant and for simplicity we let $\phi = \phi_b = 0$.

By substituting (5.46) into (5.21)–(5.24) and retaining only terms linear in the fluctuations we obtain the following system of linear differential equations for the quadrature and population fluctuations

$$\begin{aligned} \delta\dot{\hat{x}}_a &= G\,(\delta\hat{\sigma}_3 - \delta\hat{\sigma}_2)\,\alpha_0 \\ &\quad + \sqrt{\kappa_m}\,\delta\hat{X}_{Am} + \sqrt{\kappa_l}\,\delta\hat{X}_{Al} - \sqrt{G}\delta\hat{A}_P \ , \end{aligned} \tag{5.47}$$

$$\delta\dot{\hat{y}}_a = \sqrt{\kappa_m}\,\delta\hat{Y}_{Am} + \sqrt{\kappa_l}\,\delta\hat{Y}_{Al} + \sqrt{G}\delta\hat{A}_{\bar{P}}^{-} \ , \tag{5.48}$$

$$\begin{aligned} \delta\dot{\hat{\sigma}}_1 &= -\Gamma\sqrt{1-\eta}\delta\hat{\sigma}_1 + \gamma\,\delta\hat{\sigma}_2 - \sqrt{\Gamma J_{10}\eta}\delta\hat{X}_B \\ &\quad + \sqrt{\gamma}\delta\hat{A} + \sqrt{\Gamma\,(1-\eta)}\delta\hat{A}_Q \ , \end{aligned} \tag{5.49}$$

$$\begin{aligned} \delta\dot{\hat{\sigma}}_2 &= G\,(\delta\hat{\sigma}_3 - \delta\hat{\sigma}_2)\,\alpha_0^2 + G\,(J_{30} - J_{20})\,\alpha_0\,\delta\hat{x}_a + \gamma_t\,\delta\hat{\sigma}_3 - \gamma\,\delta\hat{\sigma}_2 \\ &\quad + \sqrt{\gamma}\delta\hat{A} - \sqrt{\gamma_t}\delta\hat{A}_t - \sqrt{G}\alpha_0\,\delta\hat{A}_P \ , \end{aligned} \tag{5.50}$$

$$\begin{aligned} \delta\dot{\hat{\sigma}}_3 &= -G\,(\delta\hat{\sigma}_3 - \delta\hat{\sigma}_2)\,\alpha_0^2 \\ &\quad - G\,(J_{30} - J_{20})\,\alpha_0\,\delta\hat{x}_a - \gamma_t\,\delta\hat{\sigma}_3 + \Gamma\sqrt{1-\eta}\delta\hat{\sigma}_1 \\ &\quad + \sqrt{\Gamma J_1\eta}\delta\hat{X}_B + \sqrt{\gamma}\delta\hat{A} - \sqrt{\Gamma\,(1-\eta)}\delta\hat{A}_Q \\ &\quad + \sqrt{\gamma_t}\delta\hat{A}_t + \sqrt{G}\alpha_0\,\delta\hat{A}_P \ , \end{aligned} \tag{5.51}$$

where the amplitude quadrature fluctuations of the fields are defined by

$$\delta\hat{x}_a = \delta\hat{a} + \delta\hat{a}^{\dagger}, \quad \delta\hat{X}_{Am} = \delta\hat{A}_m + \delta\hat{A}_m^{\dagger}, \ldots\ldots \text{etc} \ , \tag{5.52}$$

while the phase quadrature fluctuations are given by

$$\delta\hat{y_a} = -i(\delta\hat{a} - \delta\hat{a}^{\dagger}) \ , \quad \delta\hat{Y}_{Am} = -i(\delta\hat{A}_m - \delta\hat{A}_m^{\dagger}), \ldots\ldots \text{etc} \ . \tag{5.53}$$

Also we have

$$\sqrt{1-\eta} = \frac{\kappa_b - 4G_p J_{10}}{\kappa_b + 4G_p J_{10}} , \tag{5.54}$$

and

$$\eta = \frac{4G_p J_{10} \kappa_b}{\left(2G_p J_{10} + \frac{1}{2}\kappa_b\right)^2} . \tag{5.55}$$

Physically, η is the efficiency with which pump light is absorbed by the lasing atoms. The boundary condition at the output mirror is [17]

$$\hat{A}_{\text{out}} = \sqrt{\kappa_m}\hat{a} - \delta\hat{A}_m , \tag{5.56}$$

where $\hat{A}_{\text{out}}$ is the laser output field. In terms of the amplitude quadrature fluctuations

$$\delta\hat{X}_{\text{out}} = \sqrt{\kappa_m}\delta\hat{x}_a - \delta\hat{X}_{Am} , \tag{5.57}$$

and similarly for the phase quadrature fluctuations

$$\delta\hat{Y}_{\text{out}} = \sqrt{\kappa_m}\delta\hat{y}_a - \delta\hat{Y}_{Am} . \tag{5.58}$$

5.1.5 Noise Spectra

In frequency space (5.47)–(5.51) can be solved for the quadrature fluctuations of the output fields using (5.57). The amplitude noise spectrum (V_{out}) is given by

$$V_{\text{out}}(\omega) = \langle|\delta X_{\text{out}}|^2\rangle \tag{5.59}$$

where the lack of carets indicate Fourier transforms. The spectrum can be solved for using the frequency space equivalents of (5.31)–(5.35), namely

$$\langle\delta\Lambda(\omega)\delta\Lambda(\omega')\rangle = \langle\hat{\sigma}_2\rangle\delta(\omega-\omega') , \tag{5.60}$$

$$\langle\delta\Lambda_t(\omega)\delta\Lambda_t(\omega')\rangle = \langle\hat{\sigma}_3\rangle\delta(\omega-\omega') , \tag{5.61}$$

$$\langle\delta\Lambda_Q(\omega)\delta\Lambda_Q(\omega')\rangle = \langle\hat{\sigma}_1+\hat{\sigma}_4\rangle\delta(\omega-\omega') , \tag{5.62}$$

$$\langle\delta\Lambda_P(\omega)\delta\Lambda_P(\omega')\rangle = \langle\hat{\sigma}_3+\hat{\sigma}_2\rangle\delta(\omega-\omega') , \tag{5.63}$$

$$\langle\delta\Lambda_P^-(\omega)\delta\Lambda_P^-(\omega')\rangle = \langle\hat{\sigma}_3+\hat{\sigma}_2\rangle\delta(\omega-\omega') . \tag{5.64}$$

The full expression for the amplitude spectrum is given in Sect. 5.7. Well above threshold with negligible losses (see conditions given for (5.41)) the solution reduces to the following expression for the amplitude noise spectrum

$$V_{\text{out}}(\omega) = \frac{\omega^2}{\omega^2+(2\kappa)^2} + \frac{(2\kappa)^2\left(\omega^2+\gamma^2\right)\left[(1-\eta)+\eta V_p\right]}{\left[2\kappa\left(\gamma+2\tilde{\Gamma}\right)-2\omega^2\right]^2+\omega^2\left(4\kappa+\gamma+2\tilde{\Gamma}\right)^2}$$
$$+\frac{(2\kappa)^2\left[\omega^2+\left(2\tilde{\Gamma}\right)^2\right]}{\left[2\kappa\left(\gamma+2\tilde{\Gamma}\right)-2\omega^2\right]^2+\omega^2\left(4\kappa+\gamma+2\tilde{\Gamma}\right)^2} , \tag{5.65}$$

where $\tilde{\Gamma} = \Gamma\sqrt{1-\eta}$ and $V_p = \langle|\delta X_B|^2\rangle$ is the amplitude noise spectrum of the pump. In Sect. 5.2 we will investigate the various ways squeezing can arise from (5.65).

We can also calculate the phase fluctuation spectrum from

$$V_{\mathrm{out}}^{-}(\omega) = \langle|\delta X_{\mathrm{out}}^{-}|^2\rangle \ . \tag{5.66}$$

The phase fluctuation spectrum can be obtained by making a phase sensitive measurement of the field with respect to a stable phase reference (e.g. using homodyne detection [13]). Our assumption of constant classical phase ($\phi(t) = 0$) will only be valid for the phase spectrum if the fluctuations around that constant phase are small. The solution is given by

$$V_{\mathrm{out}}^{-}(\omega) = \frac{\omega^2 + 2\,(2\kappa)^2}{\omega^2} \ . \tag{5.67}$$

At very high frequencies the phase spectrum tends to the quantum noise level, $V_{\mathrm{out}}^{-}(\omega) = 1$. As we move to lower and lower frequencies the phase noise becomes larger and larger, eventually diverging at zero frequency. The divergence of the phase noise spectrum at zero frequency indicates that our linearization breaks down for sufficiently small Fourier frequencies[2]. On long enough times scales (low enough frequencies) the phase drifts significantly, thus invalidating the linear approximation. This is the signature of laser phase diffusion. For our purposes it is sufficient to note that the phase quadrature fluctuations at low frequencies are very large. Thus squeezing of the amplitude fluctuations at low frequencies is allowed by the uncertainty principle.

5.1.6 Phenomenological Semi-Classical Equations

It is worth noting that provided only a linearized solution is sought then the following semi-classical set of equations contains the same information as the fully quantum mechanical ones (5.21)–(5.24)

$$\begin{aligned} \dot{\alpha} &= \frac{1}{2}G\,(J_3 - J_2)\,\alpha - \frac{1}{2}\kappa\alpha + \sqrt{\kappa_m}\delta A_m + \sqrt{\kappa_l}\delta A_l \\ &\quad - \sqrt{G(J_3 + J_2)}\delta A_p \ , \end{aligned} \tag{5.68}$$

$$\begin{aligned} \dot{J}_1 &= -\frac{4G_p\kappa_b J_1}{\left(\frac{1}{2}\kappa_b + 2G_p J_1\right)^2}B^2 - \sqrt{\gamma J_2}\delta x \\ &\quad + \sqrt{4G_p\kappa_b}\frac{\left(\frac{1}{2}\kappa_b - 2G_p J_1\right)}{\left(\frac{1}{2}\kappa_b + 2G_p J_1\right)^2}\sqrt{J_4 + J_1}B\delta x_q + \gamma J_2 \ , \end{aligned} \tag{5.69}$$

[2] With ≈ 10 mW detected power, linearization will break down in frequency ranges where noise levels exceed about 70 dB above the QNL. The intensity noise of a free-running single-mode solid state laser will typically be below this level for all practical frequencies. The phase noise, on the other hand, will typically exceed this level at frequencies below about 10 KHz.

$$\dot{J}_2 = G(J_3 - J_2)\alpha^2 - \sqrt{G(J_3+J_2)}\alpha\delta x_p + \gamma_t J_3 - \gamma J_2 + \sqrt{\gamma J_2}\delta x - \sqrt{\gamma_t J_3}\delta x_t \,, \tag{5.70}$$

$$\dot{J}_3 = -\gamma_t J_3 + G(J_3 - J_2)\alpha^2 - \sqrt{G(J_3+J_2)}\alpha\delta x_p + \frac{4G_p\kappa_b J_1}{\left(\frac{1}{2}\kappa_b + 2G_p J_1\right)^2}B^2 - \sqrt{4G_p\kappa_b}\frac{\left(\frac{1}{2}\kappa_b - 2G_p J_1\right)}{\left(\frac{1}{2}\kappa_b + 2G_p J_1\right)^2}\sqrt{J_4+J_1}B\delta x_q + \sqrt{\gamma_t J_3}\delta x_t \,. \tag{5.71}$$

These equations are equivalent to our earlier semi-classical equations ((5.36)–(5.39) with α and B taken real) except now we have added fluctuation terms associated with each of the irreversible gains and losses to represent quantum noise. These terms are zero-point, random stochastic fluctuations such that $\langle \delta A_i + \delta A_i^* \rangle = \langle \delta x_i \rangle = 0$ and $\langle \delta x_i(t)\delta x_i(t')\rangle = \delta(t-t')$ for all ith, in analogy with the properties of the vacuum. All cross-correlations are zero. If a classical linearization is performed (e.g. $\alpha(t) = \alpha_0 + \delta\alpha(t)\ldots$etc) then spectra identical to those in Sects. 5.1.5 and 5.7 can be generated by considering the classical quadrature fluctuations; e.g. $\delta\alpha + \delta\alpha^*$. Notice that the quantum noise only comes in via the fluctuation terms. This leads to the observation that for linearizable systems the quantum noise is processed classically. However this is only strictly true if coherent atomic effects can be neglected (see Sect. 5.3). One advantage to working with (5.68)–(5.71) is that it is quite simple to generalize them, by inspection, to more complex systems.

5.2 Squeezing from the Rate Equation Model

Equation (5.65) unifies all the squeezing mechanisms previously noted for this type of laser. The only noise terms still significant are those due to the vacuum field input at the mirror (first term), noise from the pump source plus dipole fluctuations which couple in due to non-unit pump absorption (second term) and spontaneous emission noise due to the decay from the lower lasing level (third term). In this section we will examine the behavior of (5.65) in various limits, highlighting the various squeezing mechanisms.

5.2.1 The Quantum Noise Limited Laser

First we will retrieve the standard result that well above threshold the intensity noise of the laser is at the QNL. Hence we consider the situation in which the decay from the lower lasing level is much faster than the cavity decay rate and the pump rate, i.e. $\gamma \gg \Gamma, \kappa$. This suffices to satisfy assumptions (ii) and (iii) as given in the introduction. Equation (5.65) then reduces to

$$S_{\text{out}} = \frac{\omega^2}{\omega^2+\kappa^2} + \frac{\kappa^2}{\omega^2+\kappa^2}\left[(1-\eta) + \eta S_p\right] \,. \tag{5.72}$$

The noise contribution from the lower lasing level decay is now negligible. If the pump is quantum noise limited, (as per assumption (i) in the introduction) i.e. $S_p = 1$, the first and second terms add up to 1, for all frequencies, i.e. the output is quantum noise limited.

5.2.2 Regularized Pumping

It is important to note that the QNL result for (5.72) has different origins depending on the frequency. At low frequencies it is due to the quantum noise of the pump whilst at high frequencies it is due to vacuum fluctuations reflecting off the output mirror. It was first observed by Sokolov et al. [4] that if the pump spectrum could be made sub-Poissonian then the laser output would show amplitude squeezing at low frequencies. The maximum squeezing occurs at zero frequency and is equal to the variance of the absorbed part of the pump ($S_{ab} = ((1-\eta) + \eta S_p))$. The roll-off of squeezing with frequency is a Lorentzian with a half-width of the cavity decay rate (κ).

A problem immediately comes to mind in that the phase noise spectrum (5.67) is independent of the pump statistics. Normally one expects squeezing of one quadrature to be associated with back action producing excess noise on the other quadrature thus ensuring no violation of the uncertainty principle. However in the case of the laser, phase diffusion ensures that sufficient phase noise to maintain the uncertainty relation is always present. This can be shown explicitly by calculating the product of the phase and amplitude spectra for the case of a completely absorbed, regular pump ($S_p = 0, \eta = 1$). The result is

$$S_{\text{out}} V_{\text{out}}^{-} = 1 + \frac{\kappa^2}{\omega^2 + \kappa^2} , \tag{5.73}$$

which is always greater than one.

It was shown by Yamamoto and co-workers in theory [5] and then experimentally [6] that squeezing could be obtained from semi-conductor diode lasers using this principle. The pump for a diode laser is an electric current which can easily be made sub-Poissonian. Under carefully controlled conditions (5.72) correctly describes the output fluctuations of the diode laser, hence amplitude squeezing is seen at frequencies within the linewidth of the laser. In recent years squeezing has been observed from various types of diode lasers. Attempts have also been made (so far unsuccessfully) to obtain squeezing from solid-state lasers by using squeezed diode lasers as pump sources.

5.2.3 Rate-Matching

We now consider squeezing mechanisms which are intrinsic to the laser medium rather than being imposed externally. The first of these is called rate-matching or dynamic pump suppression, first described by Khazanov et al. [7]. Rate matching occurs when the pump rate is comparable in speed to the decay

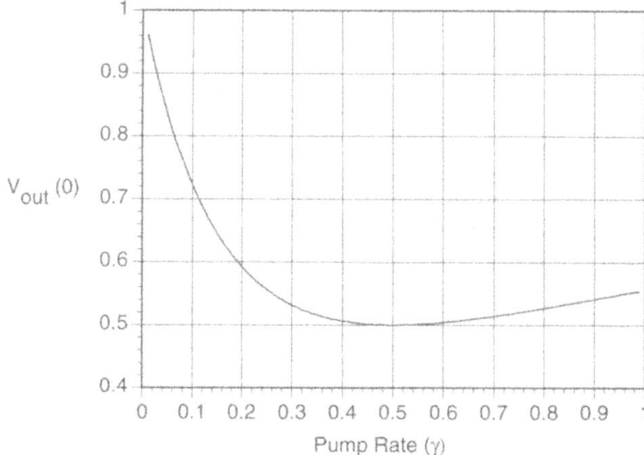

Fig. 5.2. Plot of zero frequency amplitude noise versus pump rate showing squeezing due to rate matching. The pump rate is normalized such that $\gamma = 1$

rate from the lower lasing level. Hence we now consider the limit of 5.65 in which both the pump and lower lasing level decay rate are much faster than the cavity decay rate. We also assume a coherent pump ($S_p = 1$) with low absorption efficiency ($\eta \ll 1$). Equation (5.65) reduces to

$$S_{\text{out}} = \frac{\omega^2}{\omega^2 + \kappa^2} + \frac{\kappa^2 \gamma^2}{(\kappa^2 + \omega^2)(\gamma + 2\Gamma)^2} + \frac{\kappa^2 (2\Gamma)^2}{(\kappa^2 + \omega^2)(\gamma + 2\Gamma)^2} \,. \tag{5.74}$$

Now both the pump and lower lasing level decay noise contribute to the output noise (second and third terms respectively). Surprisingly, though, the division of the noise between these two independent processes actually reduces the total noise contributed [9]. This leads to up to 50% squeezing in the output at low frequencies when the pump and decay rates are "matched" ($\gamma = 2\Gamma$). As in the regularly pumped case the roll-off of squeezing with frequency is Lorentzian. Figure 5.2 shows the dependence of squeezing at low frequency on the pump rate.

The rate-matching mechanism can be understood from a simple statistical argument about the regularity with which atoms are pumped into their upper lasing levels. First we establish a relationship between the variance in the time it takes to place an atom in its upper lasing level and the photon number variance of the output. Under the assumptions used to obtain (5.74) (i.e. that stimulated emission is very rapid and spontaneous emission out of the upper lasing level can be neglected) the arrival of an atom in the upper lasing level results instantaneously in the emission of a photon and the de-excitation of the atom to its lower lasing level. The mean number of photons, $\bar{n}$, that arrive at the output in a time T, which is long compared to the cavity lifetime, is thus given by $\bar{n} = N(T/\bar{t})$ where $\bar{t}$ is the mean time it takes for the atom to be promoted from its lower lasing level to its upper lasing level. The standard deviations in $\bar{n}$ and $\bar{t}$ are thus related by

$$\left(\frac{\Delta\bar{n}}{\bar{n}}\right)^2 = \left(\frac{\Delta\bar{t}}{\bar{t}}\right)^2 . \tag{5.75}$$

The standard deviation in the mean of $\bar{n}$ arrival times, $\Delta\bar{t}$, is related to the standard deviation of one arrival time, Δt, by

$$\Delta\bar{t} = \frac{\Delta t}{\sqrt{\bar{n}}} . \tag{5.76}$$

Under the linearization approximation the amplitude spectrum at zero frequency is equal to the long time photon number variance ($\Delta\bar{n}^2$) normalized to the mean photon number ($\bar{n}$). Hence, using (5.75) and (5.76), we can write

$$S(0) = \frac{\Delta\bar{n}^2}{\bar{n}} = \frac{\Delta t^2}{\bar{t}^2} . \tag{5.77}$$

In our 3-level laser there are two steps involved in returning the atom to its upper lasing level; a spontaneous decay from the lower lasing level to the ground state at rate γ and a pump from the ground state to the upper lasing level at rate Γ. In the absence of stimulated absorption the mean time to return to the upper lasing level would just be the sum of the mean lifetimes of the lower lasing level and the ground state, i.e. $1/\gamma + 1/\Gamma$. The effect of stimulated absorption is to increase the lifetime of the lower lasing level as the atom experiences a number of absorption and emission events before decaying to the ground state. In the limit that spontaneous decay between the lasing levels is negligible this effect doubles the lifetime of the lower lasing level. Thus $\bar{t} = 2/\gamma + 1/\Gamma$. For a QNL pump the pump and the lower lasing level decay are independent Poisson processes, i.e. their standard deviations are equal to their means. Using the standard rules for adding independent noises we obtain $\Delta t = \sqrt{(2/\gamma)^2 + (1/\Gamma)^2}$ and thus from (5.77)

$$S(0) = \frac{(2/\gamma)^2 + (1/\Gamma)^2}{(2/\gamma + 1/\Gamma)^2} . \tag{5.78}$$

This result is equal to that of (5.74) evaluated at zero frequency. Hence we see that the physical mechanism for the squeezing is the reduction in percentage noise introduced by two independent transfers of similar rates as opposed to one. A maximum reduction of a half is achieved when the effective rates are the same. The effect is analogous to the reduction in random measurement error achieved through multiple similar measurements.

This simple argument is readily generalized to more matched rates. For an $r + 3$-level laser

$$S(0) = \frac{(2/\gamma_L)^2 + (1/\gamma_1)^2 + \ldots + (1/\gamma_r)^2 + (1/\Gamma)^2}{(2/\gamma + 1/\gamma_1 + \ldots + 1/\gamma_r + 1/\Gamma)^2} , \tag{5.79}$$

where γ_L is the decay rate out of the lower lasing level, Γ is the pump rate and $\gamma_1 \ldots \gamma_r$ are the other decay rates of the atom. The rates are matched for optimum noise reduction when $\Gamma = \gamma_1 = \ldots = \gamma_r = 0.5\gamma_L$. The minimum value of $S(0)$ is then $1/(r+2)$. Almost perfect squeezing is obtained as the number of matched rates is increased. This result was first obtained by Ritsch et al. [9].

5.2.4 Inversion Filtering

The division of the low frequency noise between the pump and the lower lasing level decay noise can also occur in a frequency dependent way, even if the pump rate is much slower than the decay rate. Consider the situation in which the cavity decay rate is faster than the lower lasing level decay rate but both are much faster than the pump rate, i.e. $\kappa > \gamma \gg \Gamma$. Now (5.65) reduces to

$$S_{\mathrm{out}} = \frac{\omega^2}{\omega^2+\kappa^2} + \frac{\kappa^2\left(\omega^2+\gamma^2\right)}{\left(\kappa\gamma-2\omega^2\right)^2+\omega^2\left(2\kappa+\gamma\right)^2} + \frac{\kappa^2\left[\omega^2+\left(2\Gamma\right)^2\right]}{\left(\kappa\gamma-2\omega^2\right)^2+\omega^2\left(2\kappa+\gamma\right)^2} \,. \tag{5.80}$$

Under this condition the laser can be considered an open system, i.e. the ground level is an undepleted reservoir (as for (5.72)). As for the rate matched discussion we have assumed a QNL pump with negligible pump absorption. The resonance in the denominator of (5.80) at

$$\omega_s = \sqrt{\frac{1}{2}\kappa\gamma} \tag{5.81}$$

is now inside the cavity linewidth. At frequencies close to the resonance, noise is divided between contributions from the pump and the lower level decay rate in much the same way as when the rates are matched. In the limit that $\kappa \gg \gamma$, 50% squeezing is produced at frequencies close to ω_s. The physical mechanism behind this is a frequency filtering effect due to the storage of atoms in their lasing levels for times long compared to the cavity life time. Normally the roll-off of atomic noise sources with frequency is due to the cavity line width. Similarly the roll on of reflected vacuum noise is also determined by the cavity linewidth. As the two noise sources have inverse frequency dependences they end up adding to a constant, one, thus giving the QNL laser result. This was discussed in Sect. 5.2.1 and is shown as the dotted lines in Fig. 5.3.

However if we allow the life time of the lower lasing level to be longer than that of the cavity there is an additional filtering of the atomic noise sources whilst the roll-on of the reflected vacuum noise remains unchanged. This leads to the dip below the QNL as indicated in Fig. 5.3.

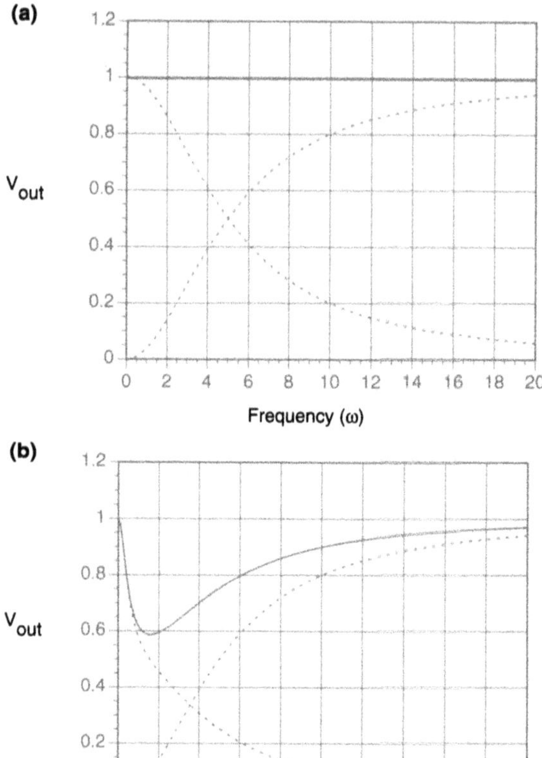

Fig. 5.3. Plot of amplitude noise spectra for conditions **(a)** $\gamma \gg \kappa = 5 \gg \Gamma$ and **(b)** $\kappa = 5, \gamma = 1 \gg \Gamma$. Dotted lines show the contributions to the spectrum from atomic noise sources (rolling off from zero frequency) and the vacuum entering through the output mirror (rolling on from zero frequency)

5.2.5 Squeezing Efficiency

We can consider the laser as a passive device which takes in light at the pump frequency and emits it at the laser frequency. An interesting question to ask is; what is the maximum conversion efficiency from QNL input to squeezed output that can be achieved using an intrinsic effect such as rate matching? This is an important question because the sensitivity with which a measurement can be made with light depends on both the intensity and the noise properties of the light. Hence, if all else was equal, one would not bother putting light through a "squeezer" if the efficiency was so low as to negate the sensitivity gain due to reduced noise. In some squeezing systems low conversion efficiency is indeed intrinsic to the production of strong squeezing [30–32]. In our previous discussion of rate matching we assumed that the pump was undepleted, ie. very low conversion efficiency. We now relax that assumption.

Under the assumptions used to obtain (5.74) but now not assuming low pump absorption we obtain

$$S_{\text{out}}(\omega) = \frac{\omega^2}{\omega^2+\kappa^2} + \frac{\kappa^2\left(\omega^2+\gamma^2\right)}{\left(\kappa^2+\omega^2\right)\left(\gamma+2\Gamma\sqrt{1-\eta}\right)^2} + \frac{\kappa^2\left(\omega^2+\left(2\Gamma\sqrt{1-\eta}\right)^2\right)}{\left(\kappa^2+\omega^2\right)\left(\gamma+2\Gamma\sqrt{1-\eta}\right)^2} . \tag{5.82}$$

With significant absorption the normal matching condition becomes

$$2\Gamma = \frac{\gamma}{\sqrt{1-\eta}} . \tag{5.83}$$

This would seem to imply that a higher pump power is required to reach matching, however the increased efficiency of the pumping process means that actually a lower pump flux is required. This can be seen by using the matching condition ($\gamma = 2\Gamma$) in (5.42) to obtain the following expression for the pump flux required to reach matching as a function of the pump absorption

$$n_{bm} = \frac{\gamma}{2\eta\left(1+\sqrt{1-\eta}\right)} . \tag{5.84}$$

At high absorption efficiency ($\eta \approx 1$) the pump flux required is $\approx \gamma/2$. In principle 50% squeezing can be achieved for arbitrarily high absorption efficiency. However in practice the range of pump fluxes for which squeezing is appreciable becomes prohibitively narrow for very high absorption efficiency. This effect is illustrated in Fig. 5.4, where (5.82) is plotted using (5.42), (5.55) and the semi-classical solutions (Sect. 5.6) to correctly relate the pump flux and efficiency to the pump rate. The dash-dotted line has efficiency of about 50% at maximum squeezing, the dashed line 75% and the solid line 96%. Notice that quite high efficiency can be comfortably achieved. This ability to increase the absolute sensitivity of a beam of light is important and is one clear advantage of rate matching over other types of squeezing.

These results can be generalized to greater numbers of matched rates with similar results. Generalizing (5.82) to a system with three rates of similar magnitude we obtain (at zero frequency)

$$S_{\text{out}}(0) = \frac{\left(\gamma_1\gamma\right)^2}{\left(\gamma\gamma_1 \pm 2\Gamma\gamma_1\sqrt{1-\eta}+\Gamma\gamma\sqrt{1-\eta}\right)^2}\left(1+\eta\left(V_p-1\right)\right) + \frac{\left(2\Gamma\gamma_1\sqrt{1-\eta}\right)^2+\left(\Gamma\gamma\sqrt{1-\eta}\right)^2}{\left(\gamma\gamma_1 \pm 2\Gamma\gamma_1\sqrt{1-\eta}+\Gamma\gamma\sqrt{1-\eta}\right)^2} , \tag{5.85}$$

where γ_1 is the rate of the new transition. The maximum squeezing obtainable is increased to 67%, achieved when the matching condition $2\Gamma\sqrt{1-\eta} = 2\gamma_1 = \gamma$ is satisfied.

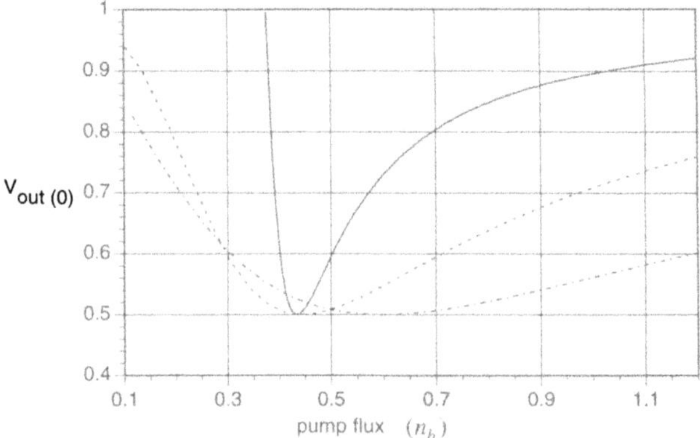

Fig. 5.4. Plot of zero frequency amplitude noise versus pump flux showing the effect of different pump efficiencies. The dashed dotted curve is for $\kappa_b = 100, \eta \approx 50\%$ at maximum squeezing, the dashed curve is for $\kappa_b = 40, \eta \approx 75\%$ at maximum squeezing and the solid curve is for $\kappa_b = 10, \eta \approx 96\%$ at maximum squeezing. Other parameters are $G = 2850$, $\kappa = 4.0$, $\gamma_t = .0001$, $G_p = 20$, normalized such that $\gamma = 1$

5.2.6 Squeezing Under Non-Ideal Conditions

We will now discuss the effect of non-ideal conditions on the squeezing. The optimum design for laser squeezing experiments depends on the particular squeezing mechanism to be exploited as well as many practical considerations dependent on the actual type of laser to be used. Such considerations are beyond the scope of this discussion. However some general conditions must be met which are discussed briefly in the following.

The basic trade-off that must be made is between minimizing intra-cavity loss whilst maximizing how far above threshold the laser is operating. The two are connected via the reflectivity of the output mirror. Higher finesse cavities have lower thresholds but also tend to have greater percentage losses. The following expression for the threshold pump rate can be found from the semi-classical solution (Sect. 5.6)

$$\Gamma_t = \frac{\frac{1}{2}\kappa\gamma_t\gamma/G}{\gamma\left(1 - \frac{\kappa}{2G}\right) - \gamma_t\left(1 + \frac{\kappa}{2G}\right)}\,, \tag{5.86}$$

which, all else equal, gets smaller as the coupling through the output mirror κ_m, and hence the total cavity loss $\kappa = \kappa_m + \kappa_l$, is decreased. On the other hand the efficiency of the laser due to intra-cavity losses is given by

$$\varepsilon = \frac{\kappa_m}{\kappa_m + \kappa_l}\,, \tag{5.87}$$

which also decreases as the output coupling is made smaller.

The effect of decreasing efficiency on the squeezing is given by the usual loss formula

$$S_{\mathrm{out}} = \varepsilon S_O + 1 - \varepsilon \ , \tag{5.88}$$

where S_O is the squeezing which would be achieved in the absence of intra-cavity loss. As $\varepsilon \to 0$, $S_{\mathrm{out}} \to 1$ and the squeezing is lost. The effect of lowering the threshold is two fold. Firstly spontaneous emission across the lasing levels introduces large amounts of excess noise close to threshold which will destroy squeezing. This noise falls off rapidly as threshold is reduced. Secondly for some lasers (particularly solid-state lasers) the squeezing linewidth is determined by the atomic dynamics close to threshold. This can be as narrow as kHz. This is a problem as technical noise, both in the laser and its pump, may be dominant at low frequencies. As threshold is reduced the linewidth broadens (gain broadening) until very far above threshold it becomes limited by the cavity linewidth (as assumed earlier).

To illustrate these effects we consider a laser with parameters typical of small solid-state lasers. We assume the lower lasing level decay is a hundred times that of the upper lasing level and that we have matched the pump ($\gamma = 100\gamma_t = 0.5\Gamma$). Intra-cavity losses are included. We consider the situation of a QNL pump spectrum (dashed lines) and that of a pump spectrum which has excess noise at low frequency (solid lines). In Fig. 5.5 we plot the noise spectra for these conditions whilst changing the reflectivity of the output mirror.

In Fig. 5.5a the output coupling is much greater than the loss and hence we are quite close to threshold. Very little squeezing is lost due to intra-cavity loss however spontaneous emission noise has significantly reduced the squeezing at zero frequency and the squeezing linewidth is so narrow that no squeezing is seen when pump noise is added. In Fig. 5.5b the output coupling has been decreased but is still significantly larger than the loss. A small amount of squeezing is lost due to optical loss but spontaneous emission noise is now negligible and the squeezing linewidth is sufficient that reasonable squeezing can be seen in the presence of pump noise. Finally in Fig. 5.5c the output coupling has been decreased till it is the same as the loss. Now there is a significant decrease in squeezing due to optical losses. On the other hand the squeezing linewidth is now much wider.

Similar effects to those illustrated in Fig. 5.5 will occur for regularized pumping. For inversion filtering the problem is more difficult as gain broadening must make the linewidth wider than ω_s whilst maintaining the relationship $\kappa \gg \gamma$. This will generally mean a very high pump rate is a prerequisite.

5.3 Squeezing from Coherent Effects

Up until now we have only considered the situation in which the dynamics of the atomic coherences have been adiabatically eliminated (see Sect. 5.1.3).

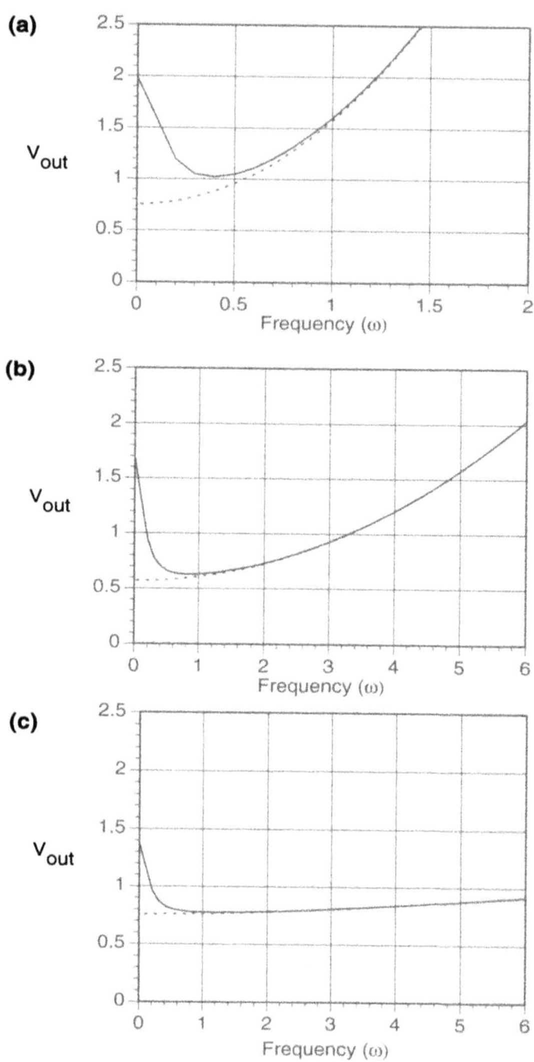

Fig. 5.5. Plot of amplitude noise spectra under non-ideal conditions showing dependence of squeezing on output coupling. Output coupling is **(a)** $\kappa_m = 500$, **(b)** $\kappa_m = 50$ and **(c)** $\kappa_m = 5$. Other parameters are $G = 2850, \kappa_l = 5, \gamma_t = .01, \Gamma = 0.5(\eta \ll 1)$. The pump spectrum is a Lorentzian with a half width of 0.14. All parameters are normalized such that $\gamma = 1$

This means that coherent atomic effects have been neglected (assumption (iv)). When the coherences are not eliminated new phenomena become possible. The effect of coherence between the pump levels was studied by Ralph and Savage [10] and found to markedly improve the squeezing. Even greater squeezing was found by Ritsch et al. [11] and other authors [33] when coherence between the pump and lasing levels is allowed, such as occurs in Raman lasers. In the following we will illustrate the new physics that arises by considering the simplest case, that of coherence between the pump levels. In practice coherent effects are only likely to be important in far infra-red gas lasers, cryogenically cooled solid-state lasers and lasers with very low active atom numbers.

5.3.1 Extending the Laser Model

We can allow coherence between the pump levels to become important by relaxing the assumptions that the decay rate from the upper pump level (γ_f) and the rate of phase decay of the pump coherence (γ_Q) are very rapid (assumptions (ii) and (iii) Sect. 5.1.3). This means that the pump coherence ($\hat{\sigma}_{14}$) and the upper pump level ($\hat{\sigma}_4$) are no longer strongly damped and cannot be adiabatically eliminated. It also means that pumping will no longer be uni-directional as some of the upper pump level population will be pumped back down to the ground level. Working from (5.13)–(5.20), we proceed as in Sect. 5.1.3 but we eliminate only the lasing coherence and the the pump field. The semi-classical equations are now

$$\dot{\alpha} = G\left(J_3 - J_2\right)\alpha - \frac{1}{2}\kappa\alpha \ , \tag{5.89}$$

$$\dot{J}_{14} = \left(\left(\frac{2g_p}{\kappa_b}J_{14} + \sqrt{\frac{4}{\kappa_b}}B\right)g_p\left(J_4 - J_1\right)\right) - \left(\frac{\gamma_f}{2} + \gamma_Q\right)J_{14} \ , \tag{5.90}$$

$$\dot{J}_1 = \left(\frac{2g_p}{\kappa_b}J_{14} + \sqrt{\frac{4}{\kappa_b}}B\right)2g_pJ_{14} + \gamma J_2 \ , \tag{5.91}$$

$$\dot{J}_2 = G\left(J_3 - J_2\right)\left(2\alpha^2\right) + \gamma_t J_3 - \gamma J_2 \ , \tag{5.92}$$

$$\dot{J}_3 = -G\left(J_3 - J_2\right)\left(2\alpha^2\right) - \gamma_t J_3 + \gamma_f J_4 \ , \tag{5.93}$$

$$\dot{J}_4 = -\left(\frac{2g_p}{\kappa_b}J_{14} + \sqrt{\frac{4}{\kappa_b}}B\right)2g_pJ_{14} - \gamma_f J_4 \ , \tag{5.94}$$

where $\langle\hat{\sigma}_{14}\rangle = J_{14}$ is the expectation value of the pump coherence. As before we consider conditions optimum for squeezing, that is negligible losses and stimulated emission the dominant rate (see preamble to (5.41)). Under these conditions the photon number is given by

$$n = \frac{N\gamma\gamma_f\Gamma_e}{\kappa\left(\gamma\gamma_f + 2\Gamma_e\gamma_f + 2\Gamma_e\gamma\right)} \ , \tag{5.95}$$

where

$$\Gamma_e \cong \frac{16G_p}{\kappa_b}B^2 \tag{5.96}$$

provided we assume the pump efficiency is low, i.e.

$$\eta \cong \frac{16G_p}{\kappa_b}\left(J_1 - J_4\right) \ll 1 \ . \tag{5.97}$$

The equations of motion for the linearized fluctuations are given by

$$\delta\dot{\hat{x}}_a = G\left(\delta\hat{\sigma}_3 - \delta\hat{\sigma}_2\right)\alpha + \sqrt{\kappa_m}\,\delta\hat{X}_{Am} + \sqrt{\kappa_l}\,\delta\hat{X}_{Al} - \sqrt{G}\delta\hat{\Lambda}_p \,, \tag{5.98}$$

$$\begin{aligned}\delta\dot{\hat{x}}_{14} = {} & \left(\frac{2g_p^2}{\kappa_b}\left(J_4 - J_1\right) - \frac{\gamma_f}{2} + \gamma_Q\right)\delta\hat{x}_{14} \\ & + 2g_p\left(\frac{2g_p}{\kappa_b}J_{14} + \sqrt{\frac{4}{\kappa_b}}\beta\right)\left(\delta\hat{\sigma}_4 - \delta\hat{\sigma}_1\right) \\ & + g_p\sqrt{\frac{4}{\kappa_b}}\left(J_4 - J_1\right)\delta\hat{X}_B + \sqrt{\gamma_f}\delta\hat{\Lambda}_{f(i)} \\ & - \sqrt{\gamma}\delta\hat{\Lambda}_{(i)} + \sqrt{\gamma_Q}\delta\hat{\Lambda}_q \,,\end{aligned} \tag{5.99}$$

$$\delta\dot{\hat{\sigma}}_1 = \left(\frac{4g_p^2}{\kappa_b}J_{14} + g_p\sqrt{\frac{4}{\kappa_b}}\beta\right)\delta\hat{x}_{14} + g_p\sqrt{\frac{4}{\kappa_b}}J_{14}\delta\hat{X}_B + \gamma\delta\hat{\sigma}_2 - \sqrt{\gamma}\delta\hat{\Lambda} \,, \tag{5.100}$$

$$\delta\dot{\hat{\sigma}}_2 = G\left(\delta\hat{\sigma}_3 - \delta\hat{\sigma}_2\right)\alpha^2 + G\left(J_3 - J_2\right)\alpha\,\delta\hat{x}_a + \gamma_t\,\delta\hat{\sigma}_3 - \gamma\,\delta\hat{\sigma}_2 + \sqrt{\gamma}\delta\hat{\Lambda} - \sqrt{\gamma_t}\delta\hat{\Lambda}_t - \sqrt{G}\alpha\,\delta\hat{\Lambda}_p \,, \tag{5.101}$$

$$\delta\dot{\hat{\sigma}}_3 = -G\left(\delta\hat{\sigma}_3 - \delta\hat{\sigma}_2\right)\alpha^2 - G\left(J_3 - J_2\right)\alpha\,\delta\hat{x}_a - \gamma_t\,\delta\hat{\sigma}_3 + \gamma_f\delta\hat{\sigma}_4 + \delta\hat{\Lambda}_{f(ii)} + \sqrt{\gamma_t}\delta\hat{\Lambda}_t + \sqrt{G}\alpha\,\delta\hat{\Lambda}_p \,, \tag{5.102}$$

$$\delta\dot{\hat{\sigma}}_4 = -\left(\frac{4g_p^2}{\kappa_b}J_{14} + g_p\sqrt{\frac{4}{\kappa_b}}\beta\right)\delta\hat{x}_{14} - g_p\sqrt{\frac{4}{\kappa_b}}J_{14}\delta\hat{X}_B - \gamma_f\delta\hat{\sigma}_4 - \sqrt{\gamma_f}\delta\hat{\Lambda}_{f(ii)} \,, \tag{5.103}$$

where $\delta\hat{x}_{14} = \delta\hat{\sigma}_{14} + \delta\hat{\sigma}_{14}^{+}$ and the noise terms are as given in (4.47)–(5.29) plus

$$\delta\hat{\Lambda}_{(i)} = \delta\hat{C}^{\dagger}\hat{\sigma}_{24}^{+} + \hat{\sigma}_{24}\delta\hat{C} \,, \tag{5.104}$$

$$\delta\hat{\Lambda}_{f(i)} = \delta\hat{C}_f^{\dagger}\hat{\sigma}_{13}^{+} + \hat{\sigma}_{13}\delta\hat{C}_f \,, \tag{5.105}$$

$$\delta\hat{\Lambda}_{f(ii)} = \delta\hat{C}_f^{\dagger}\hat{\sigma}_{34} + \hat{\sigma}_{34}^{+}\delta\hat{C}_f \,. \tag{5.106}$$

The second order expectation values are given in (5.31), (5.32) and (5.34) along with

$$\langle\delta\hat{\Lambda}_{(i)}(t)\delta\hat{\Lambda}_{(i)}(t')\rangle = \langle\hat{\sigma}_2\rangle\delta(t - t') \,, \tag{5.107}$$

$$\langle\delta\hat{\Lambda}(t)_{f(i)}\delta\hat{\Lambda}_{f(i)}(t')\rangle = \langle\hat{\sigma}_1\rangle\delta(t - t') \,, \tag{5.108}$$

$$\langle\delta\hat{\Lambda}(t)_{f(ii)}\delta\hat{\Lambda}_{f(ii)}(t')\rangle = \langle\hat{\sigma}_4\rangle\delta(t - t') \,, \tag{5.109}$$

$$\begin{aligned}\langle\delta\hat{\Lambda}(t)_{f(i)}\delta\hat{\Lambda}_{f(ii)}(t')\rangle &= \langle\delta\hat{\Lambda}(t)_{f(ii)}\delta\hat{\Lambda}_{f(i)}(t')\rangle \\ &= \langle\hat{\sigma}_{14}\rangle\delta(t - t') \,.\end{aligned} \tag{5.110}$$

All other expectation values are zero.

5.3.2 Squeezing from Coherent Pumping

We make the same assumptions used to obtain the rate matched squeezing spectra (see Sect. 5.2.3) that the laser is well above threshold with negligible losses, low absorption efficiency and narrow cavity linewidth. If we set the lattice induced phase damping to zero ($\gamma_Q = 0$) we obtain the following expression for the amplitude noise spectrum of the output field by following the procedure outlined in Sect. 5.1.5

$$\begin{aligned} S_{\text{out}} &= \frac{\omega^2}{\omega^2+\kappa^2} \\ &+ \frac{\kappa^2}{\omega^2+\kappa^2} \left\{ \frac{(1/\Gamma_e)^2 + (2/\gamma)^2 + (2/\gamma_f)^2 + 2/(\Gamma_e\gamma_f)}{(1/\Gamma_e + 2/\gamma + 2/\gamma_f)^2} \right. \\ &\left. - \frac{4\left[\left((1/\Gamma_e)^2 + 2/(\Gamma_e\gamma_f)\right)\left(1/(\gamma_f^2 + \Gamma_e\gamma_f)\right)\right]^{\frac{1}{2}}}{(1/\Gamma_e + 2/\gamma + 2/\gamma_f)^2} \right\} . \end{aligned} \quad (5.111)$$

A minimum value of $S_{\text{out}} = 0.2$ is reached at low frequency with the matching condition $\Gamma_e = 0.5\gamma_f = 0.25\gamma$ satisfied. Looking at (5.79) we find that for three matched incoherent rates we expect a minimum spectral variance 0.33. Hence there is a significant increase in the squeezing. Also the matching condition is different. The improvement is more pronounced if we assume $\gamma \gg \Gamma_e, \gamma_f$ such that there are only two matched rates. For two incoherent rates we expect from (5.79) a minimum of 0.5 when $\Gamma = \gamma_f$. But for a coherent pump we find from (5.111) a minimum of 0.25 when $\Gamma_e = 0.5\gamma_f$, a 3dB improvement. On the other hand if $\gamma_f \gg \Gamma_e, \gamma$ then we return to the case where the pumping is effectively incoherent and (5.111) reduces to (5.74).

To understand the origin of the increased squeezing we take one step back from (5.111) to the expression for the Fourier transformed amplitude fluctuations of the output field at zero frequency

$$\begin{aligned} \delta X_{\text{out}} &= \frac{2/(\gamma\sqrt{J_2})\,\delta\Lambda + 1/(\sqrt{\Gamma_e\gamma_f J_2})\,\delta\Lambda_{(i)}}{1/\Gamma_e + 2/\gamma + 2/\gamma_f} \\ &+ \frac{\sqrt{(1/\Gamma_e + 1/\gamma_f)/(\Gamma_e J_1)}\,\delta\Lambda_{f(i)} + 2/(\gamma_f\sqrt{J_4})\,\delta\Lambda_{f(ii)}}{1/\Gamma_e + 2/\gamma + 2/\gamma_f} . \end{aligned} \quad (5.112)$$

From (5.112) we can see the origin of the various contributions to the spectrum (5.111). The first and fourth terms come from fluctuations in the populations due to spontaneous emission. The second and third terms come from fluctuations in the dipole. Notice though that the dipole fluctuations are due to the same spontaneous emission terms as the population fluctuations (see (5.99) and (5.104)–(5.106)) and hence have a cross correlation (5.110). That is the physical process causing the decay of the coherence is the same as that causing the decay of the upper pump level. Let us try to "guess" the solution if the damping of the pump coherence was dominated by lattice or collisionally induced phase damping and hence the pumping was incoherent. Then the dipole and population fluctuations would have different sources. In particular the noise terms $\delta\Lambda_{f(i)}$ and $\delta\Lambda_{f(ii)}$ would have their origins in physically distinct vacuum noise. Hence there would be no cross correlation between them ($\langle\delta\Lambda_{f(i)}\delta\Lambda_{f(ii)}\rangle = 0$). Taking the self correlations unchanged (5.31)–(5.109), we can obtain the following expression for the zero frequency variance

$$S(0) = \frac{(2/\gamma)^2 + (1/\Gamma_e)^2 + (2/\gamma_f)^2 + 2/(\Gamma_e\gamma_f)}{(1/\Gamma_e + 2/\gamma + 2/\gamma_f)^2} . \quad (5.113)$$

This is indeed the correct expression for an incoherently pumped four-level laser in which back pumping from the upper pump level back to the ground state is included [29] as can be confirmed by calculating from (5.98)–(5.103) in the limit $\gamma_Q \gg \gamma_f$. The minimum variance that can be achieved is about 0.43 when $\Gamma_e \cong 0.5\gamma_f \cong 0.75\gamma$. This is better than the incoherent 3-level but not as good as a uni-directional 4-level system. Thus the source of the increased squeezing is the cross correlation between the noise terms $\delta\Lambda_{f(i)}$ and $\delta\Lambda_{f(ii)}$ in (5.110). In fact they are anti-correlated, so that a cancellation of noise can occur resulting in the additional squeezing. Including the cross correlation and using the semi-classical relationship

$$\frac{J_{14}}{\sqrt{J_4 J_1}} = -\sqrt{\frac{\gamma_f}{\Gamma_e + \gamma_f}} , \quad (5.114)$$

we can obtain the zero frequency limit of (5.111) from (5.112). The cross-correlation term depends on an atomic operator product whose expectation value can not be factorized, i.e. $\langle\hat{A}\hat{B}\rangle \neq \langle\hat{A}\rangle\langle\hat{B}\rangle$. This means that it is not possible to write simple equivalent semi-classical equations as we did for the incoherent case in Sect. 5.1.5.

5.4 Conclusion

We have presented a fully quantum mechanical theory of a CW four-level laser and solved it under conditions typical of most lasers to obtain analytic expressions for the amplitude and phase quadrature noise spectra. We have investigated the various amplitude squeezing mechanisms that can occur. We can summarize the characteristics of the different types of squeezing as follows: (i) Regularized pumping: Special requirements; sub-Poissonian pump source. Maximum squeezing; only limited by how sub-Poissonian the pump is and the level of intra-cavity loss. Comments; squeezing occurs once the laser is more than about 5 times above threshold, the squeezing linewidth is gain broadened as the laser goes further above threshold eventually being limited by the laser cavity linewidth. (ii) Rate Matching: Special requirements; pump rate similar to one or more of the spontaneous atomic transitions. Maximum squeezing; ideally given by $V(0) = 1/l$ where l is the number of atomic levels with matched transition rates but also limited by intracavity losses. Comments; threshold conditions as for regularized pumping. Squeezing efficiency can be very high, i.e. a coherent state pump can be efficiently converted to a squeezed state laser output. (iii) Inversion Filtering: Special requirements; cavity decay rate is much faster than lower lasing level decay rate. Maximum squeezing; ideally 50% but also limited by intra-cavity loss. Comments; maximum squeezing occurs at a frequency given by the square root of the cavity decay rate times the lower lasing level decay rate ($\omega_s = \sqrt{\kappa\gamma/2}$). Squeezing will not occur until gain broadening is much greater than ω_s, which may be very far above threshold.

We also briefly discussed squeezing mechanisms when coherent atomic effects are important, looking at the example of coherent pumping. We showed that increased squeezing occurs due to the cancellation of anti-correlated noise.

Acknowledgments

I would like to thank C.M. Savage, H.-A. Bachor, B.C. Buchler and P.A. Roos for useful discussions. This work was supported by the Australian Research Council.

5.5 Expectation Values

As an example of the calculations used to obtain (5.31)–(5.34) we here calculate the value of $\langle \delta\hat{A}(t)\delta\hat{A}(t')\rangle$. Substituting (4.47), we get

$$\begin{aligned} \langle \delta\hat{A}(t)\delta\hat{A}(t')\rangle = \langle & \delta\hat{C}^\dagger(t)\sigma_{12}\delta\hat{C}^\dagger(t')\sigma_{12} + \delta\hat{C}^\dagger(t)\sigma_{12}\sigma_{12}^+\delta\hat{C}(t') \\ & + \sigma_{12}^+\delta\hat{C}(t)\sigma_{12}^+\delta\hat{C}(t') + \sigma_{12}^+\delta\hat{C}(t)\delta\hat{C}^\dagger(t')\sigma_{12}\rangle \; . \end{aligned} \tag{5.115}$$

Because $\delta\hat{C}$ is a vacuum field it has the property $\langle\delta\hat{C}\rangle = \langle\delta\hat{C}^\dagger\rangle = 0$, so we are left with

$$\langle \delta\hat{A}(t)\delta\hat{A}(t')\rangle = \langle \sigma_{12}^+\delta\hat{C}(t)\delta\hat{C}^\dagger(t')\sigma_{12}\rangle \; . \tag{5.116}$$

Using the normal boson commutator $[\hat{O}(t), \hat{O}^\dagger(t')] = \delta(t-t')$, we get

$$\begin{aligned} \langle \delta\hat{A}(t)\delta\hat{A}(t')\rangle = & \langle \sigma_{12}^+\sigma_{12}\rangle\delta(t-t') \\ & + \langle \sigma_{12}^+\delta\hat{C}^\dagger(t)\delta\hat{C}(t')\sigma_{12}\rangle \; . \end{aligned} \tag{5.117}$$

In general system operators do not commute with the vacuum field inputs so we must be careful in further simplifying (5.117). Using the notation of (5.9) we can write a general expression for the commutation relations as [17]

$$\left[\hat{O}, \hat{A}_j\right] = -\frac{\sqrt{\gamma_j}}{2}\left[\hat{O}, \hat{a}_j\right] \; , \tag{5.118}$$

$$\left[\hat{O}, \hat{A}_j^\dagger\right] = -\frac{\sqrt{\gamma_j}}{2}\left[\hat{O}, \hat{a}_j^\dagger\right] . \tag{5.119}$$

Here we require the commutator

$$\left[\hat{\sigma}_{12}, \delta\hat{C}\right] = -\frac{\sqrt{\gamma}}{2}\left[\hat{\sigma}_{12}, \hat{\sigma}_{12}\right] = 0 , \tag{5.120}$$

so the second term in (5.117) is zero. Using the atomic operator relation $\hat{\sigma}_{ij}^{+}\hat{\sigma}_{ij} = \hat{\sigma}_j$, we obtain

$$\langle \delta\hat{\Lambda}(t)\delta\hat{\Lambda}(t')\rangle = \langle \sigma_{12}^{+}\sigma_{12}\rangle\delta(t-t') = \langle\hat{\sigma}_2\rangle\delta(t-t') , \tag{5.121}$$

as required.

Note that sometimes constant terms can arise when deriving the expectation values. These transform to δ function spikes at DC in Fourier space and as such can be neglected.

5.6 Semi-Classical Solutions

The full solution for the intra-cavity photon number is

$$n = \frac{N[\Gamma(\gamma-\gamma_t) - \frac{1}{2}(\gamma\Gamma + \Gamma\gamma_t + \gamma_t\gamma)\kappa/G]}{\kappa(\gamma+2\Gamma)} . \tag{5.122}$$

For low conversion efficiency or small depletion of the ground state Γ can be taken proportional to the pump flux n_b. Otherwise (5.40) and (5.55) must be used with the following solution for the ground state to correctly relate pump flux and efficiency to the pump rate.

$$J_1 = \frac{N(Gn+\gamma_t)\gamma}{Gn(\gamma+2\Gamma) + \gamma_t(\gamma+\Gamma) + \gamma\Gamma} . \tag{5.123}$$

5.7 Noise Spectrum

The full solution for the amplitude noise spectrum is given by

$$\begin{aligned} V_{\text{out}}(\omega) = \Big\{ & [\kappa_m - i\omega - F1(\omega)]^2 \\ & + \kappa_m \Gamma J_1 [F2(\omega)]^2 (1 - \eta(V_p(\omega) - 1)) \\ & + \kappa_m \gamma J_2 [F3(\omega)]^2 + \kappa_m \gamma_t J_3 [F4(\omega)]^2 \\ & + \kappa_m G (J_3 + J_2) [1 - \alpha F4(\omega)]^2 \\ & + 2\kappa_m \kappa_l \Big\} / [i\omega + F1(\omega)]^2 , \end{aligned} \tag{5.124}$$

where

$$F1(\omega) = \frac{G^2\alpha^2 (J_3 - J_2)(2i\omega + \gamma + 2\Gamma)}{(i\omega + \Gamma)(i\omega + \gamma + 2G\alpha^2 + \gamma_t) + \gamma (G\alpha^2 + \gamma_t)},$$
$$F2(\omega) = \frac{G\alpha (i\omega + \gamma - \gamma_t)}{(i\omega + \Gamma)(i\omega + \gamma + 2G\alpha^2 + \gamma_t) + \gamma (G\alpha^2 + \gamma_t)},$$
$$F3(\omega) = \frac{G\alpha (i\omega + 2\Gamma + \gamma_t)}{(i\omega + \Gamma)(i\omega + \gamma + 2G\alpha^2 + \gamma_t) + \gamma (G\alpha^2 + \gamma_t)},$$
$$F4(\omega) = \frac{G\alpha (i\omega + 2\Gamma + \gamma)}{(i\omega + \Gamma)(i\omega + \gamma + 2G\alpha^2 + \gamma_t) + \gamma (G\alpha^2 + \gamma_t)}. \quad (5.125)$$

References

1. R.J. Glauber, Phys. Rev. **131**, 109 (1963)
2. A.W. Smith and J.A. Armstrong, Phys. Rev. Lett. **16**, 1169 (1966); C. Freed and H.A. Haus, IEEE Journal of Quantum Electronics, **QE-2**, 190 (1966)
3. R. Loudon, *Quantum Theory of Light* (Oxford University Press, Oxford 1973)
4. Y.M. Golubev and I.V. Sokolov, Sov. Phys. JETP **60**, 234 (1984)
5. Y. Yamamoto, S. Machida and O. Nilsson, Phys. Rev. A **34**, 4025 (1986)
6. Y. Yamamoto and S. Machida, Phys. Rev. A **35**, 5114 (1987)
7. A.M. Khazanov, G.A. Koganov and E.P. Gordov, Phys. Rev. A **42**, 3065 (1990)
8. T.C. Ralph and C.M. Savage, Opt. Lett. **16**, 1113 (1991)
9. H. Ritsch, P. Zoller, C.W. Gardiner and D.F. Walls, Phys. Rev. A **44**, 3361 (1991)
10. T.C. Ralph and C.M. Savage, Phys. Rev. A **44**, 7809 (1991)
11. H. Ritsch, M.A.M. Marte and P. Zoller, Europhys. Lett. **19**, 7 (1992)
12. M.I. Kolobov, E. Giacobino, C. Fabre and L. Davidovich, Phys. Rev. A **47**, 1431 (1993)
13. D.F. Walls and G.J. Milburn, *Quantum Optics*, (Springer, Berlin Heidelberg New York 1994)
14. M.O. Scully and W.E. Lamb, Phys. Rev. **159**, 208 (1967)
15. H. Haken, *Laser Theory*, reproduction from *Handbuch der Phisik* (Springer, Berlin Heidelberg New York 1984)
16. M. Lax and W.H. Louisell, Phys. Rev. **185**, 568 (1969)
17. C.W. Gardiner and M.J. Collett, Phys. Rev. A **31**, 3761 (1985)
18. C. Fabre, Phys. Rep. **219**, 215 (1992)
19. T.C. Ralph, C.C. Harb and H.A. Bachor, Phys. Rev. A **54**, 4359 (1996)
20. D.D. Marcenac and J.E. Carroll, IEEE Journal of Quant.Elec. **30**, 2064 (1994)
21. B. Tromburg, H.E. Lassen and H. Olesen, IEEE Journal of Quant. Elec. **30**, 939 (1994)
22. M. Yamashita, S. Machida, T. Mukai and O. Nillson, Opt. Lett. **22**, 534 (1997)
23. C.M. Savage and T.C. Ralph, Phys. Rev. A **46**, 2803 (1992)
24. C.H. Henry, IEEE J.Quant.Elec. **18**, 259 (1982)
25. A. Karlsson and G. Bjork, Phys. Rev. A **44**, 7669 (1991)
26. A. Eschmann and C.W. Gardiner, Phys. Rev. A **54**, 760 (1996)
27. F. Marin, A. Bramati, E. Giacobino, T.C. Zhang, J.Ph. Poizat, J.F. Roch and P. Grangier, Phys. Rev. Lett. **75**, 4606 (1995)

28. H. Wang, M.J. Freeman and D.G. Steel, Phys. Rev. Lett. **71**, 3951 (1993)
29. T.C. Ralph, Phys. Rev. A **49**, 4979 (1994)
30. P. Kinsler, M. Fernee and P.D. Drummond, Phys. Rev. A **48**, 3310 (1993)
31. K.M. Gheri, D.F. Walls and M.A. Marte, Phys. Rev. A **50**, 1871 (1994)
32. R. Paschotta. M. Collett, P. Kurz, K. Fiedler, H.A. Bachor and J. Mlynek, Phys. Rev. Lett. **72**, 3807 (1994)
33. K.M. Gheri and D.F. Walls, Phys. Rev. Lett. **68**, 3428 (1994)
34. T.C. Ralph, PhD Thesis (Australian National University 1993)

6 Squeezing and Feedback

H. M. Wiseman

In its broadest conception, feedback could be defined to be any mechanism by which a system acts upon itself, via an intermediate system. This definition would classify, for example, a nonlinear refractive index as feedback on a beam of light. The light polarizes the medium in which it is propagating, which then affects the propagation of the light in that medium. If one is interested in the light as a quantum system, then such feedback can be modeled by modifying the Hamiltonian for the field. This is a well-known mechanism for generating squeezed states of light [1].

In this chapter I am concerned with a different concept of feedback, in which the intermediate system is external to the system of interest. That is to say, the system is an *open* system, in constant interaction with its environment. Feedback will occur if the change in the environment due to the system is significant in affecting the system's dynamics. Since an environment is by definition large compared to the system, it is usual for this feedback to be ignored. This is the essence of the Born, or perturbative, approach to open systems [2]. However, the system's environment can be deliberately engineered so that the feedback is important.

One obvious way to achieve this is for the environment to include a measurement apparatus which detects the influence of the system on its surroundings, a device to amplify this measurement, and a mechanism by which this amplified signal controls the dynamics of the system. It is also possible to engineer a feedback mechanism which does not involve a measurement device, but rather some more direct form of back-coupling from the environment to the system. In classical mechanics it is a moot point whether a device is designated a "measurement apparatus". However in quantum mechanics the special role of measurement implies that the direct back-coupling may be quite distinct from feedback via measurement [3]. This is one reason why a peculiarly quantum theory of feedback is necessary, and interesting. In this chapter I will be concerned only with measurement-based feedback.

The history of feedback in quantum optics goes back to the observation of sub-shot-noise fluctuations in an in-loop photocurrent in the mid 1980's by two groups [4,5]. The theory of this phenomenon was soon addressed by Yamamoto and co-workers [6,7], and Shapiro et al. [8]. The central question they were addressing was whether this feedback was producing real squeezing, a question whose answer is not as straightforward as might be thought. These treatments were based in the Heisenberg picture and used quantum Langevin equations [2] where necessary to describe the evolution of system operators. They treated the quantum noise only within a linearized approximation. Although this approximation is probably valid for all quantum optical feedback experiments performed so far, it would not be valid in the "deep quantum" regime involving few photons and non-perturbative couplings, as is being explored in the so-called "cavity quantum electrodynamics" experiments [9].

More recently an alternative approach to quantum feedback has been proposed by myself and Milburn [10,11], and developed fully by myself [12]. This is based on the theory of quantum trajectories [13–15], which is an application of quantum measurement theory to continuously monitored open quantum systems. By treating the measurement explicitly, this theory translates the quantum noise of the bath into classical noise in the record of detections. It can be shown to be equivalent to an exact (unlinearized) quantum Langevin treatment [12]. The advantage of the quantum trajectory method is that it allows arbitrary feedback to be treated by the theory, at least numerically. A particular limit of interest is that of Markovian feedback, in which the feedback dynamics can be modeled using a master equation. This result was not obtained by the authors using the quantum Langevin treatment.

A third approach [16] to feedback in quantum optics is to use the Glauber –Sudarshan P function [17–19], a quasi-probability distribution. In this theory, the fields are given an essentially classical description, but negative probabilities are allowed in order to take into account quantum correlations [16]. This theory is just as easy to use as the quantum Langevin or quantum trajectory theories when the system dynamics which can be linearized. However, like the quantum Langevin approach, it is usually intractable when the linearization approximation cannot be made. I will not discuss this theory further in this chapter.

A fourth approach is to treat the electromagnetic field as a stream of point-like particles (photons) traveling at the speed of light. This is essentially a classical approach, which cannot describe phase properties of the fields, but which is adequate if one is interested only in intensity statistics. Formally, the in-loop photon arrivals become a self-excited classical point process. This theory was used by Shapiro et al.[8] in addition to their quantum operator theory. Similar ideas have subsequently been used by other authors [20,21]. Like the other three approaches mentioned above this approach is easily applicable to linearized systems, but unlike them it does not give a full quantum description of the in-loop field. Again, I will not discuss this theory further in this chapter.

In the remainder of this chapter I have alternated 'theory' sections, which introduce the mathematical apparatus necessary for describing quantum feedback, with 'application' sections, which use the theory to investigate squeezing, and, where appropriate, discuss experimental results. In both of these parallel streams the material is presented in roughly the order in which it was developed, but the two streams are not synchronous.

First I introduce continuum fields, and then show how linearization allows feedback onto those fields to be treated analytically, yielding noise spectra for in-loop and out-of-loop measurements. Next I discuss the interaction of continuum fields with a localized quantum system, giving rise to quantum Langevin equations for system operators. This theory is used to describe nonlinear measurements (such as QND measurements) of continuum fields, and feedback based on the results of these measurements. An alternative to the quantum Langevin description is one based on quantum trajectories. This is most useful for illuminating feedback onto the localized systems, and I use it to investigate intracavity squeezing. In the Markovian limit the quantum trajectory picture of feedback allows one to derive a feedback master equation. This is of most use for describing dynamics which cannot be linearized, such as that of a strongly driven two-level atom. This turns out to be precisely what is needed to revisit the question of in-loop squeezing in terms of what the atom 'sees'.

6.1 Continuum Fields

6.1.1 Canonical Quantization

Let the fundamental field be the vector potential $\mathbf{A}(\mathbf{r},t)$ in the Coulomb gauge

$$\nabla \cdot \mathbf{A}(\mathbf{r},t) = 0 \ . \tag{6.1}$$

The free Lagrangian density for this field is [22]

$$\pounds = \tfrac{1}{2}\varepsilon_0 \left(\mathbf{E}^2 - c^2\mathbf{B}^2\right) \ , \tag{6.2}$$

where ε_0 is the permittivity of free space and c is the speed of light. The electric $\mathbf{E}$ and magnetic $\mathbf{B}$ fields are defined by

$$\mathbf{E} = -\dot{\mathbf{A}} \ ; \qquad \mathbf{B} = \nabla \times \mathbf{A} \ . \tag{6.3}$$

From (6.2), the canonical field to $\mathbf{A}$ is $-\varepsilon_0\mathbf{E}$. Thus, in quantizing the field, these obey the canonical commutation relations

$$[A_j(\mathbf{r},t), E_k(\mathbf{r}',t)] = -i\frac{\hbar}{\varepsilon_0}\delta_{jk}\delta^3_\perp(\mathbf{r}-\mathbf{r}') \ . \tag{6.4}$$

Here $\delta^3_\perp$ denotes a three-dimensional *transverse* Dirac delta-function, which is necessary to be compatible with the constraint of (6.1) [22]. Note that the Heisenberg picture operators in the canonical commutation relations are at equal times. In the Schrödinger picture, the same relations hold, but the time argument is omitted. The Euler–Lagrange (which is also the Heisenberg) equation of motion from (6.2) is the wave equation

$$\ddot{\mathbf{A}} = c^2 \nabla^2 \mathbf{A} \ . \tag{6.5}$$

Now consider the case of a beam of polarized light. That is to say, consider only one component A of $\mathbf{A}$ and let its spatial variation be confined to one direction, say z. This simplifies the analysis, and is also appropriate for determining the inputs and outputs of a quantum optical cavity. In reality, the transverse spatial extent of the beam would be confined to some area Λ which is determined by the area of the optical components involved [2]. However, as long as the x and y extensions are much greater than a wavelength, the beam can be approximated by plane waves. The appropriate wave equation is

$$\ddot{A} = c^2 \partial_z^2 A \ , \tag{6.6}$$

of which I am interested only in the forward propagating solutions

$$A(z, t + t') = A(z - ct', t) \ . \tag{6.7}$$

If the field is reflected off a cavity mirror (say at $z = 0$) then the direction of z will change at the point of reflection. This is why only one direction of propagation need be considered. The field for $z < 0$ is incoming and that for $z > 0$ is outgoing. The canonical commutation relation is now

$$[A(z,t), E(z',t)] = -i \frac{\hbar}{\varepsilon_0 \Lambda} \delta(z - z') \ . \tag{6.8}$$

Solutions for A and E satisfying the wave equation (6.6) can be constructed using the annihilation and creation operators for the modes of frequency ω, which satisfy

$$[a(\omega), a^\dagger(\omega')] = \delta(\omega - \omega') \ . \tag{6.9}$$

They are

$$A(z) = \sqrt{\frac{\hbar}{\varepsilon_0 \Lambda 2\pi c}} \int_0^\infty d\omega \frac{1}{\sqrt{2\omega}} \{a(\omega) \exp[-i\omega(t - z/c)] + \text{H.c.}\} \ , \tag{6.10}$$

$$E(z) = \sqrt{\frac{\hbar}{\varepsilon_0 \Lambda 2\pi c}} \int_0^\infty d\omega \sqrt{\frac{\omega}{2}} \{ia(\omega) \exp[-i\omega(t - z/c)] + \text{H.c.}\} \ . \tag{6.11}$$

This expression for the fields in terms of annihilation and creation operators for a continuum of modes defines the sense in which they are composed of photons of definite frequency. However, this sense is quite unlike the naive picture of a beam of light made up of (possibly different frequencies of) photons, hurtling through space at the speed of light. Each mode is spread over all space, so there is no way in which a photon, as an excitation of such a mode, can move at all. To define an annihilation operator $b(z,t)$ for a localized photon of a particular frequency, it would be necessary to sum many different mode operators. Such operators can be defined, with slight variations in the details of the definition [14,23]. The various definitions are effectively equivalent in application to quantum optical problems. The authors of [14,23] construct the localized annihilation operator from the mode annihilation operators $a(\omega)$. Here, just for variation, I am introducing a different definition for $b(z,t)$, constructed from the original fields in space-time, $A(z,t)$ and $E(z,t)$.

As established above, A and $-\varepsilon_0 E$ are canonically conjugate variables at each point in space-time. Motivated by the analogy with position and momentum, a local annihilation operator for an oscillator of angular frequency ω_0 can be defined as

$$b(z,t) = \exp[i\omega_0(t - z/c)]\sqrt{\frac{\Lambda c}{\hbar}}\left[\sqrt{\frac{\omega_0\varepsilon_0}{2}}A(z,t) - i\sqrt{\frac{\varepsilon_0}{2\omega_0}}E(z,t)\right] . \tag{6.12}$$

In terms of the mode operators, $b(z,t)$ is given by

$$\begin{aligned} b(z,t) = \frac{1}{\sqrt{2\pi}}\int_0^\infty d\omega \Bigg\{ & \frac{\omega_0+\omega}{2\sqrt{\omega_0\omega}}a(\omega)\exp[i(\omega_0-\omega)(t-z/c)] \\ & + \frac{\omega_0-\omega}{2\sqrt{\omega_0\omega}}a^\dagger(\omega)\exp[i(\omega_0+\omega)(t-z/c)]\Bigg\} . \end{aligned} \tag{6.13}$$

If the beam contains only photons of a frequency near ω_0 then it is apparent from (6.13) that we can approximate $b(z,t)$ by

$$b(z,t) \approx \frac{1}{\sqrt{2\pi}}\int_0^\infty d\omega\, a(\omega)\exp[i(\omega_0-\omega)(t-z/c)] . \tag{6.14}$$

From this we can calculate

$$[b(z,t), b^\dagger(z',t)] \approx \frac{1}{2\pi}\int_0^\infty d\omega \exp[i(\omega_0-\omega)(z'-z)/c] \tag{6.15}$$

$$\approx c\delta(z-z') , \tag{6.16}$$

which explains the choice of normalization in (6.12). Note that the approximations involved here apply only if all of the light is at a frequency close to ω_0. Thus, the width of the Dirac δ function in (6.16) should be understood to be much greater than a wavelength of light $2\pi c/\omega_0$.

The rotating exponential $\exp[i\omega_0(t - z/c)]$ means that a "localized photon" of frequency ω_0 has a slowly varying annihilation operator $b(z,t)$. This operator also has the same property as the vector potential (6.7), obeying

$$b(z, t+t') = b(z-ct', t) \tag{6.17}$$

in free space. If only frequencies near ω_0 are significantly excited, then the time-flux of energy can be easily seen to be

$$W(z,t) \simeq \hbar\omega_0 b^\dagger(z,t)b(z,t) . \tag{6.18}$$

Thus, the annihilation operator $b(z,t)$ conforms to one's naive expectations, with $b^\dagger(z,t)b(z,t)$ being the photon flux (photons per unit time) passing z at time t.

6.1.2 Photodetection

From the above discussion it should be apparent that it is not sensible to talk about a photodetector for photons of frequency ω_0 which has a response time comparable to or smaller than ω_0^{-1}. Therefore for practical purposes a photodetector is equivalent to an energy flux meter. In either case, as long as we are not interested in times comparable to ω_0^{-1}, we can assume that the signal produced by an ideal photodetector at position z_1 is given by the operator

$$I(t) = b_1^\dagger(t)b_1(t) \; , \tag{6.19}$$

where $b_1(t) \equiv b(z_1,t)$. Here I have ignored any factors of electric charge *etc.* which are sometimes included but which are actually nominal.

In experiments involving lasers, it is often the case (or at least it is harmless to assume [24]) that $b_1(t)$ has a mean amplitude $\beta = \langle b_1(t)\rangle$. Without loss of generality, I will take β to be real. In all that follows I will also assume that we are considering stationary statistics. That is, we are taking the long time limit of a system with a stationary state. In that case, *only if* the correlations of interest in the intensity of the beam of light have a characteristic time satisfying

$$\tau_{\text{corr}} \gg |\beta|^{-2} \; , \tag{6.20}$$

is it permissible to linearize (6.19). This means approximating it by

$$I(t) = \beta^2 + \delta I(t) = \beta^2 + \beta X_1(t) \; , \tag{6.21}$$

where

$$X_1(t) = b_1(t) + b_1^\dagger(t) - 2\beta \tag{6.22}$$

is the amplitude quadrature fluctuation operator for the continuum field. For the linearization to be valid the fluctuations must be small as well as slow:

$$\langle X_1(t+\tau)X_1(t)\rangle \ll \beta^2 \quad \text{for} \quad \tau \sim \tau_{\text{corr}} \; . \tag{6.23}$$

It is useful also to define the phase quadrature fluctuation operator

$$Y_1(t) = -ib_1(t) + ib_1^\dagger(t) \; . \tag{6.24}$$

For free fields, which obey (6.17), these obey the commutation relations

$$[X_1(t), Y_1(t')] = 2i\delta(t-t') \; . \tag{6.25}$$

If we define the Fourier transformed operator

$$\tilde{X}_1(\omega) = \int_{-\infty}^{\infty} dt X_1(t) e^{-i\omega t} \tag{6.26}$$

and similarly for $\tilde{Y}_1(\omega)$ then

$$[\tilde{X}_1(\omega), \tilde{Y}_1(\omega')] = 4\pi i \delta(\omega + \omega') \ . \tag{6.27}$$

For stationary statistics as we are considering, $\langle X_1(t) X_1(t') \rangle$ is a function of $t - t'$ only. From this it follows that

$$\langle \tilde{X}_1(\omega) \tilde{X}_1(\omega') \rangle \propto \delta(\omega + \omega') \ . \tag{6.28}$$

Because of the singularities in equations (6.27) and (6.28), to obtain a finite uncertainty relation it is more useful to consider the spectrum

$$S_1^X(\omega) = \frac{1}{2\pi} \int_{-\infty}^{\infty} \langle \tilde{X}_1(\omega) \tilde{X}_1(-\omega') \rangle d\omega' \tag{6.29}$$

$$= \int_{-\infty}^{\infty} e^{-i\omega t} \langle X(t) X(0) \rangle dt \ = \ \langle \tilde{X}_1(\omega) X_1(0) \rangle \ . \tag{6.30}$$

Then it can be shown that for a stationary free field [8],

$$S_1^X(\omega) S_1^Y(\omega) \geq 1 \ . \tag{6.31}$$

From this it is obvious that a coherent continuum field [17–19] is one such that for all ω

$$S_1^Q(\omega) = 1 \ , \tag{6.32}$$

where $Q = X$ or Y (or any intermediate quadrature). This is known as the *standard quantum limit* or *shot-noise limit*. A *squeezed* continuum field is one such that, for some ω and some Q,

$$S_1^Q(\omega) < 1 \ . \tag{6.33}$$

The physical significance of $S_1^X(\omega)$ is apparent from (6.21): it can be experimentally determined as

$$S_1^X(\omega) = \langle I(t) \rangle^{-1} \int_{-\infty}^{\infty} e^{-i\omega t} \langle I(t), I(0) \rangle dt \ , \tag{6.34}$$

where $\langle A, B \rangle = \langle AB \rangle - \langle A \rangle \langle B \rangle$. In fact it is possible to determine $S_1^Q(\omega)$ for any quadrature Q in a similar way. Putting the field of interest through a low-reflectivity beam splitter, while reflecting another field with a large coherent amplitude off the same beam splitter, a coherent amplitude can be added to the beam of interest. If this contribution is sufficiently large it will dominate the total coherent amplitude of the beam. Since the added component can have any chosen phase with respect to the original beam, the new linearized intensity fluctuation operator will be proportional to any chosen quadrature fluctuation operator. This technique is known as homodyne detection. In practice, balanced homodyne detection using a 50–50 beam splitter and two photodetectors is preferable, but the principle is the same [25].

6.2 In-Loop "Squeezing"

6.2.1 Description of the Device

The simplest form of quantum optical feedback is shown in Fig. 6.1. This was the scheme considered by Shapiro et al. [8]. In our notation, we begin with a field $b_0 = b(z_0, t)$ as shown in the diagram. We will take this field to have stationary statistics with mean amplitude β and fluctuations

$$\tfrac{1}{2}[X_0(t) + iY_0(t)] = b_0(t) - \beta \tag{6.35}$$

characterized by arbitrary spectra $S_0^X(\omega), S_0^Y(\omega)$. This field is then passed through a beam splitter of transmittance $\eta_1(t)$. By unitarity, the diminution in the transmitted field by a factor $\sqrt{\eta_1(t)}$ must be accompanied by the addition of vacuum noise from the other port of the beam splitter [1]. The transmitted field is

$$b_1(t) = \sqrt{\eta_1(t-\tau_1)}\, b_0(t-\tau_1) + \sqrt{\bar{\eta}_1(t-\tau_1)}\, \nu(t-\tau_1)\ . \tag{6.36}$$

Here $\tau_1 = (z_1 - z_0)/c$ and I am using the notation

$$\bar{\eta} \equiv 1 - \eta\ . \tag{6.37}$$

The operator $\nu(t)$ represents the vacuum fluctuations. The vacuum is special case of a coherent continuum field of vanishing amplitude $\langle \nu(t) \rangle = 0$, and so is completely characterized by its spectrum

$$S_\nu^Q(\omega) = 1\ . \tag{6.38}$$

Since the vacuum fluctuations are uncorrelated with any other field, and have stationary statistics, the phase and time arguments for $\nu(t)$ are arbitrary.

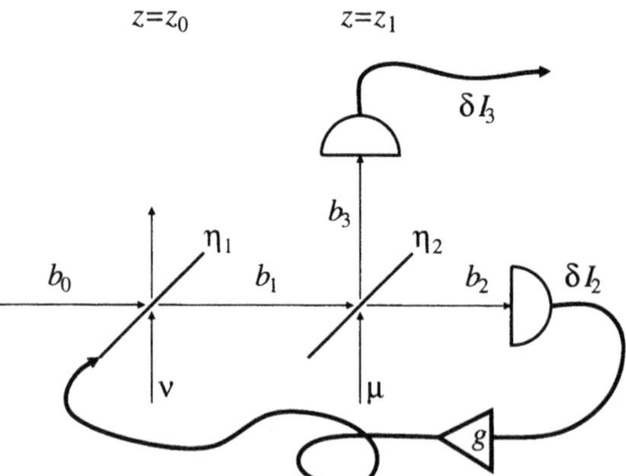

Fig. 6.1. Diagram for a traveling-wave feedback experiment. Traveling fields are denoted b and photocurrent fluctuations δI. The first beam splitter transmittance η_1 is variable, the second η_2 fixed. The two vacuum field inputs are denoted ν and μ

The beam-splitter transmittance $\eta_1(t)$ in (6.36) is time-dependent. This time-dependence can be achieved experimentally by a number of means. For example, if the incoming beam is elliptically polarized then an electro-optic modulator (a device with a refractive index controlled by a current) will alter the orientation of the ellipse. A polarization-sensitive beam splitter will then control the amount of the light which is transmitted, as done, for example, in [26]. As the reader will no doubt have anticipated, the current used to control the electro-optic modulator can be derived from a later detection of the light beam, giving rise to feedback. Writing $\eta_1(t) = \bar{\eta}_1 + \delta\eta_1(t)$, and assuming that the modulation of the transmittance is small ($\delta\eta_1(t) \ll \eta_1, \bar{\eta}_1$), one can write

$$\sqrt{\eta_1(t)} = \sqrt{\eta_1} + (1/\sqrt{\eta_1})\tfrac{1}{2}\delta\eta_1(t) \ . \tag{6.39}$$

Continuing to follow the path of the beam in Fig. 6.1, it now enters a second beam-splitter of constant transmittance η_2. The transmitted beam annihilation operator

$$b_2(t) = \sqrt{\eta_2}\, b_1(t-\tau_2) + \sqrt{\bar{\eta}_2}\, \mu(t-\tau_2) \ , \tag{6.40}$$

where $\tau_2 = (z_2 - z_1)/c$ and $\mu(t)$ represents vacuum fluctuations like $\nu(t)$. The reflected beam operator is

$$b_3(t) = \sqrt{\bar{\eta}_2}\, b_1(t-\tau_2) - \sqrt{\eta_2}\, \mu(t-\tau_2) \ . \tag{6.41}$$

Using the approximation (6.39), the linearized quadrature fluctuation operators for b_2 are

$$\begin{aligned} X_2(t) &= \sqrt{\eta_2\eta_1}\, X_0(t-T_2) + \sqrt{\eta_2/\eta_1}\, \beta\delta\eta_1(t-T_2) \\ &\quad + \sqrt{\eta_2\bar{\eta}_1}\, X_\nu(t-T_2) + \sqrt{\bar{\eta}_2}\, X_\mu(t-T_2) \ , \end{aligned} \tag{6.42}$$

$$Y_2(t) = \sqrt{\eta_2\eta_1}\, Y_0(t-T_2) + \sqrt{\eta_2\bar{\eta}_1}\, Y_\nu(t-T_2) + \sqrt{\bar{\eta}_2}\, Y_\mu(t-T_2) \ , \tag{6.43}$$

where $T_2 = \tau_2 + \tau_1$. Similarly for b_3 we have

$$\begin{aligned} X_3(t) &= \sqrt{\bar{\eta}_2\eta_1}\, X_0(t-T_2) + \sqrt{\bar{\eta}_2/\eta_1}\, \beta\delta\eta_1(t-T_2) \\ &\quad + \sqrt{\bar{\eta}_2\bar{\eta}_1}\, X_\nu(t-T_2) - \sqrt{\eta_2}\, X_\mu(t-T_2) \ , \end{aligned} \tag{6.44}$$

$$Y_3(t) = \sqrt{\bar{\eta}_2\eta_1}\, Y_0(t-T_2) + \sqrt{\bar{\eta}_2\bar{\eta}_1}\, Y_\nu(t-T_2) - \sqrt{\eta_2}\, Y_\mu(t-T_2) \ . \tag{6.45}$$

The mean fields for b_2 and b_3 are $\sqrt{\eta_1\eta_2}\,\beta$ and $\sqrt{\eta_1\bar{\eta}_2}\,\beta$ respectively. Thus, if these fields are incident upon photodetectors, the respective linearized photocurrent fluctuations are, as explained in Sect. 6.1.2,

$$\delta I_2(t) = \sqrt{\eta_1\eta_2}\, \beta\, X_2(t) \ , \tag{6.46}$$

$$\delta I_3(t) = \sqrt{\eta_1\bar{\eta}_2}\, \beta\, X_3(t) \ . \tag{6.47}$$

Here I have assumed perfect efficiency detectors. To model inefficient detectors it is necessary to add further beam splitters, with transmittance equal to the detection efficiency, in front of the detectors. The effect of this has been considered in detail in [26,27].

Having obtained an expression for $\delta I_2(t)$ we are now in a position to follow the next stage in Fig. 6.1 and complete the feedback loop. We set the modulation in the transmittance of the first beam-splitter to be

$$\delta\eta_1(t) = \frac{g}{\eta_2\beta^2}\int_0^\infty h(t')\delta I_2(t-\tau_0-t')dt' , \tag{6.48}$$

where g is a dimensionless parameter representing the low-frequency gain of the feedback loop. The response of the feedback loop, including the electro-optic elements, is assumed to be linear for small fluctuations and is characterized by the electronic delay time τ_0 and the response function $h(t')$, which satisfies $h(t)=0$ for $t<0$, $h(t)\geq 0$ for $t>0$ and $\int_0^\infty h(t')dt'=1$.

6.2.2 Stability

Clearly the feedback can only affect the amplitude quadrature X. Putting (6.48) into (6.42) yields

$$\begin{aligned} X_2(t) = {}& \sqrt{\eta_2\eta_1}\,X_0(t-T_2) + g\int_0^\infty h(t')X_2(t-T-t')dt' \\ & + \sqrt{\eta_2\bar{\eta}_1}\,X_\nu(t-T_2) + \sqrt{\bar{\eta}_2}\,X_\mu(t-T_2) , \end{aligned} \tag{6.49}$$

where $T=\tau_0+T_2=\tau_0+\tau_1+\tau_2$. This is easy to solve in Fourier space, providing that X_2 is a stationary stochastic process. This will only be the case if the feedback is stable. Using standard feedback and control theory [28], the Nyquist stability criterion is

$$\mathrm{Re}[s] < 0 , \tag{6.50}$$

where s is any solution of the characteristic equation

$$1 - gH(s)\exp(-sT) = 0 , \tag{6.51}$$

where $H(s)$ denotes the Laplace transform $H(s)=\int_0^\infty dt\exp(-st)h(t)$.

First I show that a *sufficient* condition for stability is $|g|<1$. Looking for instability, assume that $\mathrm{Re}[s]>0$. Then

$$\left|H(s)e^{-sT}\right| = \left|\int_0^\infty dte^{-s(t+T)}h(t)\right| \leq \int_0^\infty dth(t) = 1 . \tag{6.52}$$

Thus under this assumption the characteristic equation cannot be satisfied for $|g|<1$, so this regime will always be stable. If $g>1$ then it is not difficult to show that there is a positive s which will solve (6.51). Thus it is a *necessary* condition to have $g<1$. If $g<-1$, the stability of the feedback depends on T and the shape of $h(t)$. However, it turns out that it is possible to have arbitrarily large negative low-frequency feedback (that is, $g\to-\infty$), for any feedback loop delay T, provided that $h(t)$ is broad enough. The price to be paid for strong low-frequency negative feedback is a reduction in the bandwidth of the feedback, the width of $|\tilde{h}(\omega)|^2$.

To see this, consider the simplest smoothing function $h(t) = \gamma \exp(-\gamma t)$. The condition for marginal stability is that there is a solution to (6.51) for $s = i\omega$. That is,

$$1 = g \exp(-i\omega T) \frac{\gamma}{\gamma + i\omega} . \tag{6.53}$$

For the imaginary part of the right-hand side to vanish we require

$$\tan \omega T = -\omega/\gamma . \tag{6.54}$$

As we will see, for large $|g|$ we will require $\gamma \ll T^{-1}$ in which case the solutions on (6.54) can be approximated by $\omega_n = (2n+1)\pi/2T$. Under the same approximation we can ignore γ compared to ω in (6.53) to get

$$1 = -|g|(-i)^{2n+1} \frac{2\gamma T}{i(2n+1)\pi} . \tag{6.55}$$

Clearly for n odd this cannot be satisfied and so the system will be stable. However for n even we have

$$1 = |g| \frac{2\gamma T}{(2n+1)\pi} , \tag{6.56}$$

which can be satisfied (indicating marginal stability). In order to avoid this for all n we require

$$\gamma < \frac{\pi}{2T|g|} \ll \frac{1}{T} , \tag{6.57}$$

where here we see that $\gamma \ll T^{-1}$ for large negative g. Now the bandwidth of the feedback is $B \simeq 2\gamma$. Thus we have finally the approximate inequality

$$B \leq \frac{\pi}{T|g|} , \tag{6.58}$$

which shows how a finite delay time T and large negative feedback $-g \gg 1$ reduces the possible bandwidth of the feedback.

6.2.3 In-Loop and Out-of-Loop Spectra

Assuming then that the feedback is stable, we can solve (6.49) for X_2 in the Fourier domain

$$\tilde{X}_2(\omega) = \exp(-i\omega T_2) \frac{\sqrt{\eta_2 \eta_1}\, \tilde{X}_0(\omega) + \sqrt{\eta_2 \bar{\eta}_1}\, \tilde{X}_\nu(\omega) + \sqrt{\bar{\eta}_2}\, \tilde{X}_\mu(\omega)}{1 - g\tilde{h}(\omega) \exp(-i\omega T)} . \tag{6.59}$$

From this the amplitude quadrature spectrum is easily found from (6.28) and (6.29) to be

$$S_2^X(\omega) = \frac{\eta_1\eta_2 S_0^X(\omega) + \eta_2\bar{\eta}_1 S_\nu^X(\omega) + \bar{\eta}_2 S_\mu^X(\omega)}{|1 - g\tilde{h}(\omega)\exp(-i\omega T)|^2}$$
$$= \frac{1 + \eta_1\eta_2[S_0^X(\omega) - 1]}{|1 - g\tilde{h}(\omega)\exp(-i\omega T)|^2} . \tag{6.60}$$

From these formulae the effect of feedback is obvious: it multiplies the amplitude quadrature spectrum at a given frequency by the factor $|1 - g\tilde{h}(\omega)\exp(-i\omega T)|^{-2}$. At low frequencies, this factor is simply $(1-g)^{-2}$, which is why the feedback was classified on this basis into positive ($g > 0$) and negative ($g < 0$) feedback. The former will increase the noise at low frequency and the latter will decrease it. However at higher frequencies, and in particular at multiples of π/T, the sign of the feedback will reverse and $g < 0$ will result in an increase in noise and vice-versa. This is shown clearly in the theoretical investigations of Shapiro et al. [8]. All of these results make perfect sense in the context of classical light signals, except that in that case we would not worry about vacuum noise. This is equivalent to assuming that the original noise is far above the shot-noise limit, so that one can replace $1 + \eta_1\eta_2[S_0^X(\omega) - 1]$ by $\eta_1\eta_2 S_0^X(\omega)$. This gives the result expected from classical signal processing: the signal is attenuated by the beam splitters and either amplified or suppressed by the feedback.

The most dramatic effect is of course for large negative feedback. For sufficiently large $-g$ it is clear that one can make

$$S_2^X(\omega) < 1 \tag{6.61}$$

for some ω. This effect has been observed experimentally many times with different systems involving feedback [4,5,7,26,27,29,30]. Without a feedback loop this sub-shot-noise photocurrent would be seen as evidence for squeezing. However, there are a number of reasons to be very cautious about applying the word squeezing to this phenomenon. Two of these reasons are theoretical, and are discussed in the following two sub-sections. The more practical reason relates to the out-of-loop beam b_3, which I will now discuss.

From (6.44), the X quadrature of the beam b_3 is, in the Fourier domain

$$\tilde{X}_3(\omega) = \exp(-i\omega T_2)\left[\sqrt{\bar{\eta}_2\eta_1}\,\tilde{X}_0(\omega) + \sqrt{\bar{\eta}_2\bar{\eta}_1}\,\tilde{X}_\nu(\omega) - \sqrt{\eta_2}\,\tilde{X}_\mu(\omega)\right]$$
$$+ \sqrt{\bar{\eta}_2/\eta_2}\, g\tilde{h}(\omega)\exp(-i\omega T)\tilde{X}_2(\omega) . \tag{6.62}$$

Here I have substituted for $\delta\eta_1$ in terms of X_2. Now using the above expression (6.59) gives

$$\tilde{X}_3(\omega) = \exp(-i\omega T_2)\left\{\frac{\sqrt{\bar{\eta}_2\eta_1}\,\tilde{X}_0(\omega) + \sqrt{\bar{\eta}_2\bar{\eta}_1}\,\tilde{X}_\nu(\omega)}{1 - g\tilde{h}(\omega)\exp(-i\omega T)}\right.$$

$$- \left. \frac{\sqrt{\eta_2}[1 - g\tilde{h}(\omega)\exp(-i\omega T)/\eta_2]\tilde{X}_\mu(\omega)}{1 - g\tilde{h}(\omega)\exp(-i\omega T)} \right\} . \tag{6.63}$$

This yields the spectrum

$$\begin{aligned} S_3^X(\omega) = & \frac{1 + \bar{\eta}_2\eta_1(S_0^X - 1)}{|1 - g\tilde{h}(\omega)\exp(-i\omega T)|^2} \\ & - \frac{2\mathrm{Re}[g\tilde{h}(\omega)\exp(-i\omega T)] + g^2|\tilde{h}(\omega)|^2/\eta_2}{|1 - g\tilde{h}(\omega)\exp(-i\omega T)|^2} . \end{aligned} \tag{6.64}$$

The denominator is identical to that in the in-loop case, as are the first two terms in the numerator. But there are additional terms in the numerator which indicate that there is extra noise in the out-of-loop signal.

The expression (6.64) can be rewritten as

$$S_3^X(\omega) = 1 + \frac{\bar{\eta}_2\eta_1(S_0^X - 1) + g^2|\tilde{h}(\omega)|^2\bar{\eta}_2/\eta_2}{|1 - g\tilde{h}(\omega)\exp(-i\omega T)|^2} . \tag{6.65}$$

From this it is apparent that, unless the initial beam is amplitude squeezed (that is, unless $S_0^X(\omega) < 1$ for some ω) the out-of-loop spectrum will always be greater than the shot-noise-limit of unity. In other words, it is not possible to extract the apparent squeezing in the feedback loop by using a beam splitter. In fact, in the limit of large negative feedback (which gives the greatest noise reduction in the in-loop signal), the out-of-loop amplitude spectrum approaches a constant. Considering a frequency ω such that $\tilde{h}(\omega)\exp(-i\omega T)$ is real and positive, one finds that

$$\lim_{g \to -\infty} S_3^X(\omega) = \eta_2^{-1} . \tag{6.66}$$

Thus the more light one attempts to extract from the feedback loop, the higher above shot-noise the spectrum becomes.

This result is counter to an intuition based on classical light signals, where the effect of a beam splitter is simply to split a beam so that both outputs would have the same statistics. The reason this intuition fails is precisely because this is not all that a beam splitter does; it also introduces vacuum noise which is *anti-correlated* at the two output ports. The detector for beam b_2 measures the amplitude fluctuations X_2, which are a combination of the initial fluctuations X_0, and the two vacuum fluctuations X_ν and X_μ. The first two of these are common to the beam b_3, but the last, X_μ, appears with opposite sign in X_3. As the negative feedback is turned up, the first two components are successfully suppressed, but the last is actually amplified. Note that the result in (6.66) holds no matter how large $S_0(\omega)$ is compared to unity.

6.2.4 Commutation Relations

Under normal circumstances (without a feedback loop) one would expect a sub-shot noise amplitude spectrum to imply a super-shot-noise phase spectrum. However that is not what is found from the theory presented here. Rather, the in-loop phase quadrature spectrum is unaffected by the feedback, being equal to

$$S_2^Y(\omega) = 1 + \eta_1\eta_2[S_0^Y(\omega) - 1] \,. \tag{6.67}$$

It is impossible to measure this spectrum without disturbing the feedback loop because all of the light in the b_2 beam must be incident upon the photodetector in order to measure X_2. However, it is possible to measure the phase-quadrature of the out-of-loop beam by homodyne detection. This was done in [26], which verified that this quadrature is also unaffected by the feedback, with

$$S_3^Y(\omega) = 1 + \eta_1\bar{\eta}_2[S_0^Y(\omega) - 1] \,. \tag{6.68}$$

For simplicity, consider the case where the initial beam is coherent with $S_0^X(\omega) = S_0^Y(\omega) = 1$. Then $S_2^Y(\omega) = 1$ and

$$S_2^Y(\omega)S_2^X(\omega) = |1 - g\tilde{h}(\omega)\exp(-i\omega T)|^{-2} \,. \tag{6.69}$$

This can clearly be less than unity. This represents a violation of the uncertainty relation (6.31) which follows from the commutation relations (6.27). In fact it is easy to show (as done first by Shapiro et al. [8]) from the solution (6.59) that the commutation relations (6.27) are false for the field b_2 and must be replaced by

$$[\tilde{X}_2(\omega), \tilde{Y}_2(\omega')] = \frac{4\pi i\delta(\omega + \omega')}{1 - g\tilde{h}(\omega)\exp(-i\omega T)} \,, \tag{6.70}$$

which explains how (6.69) is possible.

At first sight, this apparent violation of the canonical commutation relations would seem to be a major problem of this theory. In fact, there are no violations of the canonical commutation relations. As emphasized in Sect. 6.1.1, the canonical commutation relations (6.8) are between fields at different points in space, *at the same time*. It is only for free fields (traveling forward in space for an indefinite time) that one can replace the space difference z with a time difference $t = z/c$. Field b_3 is such a free field, as it can be detected an arbitrarily large distance away from the apparatus. Thus its quadratures at a particular point do obey the time-difference commutation relations (6.25), and the corresponding Fourier domain relations (6.27). But the field b_2 cannot travel an indefinite distance before being detected. The time from the second beam splitter to the detector τ_2 is a physical parameter in the feedback system.

For times shorter than the total feedback loop delay T it can be shown that

$$[X_2(t), Y_2(t')] = 2i\delta(t - t') \quad \text{for } |t - t'| < T \; . \tag{6.71}$$

Now the field b_2 is only in existence for a time τ_2 before it is detected. Because $\tau_2 < T$, this means that the time-difference commutation relations between different parts of field b_2 are actually preserved for any time such that those parts of the field are in existence, traveling through space toward the detector. It is only at times greater than the feedback loop delay time T that non-standard commutation relations hold. To summarize, the commutation relations between any of the fields at different spatial points always holds, but there is no reason to expect the time or frequency commutation relations to hold for an in-loop field. Without these relations, it is not clear how "squeezing" should be defined. Indeed, it has been suggested [31] that "squashed light" would be a more appropriate term for in-loop "squeezing" because the uncertainty has actually been squashed, rather than squeezed out of one quadrature and into another.

6.2.5 Semiclassical Theory

A second theoretical reason against the use of the word squeezing to describe the sub-shot-noise in-loop amplitude quadrature is that (providing beam b_0 is not squeezed), the entire apparatus can be described semiclassically. In a semiclassical description there is no noise except classical noise in the field amplitudes, and shot noise is a result of a quantum detector being driven by a classical beam of light. That such a description exists might seem surprising, given the importance of vacuum fluctuations in the explanation of the spectra in Sect. 6.2.3. However, the semiclassical explanation, as explored in [8,26,32], is even simpler.

For the fields, let us use the same symbols as before, but with a cl superscript to remind us that these are classical variables rather than operators. Then the only irreducible source of noise in the system is the shot noise at the two detectors $k = 2, 3$. This arises from the assumption that a classical light field of frequency ω_0 and time-flux of energy W^{cl} induces photo-electron emissions as a random process at rate $W^{\text{cl}}/\hbar\omega_0$. Here $\hbar$ appears as a universal phenomenological constant relating the output of photodetectors to the incoming light. In addition to this irreducible noise, the light field itself may have (classical) noise which is represented by the amplitude fluctuation variable $X^{\text{cl}}(t)$, which is scaled in the same way as the quantum operator $X(t)$ has been.

In the linearized approximation we have been working in, the Poissonian shot-noise fluctuations can be approximated by Gaussian fluctuations, giving

$$\delta I_k(t) = \sqrt{I_k}\,[X_k^{\text{cl}}(t) + \xi_k(t)] \; , \tag{6.72}$$

where I_k is the mean value of the photocurrent and $\xi_k(t)$ represents independent white-noise sources obeying

$$\langle \xi_k(t')\xi_j(t)\rangle = \delta_{jk}\delta(t-t') \ . \tag{6.73}$$

In our case we have $\sqrt{I_2} = \beta\sqrt{\eta_1\eta_2}$ and $\sqrt{I_3} = \beta\sqrt{\eta_1\bar{\eta}_2}$.

Using the above expression (6.48) for $\delta\eta$, the fluctuation in the field incident on detector 2 is

$$\begin{aligned} X_2^{\rm cl}(t) &= \sqrt{\eta_1\eta_2}\, X_0^{\rm cl}(t-T_2) \\ &\quad + g\int_0^\infty h(t')[X_2^{\rm cl}(t-T-t') + \xi_2(t-T-t')]dt' \ . \end{aligned} \tag{6.74}$$

Assuming stable feedback (the conditions are the same as before) and solving this in Fourier domain we get

$$\tilde{X}_2^{\rm cl}(\omega) = \frac{\exp(-i\omega T_2)\sqrt{\eta_2\eta_1}\,\tilde{X}_0^{\rm cl}(\omega) + g\tilde{h}(\omega)\exp(-i\omega T)\tilde{\xi}_2(\omega)}{1-g\tilde{h}(\omega)\exp(-i\omega T)} \ . \tag{6.75}$$

But the photocurrent noise itself is proportional to

$$\tilde{X}_2^{\rm cl}(\omega) + \tilde{\xi}_2(\omega) = \frac{\exp(-i\omega T_2)\sqrt{\eta_2\eta_1}\,\tilde{X}_0^{\rm cl}(\omega) + \tilde{\xi}_2(\omega)}{1-g\tilde{h}(\omega)\exp(-i\omega T)} \ . \tag{6.76}$$

From (6.34) we find the spectrum

$$S_2^X(\omega) = \frac{1+\eta_2\eta_1[S_0^X(\omega)-1]}{|1-g\tilde{h}(\omega)\exp(-i\omega T)|^2} \ , \tag{6.77}$$

where I have defined

$$S_0^X(\omega) = 1 + \langle X_0^{\rm cl}(0)\tilde{X}_0^{\rm cl}(\omega)\rangle \tag{6.78}$$

as the spectrum which would be observed with no feedback and no beam splitters. With this identification, the expression (6.77) is identical with (6.60) derived using a quantized field.

In this formulation, the sub-shot noise of the in-loop current is no surprise at all. It is simply a result of feeding back an amplified negative version of the shot noise randomly produced at the detector so that the current at a later time will be anti-correlated with itself. The low noise is in the photocurrent only, not in the light beams which here are all completely classical. This explains why the out-of-loop detector does not produce a sub-shot noise signal. The photocurrent noise at that detector is proportional to

$$\begin{aligned} \xi_3(t) + X_3^{\rm cl}(t) &= \xi_3(t) + \sqrt{\eta_1\bar{\eta}_2}\, X_0^{\rm cl}(t-T_2) + g\sqrt{\bar{\eta}_2/\eta_2} \\ &\quad \times \int_0^\infty h(t')[X_2^{\rm cl}(t-T-t') + \xi_2(t-T-t')]dt' \ . \end{aligned} \tag{6.79}$$

Since ξ_3 is independent of everything else, it is noise which cannot be reduced by the feedback. Only the classical noise $X_0^{\rm cl}$ can be reduced, and that at the expense of introducing the extra uncorrelated noise ξ_2. Calculating the spectrum $S_X^3(\omega)$ again gives the same answer as the quantum theory.

6.2.6 QND Measurements of In-Loop Beams

On the basis of the above semiclassical theory it might be thought that all of the calculations of noise spectra made in this section relate only to the noise of photocurrents and say nothing about the noise properties of the light beams themselves. While this appears to be the case from the semiclassical theory, it is not really true, as can be seen from considering quantum non-demolition (QND) measurements [1]. In this context, a QND detector is one which can measure the intensity of light without absorbing it. Such devices cannot be described by semiclassical theory, which shows that this theory is not complete and hence cannot be expected to provide the correct intuition about the state of the light itself.

A specific model for a QND device will be considered in Sect. 6.4.1. Here we simply assume that a perfect QND device can measure the amplitude quadrature X of a continuum field without disturbing it beyond the necessary back-action from the Heisenberg uncertainty principle. In other words, the QND device should give a read-out at time t which can be represented by the operator $X(t)$. The correlations of this read-out will thus reproduce the correlations of $X(t)$. For a perfect QND measurement of X_2 and X_3, the spectrum will reproduce those of the conventional (demolition) photodetectors which measure these beams. This confirms that these detectors (assumed perfect) are indeed recording the true quantum fluctuations of the light impinging upon them.

What is more interesting is to consider a QND measurement on X_1. That is because the set up in Fig. 6.1 is equivalent (as mentioned above) to a set up without the second beam splitter, but instead with an in-loop photodetector with efficiency η_2. In this version, the beams b_2 and b_3 do not physically exist. Rather, b_1 is the in-loop beam and X_2 is the operator for the noise in the photocurrent produced by the detector. As shown above, this operator can have vanishing noise at low frequencies for $g \to -\infty$. However, this is not reflected in the noise in the in-loop beam, as recorded by our hypothetical QND device. Following the methods of Sect. 6.2.3, the spectrum of X_1 is

$$S_1^X(\omega) = \frac{1 + \eta_1[S_0(\omega) - 1] + g^2|\tilde{h}(\omega)|^2 \bar{\eta}_2/\eta_2}{|1 - g\tilde{h}(\omega)\exp(-i\omega T)|^2} . \tag{6.80}$$

In the limit $g \to -\infty$, this becomes at low frequencies

$$S_1^X(0) \to \frac{1 - \eta_2}{\eta_2} , \tag{6.81}$$

which is not zero for any detection efficiency η_2 finitely less than one. Indeed, for $\eta_2 < 0.5$ it is above shot-noise.

The reason that the in-loop amplitude quadrature spectrum is not reduced to zero for large negative feedback is that the feedback loop is feeding back noise $X_\mu(t)$ in the photocurrent fluctuation operator $X_2(t)$ which is independent of the fluctuations in the amplitude quadrature $X_1(t)$ of the in-loop light. The smaller η_2 is, the larger the amount of extraneous noise in the photocurrent and the larger the noise introduced into the in-loop light. In order to minimize the low-frequency noise in the in-loop light, there is an optimal feedback gain. In the case $S_0(\omega) = 1$ (a coherent input), this is given by

$$g_{\rm opt} = -\frac{\eta_2}{1-\eta_2} , \tag{6.82}$$

giving a minimum in-loop low-frequency noise spectrum

$$S_1^X(0)_{\rm min} = 1 - \eta_2 . \tag{6.83}$$

The fact that the detection efficiency does matter in the attainable in-loop squeezing shows that these are true quantum fluctuations.

6.2.7 A Squeezed Input

Although the feedback device discussed in this section cannot produce a free squeezed beam, it is nevertheless useful for reducing classical noise in the output beam b_3. It is easy to verify the result of [26] that if one wishes to reduce classical noise $S_0^X(\omega)-1$ at a particular frequency ω, then the optimum feedback is such that

$$g\tilde{h}(\omega)\exp(-i\omega T) = -\eta_1\eta_2[S_0^X(\omega) - 1] . \tag{6.84}$$

This gives the lowest noise level in the amplitude of b_3 at that frequency

$$S_3^X(\omega)_{\rm opt} = 1 + \frac{\bar{\eta}_2\eta_1[S_0^X(\omega)-1]}{1+\eta_2\eta_1[S_0^X(\omega)-1]} . \tag{6.85}$$

For large classical noise we have feedback proportional to $S_0^X(\omega)$ and an optimal noise value of $1/\eta_2$, as this approaches the limit of (6.66). The interesting regime [26] is the opposite one, where $S_0^X(\omega)-1$ is small, or even negative. This last case corresponds to to a squeezed input beam. Putting squeezing through a beam splitter reduces the squeezing in both output beams. In this case, with no feedback the residual squeezing in beam b_3 would be

$$S_3^X(\omega)_{\rm no} = 1 + \bar{\eta}_2\eta_1[S_0^X(\omega) - 1] , \tag{6.86}$$

which is closer to unity than $S_0^X(\omega)$. The optimal feedback (the purpose of which is to reduce noise) is, according to (6.84), positive. That is to say, destabilizing feedback actually puts back in beam b_3 some of the squeezing lost through the beam splitter. Since the required round-loop gain (6.84) is less than unity, the feedback loop remains stable.

This result highlights the nonclassical nature of squeezed fluctuations. When an amplitude squeezed beam strikes a beam splitter, the intensity at one output port is anticorrelated with that at the other, hence the need for positive feedback. Preliminary observations of this effect were reported in [33]. Of course the feedback can never put more squeezing into the beam than was present at the start. That is, $S_3^X(\omega)_{\rm opt}$ always lies between $S_0^X(\omega)$ and $S_3^X(\omega)_{\rm no}$. However, if we take the limit $\eta_1 \to 1$ and $S_0^X(\omega) \to 0$ (perfect squeezing to begin with) then all of this squeezing can be recovered, for any η_2. This effect might even be of practical use for attenuating highly squeezed sources, such as those produced by laser diodes, while retaining most of the squeezing [33].

6.3 Quantum Langevin Equations

In the preceding section, only continuum fields were considered, with all optical and electro-optical devices being treated as classical (i.e. deterministic) systems. Often one wishes to consider feedback onto quantum systems which may introduce extra noise terms into the equations. To do this one needs a theory for describing the dynamics of localized quantum systems interacting with continuum fields. In the optical regime this theory was put on a rigorous foundation by Gardiner and Collett [23]. This theory is based on an electric-dipole coupling, and involves key approximations which rely on a rapid oscillation (at optical frequency ω_0) of the system dipole due to the system Hamiltonian H_0. Let that dipole, in the interaction picture of H_0, be proportional to

$$c(t)\exp(-i\omega_0 t) + c^\dagger(t)\exp(i\omega_0 t) \ . \tag{6.87}$$

Here c and $c^\dagger$ are dimensionless Heisenberg-picture system operators, normalized so that $cc^\dagger|0\rangle = |0\rangle$, where $|0\rangle$ is the ground state of H_0.

Let the system be located at $z = 0$ and consider an incoming ($z < 0$) and outgoing ($z > 0$) continuum field with operator $b(z,t)$ as defined previously. Then, under the rotating wave approximation [2], the dipole coupling to the field at $z = 0$ can be modeled by the Hamiltonian

$$V(t) = i\sqrt{\gamma}[c(t)b^\dagger(0,t) - b(0,t)c^\dagger(t)] \ . \tag{6.88}$$

Here $\hbar = 1$ and $\gamma \ll \omega_0$ is the characteristic decay rate of the system. For a general interaction one would have to use both space and time arguments for the continuum field. However, the coupling considered here is strictly local. Both before ($z < 0$) and after ($z > 0$) the interaction, the field still freely propagates. This means that, although it is necessary to use a spatial as well as a temporal argument, the spatial argument need only have two values: *before* and *after*. In order to conform with pre-established usage [23], these values will be called "in" and "out". The input field is defined (in the Heisenberg picture) as

$$b_{\rm in}(t) = b(0^-,t) \ , \tag{6.89}$$

and the output field as

$$b_{\rm out}(t) = b(0^+, t) \ . \tag{6.90}$$

Both of these fields obey the commutation relations

$$[b(t), b^\dagger(t')] = \delta(t - t') \ , \tag{6.91}$$

at least for time differences shorter than the delay time in any feedback loop.

The interaction of the external field with the system at time t is described by the Hamiltonian (6.88). It is necessary to be careful in dealing with this because of the singular commutation relations (6.91). A convenient way is to use the quantum Itô stochastic calculus [23]. In the Heisenberg picture, the infinitesimal unitary transformation pertaining to the interaction at time t is

$$U(t, t + dt) = \exp\left[\sqrt{\gamma}c(t)dB_{\rm in}^\dagger(t) - \sqrt{\gamma}c(t)^\dagger dB_{\rm in}(t) - iH(t)dt\right] \ , \tag{6.92}$$

where H is a system Hamiltonian representing any small perturbation on top of the free energy (such as classical driving), and

$$dB_{\rm in}(t) = b_{\rm in}(t)dt \tag{6.93}$$

is the quantum analogue of the Wiener increment [23]. Note that the field operator in the unitary transformation (6.92) is the bath immediately before it interacts with the system, rather than the bath at the instant it is interacting with the system. This is the appropriate operator if one treats the stochastic increment $dB_{\rm in}(t)$ in the Itô sense. This means that (6.92) must be expanded to second order in the increment. For an input in the vacuum state, the operator $dB_{\rm in}$ has a vanishing first order moment, and a single nonvanishing second-order moment [23]

$$dB_{\rm in}(t)dB_{\rm in}^\dagger(t) = dt \ . \tag{6.94}$$

This could be thought of as vacuum noise.

Applying the unitary transformation (6.92) to an arbitrary system operator $s(t)$ yields

$$\begin{aligned} s(t + dt) &= U^\dagger(t, t + dt)s(t)U(t, t + dt) \\ &= s + \gamma\left(c^\dagger sc - \tfrac{1}{2}c^\dagger cs - \tfrac{1}{2}sc^\dagger c\right) dt \\ &\quad - [-iHdt + \sqrt{\gamma}dB_{\rm in}^\dagger(t)c - \sqrt{\gamma}c^\dagger dB_{\rm in}(t), s] \ , \end{aligned} \tag{6.95}$$

where all system operators in (6.95) have the time argument t. This is what I will refer to as a quantum Langevin equation (QLE) for s. Because $dB_{\rm in}(t)$ is the bath operator before it interacts with the system, it is independent of the system operator $s(t)$. Hence one can derive

$$\langle\dot{s}\rangle = \left\langle\left[\gamma\left(c^\dagger sc - \tfrac{1}{2}c^\dagger cs - \tfrac{1}{2}sc^\dagger c\right) + i[H, s]\right]\right\rangle \ . \tag{6.96}$$

From this it apparent that there is a Schrödinger picture representation of the system dynamics. In this picture

$$\langle \dot{s}(t) \rangle = \mathrm{Tr}[\dot{\rho}(t) s] \ , \tag{6.97}$$

where $\rho(t)$ is the state matrix for the system which evidently obeys

$$\dot{\rho} = \gamma \mathcal{D}[c]\rho - i[H, \rho] \ , \tag{6.98}$$

where for arbitrary operators A and B

$$\mathcal{D}[A]B = ABA^\dagger - \tfrac{1}{2}A^\dagger AB - \tfrac{1}{2}BA^\dagger A \ . \tag{6.99}$$

An equation of the form (6.98) is known as a master equation. I will discuss the master equation further in Sect. 6.5.1.

Although the noise terms in (6.95) do not contribute to (6.96), they are necessary in order for (6.95) to be a valid Heisenberg equation of motion. If they are omitted then the commutation relations of the system will not be preserved. As well as giving the evolution of the system, (6.92) transforms the input field operator into the output field operator:

$$b_{\mathrm{out}}(t) = U^\dagger(t, t+dt) b_{\mathrm{in}}(t) U(t, t+dt) = b_{\mathrm{in}}(t) + \sqrt{\gamma} c(t) \ . \tag{6.100}$$

Assuming that $b_{\mathrm{out}}(t)$ has a large coherent amplitude then one can then derive an expression for its linearized intensity fluctuations, which are proportional to

$$X_{\mathrm{out}}(t) = X_{\mathrm{in}}(t) + \sqrt{\gamma}\delta x(t) \ , \tag{6.101}$$

where $X_{\mathrm{out}}(t)$ is the fluctuation quadrature operator for the output field as usual, and where $\delta x = x - \langle x \rangle_{\mathrm{ss}}$, where $x = c + c^\dagger$ is a system quadrature with mean steady-state value $\langle x \rangle_{\mathrm{ss}}$.

6.4 Feedback Based on Nonlinear Measurements

6.4.1 QND Measurements

Section 6.2 showed that it was not possible to create squeezed light in the conventional sense using ordinary photodetection and linear feedback. Although the quantum theory appeared to show that the light which fell on the detector in the feedback loop was sub-shot noise, this could not be extracted because it was demolished by the detector. An obvious way around this would be to use a quantum non-demolition quadrature (QND) detector. One way to achieve this is for two fields of different frequency to interact via a nonlinear refractive index. In order to obtain a large nonlinearity, large intensities are required. It is easiest to build up large intensities by using a resonant cavity. To describe this requires the theory of quantum Langevin equations just presented.

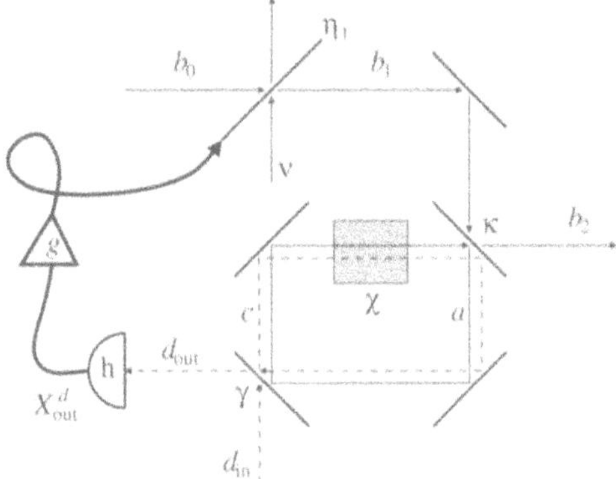

Fig. 6.2. Diagram for a traveling-wave feedback experiment based on a QND measurement. Traveling fields are denoted b and d. The first beam splitter transmittance η_1 is variable. A cavity (drawn as a ring cavity for convenience) supports two modes, a (solid line) and c (dashed line). The decay rates for these two modes are κ and γ respectively. They are coupled by a nonlinear optical process indicated by the crystal labeled χ. The perfect homodyne detection at the detector labeled h yields a photocurrent proportional to $X^d_{\text{out}} = d_{\text{out}} + d^\dagger_{\text{out}}$

Consider the apparatus shown in Fig. 6.2. The purpose of the detector in the feedback loop is to make a QND measurement of the quadrature X^b_{in} of the field b_{in}. This fields drives a cavity mode with decay rate κ described by annihilation operator a. This mode is coupled to a second mode with annihilation operator c and decay rate γ. Ignoring practicalities, I will take the coupling between the two modes to be the ideal QND coupling, as considered initially by Hillery and Scully [34] and Yurke [35], namely

$$H = \frac{\chi}{2} x^a y^c \,, \tag{6.102}$$

where

$$x^a = a + a^\dagger \,, \qquad y^c = -ic + ic^\dagger \,. \tag{6.103}$$

As described in [36], this Hamiltonian could in principle be realized by a crystal with a $\chi^{(2)}$ nonlinearity, combined with other processes. This Hamiltonian commutes with the x^a quadrature of mode a, and causes this to drive the x^c quadrature of mode c. Thus measuring the X^d_{out} quadrature of the output field d_{out} from mode c will give a QND measurement of $a + a^\dagger$, which is approximately a QND measurement of X^b_{in}.

Following the analysis of [36], the quantum Langevin equations for the quadrature operators are

$$\dot{x}^a = -\frac{\kappa}{2} x^a - \sqrt{\kappa}\, X^b_{\text{in}} \,, \tag{6.104}$$

$$\dot{x}^c = -\frac{\gamma}{2} x^c - \sqrt{\gamma}\, X^d_{\text{in}} + \chi x^a \,. \tag{6.105}$$

These can be solved in the frequency domain as

$$\tilde{x}^a(\omega) = -\frac{\sqrt{\kappa}\,\tilde{X}^b_{\rm in}(\omega)}{\kappa/2 + i\omega}\,, \tag{6.106}$$

$$\tilde{x}^c(\omega) = -\frac{\sqrt{\gamma}\,\tilde{X}^d_{\rm in}(\omega) + \sqrt{\kappa}\chi\tilde{X}^b_{\rm in}(\omega)/(\kappa/2 + i\omega)}{\gamma/2 + i\omega}\,. \tag{6.107}$$

The quadrature of the output field $d_{\rm out} = d_{\rm in} + \sqrt{\gamma}c$ is therefore

$$\tilde{X}^d_{\rm out}(\omega) = -\frac{\gamma\kappa Q\,\tilde{X}^b_{\rm in}(\omega)/(\kappa + 2i\omega) + (\gamma - 2i\omega)\tilde{X}^d_{\rm in}(\omega)}{\gamma + 2i\omega}\,, \tag{6.108}$$

where I have defined a quality factor for the measurement

$$Q = 4\chi/\sqrt{\gamma\kappa}\,. \tag{6.109}$$

In the limits $\omega \ll \kappa, \gamma$ and $Q \gg 1$ we have

$$\tilde{X}^d_{\rm out}(\omega) \simeq -Q\tilde{X}^b_{\rm in}(\omega)\,, \tag{6.110}$$

which shows that a measurement (by homodyne detection) of the X quadrature of $d_{\rm out}$ does indeed effect a measurement of the low frequency variation in $X^b_{\rm in}$.

To see that this measurement is a QND measurement we have to calculate the statistics of the output field from mode a, that is $b_{\rm out}$. From the solution (6.106) we find

$$\tilde{X}^b_{\rm out}(\omega) = -\frac{\kappa/2 - i\omega}{\kappa/2 + i\omega}\tilde{X}^b_{\rm in}(\omega)\,. \tag{6.111}$$

That is, apart from an irrelevant phase factor the output field is identical to the input field, as required for a QND measurement. Of course we cannot expect the other quadrature to remain unaffected, because of the uncertainty principle. Indeed we find

$$\tilde{Y}^b_{\rm out}(\omega) = -\frac{(\kappa - 2i\omega)}{\kappa + 2i\omega}\tilde{Y}^b_{\rm in}(\omega) + \frac{Q\gamma\kappa\,\tilde{Y}^d_{\rm in}(\omega)/(\gamma + 2i\omega)}{\kappa + 2i\omega}\,, \tag{6.112}$$

which shows that noise has been added to $Y_{\rm in}$. Indeed, in the good measurement limit which gave the result (6.110) we find the phase quadrature output to be dominated by noise:

$$\tilde{Y}^b_{\rm out}(\omega) \simeq Q\tilde{Y}^d_{\rm in}(\omega)\,. \tag{6.113}$$

6.4.2 QND-Based Feedback

We know wish to show how a QND measurement, such as that just considered, can be used to produce squeezing via feedback. The physical details of how the feedback can be achieved are as outlined in Sect. 6.2.1. In particular, in the limit that the transmittance η_1 of the modulated beam splitter goes to unity, the effect of the modulation is simply to add an arbitrary signal to the amplitude quadrature of the controlled beam. That is, the modulated beam can be taken to be

$$b_1(t) = b_0(t - \tau_1) + \beta\delta\eta_1(t - \tau_1) \ , \tag{6.114}$$

where b_0 is the beam incident on the modulated beam splitter, as in Sect. 6.2.1. In the present case, $b_1(t)$ is then fed into the QND device, as shown in Fig. 6.2, so

$$b_{\text{in}}(t) = b_1(t) \ , \tag{6.115}$$

and the modulation is controlled by the photocurrent from a (assumed perfect) homodyne measurement of X^d_{out}:

$$\delta\eta_1(t) = \frac{g}{-Q\beta}\int_0^\infty h(s)X^d_{\text{out}}(t - \tau_0 - s)ds \ . \tag{6.116}$$

Here τ_0 is the delay time in the feedback loop, including the time of flight from the cavity for mode c to the homodyne detector, and $h(s)$ is as before.

Substituting (6.114)–(6.116) into the results of the preceding subsection we find

$$\tilde{X}^b_{\text{out}}(\omega) = -\frac{\kappa - 2i\omega}{\kappa + 2i\omega}\frac{\tilde{X}^b_{\text{in}}(\omega) + \tilde{X}^d_{\text{in}}(\omega)g\tilde{h}(\omega)Q^{-1}\frac{\gamma-2i\omega}{\gamma+2i\omega}}{1 - g\tilde{p}(\omega)\tilde{h}(\omega)\exp(-i\omega T)} \ , \tag{6.117}$$

where $T = \tau_1 + \tau_0$ is the total round-trip delay time, and

$$\tilde{p}(\omega) = \frac{\gamma\kappa}{(\kappa + 2i\omega)(\gamma + 2i\omega)} \tag{6.118}$$

represents the frequency response of the two cavities. If we assume that the field d_{in} is in the vacuum state then we can evaluate the spectrum of amplitude fluctuations in X^b_{out} to be

$$S^X_{\text{out}}(\omega) = \frac{S^X_{\text{in}}(\omega) + g^2Q^{-2}}{|1 - g\tilde{p}(\omega)\tilde{h}(\omega)\exp(-i\omega T)|^2} \ . \tag{6.119}$$

Clearly in the limit $Q \to \infty$, the added noise term in the amplitude spectrum can be ignored. Then for sufficiently large negative g the feedback will produce a sub-shot noise spectrum. Note the difference between this case and that of Sect. 6.2. Here the squeezed light is not part of the feedback loop; it is a free beam. The ultimate limit to how squeezed the beam can be is caused by the noise in the measurement. In the limit $g \to -\infty$ we find

$$S^X_{\text{out}}(\omega)_{\text{min}} = |Q\tilde{p}(\omega)\tilde{h}(\omega)|^{-2} \ . \tag{6.120}$$

Since b_{out} is a free field, not part of any feedback loop, it should obey the standard commutation relations. This is the case, as can be verified from the expression (6.112) for $Y_{\rm out}^{b}$ (which is unaffected by the feedback). Consequently, the spectrum for the phase quadrature

$$S_{\rm out}^{Y}(\omega) = S_{\rm in}^{Y}(\omega) + |Q\tilde{p}(\omega)|^2 \tag{6.121}$$

shows the expected increase in noise. It is not difficult to show that the uncertainty product $S_{\rm out}^{Y}(\omega)S_{\rm out}^{X}(\omega)$ is greater than or equal to unity, as required.

6.4.3 Parametric Down Conversion

The preceding section showed that feedback based on a perfect QND measurement can produce squeezing. This has never been done experimentally because of the difficulty of building a perfect QND measurement apparatus. However, it turns out that QND measurements are not the only way to produce squeezing via feedback. Any mechanism which produces correlations between the beam of interest and another beam which are more "quantum" than the correlations between the two outputs of a linear beam splitter can be the basis for producing squeezing via feedback. Such a mechanism must involve some sort of optical nonlinearity, which is why this section is entitled "Feedback based on Nonlinear Measurements".

The production of amplitude squeezing by feeding back a quantum-correlated signal was predicted [37] and observed [38] by Walker and Jakeman in 1985, using the process of parametric down-conversion. An improved feedback scheme for the same system was later used by Tapster, Rarity and Satchell [39] to obtain an inferred amplitude spectrum $S_1^X(\omega)_{\rm min} = 0.72$ over a limited frequency range.

The essential idea is as follows. A non-degenerate optical parametric oscillator (NDOPO) can be realized by pumping a crystal with a $\chi^{(2)}$ nonlinearity by a laser of frequency ω_1 and momentum $\mathbf{k}_1$. For particular choices of ω_1 the crystal can be aligned so that a pump photon can be transformed into a pair of photons each with frequencies ω_2, ω_3 and momenta $\mathbf{k}_2, \mathbf{k}_3$ with $\omega_1 = \omega_2+\omega_3$ and $\mathbf{k}_1 = \mathbf{k}_2 + \mathbf{k}_3$. A pair of down-converted photons will thus be correlated in time because they are produced from a single pump photon. On a more macroscopic level, this means that the amplitude quadrature fluctuations in the two down-converted beams are positively correlated, and the coefficient of correlation can in principle be very close to unity. In this ideal limit, measuring the intensity of beam 2 should give a readout identical to that obtained from measuring beam 3. In effect, the measurement of beam 2 is like a QND measurement of beam 3. Thus, feeding back the photocurrent from beam 2 with an negative gain should be able to reduce the noise in beam 3 below the shot-noise limit. In the limit of perfect detection, the optimum gain becomes arbitrarily large.

In [37,39] the negative feedback was effected by controlling the power of the pump laser (which controls the rate at which photon pairs are produced). This maintains the symmetry of the experiment so that the fed-back photocurrent from beam 2 has the same statistics as the photocurrent from the free beam 3. In other words, they will both be below the shot-noise, whereas without feedback they are both at best shot-noise limited. However, it is not necessary to preserve the symmetry in this way. The measured photocurrent from beam 2 can be fed *forward* to control the amplitude fluctuations in beam 3 (for example by using an electro-optic modulator as described in Sect. 6.2.1). This feedforward was realized experimentally by Mertz et al. in 1990 [40] achieving similar results to that obtained by feedback [41]. The lesson is that unless one is concerned with light inside a feedback loop, there is no essential difference between feedback and feedforward. Indeed, the squeezing produced by QND-based feedback discussed in Sect. 6.4.3 could equally well have been produced by QND-based feedforward.

6.4.4 Second Harmonic Generation

A third example of using quantum correlations between two beams to good effect in a feedback (or feedforward) loop is that of second-harmonic generation. Like parametric down conversion, this involves a $\chi^{(2)}$ crystal, but here the two low frequency modes are assumed to be degenerate, so that $\omega_2 = \omega_3$ and $\mathbf{k}_2 = \mathbf{k}_3$, and the polarizations are the same also. For simplicity, I will call the low frequency mode the red mode and the high frequency mode with frequency $\omega_1 = 2\omega_2$ the green mode. For second harmonic generation, it is the red mode which is pumped (that is, the reverse of parametric down-conversion). In this case, the pumped mode (red) is also treated as a quantum system so that the two beams are the red and green beams. The system is easily treated using quantum Langevin equations [42].

The proposal made in [43] is as follows. The first harmonic (red) resides in a good cavity and is driven by a laser at the end with very high reflectivity. The second harmonic (green), which is generated in the red cavity, is reflected at one end only (giving two passes) so that it forms a single output beam. By itself (that is, in the absence of feedback) this system produces amplitude squeezing in both the first and second harmonic [44]. The amplitude squeezing in the red can be understood to be due to two-photon absorption, which is the effect of the adiabatically eliminated green mode. Such nonlinear absorption preferentially damps large fluctuations and so reduces the variance. The antibunching in the green mode can be attributed to the fact that the creation of a green photon requires the loss of two red photons, which reduces the chances of this event re-occurring.

Without feedback, the optimum low frequency squeezing in the red mode output is

$$S_r^X(0) = \frac{2}{3} \quad \text{for} \quad \chi = \frac{1}{3} \,, \tag{6.122}$$

and that in the green mode is

$$S_g^X(0) = \frac{1}{9} \quad \text{for} \quad \chi \to \infty \,. \tag{6.123}$$

Here χ is the ratio of the nonlinear intensity loss rate to the linear intensity loss rate for the red mode. Now there are two possible ways to feed back onto the driving of the system. The first is to use the green photocurrent to control the amplitude of the red driving, in order to try to reduce the noise in the red output light. A linearized treatment gives a new minimum of

$$S_r^X(0) = \frac{1}{2} \quad \text{for} \quad \chi = 1 \ , \quad g = \frac{1}{2} \ , \tag{6.124}$$

which is significantly less than the no-feedback value (6.122). Here, g is the low frequency loop transfer function, equal to the round-loop gain of the feedback [28]. Note that it is positive, corresponding to destabilizing feedback. This is different from the cases of QND-based feedback and down-conversion-based feedback.

The other sort of feedback is to control the driving of the red mode using the detected red photocurrent, looking to enhance the squeezing in the green output. It turns out that the minimum $S_g^X(0)$ of (6.123) as $\chi \to \infty$ cannot be lowered by feedback. However, for finite values of χ, the feedback can give a definite improvement. For example, with no feedback

$$S_g^X(0) = \frac{1}{2} \quad \text{for} \quad \chi = 1 \ , \tag{6.125}$$

whereas the optimal feedback gives

$$S_g^X(0) = \frac{1}{3} \quad \text{for} \quad \chi = 1 \ , \quad g = \frac{1}{4} \ . \tag{6.126}$$

All of these results are calculated for the case of unit efficiency photodetectors. The effect of non-unit efficiency is to reduce the effectiveness of the feedback, and to alter the conditions of optimality. The general solution, including the full spectrum with an arbitrary transfer function $g\tilde{h}(\omega)$, is contained in [43].

6.5 Quantum Trajectories

6.5.1 The Master Equation

The Quantum Langevin Equation (QLE) considered in Sect. 6.3 are Heisenberg picture equations which detail the effect of the bath on the system. They also, through the input-output relations, specify the effect of the system on the bath, and hence the relation between the system evolution and the measured photocurrents. For systems which have linear QLEs, this is the easiest method to analyze their evolution. However, even some simple systems (such as a two-level atom) do not have linear QLEs. Also, sometimes a better intuition about feedback can be gained by working in the Schrödinger picture rather than the Heisenberg picture.

If one is interested only in the evolution of the system then there is a simple Schrödinger picture equivalent to the QLE. This is the quantum master equation, an example of which was already derived as (6.98) from the corresponding QLE. Unlike a QLE, a master equation is always a linear equation. It is possible to derive the master equation directly from the system-bath coupling without deriving the QLE, simply by tracing over the bath. I present this derivation below, because it shows clearly that this step (ignoring the bath) is not an essential one. If instead of ignoring the bath, one measures it, then one obtains a quantum trajectory equation. This is a different sort of Schrödinger picture equivalent to the QLE which is more general than the master equation, as it can also relate the system evolution to the measured photocurrents. This is necessary if one is to consider feeding back the photocurrent to alter the dynamics of the system.

For convenience, I will work in the interaction picture, rather than the Schrödinger picture, so that the free evolution causing the oscillation of the dipole can be ignored. The interaction between the system and the bath is now given by

$$V(t) = i\hbar\sqrt{\gamma}\,[b^\dagger(t)c - c^\dagger b(t)]\ , \tag{6.127}$$

where c is a slowly-varying interaction picture operator, and $b(t)$ is also a slowly-varying interaction picture bath operator. The time-dependence is maintained for the bath operator, because the free Hamiltonian of the bath causes propagation at the speed of light, so a new part of the bath interacts with the system at each new point in time. These parts are labeled by the time of interaction t. Thus, $b(t)$ is an operator in the Hilbert space for a particular part of the bath. Each part has its own state matrix $\mu(t)$. For the incoming field to be a bath requires that its total state matrix be the direct product of the state matrices of the parts [14]. That is to say, the temporally separate parts of the bath must be unentangled.

Let the system at time t be known to be $\rho(t)$. Thus the initial state of the system and (relevant part of the) bath at time t is

$$R(t) = \mu(t) \otimes \rho(t)\ . \tag{6.128}$$

The infinitesimally evolved state is

$$R(t+dt) = U(t,t+dt)[\mu(t)\otimes\rho(t)]U^\dagger(t,t+dt)\ , \tag{6.129}$$

where

$$U(t,t+dt) = \exp\left[\sqrt{\gamma}\,dB^\dagger(t)c - c^\dagger\sqrt{\gamma}\,dB(t) - iH dt\right]\ , \tag{6.130}$$

and H is the system Hamiltonian in the interaction picture. If the input bath is in the vacuum state then all of the first and second order moments of $dB(t)$ vanish except, as in (6.94),

$$\mathrm{Tr}[dB(t)dB^\dagger(t)\mu(t)] = dt\ . \tag{6.131}$$

Thus, it is necessary to expand some of the terms in $U(t, t+dt)$ to second order. The result for $R(t+dt)$ is

$$\begin{aligned}&\mu(t)\otimes\rho(t)+\sqrt{\gamma}\left[dB^\dagger(t)c-c^\dagger dB(t),\mu(t)\otimes\rho(t)\right]\\&+\gamma\left\{dB^\dagger(t)\mu(t)dB(t)\otimes c\rho(t)c^\dagger-\tfrac{1}{2}dB(t)dB^\dagger(t)\mu(t)\otimes c^\dagger c\rho\right.\\&\left.-\tfrac{1}{2}\mu(t)dB(t)dB^\dagger(t)\otimes\rho c^\dagger c\right\}-i[H dt,\mu(t)\otimes\rho(t)]\ .\end{aligned}\tag{6.132}$$

The infinitesimally evolved reduced state matrix for the system is given by

$$\rho(t+dt)=\mathrm{Tr}_\mu[R(t+dt)]\ .\tag{6.133}$$

Taking the trace over the bath state in (6.132) yields the master equation (6.98):

$$\dot\rho=\gamma\mathcal{D}[c]\rho-i[H,\rho]\ .\tag{6.134}$$

The first term here represents irreversible evolution (damping in this case), and is of the unique form for such evolution as proved by Lindblad [45]. The second term represents unitary evolution which is of course reversible.

6.5.2 Photon Counting

To consider measuring the output bath it is useful to explicitly write down its state. Before interacting with the system, the input bath state is $\mu(t)=|0\rangle\langle 0|$ for all time t, where $|0\rangle$ is the lowest eigenstate for $a^\dagger a$. Here, $a=\sqrt{dt}\,b_0(t)=dB_0(t)/\sqrt{dt}$, so that $a^\dagger a$ has integer eigenvalues representing the number of photons arriving in the interval of time $[t,t+dt)$. Let the state of the system at time t be $\rho(t)$, independent of $\mu(t)$. The entangled state after the interaction of duration dt is, from (6.132)

$$\begin{aligned}R(t+dt)=&|0\rangle\langle 0|\otimes\rho(t)+\sqrt{\gamma dt}\left[|1\rangle\langle 0|\otimes c\rho(t)+|0\rangle\langle 1|\otimes\rho(t)c^\dagger\right]\\&-idt|0\rangle\langle 0|\otimes[H,\rho(t)]+\ \gamma dt\left\{|1\rangle\langle 1|\otimes c\rho(t)c^\dagger\right.\\&\left.-\tfrac{1}{2}|0\rangle\langle 0|\otimes\left[c^\dagger c\rho(t)+\rho(t)c^\dagger c\right]\right\}\ .\end{aligned}\tag{6.135}$$

The free dynamics of the electromagnetic field will now remove the bath state from the system, so that the entanglement in (6.135) will be maintained. If the outgoing bath is ignored, then the system obeys the master equation (6.134).

To obtain information about the system, the outgoing field must be measured. The obvious measurement to consider is photon counting. This I will model by projecting the field states onto eigenstates of $a^\dagger a$. It is possible to consider specific models for photon detectors, usually based on atomic systems [2]. However, these simply remove the measurement step, where possibilities become actualities, one step further along the von Neumann chain [46]. There is little which results from such models which cannot be achieved by including losses and convolving the classical photocurrent with an empirically derived detector response function. Therefore, I will simply use projection measurement operators

$$P_0 = |0\rangle\langle 0| \; ; \quad P_1 = |1\rangle\langle 1| \; . \tag{6.136}$$

If there is a null count then the unnormalized state matrix of the system and bath (whose norm gives the probability of this outcome) is given by

$$\tilde{R}_0(t+dt) = P_0 R(t+dt) P_0 = P_0 \otimes \tilde{\rho}_0(t+dt) \; , \tag{6.137}$$

where

$$\tilde{\rho}_0(t+dt) = \rho(t) - \gamma \tfrac{1}{2}\{c^\dagger c, \rho(t)\}dt - i[H, \rho(t)]dt \; . \tag{6.138}$$

This has a norm only infinitesimally different from one, so for almost all time intervals no photons are detected in the output.

If a photon is detected, then $\tilde{R}_1(t+dt) = P_1 \otimes \tilde{\rho}_1(t+dt)$. That is, the system jumps into the unnormalized conditioned state

$$\tilde{\rho}_1(t+dt) = \gamma c \rho(t) c^\dagger dt \; . \tag{6.139}$$

The norm of this state matrix is equal to the probability of a detection occurring in the interval $[t, t+dt)$, and is equal to $\gamma\langle c^\dagger c\rangle dt$. Clearly, the unconditioned master equation evolution (6.134) is retrieved by averaging over the two possible results

$$\rho(t+dt) = \tilde{\rho}_0(t+dt) + \tilde{\rho}_1(t+dt) \; . \tag{6.140}$$

It is useful and elegant to reformulate this evolution in the form of an explicitly stochastic evolution equation. This equation of motion specifies the quantum trajectory of the system. Since the measurement result is a point process, it can be represented by a random variable $dN_c(t)$ representing the increment (either zero or one) in the photon count in the interval $[t, t+dt)$. It is formally defined by

$$\mathrm{E}[dN_c(t)] = \mathrm{Tr}[c^\dagger c \rho_c(t)]\gamma dt \; , \tag{6.141}$$

$$dN_c(t)^2 = dN_c(t) \; . \tag{6.142}$$

Here the subscript c indicates that the quantity to which it is attached is conditioned on previous measurement results, arbitrarily far back in time. The conditioned state matrix obeys the stochastic master equation (SME)

$$d\rho_c(t) = \left\{ dN_c(t)\mathcal{G}[c] - dt\mathcal{H}\left[iH + \tfrac{1}{2}\gamma c^\dagger c\right] \right\} \rho_c(t) \; . \tag{6.143}$$

Here, the nonlinear (in ρ) superoperators $\mathcal{G}$ and $\mathcal{H}$ are defined by

$$\mathcal{G}[r]\rho = \frac{r\rho r^\dagger}{\mathrm{Tr}[r\rho r^\dagger]} - \rho \; , \tag{6.144}$$

$$\mathcal{H}[r]\rho = r\rho + \rho r^\dagger - \mathrm{Tr}[r\rho + \rho r^\dagger]\rho \; . \tag{6.145}$$

The nonlinearity of the SME (6.143) is indicative of the fundamental nonlinearity of quantum measurements. The original master equation for $\rho(t) = \mathrm{E}[\rho_c(t)]$ can be restored simply by replacing $dN_c(t)$ in (6.143) by its ensemble average value (6.141).

Because of the assumed perfect detection, the stochastic equation for the state matrix is equivalent to a stochastic equation for the state vector. The unraveling of the master equation into a stochastic Schrödinger equation representing photon counting is the most commonly used quantum trajectory for numerical simulations [47–52]. From the point of view of measurement theory, the stochastic master equation is of more use than the stochastic Schrödinger equation because it is more transparently related to the unconditioned master equation and because it can be generalized to cope with inefficient detectors. This will be dealt with explicitly for the case of homodyne detection.

6.5.3 Homodyne Detection Theory

The unraveling of the master equation (6.134) as a quantum trajectory is not unique. Different detection schemes will result in different quantum trajectory equations. For squeezing, the most useful detection technique is homodyne detection. In the simplest configuration, the output field of the cavity, $b_{\mathrm{out}} = \nu + \sqrt{\gamma}c$, is sent through a beam splitter of transmittance η very close to one. Into the other input port of the beam splitter is injected a very strong coherent field. This has the same frequency as the system dipole, and is known as the local oscillator. The transmitted field is then represented by the operator

$$b_1 = \nu + \sqrt{\gamma}c + \beta \ , \tag{6.146}$$

where β is a complex number representing a coherent amplitude, such that $|\beta|^2/(1-\eta)$ is equal to the input photon flux of the local oscillator. The photodetection operators are then applied as above, with the annihilation operator defined as $a = b_1\sqrt{dt}$.

Let the coherent field β be real, so that the homodyne detection leads to a measurement of the x quadrature of the system dipole. Also, let us measure time in units of γ^{-1} so that this parameter disappears from our equations. Then the rate of photodetections at the (perfect) detector

$$\mathrm{E}[dN_c(t)] = \mathrm{Tr}[(\beta^2 + \beta x + c^\dagger c)\rho_c(t)]dt \ . \tag{6.147}$$

where $x = c + c^\dagger$ as previously. In the limit that β is much larger than c, this rate consists of a large constant term plus a term proportional to x, plus a small term. It is not difficult to show that the SME for the conditioned state matrix is altered from (6.143) to

$$d\rho_c(t) = \left\{dN_c(t)\mathcal{G}[c+\beta] + dt\mathcal{H}\left[-iH - \beta c - \tfrac{1}{2}c^\dagger c\right]\right\}\rho_c(t) \ . \tag{6.148}$$

The ideal limit of homodyne detection is when the local oscillator amplitude goes to infinity. In this limit, the rate of photodetections goes to infinity, but the effect of each on the system goes to zero, because the field being detected is almost entirely due to the local oscillator. Thus, it should be possible to approximate the photocurrent by a continuous function of time, and also to derive a smooth evolution equation for the system. This was done first by Carmichael [13]. A more rigorous working of the derivation is found in [53]. The result is that over a time much longer than $\beta^{-3/2}$, but much smaller than unity, the system evolution can be approximated by the SME

$$d\rho_{\rm c}(t) = -i[H, \rho_{\rm c}(t)]dt + \mathcal{D}[c]\rho_{\rm c}(t)dt + dW(t)\mathcal{H}[c]\rho_{\rm c}(t) \; . \tag{6.149}$$

Here $dW(t)$ is an infinitesimal Wiener increment [54] satisfying

$$\mathrm{E}[dW(t)] = 0 \; , \tag{6.150}$$
$$dW(t)^2 = dt \; . \tag{6.151}$$

Thus, the jump evolution of (6.148) has been replaced by diffusive evolution. Equation (6.149) is, by its derivation, an Itô stochastic master equation, where the equal-time stochastic increment $dW(t)$ is independent of the state of the system $\rho_{\rm c}(t)$. It is trivial to see that the ensemble average evolution reproduces the nonselective master equation (6.134) by eliminating the noise term.

Just as the $\beta \to \infty$ leads to continuous evolution for the state, it also changes the point process photocount into a continuous photocurrent with white noise. Removing the constant local-oscillator contribution and scaling appropriately gives

$$I_{\rm c}^{\rm hom}(t) \equiv \lim_{\delta t \to 0} \lim_{\beta \to \infty} \frac{\delta N_{\rm c}(t) - \beta^2 \delta t}{\beta \delta t} = \langle x \rangle_{\rm c}(t) + \xi(t) \; , \tag{6.152}$$

where $\xi(t) = dW(t)/dt$. It is not difficult to see that if the detector efficiency is η, the homodyne photocurrent becomes

$$I_{\rm c}^{\rm hom}(t) \equiv \lim_{\delta t \to 0} \lim_{\beta \to \infty} \frac{\delta N_{\rm c}(t) - \eta \beta^2 \delta t}{\sqrt{\eta}\, \beta \delta t} = \sqrt{\eta}\, \langle x \rangle_{\rm c}(t) + \xi(t) \; , \tag{6.153}$$

and the SME (6.149) is modified to

$$d\rho = -i[H, \rho_{\rm c}(t)]dt + \mathcal{D}[c]\rho_{\rm c}(t)dt + \sqrt{\eta} dW(t)\mathcal{H}[c]\rho_{\rm c}(t) \; . \tag{6.154}$$

6.5.4 Homodyne-Mediated Feedback

In the following section I will consider the use of feedback to produce or enhance squeezing in intracavity fields. Since squeezing is the reduction in fluctuations of one quadrature of a field, the obvious sort of feedback to consider is one using the homodyne photocurrent obtained from measuring the output field. Here I will develop the theory for describing this sort of feedback. As well as being more relevant for our purposes than feedback using the direct detection photocurrent [12], it is also somewhat easier to treat theoretically, which is why it was derived first [10,11].

In principle, the homodyne photocurrent could be subject to any sort of filtering prior to being fed back, including nonlinear filtering. However it turns out that for the applications we wish to consider, only linear filtering is desired. Also, rather than using a response function $h(t)$ I will simply take the feedback to be delayed by a time T. That means that the evolution due to the feedback can simply be written as

$$[\dot{\rho}_{\rm c}(t)]_{\rm fb} = I_{\rm c}^{\rm hom}(t-T)\mathcal{K}\rho_{\rm c}(t)/\sqrt{\eta} \; , \tag{6.155}$$

where $\mathcal{K}$ is a superoperator. Since $I_{\rm c}^{\rm hom}(t)$ may be negative, $\mathcal{K}$ must be such as to give valid evolution irrespective of the sign of time. That is to say, it must give reversible evolution with

$$\mathcal{K}\rho \equiv -i[F,\rho] \tag{6.156}$$

for some Hermitian operator F. In other words, we can represent the effect of the feedback by the Hamiltonian

$$H_{\rm fb} = F\left[\langle c+c^\dagger\rangle_{\rm c}(t-T) + \xi(t-T)/\sqrt{\eta}\right] \; . \tag{6.157}$$

Because the stochasticity in the measurement (6.154) and the feedback (6.155) is Gaussian white noise, it is relatively simple to determine the effect of the feedback. Bearing in mind that the feedback must act after the measurement, and that (6.155) must be interpreted as a Stratonovich equation [11], the result for the total conditioned evolution of the system is

$$\begin{aligned}\rho_{\rm c}(t+dt) = &\left\{1 + \mathcal{K}[\langle c+c^\dagger\rangle_{\rm c}(t-T)dt + dW(t-T)/\sqrt{\eta}] + \frac{1}{2\eta}\mathcal{K}^2 dt\right\} \\ &\times\left\{1 + \mathcal{H}[-iH]dt + \mathcal{D}[c]dt + \sqrt{\eta}dW(t)\mathcal{H}[c]\right\}\rho_{\rm c}(t) \; . \end{aligned} \tag{6.158}$$

For T finite, this becomes

$$\begin{aligned}d\rho_{\rm c}(t) = &\, dt\left\{\mathcal{H}[-iH] + \mathcal{D}[c] + \langle c+c^\dagger\rangle_{\rm c}(t-T)\mathcal{K} + \frac{1}{2\eta}\mathcal{K}^2\right\}\rho_{\rm c}(t) \\ &+ dW(t-T)\mathcal{K}\rho_{\rm c}(t)/\sqrt{\eta} + \sqrt{\eta}dW(t)\mathcal{H}[c]\rho_{\rm c}(t) \; . \end{aligned} \tag{6.159}$$

On the other hand, putting $T=0$ in (6.158) gives

$$\begin{aligned}d\rho_{\rm c}(t) = &\, dt\left\{-i[H,\rho_{\rm c}(t)] + \mathcal{D}[c]\rho_{\rm c}(t) - i[F, c\rho_{\rm c}(t) + \rho_{\rm c}(t)c^\dagger]\right. \\ &\left.+\, \mathcal{D}[F]\rho_{\rm c}(t)/\eta\right\} + dW(t)\mathcal{H}[\sqrt{\eta}c - iF/\sqrt{\eta}]\rho_{\rm c}(t) \; . \end{aligned} \tag{6.160}$$

6.6 Intracavity Squeezing

6.6.1 The Linear System

In order to understand the effect of quantum limited feedback on intracavity squeezing, it is useful to consider an exactly solvable system. In this section, I will mainly be following [11] in considering the case of a linear optical system, with linear feedback based on homodyne (or QND) detection. By a linear system, I mean that the equation of motion for the two quadrature operators are linear. This is approximately the case for many quantum optical systems, in the limit of large photon numbers. For specificity, I will chose a system which is exactly linear. If, as in the remainder of this chapter, one is interested in the behavior of one quadrature only (here the x quadrature), then all linear dynamics can be composed of damping, driving, and parametric driving. Damping will be assumed to be always present (as necessary to do feedback or obtain an output from the cavity) and will have rate 1. Constant linear driving simply shifts the origin away from $x = 0$, and will be ignored. Stochastic linear driving in the white noise approximation causes diffusion in the x quadrature, at a rate l. Finally, if the strength of the parametric driving ($H \sim xy$) is θ (where $\theta = 1$ would represent a degenerate parametric oscillator at threshold), then the master equation for the system is

$$\dot{\rho} = \mathcal{D}[a]\rho + \tfrac{1}{4}l\mathcal{D}[a^\dagger - a]\rho + \tfrac{1}{4}\theta[a^2 - a^{\dagger 2}, \rho] \equiv \mathcal{L}_0\rho \ , \tag{6.161}$$

where a is the annihilation operator for the cavity mode.

An alternative definition for the linearity of the x quadrature dynamics is that the marginal distribution of the Wigner function for x (which is the true probability distribution for x) obeys an Ornstein–Uhlenbeck equation. That is to say,

$$\dot{P}(x) = \left(\partial_x kx + \tfrac{1}{2}D\partial_x^2\right) P(x) \ , \tag{6.162}$$

where k and D are constants. The solution of this equation is a Gaussian with variance

$$V = \frac{D}{2k} \ . \tag{6.163}$$

For the particular master equation above (the properties of which will be denoted by the subscript 0), the drift and diffusion constants are

$$k_0 = \tfrac{1}{2}(1 + \theta) \ , \tag{6.164}$$

$$D_0 = 1 + l \ . \tag{6.165}$$

In this case, $V_0 = (1+l)/(1+\theta)$. If this is less than unity, the system exhibits squeezing of x. It is more useful to work with the normally ordered variance, which becomes negative if the x quadrature is squeezed. Here, I will denote it

$$U = V - 1 \ , \tag{6.166}$$

which for this system takes the value

$$U_0 = \frac{l - \theta}{1 + \theta} \ . \tag{6.167}$$

If the system is to stay below threshold (so that the y quadrature does not become unbounded), then the maximum value for θ is one. At this value,

$U_0 = -1/2$ when the x diffusion rate $l = 0$. Therefore the minimum value of squeezing which this linear system can attain as a stationary value is half of the theoretical minimum of $U = -1$.

6.6.2 Homodyne-Mediated Feedback

We now wish to consider the effect of homodyne-mediated feedback on the intracavity light. This is most easily understood using the quantum trajectory picture in the Markovian limit. Thus we want the stochastic master equation for the conditioned state matrix $\rho_c(t)$ (6.160)

$$d\rho_c(t) = dt\left(\mathcal{L}_0\rho_c(t) + \mathcal{K}[a\rho_c(t) + \rho_c(t)a^\dagger] + \frac{1}{2\eta}\mathcal{K}^2\rho_c(t)\right) + dW(t)\left(\sqrt{\eta}\mathcal{H}[a] + \mathcal{K}/\sqrt{\eta}\right)\rho_c(t) \,. \tag{6.168}$$

Here, $\mathcal{L}_0$ is as defined in (6.161).

The question now arises as to what to choose for the $\mathcal{K}$. Seeking to reduce the fluctuations in x suggests the feedback operator, which is related to $\mathcal{K}$ by (6.156), should be

$$F = -\lambda y/2 \,. \tag{6.169}$$

As a separate Hamiltonian, this translates a state in the negative x direction for λ positive. By controlling this Hamiltonian by the homodyne photocurrent one thus has the ability to change the statistics for x and perhaps achieve better squeezing. This Hamiltonian can be effected by driving the cavity (at a second mirror which can be assumed to have a negligible loss rate compared to the first mirror). Using this choice and changing (6.168) into a stochastic Liouville equation for the conditioned Wigner function gives

$$dP_c(x) = dt\left[\partial_x(k_0+\lambda)x + \frac{1}{2}\partial_x^2\left(D_0 + 2\lambda + \lambda^2/\eta\right)\right]P_c(x) + dW(t)\left[\sqrt{\eta}\left(x - \bar{x}_c(t) + \partial_x\right) + (\lambda/\sqrt{\eta})\partial_x\right]P_c(x) \,, \tag{6.170}$$

where $\bar{x}_c(t)$ is the mean of the distribution $P_c(x)$ and $dW(t)$ is as usual.

This equation is obviously no longer a simple Ornstein–Uhlenbeck equation. Nevertheless, it still has a Gaussian as an exact solution, as can be shown by direct substitution. The mean $\bar{x}_c$ and variance V_c of the conditioned Gaussian distribution are found to obey

$$\dot{\bar{x}}_c = -(k_0+\lambda)\bar{x}_c + \xi(t)\left[\sqrt{\eta}\left(V_c - 1\right) - (\lambda/\sqrt{\eta})\right] \,, \tag{6.171}$$

$$\dot{V}_c = -2k_0V_c + D_0 - \eta\left(V_c - 1\right)^2 \,. \tag{6.172}$$

Two points about the evolution equation for V_c are worth noting. It is completely deterministic (no noise terms), and it is not influenced by the presence of feedback. Furthermore, for this linear system, it is independent of $\bar{x}_c$. Thus, the stochasticity and feedback terms in the equation for the mean do not even enter that for the variance indirectly.

The equation for the conditioned variance is more simply written in terms of the conditioned normally ordered variance $U_c = V_c - 1$

$$\dot{U}_c = -2k_0 U_c - 2k_0 + D_0 - \eta U_c^2 \,. \tag{6.173}$$

On a time scale as short as a cavity lifetime, U_c will approach its stable steady-state value of

$$U_c = \eta^{-1}\left(-k_0 + \sqrt{k_0^2 + \eta(-2k_0 + D_0)}\right) \,. \tag{6.174}$$

Substituting the steady-state conditioned variance into (6.171) gives

$$\dot{\bar{x}}_c = -(k_0 + \lambda)\bar{x}_c + \xi(t)\frac{1}{\sqrt{\eta}}\left[-k_0 + \sqrt{k_0^2 + \eta(-2k_0 + D_0)} - \lambda\right] \,. \tag{6.175}$$

If one were to choose

$$\lambda = -k_0 + \sqrt{k_0^2 + \eta(-2k_0 + D_0)} = -k_0 + \sqrt{k_0^2 + 2\eta k_0 U_0} \,, \tag{6.176}$$

then there would be no noise at all in the conditioned mean and so one could set $\bar{x}_c = 0$. In other words, this value of λ is precisely the value required to minimize the unconditioned variance under feedback. When all fluctuations in the mean are suppressed, the unconditioned variance is equal to the conditioned variance.

In general, the unconditioned variance will consist of two terms, the conditioned quantum variance in x plus the classical (ensemble) average variance in the conditioned mean of x:

$$U_\lambda = U_c + \mathrm{E}[\bar{x}_c^2] \,. \tag{6.177}$$

The latter term is found from (6.175) to be

$$\mathrm{E}[\bar{x}_c^2] = \eta^{-1}\frac{1}{2(k_0 + \lambda)}\left[-(k_0 + \lambda) + \sqrt{k_0^2 + \eta(-2k_0 + D_0)}\right]^2 \,. \tag{6.178}$$

Adding (6.174) gives

$$U_\lambda = \eta^{-1}\frac{\lambda^2 + \eta(-2k_0 + D_0)}{2(k_0 + \lambda)} = (k_0 + \lambda)^{-1}\left(k_0 U_0 + \frac{\lambda^2}{2\eta}\right) \,. \tag{6.179}$$

An immediate consequence of this expression is that U_λ can only be negative if U_0 is. That is to say, classical feedback based on homodyne detection cannot *produce* intracavity squeezing. However, this does not mean that the feedback cannot *enhance* squeezing. Obviously, the best intracavity squeezing will be when $\eta = 1$, in which case the intracavity squeezing can be simply expressed as

$$U_{\min} = k_0\left(-1 + \sqrt{1 + R_0}\right) \,, \tag{6.180}$$

where $R_0 = (-2k_0 + D_0)/k_0^2 \geq -1$. It can be proven that that $U_{\min} \leq U_0$, with equality only if $\eta = 0$ or $U_0 = 0$. This result implies that the intracavity variance in x can always be reduced by classical homodyne-mediated feedback, unless it is at the classical minimum. In particular, intracavity squeezing can always be enhanced. For the parametric oscillator defined originally in (6.161), with $l = 0$, $U_{\min} = -\theta/\eta$. For $\eta = 1$, the (symmetrically ordered) x variance is $V_{\min} = 1 - \theta$. The y variance, which is unaffected by feedback, is seen from (6.161) to be $(1-\theta)^{-1}$. Thus, with perfect detection, it is possible to produce a minimum uncertainty squeezed state with arbitrarily high squeezing as $\theta \to 1$. This is not unexpected as a parametric amplifier (in an undamped cavity) also produces minimum uncertainty squeezed states. The feedback removes the noise which was added by the damping which is necessary to do the measurement used in the feedback.

The reason that this feedback cannot produce squeezing is that the conditioning of the variance according to (6.173) cannot change the sign of the normally-ordered variance U. The homodyne measurement does reduces the conditioned variance, except when it is equal to the classical minimum of 1. The more efficient the measurement, the greater the reduction. Ordinarily, this reduced variance is not evident because the measurement gives a random shift to the conditional mean of x, with the randomness arising from the shot noise of the photocurrent. By appropriately feeding back this photocurrent, it is possible to precisely counteract this shift and thus observe the conditioned variance.

If the time delay T in the feedback loop is not negligible then the counteraction will be less than perfect. It is possible to calculate this effect exactly for an arbitrary linear feedback response using the quantum trajectory theory [11]. However, it would generally be easier to return to the approach based on quantum Langevin equations [55]. For short delay $T \ll 1$, there is a simple expression for the modified normally ordered variance:

$$U_{\lambda;T} = U_\lambda(1 + \lambda T) \ . \tag{6.181}$$

For squeezed systems, with $U_0 < 0$, the optimum value of U_λ occurs for λ negative, as shown above. Thus, the time delay reduces the total squeezing by the factor $(1+\lambda T)$. On the other hand, classical noise is reduced to $U_\lambda > 0$ with λ positive, so that the total noise is increased by the factor $(1 + \lambda T)$. Overall, the time delay degrades the effectiveness of the feedback, as expected.

Note that the optimal λ of (6.176) has the same sign as U_0. That is to say, if the system produces squeezed light, then the best way to enhance the squeezing is to add a force which displaces the state in the direction of the difference between the measured photocurrent and the desired mean photocurrent. This is the opposite of what would be expected classically, and can be attributed to the effect of homodyne measurement on squeezed states. For classical statistics ($U \geq 0$), a higher than average photocurrent reading $[\xi(t) > 0]$ leads to the conditioned mean $\bar{x}_c$ increasing (except if $U = 0$ in which case the measurement has no effect). However, for nonclassical states with $U < 0$, the classical intuition fails as a positive photocurrent fluctuation causes $\bar{x}_c$ to decrease. This explains the counterintuitive negative value of λ required in squeezed systems, which naively would be thought to destabilize the system and increase fluctuations. The value of the positive feedback required (6.176) is such that the overall restoring force $k_0 + \lambda$ is still positive.

Succinctly, one can state that conditioning can be made practical by feedback. The intracavity noise reduction produced by classical feedback can be precisely as good as that produced by conditioning. This reinforces the simple explanation as to why homodyne-mediated classical feedback cannot produce nonclassical states: because homodyne detection cannot. Nonclassical feedback (such as using the photocurrent to influence nonlinear intracavity elements) may produce nonclassical states, but such elements can produce nonclassical states without feedback, so this is hardly surprising. In order to produce nonclassical states by classical feedback, it would be necessary to have a nonclassical measurement scheme. That is to say, one which does not rely on measurement of the extracavity light to procure information about the intracavity state. Intracavity measurements (in particular, quantum non-demolition measurements) are not limited by the random process of damping to the external continuum. The extra term which the measurement introduces into the nonselective master equation will not produce nonclassical states, but may allow the measurement to produce nonclassical *conditioned* states. One would thus expect that intracavity QND measurements would enable feedback to overcomes the classical limit, and I will now show that this is indeed the case.

6.6.3 QND-Mediated Feedback

The natural choice of quantum non-demolition variable is the quadrature to be squeezed, say x as before. I use the same model for a QND measurement as in Sect. 6.4.1. Mode a is coupled to mode c by the Hamiltonian in (6.102). The other dynamics of mode a are defined as before by its Liouville superoperator $\mathcal{L}_0$. The density operator for both modes thus obeys the following master equation:

$$\dot{R} = \mathcal{L}_0 R - \frac{\chi}{2}[x(c - c^\dagger), R] + \gamma\mathcal{D}[c]R \ . \tag{6.182}$$

In order to treat mode c as part of the apparatus rather than part of the system, it is necessary to eliminate its dynamics. This can be done by assuming that it is heavily damped, with γ much larger than all other rates. Then, apart from initial transients, it will have few photons and will be slaved to mode a. Following standard techniques for adiabatic elimination [53] gives the master equation for ρ, the density operator for mode a alone, as

$$\dot{\rho} = \mathcal{L}_0\rho + \Gamma\mathcal{D}[x/2]\rho \ , \tag{6.183}$$

where the measurement strength parameter is $\Gamma = 4\chi^2/\gamma$.

Now add homodyne measurement of the b mode with efficiency η. Starting from the conditioned state matrix R_c before the adiabatic elimination and following it through gives the conditioning master equation for ρ_c [11]

$$d\rho_c = dt\mathcal{L}_0\rho_c + dt\Gamma\mathcal{D}[x/2]\rho_c + \sqrt{\eta\Gamma}dW(t)\mathcal{H}[x/2]\rho_c \ . \tag{6.184}$$

Normalizing the homodyne photocurrent so that the noise is the same as in preceding sections gives

$$I_c(t) = \sqrt{\mathrm{H}}\langle x\rangle_c(t) + \xi(t) \ . \tag{6.185}$$

Here I am using H (a capital η) for $\eta\Gamma$ as the effective efficiency of the measurement. This is related to the parameter Q in Sect. 6.4.1 by $\mathrm{H} = \eta Q^2/4$. Note that this is not bounded above by unity, since it is possible for χ^2/γ to be much greater than one even with χ much less than γ. Recall that all rates are measured in units of the a mode linewidth (which was κ in Sect. 6.4.1).

The photocurrent (6.185) can be used in feedback onto the a mode just as in preceding sections. A feedback term of the form

$$[\dot\rho_c]_{\mathrm{fb}} = I_c(t-T)\mathcal{K}\rho_c/\sqrt{\mathrm{H}} \tag{6.186}$$

gives, in the limit $T \to 0$, the conditioned evolution

$$\begin{aligned} d\rho_c = {} & dt\left(\mathcal{L}_0\rho_c + \Gamma\mathcal{D}[x/2]\rho_c + \mathcal{K}\tfrac{1}{2}[x\rho_c + \rho_c x] + \frac{1}{2\mathrm{H}}\mathcal{K}^2\rho_c\right) \\ & + dW(t)\left(\sqrt{\mathrm{H}}\mathcal{H}[x/2] + \mathcal{K}/\sqrt{\mathrm{H}}\right)\rho_c \ . \end{aligned} \tag{6.187}$$

Using the same expressions as in Sect. 6.6.2 implies that the probability distribution for the x quadrature obeys

$$\begin{aligned} dP_c(x) = {} & dt\left[\partial_x(k_0 + \lambda)x + \frac{1}{2}\partial_x^2\left(D_0 + \lambda^2/\mathrm{H}\right)\right]P_c(x) \\ & + dW(t)\left[\sqrt{\mathrm{H}}[x - \bar{x}_c(t)] + (\lambda/\sqrt{\mathrm{H}})\partial_x\right]P_c(x) \ . \end{aligned} \tag{6.188}$$

The mean and variance of this conditioned distribution obey

$$\dot{\bar{x}}_c = -(k_0 + \lambda)\bar{x}_c + \xi(t)\left(\sqrt{\mathrm{H}}V_c - \lambda/\sqrt{\mathrm{H}}\right), \tag{6.189}$$

$$\dot{V}_c = -2k_0V_c + D_0 - \mathrm{H}V_c^2 \ . \tag{6.190}$$

These equations are identical to the corresponding equations for homodyne mediated feedback (6.171) and (6.172) apart from the replacement of $(V_c - 1)$ by V_c and η by H in the measurement terms. In the limit $\mathrm{H} \to \infty$, (6.190) predicts an arbitrarily small steady-state conditioned variance. This is characteristic of a good QND measurement. Of course, the quantum noise has not been eliminated but rather redistributed. For H to be large requires Γ to be large also, so that the variance in the unsqueezed quadrature is greatly increased by the measurement term in (6.183). This ensures that Heisenberg's uncertainty principle is not violated.

For this QND measurement the stationary value for V from (6.190) is

$$V_{\mathrm{c}} = \mathrm{H}^{-1}\left(-k_0 + \sqrt{k_0^2 + \mathrm{H}D_0}\right) . \tag{6.191}$$

Thus choosing the feedback strength to be

$$\lambda = -k_0 + \sqrt{k_0^2 + \mathrm{H}D_0} . \tag{6.192}$$

eliminates the stochastic element in (6.189). In this case, the stationary conditioned variance (6.191) is the minimum achievable variance. In the limit $\mathrm{H} \to \infty$, it is easy to see that $V_{\mathrm{min}} = V_{\mathrm{c}}$ approaches the theoretical minimum value of 0. That is, perfect squeezing can be produced inside the cavity by QND mediated feedback. In this limit, one requires the feedback to be very strong, with $\lambda \simeq \sqrt{\mathrm{H}D_0}$. Unlike the homodyne mediated feedback case, λ should always be positive, as in accord with classical intuition. Indeed, all of the features of QND mediated feedback conform to a classical theory of feedback with measurements of finite accuracy (related to H). The quantum nature of the feedback is manifest only in the increased fluctuations in y due to the measurement back-action not present classically.

6.6.4 Mimicking a Squeezed Bath

The application of feedback based on a QND homodyne measurement to a cavity state has also been considered by Tombesi and Vitali [56]. However, rather than directly trying to minimize the x quadrature variance of the field mode, as I have discussed above, their goal was to mimic the dynamics produced by shining a broad-band squeezed vacuum onto the cavity mirror. Being broad-band compared to the cavity mode, a squeezed vacuum input is parametrized by two numbers, N, M which change the single nonzero second-order moments of (6.94) to

$$\begin{aligned} &dB_{\mathrm{in}}^{\dagger} dB_{\mathrm{in}} = N dt \; ; \;\; dB_{\mathrm{in}} dB_{\mathrm{in}}^{\dagger} = (N+1) dt \; ; \\ &dB_{\mathrm{in}} dB_{\mathrm{in}} = (dB_{\mathrm{in}}^{\dagger} dB_{\mathrm{in}}^{\dagger})^* = M dt \; . \end{aligned} \tag{6.193}$$

Positivity of the bath state matrix requires $|M|^2 \leq N(N+1)$ [2]. The master equation resulting from such bath correlations is [2]

$$\begin{aligned} \dot{\rho} = &-i[H, \rho] + (N+1)\mathcal{D}[a] + N\mathcal{D}[a^{\dagger}] \\ &-M\tfrac{1}{2}[a^{\dagger}, [a^{\dagger}, \rho]] - M^*\tfrac{1}{2}[a, [a, \rho]] \; . \end{aligned} \tag{6.194}$$

Tombesi and Vitali show that an equation of this form can be produced using feedback based on a QND homodyne measurement and that, not surprisingly, it can can produce intracavity squeezing.

6.6.5 The Micromaser

A final example of feedback onto an intracavity state which can give nonclassical noise reduction is that of the micromaser [57]. This consists of a small microwave cavity through which a monochromatic beam of resonant two-level atoms is passed. The atomic state upon exit can be measured and the result used in feedback. The case of modifying the cavity quality factor has been considered by Liebman and Milburn [58]. Because of the nature of the Jaynes–Cummings Hamiltonian, the micromaser dynamics are complicated without feedback, and even more complicated with. However, one result is easy to explain. In the limit of short transit time the atoms (assumed all to enter in the upper state) act simply as a linear amplifier of the cavity mode. In the absence of feedback the stationary state is thermal, with a photon number variance much greater than the mean. With weak feedback, increasing the cavity damping rate whenever an outgoing atom is detected in the lower state, the photon distribution can be made sub-Poissonian, with a variance equal to half the mean. For longer transit times, the no-feedback dynamics show the effects of trapping states (where the atom undergoes an integer number of Rabi cycles in transit [57]), and the minimum stationary variance is typically as low as one quarter of the mean. Feedback can produce an arbitrary small minimum variance near a trapping state. However, this result is very sensitive to the transit time and so may be washed out by a realistic atomic velocity profile.

6.7 Feedback Master Equation

In (6.160) we have the stochastic master equation for a system undergoing homodyne measurement, with instantaneous linear feedback of the homodyne photocurrent. This is a Markovian Itô stochastic equation, so that is possible to take the ensemble average simply by removing the stochastic term. This removes all nonlinear terms from the stochastic master equation, and gives the deterministic master equation

$$\dot{\rho} = -i[H, \rho] + \mathcal{D}[c]\rho - i[F, c\rho + \rho c^\dagger] + \frac{1}{\eta}\mathcal{D}[F]\rho \,. \tag{6.195}$$

I will call this the homodyne feedback master equation. The first feedback term, linear in F, is the desired effect of the feedback which would dominate in the classical regime. The second feedback term causes diffusion in the variable conjugate to F. It can be attributed to the inevitable introduction of noise by the measurement step in the quantum-limited feedback loop. The lower the efficiency, the more noise introduced.

The homodyne feedback master equation can be rewritten in the Lindblad form [45] as

$$\dot{\rho} = -i\left[H + \tfrac{1}{2}(c^\dagger F + Fc), \rho\right] + \mathcal{D}[c - iF]\rho + \frac{1-\eta}{\eta}\mathcal{D}[F]\rho \equiv \mathcal{L}\rho \,. \tag{6.196}$$

In this arrangement, the effect of the feedback is seen to replace c by $c - iF$, and to add an extra term to the Hamiltonian, plus an extra diffusion term which vanishes for perfect detection. It is possible to derive an analogous feedback master equation for direct detection, but this is not needed for squeezing. In the limit where intensity fluctuations can be linearized as amplitude quadrature fluctuations, direct detection is essentially equivalent to homodyne detection of the amplitude quadrature.

It is very important to note that although (6.196) has the appearance of a normal master equation, one cannot simply use it in the customary way for calculating the spectrum of the photocurrent used in the feedback loop. For example the two-time correlation function of the in-loop photocurrent is not given by the standard expression:

$$\mathrm{E}[I_{\mathrm{c}}^{\mathrm{hom}}(t+\tau)I_{\mathrm{c}}^{\mathrm{hom}}(t)] \neq \eta\langle : x(t+\tau)x(t) : \rangle + \delta(\tau) \ , \tag{6.197}$$

where the normally ordered two-time correlation function for x is defined as

$$\langle : x(t+\tau)x(t) : \rangle = \mathrm{Tr}\left\{(c+c^\dagger)e^{\mathcal{L}\tau}[c\rho(t)+\rho(t)c^\dagger]\right\} \ . \tag{6.198}$$

That is because the photocurrent at time t changes the system through the feedback as well as through the conditioning. Taking this into account, it is not difficult to show [12] that the correct expression is

$$\begin{aligned}\mathrm{E}[I_{\mathrm{c}}^{\mathrm{hom}}(t+\tau)I_{\mathrm{c}}^{\mathrm{hom}}(t)] = \eta\mathrm{Tr}\left\{(c+c^\dagger)e^{\mathcal{L}\tau}[(c-iF/\eta)\rho(t)\right.\\ \left.+\,\rho(t)(c^\dagger+iF/\eta)]\right\} + \delta(\tau) \ .\end{aligned} \tag{6.199}$$

Note that the feedback is present in both the term in square brackets and the evolution by $\mathcal{L}$ for the time τ. The former presence means that the in-loop photocurrent may have a sub-shot-noise spectrum, even if the system dynamics are classical. That is to say, for an optical system it may be possible to use an a semiclassical analysis (with classical light fields and detector shot noise) which nevertheless correctly predicts a sub-shot noise photocurrent spectrum. This is not surprising given the analogous result for free squeezing in Sect. 6.2.5.

6.8 In-Loop Squeezing Revisited

6.8.1 In-Loop Squeezing

In Sect. 6.2 it was shown that a sub-shot-noise in-loop photocurrent is not evidence for squeezing of the in-loop light in the usual sense, not least because the usual two-time commutation relations do not apply to an in-loop field. Also, a linear optical element (a beam splitter) fails to extract any squeezing from the loop, and in fact produces an above-shot-noise output. Nevertheless it was also shown that an in-loop nonlinear optical element (a QND intensity meter) agrees with the intensity statistics seen by the in-loop detector. This

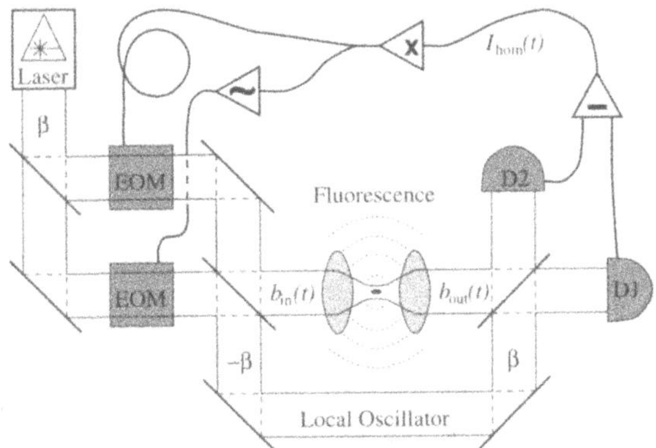

Fig. 6.3. Diagram of the experimental configuration discussed in the text. All beam splitters are 50:50. The atom is represented by the small ellipse at the focus of $b_{\rm in}(t)$. The difference $I_{\rm hom}(t)$ between the photocurrents at detectors D1 and D2 is amplified and split. The two signals (with opposite sign) are fed back to the two electro-optic modulators (EOM)

means that it is an open question as to whether the effect of the in-loop "squashed" [31] light on other nonlinear optical elements is more akin to that of squeezed light or light with classical fluctuations.

The simplest nonlinear optical element is a two-level atom. Shortly after the first observation of squeezing, Gardiner [59] made a seminal prediction regarding its effects on such an atom [60], namely that immersing an atom in broad-band squeezed light would break the equality between the transverse decay rates for the two quadratures of the atomic dipole. In particular, one decay rate could be made arbitrarily small, producing an arbitrarily narrow line in the power spectrum of the atom's fluorescence. This was seen, as the title of [59] proclaims, as a "direct effect of squeezing".

In this section I present work published recently [61], which considers the question of whether this atomic line-narrowing is characteristic only of squeezing in the conventional sense ('real squeezing'), or whether it can be produced by light which gives rise to a below shot-noise photocurrent by virtue of being part of a feedback loop ('in-loop squeezing'). The answer is that in-loop squeezing *can* do the job. In fact, the inhibition of the decay of one quadrature depends on the amount of squeezing and the quality of mode-matching to the atom in exactly the same way for in-loop squeezing as for free-field squeezing. Thus in-loop squeezing appears likely to be an important tool for future experimental investigation of the effect of low-noise light on atoms, as it is usually easier to generate than free squeezing.

Consider the apparatus shown in Fig. 6.3, but for the moment without the fluorescent atom. The Mach–Zehnder interferometer on the left hand side has two functions. First it produces a weak beam ($b_{\rm in}$) which is given by

$$b_{\rm in}(t) = \nu(t) - (i/2)(\beta e^{i\phi} - \beta e^{-i\phi}) \approx \nu(t) + \beta\phi \ . \tag{6.200}$$

Here β, assumed real, is the coherent amplitude of the laser, and $\pm\phi$, assumed small, are the phase shifts imposed by the electro-optic modulators. The operator $\nu(t)$ represents vacuum fluctuations as in Sect. 6.2. The second function of the interferometer is to produce a local oscillator beam with mean amplitude

$$-(1/2)(\beta e^{i\phi} + \beta e^{-i\phi}) = -\beta[1 + O(\phi^2)] \ . \tag{6.201}$$

With appropriate phase shifts assumed, this local oscillator is then used for making a homodyne measurement of the $X = b + b^\dagger$ quadrature of b_{out} (which, in the absence of the atom, is identical with b_{in}). If the efficiency of the detectors is ε then the homodyne photocurrent is represented by the operator [2]

$$I_{\mathrm{hom}}(t) = \sqrt{\varepsilon}\, X_{\mathrm{out}}(t) + \sqrt{1-\varepsilon}\, \xi_\varepsilon(t) \ , \tag{6.202}$$

where $\xi_\varepsilon(t)$ is a unit-norm real white noise process.

Although $S^X_{\mathrm{out}}(\omega)$ is shot-noise limited for high frequencies, it need not be for lower frequencies. In particular, the feedback loop shown can produce a spectrum below the shot-noise as shown in Sect. 6.2. The current $I_{\mathrm{hom}}(t)$ is amplified and used to control ϕ. If we set

$$\phi(t) = \frac{g}{2\beta\sqrt{\varepsilon}} \int_0^\tau h(t') I_{\mathrm{hom}}(t - T - t')dt' \ , \tag{6.203}$$

then we have a feedback loop with a low-frequency round-loop gain of g which is stable under the conditions given in Sect. 6.2.2. Solving in the Fourier domain,

$$\tilde{X}_{\mathrm{out}}(\omega) = \left[\tilde{\xi}_\nu(\omega) + g\tilde{h}(\omega)e^{-i\omega T}\sqrt{\frac{1-\varepsilon}{\varepsilon}}\, \tilde{\xi}_\varepsilon(\omega)\right] \frac{1}{1 - g\tilde{h}(\omega)e^{-i\omega T}} \ . \tag{6.204}$$

Thus X_{in} (which here equals X_{out}) has a spectrum

$$S^X_{\mathrm{in}}(\omega) = [1 + g^2|\tilde{h}(\omega)|^2(\varepsilon^{-1} - 1)]/|1 - g\tilde{h}(\omega)e^{-i\omega T}|^2 \ . \tag{6.205}$$

At a frequency $\bar{\omega}$ much less than the feedback bandwidth $\sim \tau^{-1}$, and much less than the reciprocal of the delay time T^{-1}, $\tilde{h}(\bar{\omega})\exp(-i\omega T) \simeq 1$ and the minimum noise is

$$S^X_{\mathrm{in}}(\bar{\omega})_{\mathrm{min}} = 1 - \varepsilon \ , \quad \text{for} \quad g = -\varepsilon/(1-\varepsilon) \ , \tag{6.206}$$

which is clearly below the standard quantum limit. These results reproduces those of (6.83), where here ε is playing the same role as η_2.

6.8.2 An In-Loop Atom

Returning to Fig. 6.3, we now include the two-level atom, which is assumed to be resonant to the laser. It couples strongly only to modes of the radiation field having the appropriate dipole spatial distribution [1]. However, by focusing a beam as shown in Fig. 6.3, it is possible to mode-match a significant proportion, say η, of $b_{\rm in}$ into the atom's input. Recent numerical calculations indicate that practical schemes for focusing light in free space have a limit on η of order 0.1 [62]. This suggests that in practice, a more efficient way to increase the effective η would be to couple the light into a microcavity, as in [63]. However, it is conceptually simpler to consider the free-space set up in Fig. 6.3.

The Hamiltonian of the atom in the interaction picture at time t is

$$H(t) = -i[\sqrt{\eta}\, b_{\rm in}(t) + \sqrt{1-\eta}\, \mu(t)]\sigma^\dagger(t) + \text{H.c.} \ . \tag{6.207}$$

Here $\sigma = |g\rangle\langle e|$ is the atomic lowering operator and I have set the longitudinal atomic decay rate to unity. The operator $\mu(t)$ represents an independent vacuum input. Under this coupling, the output field is found from the techniques of Sect. 6.3 to be

$$b_{\rm out}(t) = b_{\rm in}(t) + \sqrt{\eta}\, \sigma(t) \ . \tag{6.208}$$

Although it would be possible to give a description of the entire feedback loop in terms of atomic and radiation field operators [61], it is simpler to use the quantum trajectory theory of homodyne measurement as outlined in Sect. 6.5. In this theory, only the atom is treated as a quantum mechanical system with state matrix $\rho(t)$; the rest of the apparatus is considered as a complicated measurement and feedback device for the atom. The photocurrent $I_{\rm c}^{\rm hom}(t)$ is therefore a classical quantity. From (6.153) it is given by

$$I_{\rm c}^{\rm hom}(t) = \bar{I}_{\rm c}^{\rm hom}(t) + \xi^{\rm hom}(t) \ , \tag{6.209}$$

where $\xi^{\rm hom}(t)$ is local-oscillator shot noise, which in this case is the only source of noise in the whole system. From (6.202) and (6.208), the expected value $\bar{I}_{\rm c}^{\rm hom}(t)$ conditioned upon the prior photocurrent record is

$$\bar{I}_{\rm c}^{\rm hom}(t) = \sqrt{\eta\varepsilon}\, \mathrm{Tr}[\rho_{\rm c}(t)\sigma_x] + \sqrt{\varepsilon}\, 2\beta\phi_{\rm c}(t) \ . \tag{6.210}$$

Here $\phi_{\rm c}(t)$ is not set to its average value of zero because it is determined by the prior classical photocurrent via (6.203).

From the theory of Sect. 6.5, the atom will obey the following nonlinear stochastic master equation

$$d\rho_{\rm c} = dt\mathcal{D}[\sigma]\rho_{\rm c} + \sqrt{\eta\varepsilon}\, dW^{\rm hom}(t)\mathcal{H}[\sigma]\rho_{\rm c} - idt[H_{\rm fb}, \rho_{\rm c}] \ . \tag{6.211}$$

The final Hamiltonian is due to the feedback. It is identical to the term due to feedback in the fundamental atomic Hamiltonian (6.207), namely

$$H_{\rm fb}(t) = \sqrt{\eta}\, \beta\phi(t)\sigma_y \ . \tag{6.212}$$

So far, it is not a singular quantity because of the smoothing of the photocurrent in (6.203).

Now consider the limit of instantaneous feedback on the atomic time-scale, $\tau, T \ll 1$. In this limit

$$2\beta\phi(t) = gI_{\rm c}^{\rm hom}(t)/\sqrt{\varepsilon} \tag{6.213}$$

and thus we can derive

$$I_{\rm c}^{\rm hom}(t) = (1-g)^{-1}\{\xi(t) + \sqrt{\eta\varepsilon}\,{\rm Tr}[\rho_{\rm c}(t)\sigma_x]\} \ . \tag{6.214}$$

Hence from (6.212) and (6.213),

$$H_{\rm fb}(t) = \lambda\tfrac{1}{2}\sigma_y\{{\rm Tr}[\rho_{\rm c}(t)\sigma_x] + \xi(t)/\sqrt{\eta\varepsilon}\,\} \ , \tag{6.215}$$

where it is to be understood that t on the right-hand side of this equation actually stands for $t-0^+$. Since $-\infty < g < 1$, the feedback parameter λ is given by

$$\lambda = g\eta/(1-g) \ \in (-\eta, \infty) \ . \tag{6.216}$$

Taking the Markovian limit allows us to derive a deterministic master equation for the atom. Equation (6.211) no longer applies, because $H_{\rm fb}(t)$ in (6.215) is singular. However, by comparison of (6.215) with (6.157) we see that we can follow the methods of Sects. 6.5.4 and 6.7 to obtain

$$\dot\rho = \mathcal{D}[\sigma]\rho - i\lambda[\tfrac{1}{2}\sigma_y, \sigma\rho + \rho\sigma^\dagger] + \frac{\lambda^2}{\eta\varepsilon}\mathcal{D}[\tfrac{1}{2}\sigma_y]\rho \ . \tag{6.217}$$

This equation, and the following relation between λ and the in-loop squeezing

$$S_{\rm in}^X(\bar\omega) = \frac{1+g^2(\varepsilon^{-1}-1)}{(1-g)^2} = 1 + \frac{2\lambda}{\eta} + \frac{\lambda^2}{\eta^2\varepsilon} \ , \tag{6.218}$$

are the central results of this section. In (6.218), we still have $\bar\omega \ll \tau^{-1}$, but now also $\bar\omega \gg 1$. This ensures that the atomic variables (with characteristic time scale of unity) do not contribute significantly to the spectrum at $\bar\omega$, so that (6.205) is still valid.

From the master equation (6.217) it is easy to derive the following dynamical equations:

$$\begin{aligned} {\rm Tr}[\dot\rho\sigma_x] &= -\gamma_x{\rm Tr}[\rho\sigma_x] \ , & (6.219)\\ {\rm Tr}[\dot\rho\sigma_y] &= -\gamma_y{\rm Tr}[\rho\sigma_y] \ , & (6.220)\\ {\rm Tr}[\dot\rho\sigma_z] &= -\gamma_z{\rm Tr}[\rho\sigma_z] - C \ . & (6.221) \end{aligned}$$

Only the equation for σ_y is unaffected by the feedback, with $\gamma_y = 1/2$. The new decay rate for σ_x is

$$\gamma_x = \frac{1}{2}\left[1 + 2\lambda + \frac{\lambda^2}{\eta\varepsilon}\right] \ , \tag{6.222}$$

and the modified parameters for σ_z are

$$\gamma_z = \gamma_y + \gamma_x \ , \quad C = 1 + \lambda \ . \tag{6.223}$$

In steady state

$$\mathrm{Tr}[\rho_{\rm ss}\sigma_x] = \mathrm{Tr}[\rho_{\rm ss}\sigma_y] = 0 \ , \tag{6.224}$$
$$\mathrm{Tr}[\rho_{\rm ss}\sigma_z] = -1 + \lambda^2/[2\eta\varepsilon(1+\lambda) + \lambda^2] \ . \tag{6.225}$$

The most interesting of these results is that negative feedback can reduce the decay rate of the x component of the atomic dipole below its natural value of 1/2. From (6.218) it can be re-expressed as

$$\gamma_x = \tfrac{1}{2}\left[(1-\eta) + \eta S_{\rm in}^X(\bar{\omega})\right] \ . \tag{6.226}$$

This clearly shows that γ_x has two contributions: $(1-\eta)/2$ from the vacuum input and $\frac{1}{2}\eta S_{\rm in}^X(\bar{\omega})$ from the in-loop squeezed light. The greatest reduction occurs for minimum in-loop fluctuations as in (6.206), for which

$$(\gamma_x)_{\rm min} = \tfrac{1}{2}(1-\eta\varepsilon) \ , \quad \text{for} \quad \lambda = -\eta\varepsilon \ . \tag{6.227}$$

The slower decay of σ_x can be directly observed in the power spectrum of the fluorescence of the atom into the vacuum modes. This measures the photon flux per unit frequency into these modes and is defined by

$$P(\omega) = \frac{1-\eta}{2\pi}\langle\tilde{\sigma}^\dagger(-\omega)\sigma(0)\rangle_{\rm ss} \ . \tag{6.228}$$

From (6.219)–(6.221) this is easily evaluated to be

$$P(\omega) = \frac{(1-\eta)(\gamma_z - C)}{8\pi\gamma_z}\left[\frac{\gamma_x}{\gamma_x^2+\omega^2} + \frac{\gamma_y}{\gamma_y^2+\omega^2}\right] \ . \tag{6.229}$$

For the optimal squeezing ($\lambda = -\eta\varepsilon$) we have

$$P(\omega) = \frac{(1-\eta)\eta\varepsilon}{4\pi(2-\eta\varepsilon)}\left[\frac{1-\eta\varepsilon}{(1-\eta\varepsilon)^2+4\omega^2} + \frac{1}{1+4\omega^2}\right] \ . \tag{6.230}$$

This is plotted in Fig. 6.4 for $\eta = 0.8$ and $\varepsilon = 0.95$.

6.8.3 Comparison with Free Squeezing

To compare the above results with those produced by free squeezing we again assume that the mode-matching of the squeezed modes into the atom is η, and that the squeezing is broad-band compared to the atom. Assuming also that the input light is in a minimum-uncertainty state for the X and Y quadratures [1], it can be characterized by a single real number L, with

$$S_{\rm in}^X(\omega) = L = 1/S_{\rm in}^Y(\omega) \ . \tag{6.231}$$

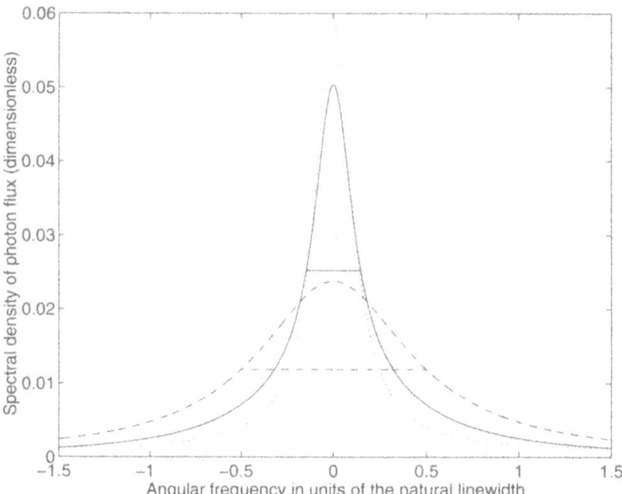

Fig. 6.4. Plot of the Power Spectrum $P(\omega)$ of the fluorescence into the vacuum modes, for in-loop squeezing (solid) and free squeezing (dotted), with mode-matching $\eta = 0.8$ and squeezing $S_{\rm in}^{X}(\bar{\omega}) = 0.05$. The linewidth for in-loop squeezing is slightly broader because the contribution from σ_y is *not* broadened in this case. The natural-width spectrum from weak thermal driving (dashed) is scaled up for comparison

This parameter is related to those of Sect. 6.6.4 by $L = 2N + 2M + 1$, where $M^2 = N(N+1)$ and M is assumed real. This yields the master equation

$$\dot{\rho} = (1-\eta)\mathcal{D}[\sigma]\rho + \frac{\eta}{4L}\mathcal{D}[(L+1)\sigma - (L-1)\sigma^{\dagger}]\rho \, , \tag{6.232}$$

which leads again to (6.219)–(6.221), but with

$$\gamma_x = \tfrac{1}{2}\left[(1-\eta) + \eta L\right] \, , \tag{6.233}$$

$$\gamma_y = \tfrac{1}{2}\left[(1-\eta) + \eta L^{-1}\right] \, , \tag{6.234}$$

$$\gamma_z = \gamma_x + \gamma_y \, , \quad C = 1 \, . \tag{6.235}$$

For $L < 1$ the decay of σ_x is again inhibited. The crucial observation to be made is that the dependence of γ_x on the degree of X quadrature squeezing of the input light is exactly the same as for in-loop squeezing, as is seen by comparing (6.231) and (6.233) with (6.226). The only difference between the two cases is that C is unaffected by the free squeezing and that γ_y is not increased by the in-loop squeezing. The latter is a direct consequence of the fact that an in-loop field is not bound by the usual two-time uncertainty relations. The free squeezing fluorescence spectrum is again given by (6.229). This is also plotted in Fig. 6.4 for $\eta = 0.8$ and $L = 0.05$. As this figure shows, the spectra are certainly not identical, but the sub-natural linewidth is much the same in both.

To conclude, line-narrowing of an atom is not a diagnostic of free squeezing. Rather, it requires only temporal anticorrelations of one quadrature of the input field (for times much shorter than the atomic lifetime) such as can be produced by a negative electro-optic feedback loop. The dependence of the line-narrowing on the input squeezing and the degree of mode-matching is the same for in-loop squeezing as for free squeezing. Because the quadrature operators of an in-loop field do not obey the usual two-time commutation relations, the reduction in noise in one quadrature does not imply an increase in noise in the other. Hence the line-narrowing of one quadrature of the atomic dipole by in-loop squeezing does not entail the line-broadening of the other quadrature. What significance this difference has in the physics of more complex atomic interactions with squeezed light [60] is a question requiring much investigation.

In-loop squeezing is generally easier to produce than free squeezing for a number of reasons. First, in-loop squeezing does not require expensive and delicate sources such as nonlinear crystals, but rather off-the-shelf electronic and electro-optical equipment. Second, the amount of squeezing is limited only by the efficiency of the photodetection. For homodyne detection, as required here, an efficiency of 95% is readily obtainable [64] and would enable in-loop squeezing of 95%. Third, in-loop squeezing can be produced at any frequency for which a coherent source is available, so experiments could be conducted on any atomic transition. The one difficulty with in-loop squeezing is that it requires a feedback loop response time much shorter than an atomic lifetime, but this would not be a problem for metastable transitions. Thus as well as giving us a better theoretical understanding of the effects on matter of light with fluctuations below the standard quantum limit, in-loop squeezing should be a practical alternative to free squeezing in the experimental investigation of these effects.

6.8.4 Other Uses of "Squashed Light"

The analysis given above is probably the most interesting application for squashed light proposed so far. However, it is worth pointing out that another conceivable application was suggested in [26]. This was in the context of a QND measurement of an the squashed in-loop quadrature. For a perfect QND device, such a measurement does not disturb the feedback loop. Now because the in-loop fluctuations are squashed, the readout from the QND device will have less noise than it would have without feedback. This is useful if the measurement is being made with the aim not of determining the in-loop intensity, but of determining the coupling constant between the in-loop light and the QND meter. For a given detector bandwidth, the squashing of intensity fluctuations would thus enable this parameter to be estimated more accurately. The improvement in accuracy would be limited only by the efficiency of the photodetectors used in the feedback loop, as in (6.206).

Another potential application for squashed light was proposed recently in [31]. Here the idea is to reduce radiation pressure fluctuations on a mirror by squashing the amplitude fluctuations. The feedback involves standard photodetection and an amplitude modulator. The point of reducing the radiation pressure fluctuations is again to allow a more accurate estimation of other parameters, namely the spectrum of the thermal noise in the mirror. In this case it turns out that in the optimal regime the fluctuations can at most be reduced by a factor of one half.

6.9 Conclusion

What then, can we say in conclusion about squeezing and feedback? The first and foremost fact is that, in the absence of any nonlinear optical elements, feedback cannot produce free squeezing. A nonlinear optical element is any element whose effect cannot be modeled by a displacement, rotation, or damping of the amplitude of the light. By free squeezing I mean squeezing whose existence can be verified by detection in a conventional (demolition) detector which is not part of the feedback loop. This includes both continuum squeezing and intracavity squeezing (which can be measured by dumping the light out of the cavity).

If there are nonlinear optical elements present then there are many interesting things one can achieve with feedback involving squeezing. First, if those elements are used to generate a squeezed beam which is split, then feedback (or indeed feedforward) onto an intensity modulator can effectively transfer amplitude squeezing electronically from the beam which is detected to the free beam. This was discussed in Sect. 6.2.7. Second, nonlinear optical elements can create quantum correlations between two beams. Then, even if the beams separately are not squeezed, feedback (or, equally effectively, feedforward) of the light from one beam can make the other beam squeezed. If they are squeezed, then feedback or feedforward can enhance this squeezing. A number of scenarios were discussed in Sect. 6.4. Third, if the nonlinear element creates squeezing within a cavity then controlling the driving of the cavity by a photocurrent derived from a homodyne measurement of the output can enhance the squeezing, as discussed in Sect. 6.6. This could be seen as a generalization of the first case discussed above, where the cavity mirror is acting as a beam splitter and the intracavity light is being continually split.

A quite separate issue is the nature of the light within a feedback loop. At first sight it appears easy to make this squeezed, as negative feedback (without the aid of any optical nonlinearity) can produce an *in-loop* photocurrent with arbitrarily low noise, regardless of the efficiency of the detector. However, the fact that this is the current from an in-loop detector means that a sub-shot-noise spectrum does not have the usual significance. First, there is a perfectly good semiclassical explanation for this phenomenon in terms of coherent field states. Second, the usual two-time commutation relations for a free field do not hold for an in-loop field. This means that there is no "squeezing" of the uncertainty from the amplitude to the phase quadrature, rather just an apparent "squashing" of the amplitude uncertainty with no effect on the phase uncertainty. Lastly, from a practical point of view, the squeezed light cannot be removed from the loop using a linear optical device. A beam splitter inserted in the in-loop beam produces a free (out-of-loop) beam with noise level above rather than below the shot noise.

Despite the differences between in-loop squashing and free squeezing, it turns out that there are similarities. An in-loop quantum non-demolition amplitude detector will respond to amplitude squashing exactly as to amplitude squeezing. For a feedback loop constructed using a perfectly efficient (demolition) detection, a perfect in-loop non-demolition detector will have a read-out identical to that of the in-loop demolition detector, and, for arbitrarily large negative feedback, this will have an arbitrarily low noise level. For non-unit efficiency of the demolition detector in the feedback loop, the two detectors will not agree and the maximum degree of squashing observed by the non-demolition detector will be limited by the demolition detector efficiency.

The degree of squashing seen by a perfect QND detector turns out to be a legitimate measure of the degree of noise reduction in the light, and this squashing can have effects on nonlinear optical devices very similar to those produced by squeezing. In particular, shining broad-band quadrature squashed light on a resonant two-level atom will cause the decay of the in-phase quadrature of the atomic dipole to be suppressed. This manifest itself as a line-narrowing in the power spectrum of the atom's fluorescence, an effect which was originally thought to be characteristic of squeezing. Moreover, the reduction in the linewidth of the in-phase quadrature depends on the degree of squashing (as measured by a QND device) in precisely the same way as it does on the degree of squeezing in the case of freely propagating light. This has important experimental implications as there are are a number of factors which make squashing easier to achieve than squeezing.

Very recently, there have been more intriguing developments in the theory of squashed light and its application to quantum spectroscopy [65]. First, I showed that it is possible to squash light simultaneously in both quadratures, and also to squeeze light in one quadrature and squash it the other. In the limit of perfect squeezing, and perfect squashing (which requires unit efficiency detectors) the in-loop fluctuations can be banished from both quadratures! An atom coupled only to this in-loop light would (to a first approximation) have its spontaneous decay completely suppressed. This surprising prediction is one more example of the continuing fruitful investigation of the relationship between feedback and squeezing.

Acknowledgment

I would like to thank Laura Thomsen for a careful reading of this manuscript. In work subsequent to the completion of this manuscript, we have generalized the work described above by considering the effect of twin-beam squashed light on a three-level atom [66]. We have also shown how feedback can be used to prepare near-minimum uncertainty spin-squeezed states [67].

References

1. D.F. Walls and G.J. Milburn, *Quantum Optics* (Springer, Berlin Heidelberg New York 1994)
2. C.W. Gardiner, *Quantum Noise* (Springer, Berlin Heidelberg New York 1991)
3. H.M. Wiseman and G.J. Milburn, Phys. Rev. A **49**, 4110 (1994)
4. J.G. Walker and E. Jakeman, Proc. Soc. Photo-Opt. Instrum. Eng. **492**, 274 (1985)
5. S. Machida and Y. Yamamoto, Opt. Commun. **55**, 219 (1985)
6. H.A. Haus and Y. Yamamoto, Phys. Rev. A **34**, 270 (1986)
7. Y. Yamamoto, N. Imoto and S. Machida, Phys. Rev. A **33**, 3243 (1986)
8. J.M. Shapiro, G. Saplakoglu, S.T. Ho, P. Kumar, B.E.A. Salech and M.C. Teich, J. Opt. Soc. Am. B **4**, 1604 (1987)
9. H.J. Carmichael, L. Tian, W. Ren and P. Alsing, in *Cavity QED*, ed. P. Berman, vol. 34 of Advances in AMO Physics (1994)
10. H.M. Wiseman and G.J. Milburn, Phys. Rev. Lett. **70**, 548 (1993)
11. H.M. Wiseman and G.J. Milburn, Phys. Rev. A **49**, 1350 (1994)
12. H.M. Wiseman, Phys. Rev. A **49**, 2133 (1994)
13. H.J. Carmichael, *An Open Systems Approach to Quantum Optics* (Springer, Berlin Heidelberg New York 1993)
14. C.W. Gardiner, A.S. Parkins and P. Zoller, Phys. Rev. A **46**, 4363 (1992)
15. H.M. Wiseman and G.J. Milburn, Phys. Rev. A **47**, 1652 (1993)
16. L. Plimak, Phys. Rev. A **50**, 2120 (1994)
17. R.J. Glauber, Phys. Rev. **130**, 2529 (1963)
18. R.J. Glauber, Phys. Rev. **131**, 2766 (1963)
19. E.C.G. Sudarshan, Phys. Rev. Lett. **10**, 277 (1963)
20. A.S. Troshin, Opt. Spektrosc. (USSR) **70**, 389 (1991)
21. A. Heidmann and J. Mertz, J. Opt. Soc. Am. B **10**, 1637 (1993)
22. C. Cohen-Tannoudji, J. Dupont Roc and G. Grynberg, *Photons and Atoms: An Introduction to Quantum Electrodynamics* (Wiley, New York 1989)
23. C.W. Gardiner and M.J. Collett, Phys. Rev. A **31**, 3761 (1985)
24. K. Mølmer, Phys. Rev. A **55**, 3195 (1996)
25. H.P. Yuen and J.H. Shapiro, IEEE Trans. IT **26**, 78 (1980)
26. M.S. Taubman, H. Wiseman, D.E. McClelland and H.A. Bachor, J. Opt. Soc. Am. B **12**, 1792 (1995)
27. A.V. Masalov, A.A. Putilin and M.V. Vasilyev, J. Mod. Opt. **41**, 1941 (1994)
28. J.J. Stefano, A.R. Subberud and I.J. Williams, *Theory and Problems of Feedback and Control Systems 2e* (Mc Graw-Hill, New York 1990)
29. Ya.A. Fananov, Opt. Spektrosc. (USSR) **70**, 392 (1991)

30. S.H. Youn, W. Jhe, J.H. Lee and J.S. Chang, J. Opt. Soc. Am. B **11**, 102 (1994)
31. B.C. Buchler, M.B. Gray, D.A. Shaddock, T.C. Ralph and D.E. McClelland, Opt. Lett. **24**, 259 (1999)
32. D.B. Khoroshko and S.Ya. Kilin, JETP **79**, 691 (1994)
33. M.S. Taubman, T.C. Ralph, A.G. White, P.K. Lam, H.M. Wiseman, D. McClelland and H.A. Bachor, Proceedings of the 12th International Congress on Lasers in Research and Engineering (Springer, Berlin Heidelberg New York 1996)
34. M. Hillery and M.O. Scully, Phys. Rev. D **25**, 3137 (1982)
35. B. Yurke, J. Opt. Soc. Am. **B2**, 732 (1985)
36. P. Alsing, G.J. Milburn and D.F. Walls Phys. Rev. A **37**, 2970 (1988)
37. E. Jakeman and J.G. Walker, Opt. Commun. **55**, 219 (1985)
38. J.G. Walker and E. Jakeman, Optica Acta **32**, 1303 (1985)
39. P.R. Tapster, J.G. Rarity and J.S. Satchell, Phys. Rev. A **37**, 2963 (1988)
40. J. Mertz, A. Heidmann, C. Fabre, E. Giacobino and S. Reynaud, Phys. Rev. Lett. **64**, 2897 (1990)
41. J. Mertz, A. Heidmann and C. Fabre, Phys. Rev. A **44**, 3329 (1991)
42. M.J. Collett and R.B. Levien, Phys. Rev. A **43**, 5068 (1991)
43. H.M. Wiseman, M.S. Taubman and H.A. Bachor, Phys. Rev. A **51** 3227 (1995)
44. R. Paschotta, M. Collett, P. Kurz, K. Fielder, H.A. Bachor and J. Mlynek, Phys. Rev. Lett. **72**, 3807 (1994)
45. G. Lindblad, Commun. Math. Phys. **48**, 199 (1976)
46. J. von Neumann, *Mathematical Foundations of Quantum Mechanics* (Springer, Berlin 1932); English translation (Princeton University Press, Princeton 1955)
47. J. Dalibard, Y. Castin and K. Mølmer, Phys. Rev. Lett. **68**, 580 (1992)
48. K. Mølmer, Y. Castin and J. Dalibard, J. Opt. Soc. Am. B **10**, 524 (1993)
49. R. Dum, P. Zoller and H. Ritsch, Phys. Rev. A **45**, 4879 (1992)
50. R. Dum, A.S. Parkins, P. Zoller and C.W. Gardiner, Phys. Rev. A **46**, 4382 (1992)
51. L. Tian and H.J. Carmichael, Phys. Rev. A **46**, R6801 (1992)
52. P. Marte, R. Dum, R. Taïeb and P. Zoller, Phys. Rev. A **47**, 1378 (1993)
53. H.M. Wiseman and G.J. Milburn, Phys. Rev. A **47**, 642 (1993)
54. C.W. Gardiner, *Handbook of Stochastic Methods* (Springer, Berlin Heidelberg New York 1985)
55. V. Giovannetti, P. Tombesi and D. Vitali, Phys. Rev. A **60**, 1549 (1999)
56. P. Tombesi and D. Vitali, Phys. Rev. A **50**, 4253 (1994)
57. P. Filipowicz, J. Javanainen and P. Meystre, Phys. Rev. A **34**, 3077 (1986)
58. A. Liebman and G.J. Milburn, Phys. Rev. A **51**, 736 (1995)
59. C.W. Gardiner, Phys. Rev. Lett. **56**, 1917 (1986)
60. Z. Ficek and P.D. Drummond, Phys. Today **50**, 34 (1997)
61. H.M. Wiseman, Phys. Rev. Lett. **81**, 3840 (1998)
62. S.J. van Enk and H.J. Kimble, Phys. Rev. A **61**, 051802 (2000)
63. D.W. Vernooy, A. Furusawa, N.Ph. Georgiades, V.S. Ilchenko and H.J. Kimble, Phys. Rev. A **57**, R2293 (1998)
64. S. Schiller, G. Breitenbach, S.F. Pereira, T. Muller and J. Mlynek, Phys. Rev. Lett. **77**, 2933 (1996)
65. H.M. Wiseman, J. Opt. B: Quant. Semiclass. Opt. **1**, 459 (1999)
66. L.K. Thomsen and H.M. Wiseman, Phys. Rev. A **64**, 043805 (2001)
67. L.K. Thomsen, S. Mancini and H.M. Wiseman, Phys. Rev. A **65**, 061801 (2002)

Part III

Applications of Quantum Squeezing

7 Communication and Measurement with Squeezed States

H. P. Yuen

In this chapter I will discuss the advantages, in principle, of using nonclassical states [1] in communication and measurement situations involving classical information transfer. Most of the discussions will be concerned with the quantum states of light, in particular, the quadrature squeezed states and number states. Thus, optical terminology will be freely employed even though the principles are generally applicable to fermions also, and gravitational wave detection by a free mass will also be treated. I shall focus on the general theoretical concepts and principles underlying such applications of nonclassical states without extensive mathematical derivations, and also no review of the physics involving these states which are covered elsewhere in this book. I shall mostly avoid precise mathematical definitions and formulations, although the treatment is as precise as most standard treatments in theoretical physics or engineering science.

It is, of course, the defining characteristic of a quadrature squeezed state that the quantum fluctuation in one of its quadrature is reduced below that of a coherent state. Let $|\alpha\rangle$ be a coherent state (CS) of an optical field mode with photon annihilation operator $\hat{a} = (\hat{x} + i\hat{y})/2$ so that

$$(\Delta x)^2 = (\Delta y)^2 = 1 \; . \tag{7.1}$$

In a two-photon coherent state (TCS) [2] $|\mu\nu\alpha\rangle$, which are the pure quadrature squeezed states, one obtains with a proper choice of quadrature $\hat{x}^\theta$

$$\left(\Delta\hat{x}^\theta\right)^2 = (|\mu| - |\nu|)^2 \; , \qquad \left(\Delta\hat{x}^{\theta+\pi/2}\right)^2 = (|\mu| + |\nu|)^2 \; . \tag{7.2}$$

Since $|\mu|^2 - |\nu|^2 = 1$, $|\mu\nu\alpha\rangle$ is a minimum uncertainty state on $\hat{x}^\theta$, $\hat{x}^{\theta+\pi/2}$,

$$\left(\Delta x^\theta\right)^2 \left(\Delta x^{\theta+\pi/2}\right)^2 = 1 \; . \tag{7.3}$$

For simplicity, let μ, ν, α be real and $|\mu - \nu| < 1$, thus

$$\langle\alpha|\hat{x}|\alpha\rangle = 2\alpha = \langle\mu\nu\alpha|\hat{x}|\mu\nu\alpha\rangle \; , \tag{7.4}$$

$$\langle\alpha|\left(\Delta\hat{x}\right)^2|\alpha\rangle > \langle\mu\nu\alpha|\left(\Delta\hat{x}\right)^2|\mu\nu\alpha\rangle \; . \tag{7.5}$$

This is often taken to mean that in the proper quadrature, a TCS is less noisy than a CS and so is better for communication and measurement. However, (7.4)–(7.5) is *not* a proper justification of such an assertion.

First of all, the states $|\mu\nu\alpha\rangle$ and $|\alpha\rangle$, which is $|\mu\nu\alpha\rangle$ with $\nu = 0$, have different energy,

$$\langle\mu\nu\alpha|a^\dagger a|\mu\nu\alpha\rangle = |\alpha|^2 + |\nu|^2 \ . \tag{7.6}$$

It is not a priori clear that if a portion of the energy associated with the mean field α is moved to increase $(\Delta y)^2$ so that $(\Delta x)^2$ is less, the overall effect is beneficial. Assuming that a signal-to-noise ratio (SNR) criterion is appropriate for the present illustration,

$$\mathrm{SNR} \equiv \frac{\langle x\rangle^2}{(\Delta x)^2} \tag{7.7}$$

it was shown [3] that TCS indeed maximizes (7.7) under the constraint of a fixed energy for an arbitrary state, $\langle a^\dagger a\rangle \leq S$, with the result

$$\mathrm{SNR}_{|\mu\nu\alpha\rangle} = 4S(S+1) \tag{7.8}$$

for $\nu = S/\sqrt{2S+1}$ as compared to $\mathrm{SNR}_{|\alpha\rangle} = 4S$. Secondly, in communication with $|\alpha\rangle$, both quadratures can be used to carry information and thus may yield a higher capacity than the use of $|\mu\nu\alpha\rangle$ with only one quadrature, which is equivalent to using half of the available bandwidth. It turns out that for the unrestricted capacity [4], and much more so for the binary signaling capacity [5], the use of $|\mu\nu\alpha\rangle$ does lead to improvement over $|\alpha\rangle$. The relevant communication concepts and further details are to be discussed in the sequel. The point here is that the advantage of $|\mu\nu\alpha\rangle$ over $|\alpha\rangle$ is not as obvious or intuitive as it may first appear. Similarly, while number states $|n\rangle$ and direct detection produce a noiseless system, it is discrete as compared to the in-principle continuum of states $|\alpha\rangle$. Again, it is not a priori clear that $|n\rangle$ would lead to a higher capacity.

The real point involving nonclassical states, I believe, is the following. Historically or typically in physics, one analyzes a given physical phenomenon and sees if it can be useful in application, whereas in engineering one often synthesizes to produce something to perform a certain function efficiently. (This opposition between analysis and synthesis is, of course, neither absolute nor pervasive in physics versus engineering.) For a long time after the laser was invented, the ideal laser state was supposed to be a coherent state, a quantum source one has to live with. Thus, all practical light sources were supposedly characterized by classical states, i.e. pure coherent states or their random superposition. However, states which are not classical, the *nonclassical states*, are clearly possible to have, at least in principle. In a synthesis or optimization approach, one would want to find out whether such states could lead to a better system for the application under consideration. Thus, the following *questions* suggest themselves in any given problem situation: What are the appropriate performance criteria and resource constraints? What are the best states or state-measurement combination one should use according to the criteria and the constraints? How much better are they compared to the conventional or standard system? The above discussion surrounding (7.7) and (7.8) furnishes an example of answers to such questions. Typically, the answer would involve quantities that are only specified mathematically, such as a TCS. If it seems worthwhile to develop such new systems, further questions on concrete physical realizations would have to be addressed. In these days of "quantum information", such questions are even more pervasive and important.

In the following section, I will review some basic concepts in classical communication, distinguish communication from detection, and discuss how physical measurement fits into both. In Sect. 7.2 the issues of quantum communication for classical information transfer will be explained. [Note that "quantum information" is entirely outside the scope of this chapter.] I will discuss the information capacity of nonclassical states, and the apparently only possible useful application of nonclassical states in fiber optic communication, to date – the use of nonclassical amplifiers and duplicators. In Sect. 7.3, I will discuss the use of nonclassical states in physical measurement problems, and the communication theoretic limit on the accuracy of measurements. In Sect. 7.4 the validity of the standard quantum limit for monitoring free-mass positions is addressed. Throughout I will try to explain the intuitive relevance of the various basic communication parameters, to highlight the main ideas with careful formulation but minimum details, and to dispel a few common misconceptions. Some results are also presented here for the first time.

7.1 Classical Communication and Measurement

7.1.1 Classical Information Transmission

For our purpose, a classical communication system can be schematically represented by Fig. 7.1. A source generates a classical quantity u, which is a member of an alphabet set U, $u \in U$, which may be discrete or continuous. Since u is generated probabilistically according to some distribution, the corresponding random variable is denoted by U. The transmitter modulates u onto a signal $X^{(\mathrm{in})}(t,u)$, which is a time-varying classical function. The channel, which usually represents all the disturbance in the system from source to destination, yields an output $X^{(\mathrm{out})}(t,u)$ statistically related to the input $X^{(\mathrm{in})}(t,u)$. The receiver processes $X^{(\mathrm{out})}(t,u)$ to produce an estimate $v \in U$ of u to satisfy the performance criteria.

If U is a finite set $\{1,\ldots,M\}$, the criterion of error probability is often employed. If U is continuous, the mean-square error between U and V is often taken as the criterion. In both cases the system is designed, subject to whatever constraints under consideration, to minimize the error or to produce

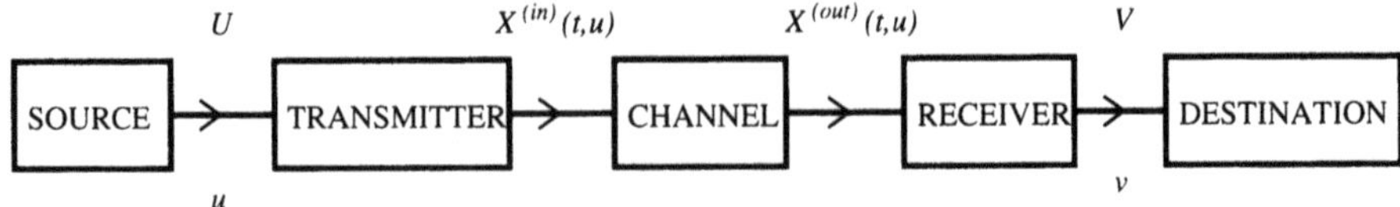

Fig. 7.1. Schematic representation of a classical communication system: for U and V capitals denote random quantities, lower case their samples but no such distinction is made for $X^{(\mathrm{in})}$ and $X^{(\mathrm{out})}$

a sufficiently small error. In a *communication* situation, one has joint design over the transmitter and receiver whereas in a *detection* situation, one is concerned only with the receiver design. Thus, in communications one may pick $X^{(\mathrm{in})}(t,u)$ to influence $X^{(\mathrm{out})}(t,u)$, and in detection one is faced with a given statistical description of $X^{(\mathrm{out})}(t,u)$. Clearly, communication is broader than detection. In the communication case it is important to deal explicitly with the time-sequential nature of the source output, with u regarded as a sequence $u_1, u_2, \ldots, u_i, \ldots$ with corresponding $X_i^{(\mathrm{in})}(t,u_i)$ and $X_i^{(\mathrm{out})}(t,u_i)$.

The system constraints in both cases are similar. The physical transmission medium (and often together with the unavoidable disturbance in the receiver structure) specifies the *channel representation*, the statistical relation between $X^{(\mathrm{in})}(t,u)$ and $X^{(\mathrm{out})}(t,u)$. Constraints on the channel typically include all the physical limitations on the transmitter, the medium, and the receiver. They usually include a power or energy limitation on $X^{(\mathrm{in})}(t,u)$, a total time T and a total bandwidth W available for transmission and reception. In addition to small error, the system objectives include moderate implementation complexity, which is not always easy to quantify, and also large data rate in the case of communication.

The concept of data rate or information rate is fundamental in communication. It is usually measured in bits per second, or bits per use which is immediately converted to bits per second when multiplied by uses per second. For a data source generating one of M equiprobable messages per T seconds, the data rate R is defined to be

$$R = (\log_2 M)/T \; . \tag{7.9}$$

This definition explicitly indicates that it is the *number of message possibilities* that characterizes the rate of a source. It immediately shows why one can have more than one bit per photon. Indeed, one can have an infinite number of bits per photon if that photon can fall into, say, one of an infinite number of different time slots. For a general statistical source, the Shannon entropy H for the source is used, in bits per use of the source or bits per source symbol. A full description of communication, information, and detection theory can be found in [6–9]. In the present treatment, only some significant relevant points would be highlighted.

The concept of data rate (7.9) already forces upon us a *fundamental discrete view of nature* in any realistic physical process. If one can assess a true continuum, or indeed a true discrete infinity (in communications the word "discrete" often means discrete and finite), one would be able to get infinite data rate, e.g. when one can distinguish the real numbers between 0 and 1 with infinite precision. In reality, a continuum can support only a finite number of bits either from unavoidable disturbance or from the laws of quantum physics. A discussion of certain points relating to this finite/infinite dichotomy can be found in [10]. Here I would like to emphasize that communication is inherently a finite (discrete, digital) process. Any continuous quantity would finally appear in some discrete fashion in actual utilization.

Not surprisingly, the desirable goals of large data rate and small error probability are in conflict with each other. It is easy to see from the law of large number that if one slows down the data rate, say by repeatedly sending the same message, one can decrease the error probability, indeed to zero asymptotically but with the data rate also going to zero. What is not obvious, but given by Shannon's channel coding theorem, is that for a fixed channel representation, there is a nonzero rate called channel capacity below which one can transmit with arbitrarily small error probability by using increasingly long codes. A (channel) code is a signaling scheme in which all the signaling symbols in a sequence over many uses are processed simultaneously, which clearly makes the implementation more complex. However, long sequences have the statistical regularity given by the probabilistic description similar to the law of large number, which, e.g. implies that in a long sequence of fair coin tosses there is roughly one half heads, compared to nothing that can be said which is applicable to a single or a few tosses. It is this *statistical regularity in long sequences* that leads to the possibility of vanishingly small error probability with a nonzero rate as given by Shannon's theorem.

The maximum such nonzero rate under whatever constraints and specifications on a channel is called the *capacity* of that constrained or specific channel, and is equal to the mutual information between the channel input and output. Referring to Fig. 7.1, we will later discuss the time-varying signal aspect but for the moment consider just channel input $X^{(\mathrm{in})}$ and output $X^{(\mathrm{out})}$ from alphabets X and Y with the channel specified statistically by the conditional probability $p(X^{(\mathrm{out})}|X^{(\mathrm{in})})$, $X^{(\mathrm{out})} \in \mathrm{Y}$, $X^{(\mathrm{in})} \in \mathrm{X}$, interpreted as a probability density or probability mass according to whether the alphabet is continuous or discrete. With an input probability $p(X^{(\mathrm{in})})$, the joint probability $p(X^{(\mathrm{out})}, X^{(\mathrm{in})}) = p(X^{(\mathrm{out})}|X^{(\mathrm{in})})p(X^{(\mathrm{in})})$ completely specifies the channel action and the mutual information $I(X;Y)$ is defined by, in the continuous case

$$I(X;Y) \equiv \int p(X^{(\mathrm{in})}, X^{(\mathrm{out})}) \log \frac{p(X^{(\mathrm{in})}|X^{(\mathrm{out})})}{p(X^{(\mathrm{in})})} dX^{(\mathrm{in})} dX^{(\mathrm{out})} , \qquad (7.10)$$

and similarly, in the discrete case,

$$I(X;Y) \equiv \sum_{X^{(\mathrm{in})}, X^{(\mathrm{out})}} p(X^{(\mathrm{in})}, X^{(\mathrm{out})}) \log \frac{p(X^{(\mathrm{in})}|X^{(\mathrm{out})})}{p(X^{(\mathrm{in})})} . \qquad (7.11)$$

The Shannon entropy $H(U)$ of a single random variable U can be defined as average self information, or

$$H(U) \equiv -\int p(u) \log p(u) du \ , \tag{7.12}$$

$$H(U) \equiv -\sum_{u} p(u) \log p(u) \ , \tag{7.13}$$

in the continuous and discrete case. Note that (7.13) is always nonnegative while (7.12) can be negative. Shannon's source-channel coding theorem and its converse [7,9,12] state that successive independent samples of a discrete U can be transmitted over a memoryless channel $p(X^{(\text{out})}|X^{(\text{in})})$ with arbitrarily small (but not exactly zero) error probability between U and V (see Fig. 7.1) if $H(U) < I(X;Y)$, and the block error $P_e \rightarrow 1$ for $H(U) > I(X;Y)$. The case $H(U) = I(X;Y)$ forms a *boundary* with P_e bounded away from zero in general unless the channel is noiseless. It is important to observe the conceptual distinction between a source output U and a channel input X, even though they may happen to be the same physical quantity. Note also that a continuous alphabet channel in reality still has a *finite* capacity and so can reliably transmit only a discrete quantity. If U is a continuous random variable, some performance criterion such as mean-square error would need to be adopted which cannot be made vanishingly small. The extent to which it can be minimized is dealt with in rate-distortion theory [7–13] discussed in Sect. 7.3. It is important to note that for a noisy channel, the use of long codes to obtain a reliable system with high rate significantly increases the system complexity, especially in the decoding operation.

The name "capacity" is usually applied to the $I(X;Y)$ maximized with respect to $p(X^{(\text{in})})$ under whatever constraints, but it is also used to refer to whatever maximum $I(X;Y)$ obtained by different restrictions on the utilization of a given channel, e.g. under discretization (usually called quantization in the communication and signal processing literature) of the input and output of a continuous channel. The point is that with various special restrictions including a fixed $p(X^{(\text{in})})$, a given channel would give rise to many other channels, each with its own "capacity". Even more proliferation occurs in the quantum case. It is essential to understand the exact conditions under which a so-called "capacity" is obtained, for it is often not a really meaningful capacity in the sense of *ultimate* capability limit on the transmission medium or system.

7.1.2 Signal, Noise and Dimensionality

I will now try to describe qualitatively the effect of noise on data rate, finally leading to the famous Shannon capacity formula for an additive white Gaussian noise (AWGN) channel which is directly applicable to squeezed states.

Let P and N be the total average signal and noise power of an AWGN channel represented by

$$X^{(\mathrm{out})}(t) = X^{(\mathrm{in})}(t) + n(t) , \tag{7.14}$$

where $n(t)$ is the white noise. Let W be the available bandwidth, i.e. the duration in frequency occupied by the signals $X^{(\mathrm{in})}(t)$. Then the optimizing input signals for capacity is a white Gaussian process with resulting

$$C = W \log(1 + \frac{P}{N}) . \tag{7.15}$$

In terms of the noise spectral density N_0, one has the famous formula

$$C = W \log(1 + \frac{P}{N_0 W}) . \tag{7.16}$$

Equation (7.15) can be derived from the mutual information expression (7.10), as given by Shannon [11] and later more rigorously in [7]. The intuitive reason why (7.15) takes the form it does, according to such a derivation, would then have to be traced through the reason why (7.10) or (7.11) provides a *general* capacity for information transmission. In the discrete case (7.11), this can be gleaned from Shannon's original proof in [11], and the continuous case may be viewed as a discrete limit as developed in [7]. However, a direct approach can be given for a Gaussian channel, also provided by Shannon [13], which explains the nature of various relevant quantities quite succinctly.

Consider the transmission and reception of a single continuous real variable in noise

$$X^{(\mathrm{out})} = X^{(\mathrm{in})} + n . \tag{7.17}$$

If $X^{(\mathrm{in})}$ is restricted to an interval of length L, an infinite number of bits per channel use is obtained in the absence of noise, $n = 0$, for any $L > 0$. If the noise n always has value in the interval $[-\Delta/2, \Delta/2]$, the number of bits per use is reduced to the finite

$$(L + \Delta)/\Delta \tag{7.18}$$

including edge effects. If X and n are independent continuous random variables with variances P and N, or standard deviations $\sqrt{P}$ and $\sqrt{N}$, a crude estimate patterning after (7.18) would suggest that the number of amplitudes that can be well distinguished, or equivalently the number of bits per use, is

$$\sim k\sqrt{(P + N)/N} , \tag{7.19}$$

where k is a small constant in the neighborhood of unity depending on how "well distinguished" is to be interpreted. We may recall earlier in this section it was mentioned that in a long sequence of independent trials, statistical regularity appears and provides deterministic features to the sequence. This kind of effect would indeed turn the approximate relation (7.19) into an exact one similar to (7.18). In the case of time-varying signals, this comes about in the long signal duration T limit as follows.

First of all, the collection of time functions of "approximate time duration" T and "approximate bandwidth" W span a linear space of dimension

$$D \sim 2TW \tag{7.20}$$

according to the Dimensionality Theorem [6], an improved version of the sampling theorem [13]. The word "approximate" above is necessary because no time function with a Fourier transform can be both strictly time-limited and strictly band-limited, but the exact definitions of "approximate" do not alter the final result (7.20) [14]. The Dimensionality Theorem (7.20), as I discussed elsewhere [15], has momentous consequences in the description of nature. Here, it cuts down, even in the absence of noise, an otherwise infinite dimensional space to a finite dimension in a realistic system where both T and W have to be finite. Thus, a signal or time function can be viewed geometrically as a point in a finite dimensional Hilbert space. (The linear space is readily given an inner product via $\int_T X_1^{(\mathrm{in})}(t) X_2^{(\mathrm{in})}(t) dt$.)

In this geometric representation, the effect of an additive noise is to add a noise vector to the signal vector. The effect of a power constraint P on $X^{(\mathrm{in})}(t)$ is to have it lie inside a D-dimensional sphere of radius $\sqrt{P}$ in the whole space R^D, the Euclidean space of dimension D. For white Gaussian processes, the coefficients of its expansion in any orthonormal basis are independent Gaussian random variables, with variances all given by the same quantity, the average power of the process [6–8,13]. The receiver looks at the received point A in the signal space, and picks the nearest signal point in Euclidean distance to A for minimizing the error probability assuming equiprobable messages. The situation is illustrated in Fig. 7.2.

Thus, in time T there are $2TW$ independent Gaussian amplitudes from (7.20), and from (7.19) the total number of well distinguished signals is

$$M = \left[k \sqrt{\frac{P+N}{N}} \right]^{2TW} . \tag{7.21}$$

The number of bits per second is, from (7.9),

$$\frac{\log_2 M}{T} = W \log_2 k^2 \frac{P+N}{N} . \tag{7.22}$$

The capacity formula (7.15) for AWGN channel follows from (7.22) with $k = 1$. It comes about more precisely as follows. As a result of the statistical regularity mentioned above, for large T the signals $X^{(\mathrm{in})}(t)$ must almost all lie on a sphere of radius $\sqrt{2TWP}$, signal plus noise $X^{(\mathrm{out})}(t)$ on a sphere of radius $\sqrt{2TW(P+N)}$, with noise $n(t)$ on a sphere of radius $\sqrt{2TWN}$ centered at the original signal point. Note that the Euclidean structure of the signal space, which is absent in the general discrete case, is crucial here

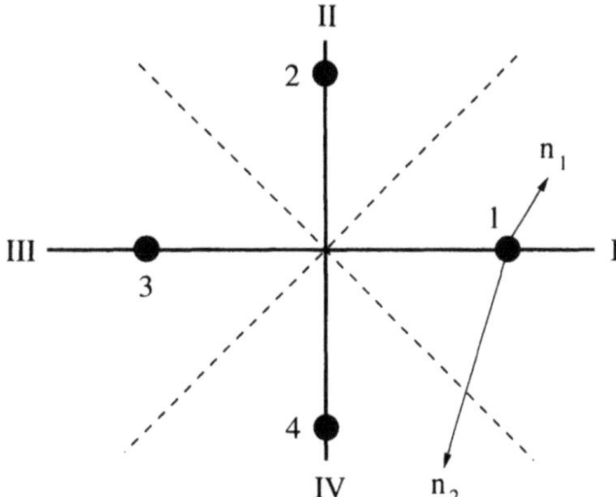

Fig. 7.2. Geometric representation of the receiver: four possible signals 1 to 4 in a 2-dimensional space with corresponding minimum distance decision regions I to IV formed by the dotted lines. The additive white Gaussian noise vector n_1 added to signal 1 would be decoded correctly, while the noise vector n_2 pushes signal 1 to the decision region IV for signal 4, and would be decoded incorrectly

– the different coordinate values of, say, $n(t)$ in the D-dimensional space are Gaussian distributed yielding the given average value DN for the norm $\int_T n^2(t)dt$ of the $n(t)$ vector, so that $n(t)$ is on a sphere of radius $\sqrt{DN}$ around the source signal point. For arbitrarily small error probability, one would want the noise spheres around different signal points to overlap arbitrarily little. A "sphere-packing" argument (see (7.35) below also) then readily establishes the converse to the coding theorem for (7.15), namely that it is impossible to transmit with arbitrarily small error probability at rates above C. For the positive statement that it is indeed possible to so transmit at rates below C, a "random coding" argument is required which in fact establishes the following amazing result: if the signals are selected at random, with probability one the resulting error probability is arbitrarily small. The dichotomy at C, for all rates R below C the block error $P_e \to 0$ for almost all long codes ($n \to \infty$) while for rates above C, $P_e \to 1$ for almost all long codes, is exactly like a phase transition. In practice, it turns out that long codes or signal sets that have enough structure to be readily described, encoded and decoded, do not approach capacity although the situation seems to be changing very recently. For more details of the above description see [6] and [13].

Besides communication of information, the problem of estimating a continuous entity is also of prime concern in this chapter. Consider a Gaussian random variable U with zero mean (or normalize it away) and variance σ^2, which is received in the form Au in Gaussian noise with a possible gain or loss A, i.e.

$$X^{(\text{out})} = Au + n \; . \tag{7.23}$$

This may arise in linear modulation, or in the estimation of U in any experiment. (The word "detection" is commonly reserved for the "estimation" of a discrete U.) If the mean-square error $\overline{\epsilon^2}$ between the estimate $V = \hat{u}(X^{(\text{out})})$ and U is to be minimized, the best estimate is given by [6,8] the conditional mean $E[U|X^{(\text{out})}]$,

$$\hat{U}_{MMSE}(X^{(\text{out})}) = \frac{X^{(\text{out})}/A}{1 + N/\sigma^2 A^2} , \tag{7.24}$$

where N is the noise variance, with resulting

$$\overline{\epsilon^2} = \frac{\sigma^2}{1 + \sigma^2 A^2/N} \equiv \sigma^2 (1 + \frac{S}{N})^{-1} \tag{7.25}$$

in terms of the a priori variance σ^2 and a signal-to-noise ratio S/N.

In the physics literature on quantum information and its applications, the criterion of mutual information is often used in place of detection or estimation error in situations (such as cryptographic eavesdropping) for which no coding is possible. Depending on the problem, the best possible outcome of such use would be a bound rather than the desired performance criterion.

7.1.3 Communication versus Measurement

The estimation/detection problem clearly parallels the problem of producing an estimate of a desired quantity u from the measured data $X^{(\text{out})}$ in a physical experiment. More generally, in an estimation problem one is given a fixed statistical specification $X^{(\text{out})}(t, u)$ and forms in general a nonlinear estimate $\hat{u}(X^{(\text{out})})$ so that a cost function $C(\hat{U}, U)$ is minimized. For mean-square error,

$$C(\hat{U}, U) = \overline{\epsilon^2} = \mathrm{E}[|U - \hat{U}|^2] , \tag{7.26}$$

where the expectation E is taken over all the random quantities involved. In a communication situation, the channel is a statistical transformation F on the input $X^{(\text{in})}(t, u)$

$$X^{(\text{out})}(t, u) = F[X^{(\text{in})}(t, u)] , \tag{7.27}$$

which reads, for an additive noise channel,

$$X^{(\text{out})}(t, u) = X^{(\text{in})}(t, u) + n(t) \tag{7.28}$$

for which one can control $X^{(\text{in})}(t, u)$ subject to the system constraints. In contrast to the estimation case, a direct optimization approach to a communication problem with joint transmitter/receiver optimization has never been developed in a useful way. Instead of asking for the optimum system for a fixed time duration T, T itself is floated as a design parameter in the development of channel encoding-decoding design subject to Shannon's coding theorems.

It can be seen that a *physical measurement* is generally *not* just an estimation problem, because $X^{(\text{out})}(t,u)$ can be influenced to some extent through the choice of the physical measurement process, although perhaps not as much as controlling $X^{(\text{in})}(t,u)$ in (7.27). In particular, it is a major part of the measurement system design to find an appropriate ***physical variable*** $X^{(\text{in})}$ to couple to the desired information parameter u to form $X^{(\text{in})}(t,u)$ for information extraction after the corruption of $X^{(\text{in})}(t,u)$ to $X^{(\text{out})}(t,u)$ by the "channel" is taken into account. However, there is usually no question of data rate in a measurement. Thus, physical measurements, which are of prime concern to us, are described somewhere between communication and detection/estimation. This situation already obtains in classical measurements, and in cases of fixed quantum states and quantum measurements. It becomes more so in quantum communications where the quantum states and quantum measurements can be freely chosen. As developed in Sect. 7.3, the feasibility of choosing quantum states moves a physical measurement problem away from being a pure estimation problem to becoming more like a communication problem, although it never fully becomes a standard communication problem.

7.2 Quantum Communication

7.2.1 Quantum versus Classical Communication

By "quantum communication" we mean more than the study of quantum effects in communication systems involving classical information transfer. Specifically, in quantum communications we are concerned with the system performance under a variety of different quantum measurements and quantum states. Referring to Fig. 7.1, the statistical specification of the channel plus transmitter, e.g. is given by a conditional probability $p(X^{(\text{out})}|u)$. The classical variable $X^{(\text{out})}$ may well be of quantum origin, say it is the eigenvalue of a quantum observable obtained in a measurement. However, as far as the analysis of this system is concerned, the fact that $p(X^{(\text{out})}|u)$ arises from quantum mechanics makes no difference, and it would proceed just like a classical communication system. If this $p(X^{(\text{out})}|u)$ arises from a quantum state $\rho(u)$ and a quantum measurement of a self-adjoint $\hat{X}^{(\text{out})}$ with eigenstates $|X^{(\text{out})}\rangle$, $p(X^{(\text{out})}|u) = \langle X^{(\text{out})}|\rho(u)|X^{(\text{out})}\rangle$, one may well ask whether other possible choices of $\rho(u)$ and observable with resulting different $p(X^{(\text{out})}|u)$ may lead to better performance. These additional freedoms of quantum measurement and quantum state selection are absent in a classical communication system. They constitute the new content of *quantum communication.*

A general quantum communication system is depicted schematically in Fig. 7.3. The channel input and output signals $\hat{X}^{(\text{in})}(t)$ and $\hat{X}^{(\text{out})}(t)$ are now field operators in quantum states $\rho(u)$ and $\rho_R(u)$.

Generally, as indicated by the Dimensionality Theorem (7.20), a finite number of modes each with two degrees of freedom, such as an optical mode

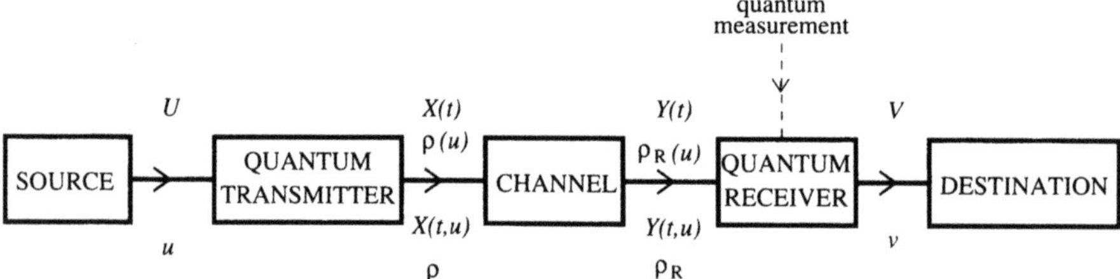

Fig. 7.3. Schematic representation of a quantum communication system: the message dependence generally enters through the state for the field, but can be put in the field operator itself in some instances

with two quadratures, would suffice so that the density operators ρ and ρ_R are well-defined. A specific classical statistical characterization of the system would result upon a choice of quantum measurement at the receiver. The most general characterization of a quantum measurement is the so-called "completely positive operation measure" with the corresponding measurement statistics given by a positive operator-valued measure (POM) [16]. Let $\hat{O}(X^{(\mathrm{out})})$ be a POM on $\hat{X}^{(\mathrm{out})}(t)$, the channel output, thus

$$\sum_{X^{(\mathrm{out})}} \hat{O}(X^{(\mathrm{out})}) = I \quad \text{or} \quad \int \hat{O}(X^{(\mathrm{out})}) dX^{(\mathrm{out})} = I \tag{7.29}$$

and each $\hat{O}(X^{(\mathrm{out})})$ is a nonnegative self-adjoint operator, with $\hat{O}(X^{(\mathrm{out})}) = |X^{(\mathrm{out})}\rangle\langle X^{(\mathrm{out})}|$ for orthogonal $|X^{(\mathrm{out})}\rangle$ in the case of a self-adjoint observable. The statistics are given by

$$p(X^{(\mathrm{out})}|u) = \mathrm{Tr}\rho_R(u)\hat{O}(X^{(\mathrm{out})}) \ . \tag{7.30}$$

The output state $\rho_R(u)$ is determined by the channel action on the input state $\rho(u)$. The additional quantum "freedoms" in quantum communication consist in the selection of $\hat{O}(X^{(\mathrm{out})})$ and $\rho(u)$. Note that it may be more convenient, as in the case of frequency modulation, to enter the information variable u in $\hat{X}^{(\mathrm{in})}(t)$ in parallel with the classical case, and specify the transmitter in terms of $\hat{X}^{(\mathrm{in})}(t,u)$ and ρ rather than $\hat{X}^{(\mathrm{in})}(t)$ and $\rho(u)$. In this formulation, the quantum state ρ and the classical modulation process are decoupled. However, if the information u enters through the quadratures, it would be necessary to use $\rho(u)$, for which the quantum state selection and modulation selection are tied together.

Historically, the serious study of optical communication began immediately after the laser was experimentally realized, for which quantum effects are clearly important as $\hbar\omega/k \sim 10^4$ K. While the evaluation of system performance went on for coherent states and the three standard measurements: direct, homodyne, and heterodyne detections, quantum communication theory in our sense was also developed. Forney [17] and Gordon [18] proposed the entropy bound for information transfer with a fixed set of states valid for arbitrary measurement, to be discussed in Sect. 7.2.3. Helstrom [19] studied the quantum measurement optimization problems in the spirit of classical detection/estimation theory, which were further developed by Holevo [20] and Yuen [21]. Each of the above three standard measurements corresponds, respectively, to the quantum measurement of photon number, single field quadrature, and joint quadratures described by a POM but not a self-adjoint operator [22]. In such work, which actually has many applications in physics [23] but will not be further discussed in this chapter, the states are fixed and the quantum measurement is selected so that the resulting classical statistical system leads to the best possible performance compared to other measurements. The issue of quantum channel representation was treated [4,24,25] and the possibility of receiver state control is suggested [2,4,25–27]. The general problem of transmitter quantum state selection was considered by Yuen [3,4], leading to the development of TCS as indicated in the introductory section. The question of optimal state influence on channel capacity was also implicit in connection with the application of the entropy bound, indicating that number states and photon counting are best for free boson fields [4,28]. For recent advances in determining the capacities and error exponents of various quantum channels by Holevo and the Hirota group, see [29–31]. For other advances including work on quantum tomography by the D'Ariano group, see [32–34]. For applications of squeezed states to quantum cryptography, see [35,36].

7.2.2 Mutual Information

The capacities, or mutual informations maximized over the input distributions, for various boson channels are discussed extensively in [37]. Here I would like to focus on five capacities for the narrowband free boson channel under an average power constraint: number state and photon counting, TCS and homodyning, coherent state and the three standard measurements. Hopefully, it would become clear within this and the next subsection that they are the most important cases capturing the essence of the situation.

For the free electromagnetic field at optical frequencies, all the current or forseeable future systems are narrowband, i.e. the available bandwidth is only a small fraction of the center frequency. Due to various facts of nature, it would be extremely difficult and inefficient to utilize photons at higher frequencies, say X-rays, in a communication situation. Thus, there is no practical significance in studying wideband photonic channels. The constraint of average power can be separated into two parts: average with respect to the statistics of the information variable U and average with respect to the quantum nature of the state ρ. In either case a peak power (or energy or power spectral density) constraint can also be applied. In the case of classical signals the peak power constraint is indeed quite meaningful and realistic, but is often hard to handle mathematically and usually avoided. In the case of quantum states, a peak energy constraint would cut off the Hilbert space of states $\mathcal{H}$ at a maximum number state eigenvalue n_m so that $\mathcal{H}$ becomes finite-dimensional. This, however, is unrealistic or at least hard to handle in so far as one considers a coherent state $|\alpha\rangle$, which has components in all $|n\rangle$, to be realizable. Some discussion on this issue is given in [10], although in its full scope it is a complicated and profound issue. Here I would advocate, if only on the ground of mathematical convenience, that energy constraint is to be applied to the quantum state average $\mathrm{Tr}(\rho a^\dagger a)$, and not to yield an n_m.

Let $P = hf_0WS$ be the available signal power of a narrowband channel of center frequency f_0 and photon numbers S per mode. The photon number capacity is [28,37,38]

$$C_{\mathrm{op}} = W[(S+1)\log(S+1) - S\log S] \; . \tag{7.31}$$

For TCS with homodyne detection [4,37],

$$C_{TCS} = W\log(1+2S) \; . \tag{7.32}$$

If both quadratures of the TCS are utilized under the same power constraint with optimized TCS-heterodyne [22] or joint quadrature measurement, it can be shown from the Kuhn–Tucker optimizality conditions of nonlinear programming that as S is increased from 1 the optimum capacity is indeed achieved through utilization of only one quadrature. The coherent state heterodyne and homodyne capacities are

$$C_{\mathrm{het}} = W\log(1+S) \; , \tag{7.33}$$

$$C_{\mathrm{hom}} = \frac{W}{2}\log(1+4S) \; . \tag{7.34}$$

Equations (7.32)–(7.34) are easily derived from (7.15) because the corresponding channels are AWGN ones – the fluctuations in a TCS or a coherent state, which is merely a classical amplitude superposed on vacuum, are behaving as independent additive Gaussian noises, and are white noises under the narrowband assumption. The coherent-state photon counting capacity C_{ph} does not have a simple closed form but is readily computed numerically [39]. These five capacities are compared numerically in Fig. 7.4 reproduced from [4], for a fairly wide bandwidth. It may be observed that C_{TCS} is always larger than C_{het} and C_{hom}, and is also larger than C_{ph} in the case of more than a fraction of a photon per mode.

However, the difference between C_{TCS} and $C_{\mathrm{het}}, C_{\mathrm{hom}}$ is not big. Indeed, C_{TCS} is less than twice C_{hom} although the SNR of TCS is the square of that for coherent states. Because the data rate for a mode goes as log(1+SNR) from (7.15), the square in SNR becomes less than a multiplicative factor of 2. The difference between C_{TCS} and C_{het} is even less, (7.32) is equivalent to doubling the signal power in C_{het} with the same bandwidth. The underlying reason can be understood as follows. In the geometric representation of signal and noise sketched in Sect. 7.1.2, it can be seen that the effect of noise is to move a given signal point away from its position. If the noise is big enough,

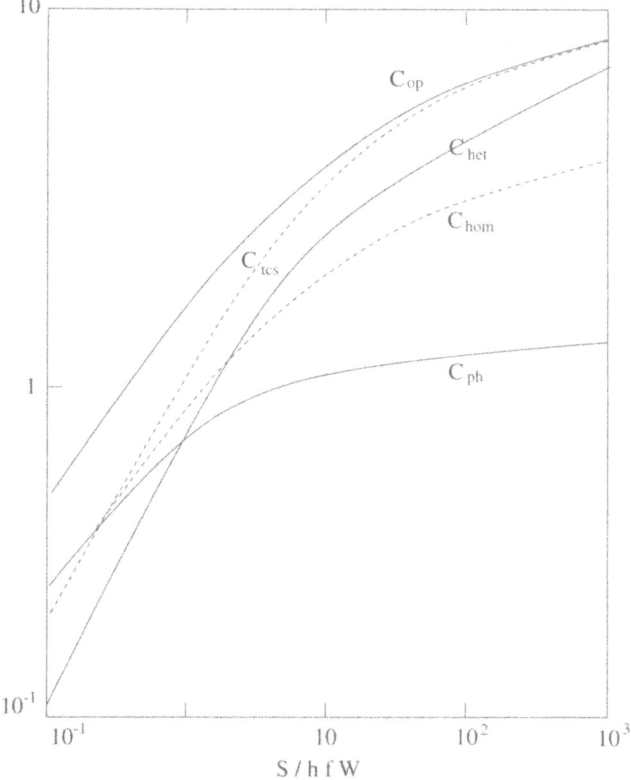

Fig. 7.4. Comparison of capacities in bits per second for $f_0 \sim 5.7 \times 10^{14}$ Hz and $W \sim 10^{14}$Hz

it would move it closer to another signal point B as compared to the original point A, and the optimum receiver would decide it is this other signal B that was transmitted, hence making an error, as illustrated in Fig. 7.2. Thus, a good system would have the signal points as far apart as possible from the viewpoint of errors, and have as many signal points as possible from the viewpoint of data rate, two conflicting goals. For a fixed dimension $D \sim 2WT$, a larger power P yields a larger sphere and the same number of M signal points can be placed further apart inside the sphere, leading to a smaller error for a fixed noise power N. Increasing W, however, is more beneficial than increasing P, thus $C_{\text{het}} > C_{\text{hom}}$ as W increases, even though C_{hom} has a bigger SNR. To see the role of W versus P, recall the discussion around (7.21) and (7.22) that one wants the noise spheres around different signal points to be almost non-overlapping to yield small error probability. As a result of this "sphere packing", the number of well distinguished signals is roughly the ratio of the signal plus noise volume to the noise volume. The volume $V_D(r)$

of a D-dimensional sphere of radius r is $B_D r^D$ for a D-dependent constant B_D, which implies

$$V_D(\sqrt{D(P+N)})/V_D(\sqrt{DN}) = \left(1+\frac{P}{N}\right)^D \tag{7.35}$$

since the radii of the signal plus noise and noise spheres is $\sqrt{2TW(P+N)}$ and $\sqrt{2TWN}$ respectively. The quantity (7.35) grows exponentially in D or W but only to a fixed power in P. This more important role of W versus P clearly manifests in (7.15) and (7.16).

Having understood why the apparent large gain in SNR given by (7.8) for TCS leads only to a small gain in capacity, the question becomes whether TCS would be significant in improving optical communications compared to coherent states. This rest of this Sect. 7.2 is devoted to a detailed examination of this issue. We may first observe that complicated coding, especially the decoding process, is required to approach capacity given by any of (7.32)–(7.34). If one looks at the error behavior of information transfer under specific simple signaling scheme, e.g. the antipodal signals discussed in [5], the full SNR square advantage may appear. That is, more restricted "capacities" than (7.32)–(7.34) may show a large advantage with TCS. In Sect. 7.2.3 we will see that the number state capacity C_{op}, which is so close to C_{TCS}, is actually the optimum rate for any states and measurements subject to the average power constraint. This capacity C_{op} can be obtained *without* the need for complicated decoding because the ideal number state channel is noiseless – there is no need to use long sequences to yield statistical regularity. Thus, the use of number states can be considered as an *alternative to channel coding*. Number states, as intensity squeezed states, have a lot of similarity to TCS in regard to their physical generation and propagation characteristics. Unfortunately, the use of such nonclassical states as information sources would *not* be advisable in practical communication systems. In addition to various problems of a more practical nature, such as phase coherence for TCS and good detectors for number states, the inevitable presence of significant loss would wipe out the advantage of nonclassical states. This issue will be treated in Sect. 7.2.4 after the following discussion of the entropy bound that established the optimality of C_{op} given by (7.31).

7.2.3 The Entropy Bound

Given a fixed set of density operators ρ_λ dependent on a discrete or continuous random variable Λ with probability (density) $p(\lambda)$, define

$$\overline{\rho} \equiv \sum_\lambda p(\lambda)\rho_\lambda \qquad \text{or} \qquad \int p(\lambda)\rho_\lambda d\lambda\ . \tag{7.36}$$

Let $S(\rho) \equiv -\mathrm{Tr}\rho \log \rho$ be the Von Neumann entropy of ρ, and let $\hat{O}(X^{(\mathrm{out})})$ be the POM giving the measurement probability. Then the mutual information between λ and $X^{(\mathrm{out})}$ is bounded by

$$I(\Lambda; \mathrm{Y}) \leq S(\overline{\rho}) - \overline{S}(\rho_\lambda) \tag{7.37}$$

$$\overline{S}(\rho_\lambda) = \sum_\lambda p(\lambda) S(\rho_\lambda) \qquad \text{or} \qquad \int p(\lambda) S(\rho_\lambda) d\lambda \; . \tag{7.38}$$

This entropy bound (7.37), first given by Forney [17] and Gordon [18], was proved for finite discrete Λ and finite dimensional Hilbert space $\mathcal{H}$ by Zador [40] and independently by Holevo [41], and general Λ and infinite dimensional $\mathcal{H}$ by Ozawa [38]. The long complicated history of this bound is outlined in [10]. Recently, the inequality in (7.37) is shown to be achievable if the measurement can be made over a long sequence of states instead of just symbol by symbol in the sequence [42,43]. However, while this may be considered to establish the capacity of a quantum channel defined by the mapping $\lambda \mapsto \rho_\lambda$, such a specification of a quantum channel is neither general nor practical. The main reason is that there is no way to tell whether the particular map $\lambda \mapsto \rho_\lambda$ is optimum under the constraint of the problem. As we have emphasized in Sect. 7.2.1, a general quantum communication problem involves *both* the choice of states and measurements. Under an average energy constraint for a single mode

$$\sum_\lambda p(\lambda) \mathrm{Tr}\left[\rho_\lambda a^\dagger a\right] \leq S \tag{7.39}$$

one cannot tell a priori what the optimal $\lambda \mapsto \rho_\lambda$ should be, even if one assumes all ρ_λ are coherent states.

On the other hand, the bound (7.37) in its full generality readily shows [38] that for a single boson mode under (7.39), the maximum $I(\Lambda; \mathrm{Y})$ is achieved by taking λ and $X^{(\mathrm{out})}$ as a non-negative integer, with number states $\rho_{\lambda=n} = |n\rangle\langle n|$ and the I value given by (7.31) for $W = 1$. The wideband capacity can be similarly derived [38]. Note that apart from showing that more general processing such as feedback would not increase I, this joint optimization over state/measurement and modulation (the map $\lambda \mapsto \rho_\lambda$) demodulation (the map $X^{(\mathrm{out})} \mapsto \lambda$) does establish that (7.31) is the *ultimate* quantum limit on the possible rate of information transfer for a boson mode of average energy S.

7.2.4 Effect of Loss

The effect of linear loss on a mode can be represented as a transformation on the modal photon annihilation operator [2,4,44–46],

$$b = \eta^{\frac{1}{2}} a + (1 - \eta)^{\frac{1}{2}} d \; , \tag{7.40}$$

where η is the transmittance and the d-mode is in vacuum. Note that the effect of quantum efficiency on a detector can be so represented as well. It follows immediately from (7.40) that the resulting quadrature fluctuation in b has a floor level $(1-\eta)/4$, which is essentially the coherent state noise level for $\eta \ll 1$. Similarly for a number state, the b-mode photon number fluctuation

$$\langle \Delta N_b^2 \rangle = \eta^2 \langle \Delta N_a^2 \rangle + \eta(1-\eta)\langle N_a \rangle \tag{7.41}$$

contains a partition noise equal to the mean $\langle N_b \rangle = \eta \langle N_a \rangle$ for $\eta \ll 1$, washing away the sub-Poissonian character of the a-mode. Generally, one can readily show from (7.40) that the state ρ_b is very close to a coherent state of mean $\eta^{1/2}\langle a \rangle$ for large loss, thus any nonclassical state becomes essentially classical.

The implication of this fact on the utility of nonclassical states is profound, especially in engineering applications where significant loss is usually present, e.g. in fiber optic communications. Unless a special environment is created [4] to compensate for the squeezing or nonclassical effect in the presence of loss, there is no way to keep a nonclassical state at the reception end. While this is possible in principle, it seems that is not worth the trouble. Even in scientific experiments or in the process of nonclassical state generation, loss places a severe limit on the amount of squeezing obtainable. The *sensitivity* of nonclassical states to loss and interference would place strenuous requirements on all the system components, making any such system extremely difficult to implement. To me, a similar kind of argument leads to a similar implication in the field of quantum information.

Given the close value of C_{TCS} to C_{het} in (7.32)–(7.33) in the absence of loss, it should be clear that there is hardly any advantage left in the presence of loss. While the ultimate quantum capacity C_ℓ of a lossy channel is not known, an upper bound on C_ℓ can be easily derived. Under an average energy constraint S and loss η, equation (7.31) for C_{op} with S replaced by ηS would provide a bound on C_ℓ. The gap between C_{op} and C_{het} with ηS is the largest gain, probably not actually achievable, that one can possibly obtain with nonclassical states in a lossy channel. The smallness of this gap, as seen from Fig. 7.4, shows that there is little significance in pursuing quantum communications with nonclassical sources in practice, a conclusion I drew over twenty years ago.

7.2.5 Quantum Amplifiers and Duplicators

Not all is lost, however. As to be presently explained, the use of novel quantum amplifiers and related devices on coherent state sources can lead to a number of significant communication applications not possible with the usual phase-insensitive linear amplifier (PIA). A characteristic feature of these novel devices is that their outputs are often nonclassical states for coherent state inputs, even though it is not the nonclassical nature of these states that is relevant in the application.

Table 7.1. Abbreviations used in the Table 7.2 and text

CS	coherent state
NS	number state
TCS	two-photon coherent state
PCS	phase coherent state
PIA	phase-insensiive linear amplifier
PSA	phase-sensitive linear amplifier
PNA	photon-number amplifier
QPA	quantum phase amplifier
PND	photon number duplicator
POA	photon on-off amplifier
QND	quantum nondemolition measurements
POM	positive operator-valued measure
SQL	standard quantum limit

Corresponding to the three standard quantum measurements are three quantum amplifiers, the photon number amplifier (PNA), the phase-sensitive linear amplifier (PSA), and the PIA. If b and a are the output and input modal photon annihilation operator of the amplifier, these three amplifiers can be represented as [45–48], with a power gain $G > 1$,

$$\text{PIA} \quad b = G^{1/2}a + (G-1)^{1/2}v^\dagger \ , \qquad [v, v^\dagger] = 1 \ , \tag{7.42}$$

$$\text{PSA} \quad b_1 = G^{1/2}a_1 \ , \qquad b_2 = G^{-1/2}a_2 \ , \tag{7.43}$$

$$\text{PNA} \quad b^\dagger b = G a^\dagger a \ , \quad G \quad \text{integer} \ . \tag{7.44}$$

A fourth quantum phase amplifier [49,50] is

$$\text{QPA} \quad e_+ = e_+^G \ , \quad e_+ \equiv (a^\dagger a + 1)^{\frac{1}{2}} a^\dagger \ , \tag{7.45}$$

which is related to the ideal phase measurement [19,21,51] described by a POM involving the Susskind–Glogower states and corresponding phase-coherent states [51].

Table 7.1 lists the abbreviations used in the Table 7.2 and text. Table 7.2 summarizes the different types of amplifiers and duplicators. The column on states merely emphasizes that the nature of these states would be preserved only by the corresponding amplifier, not that the amplifier is noiseless only for those states.

In (7.42) and (7.44), the photon operator b has to be defined on two modes. For a fuller discussion of these amplifiers, see [48] and [49] which also contains an extensive treatment of duplicators to be discussed later in this section. The main point about (7.42)–(7.44) is that the amplifier output of each is, for the corresponding measurement, a perfect "noiseless" scaled (amplified) version of the input for *arbitrary* input state, i.e. they are noiseless amplifiers for

Table 7.2. Types of quantum amplifier

Detection	Amplifier	States	Duplicators
heterodyne	PIA	CS	BQD
homodyne	PSA	TCS	SQD
direct	PNA	NS	PND
phase(ideal)	QPA	PCS	QPD

the corresponding detection scheme. Thus, the often found statement that quantum amplifier necessarily introduces noise, say in the sense of having a noise figure $F \equiv SNR_a/SNR_b > 1$, is *wrong*. As summarized in Table 1, if the proper amplifier matching the measurement is used there is no additional noise ideally, similar to the classical case. All the noise then arises *inherently* from the quantum nature of the input. (This is also true in both balanced and unbalanced homodyne/heterodyne detection for which the effective amplifier, the local oscillator, introduces no noise in the high gain limit. See [52]. It is a pervasive misconception that the noise in homodyne/heterodyne detection is local-oscillator shot noise.) Without going into a detailed exposition, this is actually clear intuitively from the basic principles of quantum physics. When you fix a measurement, the situation is classical for any given state as discussed in Sect. 7.2.1 on quantum vs. classical communication, *in the sense that* a fixed probabilistic description is obtained. The situation is a little more subtle in the case of POM rather than self-adjoint operator, but can be understood by analyzing the POM as commuting self-adjoint operators measurement on an extended space which can always be done [20].

The generation mechanism of PSA is identical to quadrature squeezing, which, being piecewise linear, is not exactly a nonlinear effect. On the other hand, PNA, QPA and the duplicators involve truly nonlinear quantum effects [47–50] which would not be discussed here. None of these new quantum devices except PSA has been successfully demonstrated experimentally in a useful manner.

At this point, I would like to address a confusing point about the capability of amplifiers. It is often stated that an amplifier at the receiver could improve the receiver performance. The optimum receiver performance is determined by the specification $\{\hat{X}^{(\mathrm{out})}(t), \rho(u)\}$ in Fig. 7.3. Nothing, and certainly no amplifier, can ever improve that as a matter of tautology. What can be improved is a specific receiver structure that does not lead to the optimum performance. In such a case, the use of an amplifier or some other device may improve the suboptimal receiver performance. This point is related to, but different from, the so-called data processing theorem [7] in information theory which shows that no processing can increase the information transfer over a channel.

The above amplifiers can be used as pre-amplifiers to suppress subsequent receiver noise in the corresponding detections, in either engineering or scientific applications. They can also be used to advantage [53] in the attempt to create a transparent optical local area network. For such a purpose, however, the duplicators [46–48,54] would be perfect. A photon number duplicator (PND) is a device with one input a in state ρ_a and two outputs b, c such that each of the output photon counting statistics is the same as that of the input

$$\langle n|\rho_a|n\rangle = \langle n|\rho_b|n\rangle = \langle n|\rho_c|n\rangle \ . \tag{7.46}$$

Typically, the output photon counts for the b and c modes are perfectly correlated, thus PND also provides a perfect realization of a photon number quantum nondemolition measurement (QND) with only a finite energy [47]. Single and double quadrature duplicators can be similarly described.

The amplifiers can be used as line amplifiers in long distance optical fiber communications. For example, the use of PSA not only improves the SNR by a factor of 2 for coherent state sources in a long amplifier chain, it also significantly reduces the Gordon–Haus soliton timing error [55]. Considerable experimental progress [56] has been made on such possible application, but the required phase coherence renders it impractical. For on-off signals, the use of PNA leads to the following error probability

$$P_e = \frac{1}{2}\exp\{-S[1 - f_n(G)]\} \ , \tag{7.47}$$

where the functions $f_n(G)$ obey the recurrence relation

$$f_{n+1}(G) = (1 - G^{-1})^G[1 + (G-1)^{-1} f_n(G)]^G \tag{7.48}$$

with $f_0(G) = 0$. Equations (7.47)–(7.48) apply to a chain of n amplifiers of gain G and loss G^{-1} between two adjacent amplifiers, assuming direct detection. In Fig. 7.5, this error exponent $1 - f_n(G)$ is compared with that of the PIA line, $1/(4n)$, obtained under the Gaussian approximation for direct detection.

As can be seen in the figure, even more improvement, in fact the optimum improvement, is obtained with the use of a photon on-off amplifier [57] (POA) tailored for the situation. In the state description, a POA acts on two modes but for the input mode it reads

$$\text{POA} \quad \begin{aligned} |0\rangle &\mapsto |0\rangle \\ |1\rangle &\mapsto |\alpha\rangle \\ &\ \vdots \\ |n\rangle &\mapsto |\alpha\rangle \\ &\ \vdots \end{aligned} \tag{7.49}$$

where $|\alpha\rangle, |0\rangle$ are the two on-off coherent states, $S = |\alpha|^2$. The resulting error probability is

$$P_e = \frac{1}{2}[1 - (1 - e^{-S})^n] \ , \tag{7.50}$$

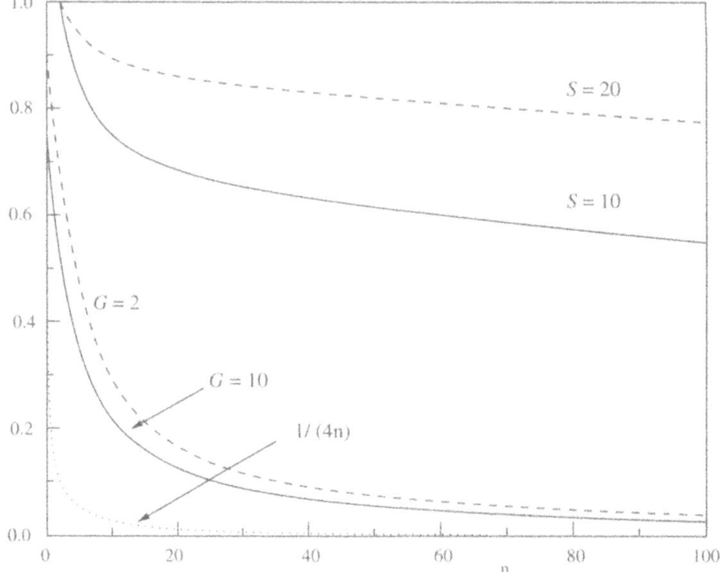

Fig. 7.5. Comparison of error exponents $S^{-1}\ln 2P_e$ as a function of stages n – the PIA line exponent is independent of S and G, the PNA exponent is independent of S and the POA exponent is independent of G

which is the same as that obtained by a repeater, i.e. by direct detection and retransmission at each of the n stages. In general, it is possible to write down a *perfect* quantum amplifier for any given signaling and detection scheme which performs as well as a repeater, although the actual installation of POA or any such amplifiers in a long line would entail the loss of flexibility, as compared to PNA, for adapting to other signaling schemes.

Quantum amplifiers are also useful in quantum cryptography [48]. A major problem of the quantum cryptographic schemes is that they cannot be amplified to compensate for the loss without disrupting the operation of the scheme. In [58] a new quantum cryptographic scheme is introduced that allows amplification, which greatly extends the distance over a fiber for which the scheme works.

7.3 Ultimate Limit on Measurement Accuracy

7.3.1 Measurement System and Ultimate Performance

In this section, the question is addressed on the ultimate, quantum as well as classical, limit on the measurement accuracy obtainable with various measurement systems. The optimum performance ideally achievable with a measurement system is of course an important piece of design information, but more importantly I would like to assess the potential of such systems, and ways to realize them in principle, in order to explore the feasibility of developing ultrahigh precision measurement systems important in many applications, especially scientific ones. My approach [58] is based on the communication characterization of measurement discussed so far, especially in Sect. 7.1.3, while adopting quantum and classical communication theory to provide the answers. Since the correspondence between communication and measurement is not exactly isomorphic, we will find that it is possible to obtain limits on the measurement accuracy, but not always possible to be assured that those limits are attainable. Indeed, even if the correspondence is perfect, there are still additional questions, such as what systems are actually available, that would resist a complete mathematical characterization in the forseeable future. Nevertheless, as to be discussed presently, some of the results obtained are somewhat surprising, and also promising. In the next Sect. 7.3.2 the rate distortion limit in classical communication theory will be explained, and in Sect. 7.3.3 the corresponding quantum limits will be presented. Here I would like to first highlight the results and their implications.

The final error in a measurement system may depend, even in principle excluding nonideal environmental perturbations, on more than a single source or variety. For example, in the detection of very weak gravitational radiation by a Michelson interferometer, the radiation pressure error needs to be added to the photon detection error to form the total error. The application of squeezed states in this situation is treated elsewhere in this book and would not be discussed. Here, the general theory would be illustrated only with a measurement medium or channel that can be characterized as a free boson field, so that the results in Sect. 7.2 may be utilized. The general approach, however, is applicable to any specific measurement system.

Consider the problem of estimating a parameter U with Gaussian density $p_G(u)$ of zero mean and variance σ^2 via a single mode optical field of average energy S. While the optimization of (7.7) yields TCS as the solution, two choices have already been fixed in advance: the parameter u is to be modulated into the mean α_1 of the state, and homodyning or measurement of α_1 is to be performed. If one relaxes these two conditions in accordance with the general quantum communication approach of Sect. 7.2.1, one may pick a state $\rho(u)$ subject to

$$\int \mathrm{d}u p_G(u) \mathrm{Tr}\left[\rho(u) a^\dagger a\right] \leq S \tag{7.51}$$

and a general measurement represented by the POM $\hat{O}(y)$, so that the mean-square error $\overline{\epsilon^2}$ between y and u is minimized. It is not clear at all that the combination of linear modulation, TCS and homodyne is the optimum solution or yields a near optimum performance to this problem. As the following development shows, a lower bound for the root-mean-square error $\delta u \equiv (\overline{\epsilon^2})^{1/2}$ under (7.51) can be derived

$$\delta u \geq \sigma \frac{S^S}{(S+1)^{S+1}} \sim \frac{\sigma}{eS} \ , \quad S \gg 1 \ , \tag{7.52}$$

which is very close to the TCS linear modulation performance,

$$\delta u^{TCS} = \frac{\sigma}{2S+1} \sim \frac{\sigma}{2S} \ , \quad S \gg 1 \ . \tag{7.53}$$

Partly because it is not even clear whether the lower bound (7.52) can indeed be achieved, one would ordinarily be quite satisfied with the difference between 1/2 and e^{-1} and stop looking for another system unless the TCS system is not practical for whatever reason. One may say the linear TCS system is essentially optimum. The corresponding coherent state performance is $\delta u' \sim \sigma/\sqrt{S}$.

For a uniformly distributed phase parameter $\phi \in [-\pi, \pi)$, the corresponding lower bound for the root-mean-square error is

$$\delta\phi = \lambda \frac{S^S}{(S+1)^{S+1}} \sim \frac{\lambda}{eS} \ , \quad S \gg 1 \ , \tag{7.54}$$

where $\lambda \sim 1.35$. This single-mode $1/S$ behavior, improved over the $1/\sqrt{S}$ dependence for coherent state, has been obtained previously for two different concrete systems utilizing TCS [60,61] and number states [62]. This should not be surprising given the closeness between the number state and TCS capacities, (7.31) and (7.32).

In the case of a narrowband field with $m = D/2 = WT$ modes but the same total energy or photon number S, the lower bound for the measurement of a Gaussian U is

$$\delta u \geq \sigma \frac{S^S m^m}{(S+m)^{S+m}} \tag{7.55}$$

$$\rightarrow (1 + \frac{m}{S})^{-S} e^{-S} \ , \quad m \gg 1 \ . \tag{7.56}$$

For the uniform phase ϕ, $\delta\phi$ is given by (7.55) with σ replaced by λ as in (7.52) and (7.54). Note that (7.56) goes to zero as $D \to \infty$. This is because infinite capacity is obtained when the narrowband expression is extrapolated to infinite bandwidth, which is not physically meaningful as the hf dependence becomes important [21,37]. The capacity is finite when such dependency is taken into account [28,37,38].

The result (7.55) or (7.56) is rather remarkable. For the same total energy S spread over a large number of modes, the performance can be improved from $1/S$ to $\exp(-S)$ assuming the bound can be approached. As a communication limit for which one can control the modulation, this is certainly the case. Intuitively, it can be traced to the relative importance of D or W over P or S as discussed in Sects. 7.1.1 and 7.2.2. As the quantum state affects only the SNR and not D, one may expect such behavior for coherent states also, which is indeed born out to be the case as developed in Sect. 7.3.3. Even in the measurement situation, the improvement of a fixed energy spread over many modes is a real one, to be demonstrated for a concrete frequency modulation scheme also developed in Sect. 7.3.3.

(An aside between multimode and single mode results. It is sometimes said that two-mode squeezing is basically different from single mode squeezing because two modes are needed. However, by a simple modal transformation equivalent to removing cross terms in the multimode Hamiltonian, two-mode squeezing can be reduced to single mode squeezing, that is, in the multimode situation one picks the right mode that yields squeezing. The general theory is given in [63,64], which indeed was what led to the prediction of squeezing in degenerate four-wave mixing [65].)

7.3.2 Classical Rate-Distortion Limit

The rate-distortion function $R(d)$ of a random variable U was introduced by Shannon [11], with by now a very extensive literature. Here we consider just continuous U with density function $p(u)$ although discrete U works the same. For a distortion measure $d(u, v)$ between u and v, such as $|u - v|^2$ or $|u - v|$, the average distortion is

$$E[d(U, V)] \equiv \int d(u, v)p(u)p(v|u)dudv \ . \tag{7.57}$$

The rate distortion function $R(d)$ of u is defined to be the minimum mutual information

$$R(d) \equiv \min_{E[d(U,V)]\leq d} I(U; V) \tag{7.58}$$

over all possible choices of $p(v|u)$ subject to the constraint that $E[d(U, V)]$ is less than or equal to a given level $d \geq 0$. One may think of V as a data-compressed version of $U - V$ represents U with an average distortion d, thus it takes less bits to represent V than U for $d > 0$. Shannon's source coding theorem with a fidelity criterion and its converse [7,9,12] state that a source variable U can be asymptotically represented with an average distortion d if and only if at least $R(d)$ bits per source symbol is provided. Similar to channel coding, long sequence encoding and decoding that ensures statistical regularity are in general required to achieve such minimum in the asymptotic limit. Nevertheless, roughly speaking $R(d)$ is the minimum number of information bits per symbol required to represent a source with an average distortion d per symbol.

The channel capacity C may be written as a function $C(\beta)$ where β denote the resource parameters available, including power and bandwidth, as well as other characteristics of the channel such as noise power. Referring to Fig. 7.1, the question arises on the minimum distortion d one can obtain for transmitting a source variable U over a channel with capacity $C(\beta)$. The answer is provided by Shannon's joint source-channel coding theorem [7,9,12] in the so-called rate distortion limit or rate distortion bound. Recall that roughly speaking, the channel coding theorem says that $C(\beta)$ is the maximum number of information bits one can transfer error-free over a channel with parameters β. By combining the source and channel coding theorem, one has $C(\beta) \geq R(d)$ so that, since R is a monotone decreasing function of d,

$$d \geq R^{-1}C(\beta) . \tag{7.59}$$

Intuitively, this works as a lower bound as a consequence of the converse to the coding theorem because otherwise one can transmit more than the capacity rate or compress smaller than the source rate. The positive coding theorem assures that the bound may be approached arbitrarily closely in a communications situation.

Even though the rate distortion bound (7.59) is generally achieved only with source and channel coding, it is occasionally achieved without any coding or nonlinear modulation. Consider the transmission of a zero-mean Gaussian U of variance σ^2 under with the mean-square error criterion, $d(u,v) = |u-v|^2$, over an additive Gaussian noise channel

$$X^{(\text{out})} = X^{(\text{in})} + n \tag{7.60}$$

with noise variance N. Then [7,9,11–13] for U

$$\begin{aligned} R_u(d) &= \frac{1}{2}\log(\sigma^2/d) , \qquad & 0 \leq d \leq \sigma^2 , \\ &= 0 , & d \geq \sigma^2 , \end{aligned} \tag{7.61}$$

and

$$C(S) = \frac{1}{2}\log(1+\frac{S}{N}) \tag{7.62}$$

under $E[X^2] \leq S$. The rate distortion limit (7.59) becomes

$$d \geq \sigma^2(1+\frac{S}{N})^{-1} . \tag{7.63}$$

If one sends U as X over the channel as in (7.23) so that $S = A^2\sigma^2$, and use the estimate (7.24), the resulting mean-square error (7.25) is exactly the lower limit (7.63). This shows that for this problem of transmitting a Gaussian parameter matched, in per use or per symbol to a Gaussian noise channel, even in the full generality of Fig. 7.1 there is *nothing* that can do better than linear modulation-demodulation! Indeed, no way other than through $R(d)$ has ever been successfully employed to show the optimality of linear modulation in this problem.

The rate distortion function of the uniform phase variable ϕ is difficult to evaluate exactly. However, the Shannon upper and lower bounds [7,11] on $R(d)$ for ϕ differs only by about 0.3 bit per symbol. Thus the upper bound

$$R_\phi(d) \leq \log\frac{1.35}{\sqrt{d}} \tag{7.64}$$

would be used for $R_\phi(d)$.

There are two complications in the application of the rate distortion bound to measurement problems. The first arises from the fact that in a measurement one has little or no room for source coding as the parameter U is usually out of one's control before modulating onto the physical channel input variable $X^{(\text{in})}$. Thus, while (7.59) remains a limit, in general there may be no way to approach it. One may try to replace $R(d)$ of (7.61) by some realistic $R(d)$ obtained with whatever one can do to U, but contrary to what is stated in [56], it is not clear how such $R(d)$ may be evaluated. On the other hand, from experience the $R(d)$ for a Gaussian parameter with different encoding criteria vary little, and so the exact form of $R(d)$ is not expected to make any major difference in the final result (7.59).

The second problem is, I believe, more serious and closer to the heart of the matter. It arises because no channel coding may be employed in a typical measurement situation. One can similarly try to replace $C(\beta)$ by a mutual information $J(\beta)$ incorporating the realistic limitations and freedom, which again seems hard to do. This is an essential connection, however, because the modulation of U into X represents how one physically couples U into the measurement medium in the measurement system. It makes a difference whether the optical field couples to U via an interferometer configuration or a source impressing configuration, and e.g. whether the frequency or the amplitude is modulated by U. In any case, if such a meaningful $J(\beta)$ can be obtained, then a measurement rate distortion limit can be obtained from $J(\beta) \geq R(d)$ in the form similar to (7.59)

$$d \geq R^{-1} J(\beta) \, . \tag{7.65}$$

7.3.3 Ultimate Quantum Measurement System Limit

By combining the above rate distortion theory with classical capacities replaced by quantum capacities, one obtains quantum rate distortion limits for a general quantum system of Fig. 7.3, taking into account all the freedom of classical modulation-demodulation and quantum measurement as well as state selections. Note that the uncertainty principles are far from sufficient to determine such ultimate limits. In the original form they are merely restrictions on quantum states, and even in their extended form [20] they do not account for the many freedom represented in Fig. 7.3. A few more remarks on this may be found in [10] and [59].

From (7.61) and the single mode version of (7.32), one finds (7.52) using the optimum number state capacity. With C_{TCS} of (7.32), one finds the same δu^{TCS} as (7.53) for linear modulation without coding! This is *exactly* the situation around (7.25) and (7.63) pointed out above. For C_{het} of (7.30), one obtains

$$\delta u^{CS} = \frac{\sigma}{S+1} \, , \tag{7.66}$$

which may be compared to

$$\delta u' \sim \frac{\sigma}{\sqrt{S}} \tag{7.67}$$

obtained in coherent state systems without coding or nonlinear modulation. The reason why coding or nonlinear modulation is necessary in the coherent state case is that bandwidth expansion, two quadratures in a coherent state versus the single real parameter to be estimated, has to be utilized. Thus, apart from a gain on δu by a factor of 2, *the use of TCS for measurement is essentially the same as coding on a coherent state system* as far as the performance goes, a rather unexpected result.

For the uniform phase parameter, one finds (7.54) from (7.31) and (7.64), and similarly

$$\delta\phi^{TCS} \sim \frac{1}{2S}\,, \qquad S \gg 1\,, \tag{7.68}$$

$$\delta\phi^{CS} \sim \frac{1}{S}\,, \qquad S \gg 1\,. \tag{7.69}$$

Again, the $1/S$ behavior can be obtained without coding on TCS or number state systems, while coherent state systems without coding yields

$$\delta\phi' \sim 1/\sqrt{S}\,. \tag{7.70}$$

In the multimode narrowband situation, one similarly obtains (7.55) for the optimum case, with $m = D/2 = WT$, and

$$\delta u^{TCS} = \sigma(1 + \frac{2S}{m})^{-m}\,, \tag{7.71}$$

$$\delta u^{CS} = \sigma(1 + \frac{S}{m})^{-m}\,. \tag{7.72}$$

Similarly results for ϕ can be written down with σ replaced by λ as in (7.52) and (7.54). From (7.71) and (7.72),

$$\delta u^{TCS} \to \sigma e^{-2S}, \qquad \delta u^{CS} \to \sigma e^{-S}, \qquad D \to \infty \tag{7.73}$$

an exponential decrease in S versus $1/S$ in the single mode case. Even though the narrowband assumption is violated in $D \to \infty$, the exponential improvement is real from (7.55) or (7.71)–(7.72) as D can be very large at optical frequencies.

To show that multimode system is indeed better for measurement for the same total energy, consider the following pulse frequency modulation scheme

$$X^{(\mathrm{in})}(t,\phi) = \sqrt{\frac{2S}{T}}\sin(w_0 + \beta\phi)t\,, \qquad 0 \le t \le T\,, \tag{7.74}$$

where β is a known fixed constant and S the total energy in the signal. Classically, it is known [8] that in the presence of additive white Gaussian noise, the use of (7.74) and corresponding nonlinear demodulation lead to a decrease of root-mean-square error $\delta\phi$ by a factor $\sim 1/m$ compared to the linear modulation case, when a threshold constraint involving S, T, β and the noise variance N_0 is satisfied which occurs for sufficiently large D or S. If (7.74) is used in either a coherent state-heterodyne or TCS-homodyne systems, one would obtain

$$\delta\phi^{TCS} \sim \frac{1}{mS}, \quad \delta\phi^{CS} \sim \frac{1}{m\sqrt{S}} \tag{7.75}$$

compared to (7.68) and (7.70). While showing the importance of bandwidth, the net gain $1/m$ is in itself already significant as D is large.

7.4 Position Monitoring with Contractive States

As a final application of squeezed states, we discuss the problem of repeatedly measuring the position of a free mass for which the state after each measurement is important as it determines the state at the next measurement instant. This feature makes the problem, relevant to gravitational-wave interferometers treated elsewhere in this book, quite different from the other ones we have discussed so far in this chapter, for which all the information can be extracted from the system by one measurement. There is still considerable confusion in the literature on the validity of the so-called "standard quantum limit" (SQL) on how small the position fluctuation $\langle \Delta\hat{X}^2(t)\rangle$ can be obtained in a sequence of position measurements, although the issues in principle have been cleared up entirely over ten years ago. Perhaps this is partly because some of the following clarification never appeared in print.

The SQL states that [66,67] if a position measurement is made at $t = 0$, the fluctuation at $t > 0$ is at least

$$\langle \Delta\hat{X}^2(t)\rangle_{SQL} = \hbar t/m \ , \tag{7.76}$$

where m is the mass of a fermion. The derivation of (7.76), however, was incorrectly taken to be universally valid as a consequence of the Uncertainty Principle, and it was concluded that the free mass position is not a "QND observable" – namely, that the disturbance to the system from the first position measurement demolishes the possibility of an accurate second measurement after an interval of free evolution. To delineate how a position monitoring scheme works, consider the monitoring of weak classical forces $f_1(t)$ and $f_2(t)$ coupled linearly to a free mass with position $\hat{X}$ and momentum $\hat{P}$

$$H_I = f_1(t)\hat{X} + f_2(t)\hat{P} \ . \tag{7.77}$$

In the Heisenberg picture,

$$\hat{X}(t) = \hat{X}(0) + \hat{P}(0)t/m + \int_0^t f_2(t')dt' - \int_0^t dt' \int_0^{t'} d\tau f_1(\tau)/m \ , \tag{7.78}$$

$$\hat{P}(t) = \hat{P}(0) - \int_0^t f_1(t')dt' \ . \tag{7.79}$$

Typically, $f_2 = 0$ and $\hat{X}$ is more readily measurable than $\hat{P}$ in practice. From (7.78), information on $f_1(t)$ can be obtained by measurements on $\hat{X}(t)$ at different times.

If a position measurement at $t = 0$ is made in the sense of Pauli's first-kind measurement [68], the position eigenstates $|X\rangle$ is to be used to compute the measurement probability $p(X)$ and the state of the mass after the measurement with a reading X' is $|X'\rangle$. Thus, first-kind measurement of a self-adjoint operator is one for which the Von Neumann projection postulate applies. From (7.79), the position fluctuation $\langle \Delta\hat{X}^2(t)\rangle$ is often concluded to be infinite, because the "back-action" causes $\langle \Delta^2\hat{P}(0)\rangle = \infty$ with $\langle \Delta\hat{X}^2(0)\rangle = 0$ from the Uncertainty Principle. Since $\langle \hat{P}^2(0)\rangle = \langle \Delta^2\hat{P}(0)\rangle + \langle \hat{P}(0)\rangle^2$, an infinite average energy is obtained for the mass in a position eigenstate $|X\rangle$, thus one can actually only make "approximate" position measurements which are generally described by POM as far as the measurement statistics goes, with $\langle \Delta\hat{X}^2(0)\rangle > 0$. In any event, it was concluded that whatever position measurement is used the Uncertainty Principle implies the SQL (7.76).

In [69], it was pointed out that this conclusion is not valid from (7.78) when $\langle \Delta\hat{X}(0)\Delta\hat{P}(0) + \Delta\hat{P}(0)\Delta\hat{X}(0)\rangle$ is negative. It was also pointed out that $\langle \Delta\hat{X}^2(t)\rangle$ can be arranged to be as small as desired at any $t > 0$ if the state after measurement is left in a "contractive state" $|\mu\nu\alpha\omega\rangle$, which is a TCS $|\mu\nu\alpha\rangle$ with the frequency ω put back explicitly and the parameters μ, ν, ω chosen appropriately so that the "generalized minimum uncertainty wave packet" $\langle X|\mu\nu\alpha\omega\rangle$ contracts rather than spreads in t up to a desired measurement time. It was observed that measurements of the second kind, in particular a class of measurements formally described by Gordon and Louisell, may be used to beat the SQL. Specifically, the measurement described by [68,69]

$$|\mu\nu\alpha\omega\rangle\langle\mu'\nu'\alpha\omega| \tag{7.80}$$

would work, where $|\mu'\nu'\alpha\omega\rangle$ is used to compute the measurement probability with reading $\alpha = \alpha_1 + i\alpha_2$,

$$\alpha_1 = x(m\omega/2\hbar)^{1/2} \ , \qquad \alpha_2 = p/(2\hbar m\omega)^{\frac{1}{2}} \ , \tag{7.81}$$

which may be considered a joint approximate measurement of $\hat{X}$ and $\hat{P}$ similar to TCS-heterodyne [22], and $|\mu\nu\alpha\omega\rangle$ is the state after measurement of reading α arranged to be a contractive state for the next measurement. The position measurement would be sharp if $\langle \Delta\hat{X}^2\rangle \sim |\mu' - \nu'|^2 \to 0$, while μ, ν, ω are chosen so that the mass state has a sharply defined position at the next measurement instant.

Two criticisms were made on the success of this approach to beat the SQL. First, it was pointed out that it was not clear a measurement described as in (7.80) is realizable in principle. A quantum measurement realization can be described by the coupling of a "probe" to the system with commuting self-adjoint operators being measured on the probe, and with all the quantities computed by the usual rules of quantum mechanics (without the need for the projection postulate as emphasized by Ozawa [16].) While two realizations were produced [71], they were criticized on the ground that the probe-system interaction Hamiltonians H_I are time-dependent and so are equivalent to "state preparation". While these realizations are actually quite different from the state preparations that were discussed and are in fact full-fledged quantum measurement realizations in accordance with standard quantum measurement theory, the situation is resolved beyond dispute when a time-independent H_I was found [72] for realizing (7.80). More significantly, Ozawa [16,73] has obtained a complete characterization of quantum measurement including the state after measurement in the concept of a completely positive operation measure mentioned in Sect. 7.2.1, and he showed that any Gordon–Louisell measurement representation, in which a complete but not necessarily orthogonal set of states is used to yield the measurement statistics and the state after measurement depends only the measured value, is indeed realizable.

To discuss the second criticism, one needs to examine more closely how the measurement scheme based on (7.80) actually works. Let α' be the reading at $t = 0$ so that the state at $t = 0+$ is $|\mu\nu\alpha'\omega\rangle$. After another time t, the free mass is in state $|\mu_t\nu_t\alpha'_t\omega\rangle$ with $|\mu_t - \nu_t| \to 0$. From (7.78)–(7.79) with $f_2 = 0$, the value α'_t is given by

$$\alpha'_{t1} = \alpha'_1 + \alpha'_2 t/m - \int_0^t dt' \int_0^{t'} d\tau f_1(\tau)/m \, , \tag{7.82}$$

$$\alpha'_{t2} = \alpha'_2 - \int_0^t f_1(\tau) d\tau \, . \tag{7.83}$$

Equations (7.82) and (7.83) provide the average of the reading α'' at t, which can be represented by

$$\alpha''_1 = \alpha'_1 + \alpha'_2 t/m - \int_0^t dt' \int_0^{t'} d\tau f_1(\tau)/m + n_1 \, , \tag{7.84}$$

$$\alpha''_2 = \alpha'_{t2} + n_2 \, , \tag{7.85}$$

where the fluctuation of n_1 is vanishingly small from $|\mu_t - \nu_t| \to 0$ while the noise n_2 is big. From (7.84), one may use the α''_1 reading to estimate f_1 after it is subtracted from the value of $\alpha'_1 + \alpha'_2 t/m$ known at time t. The reading α''_2 is also taken so that it could be used for the subtraction at the next measurement, although it is not used for estimating f_1 as it is noisy and helps little. It is clear that this scheme beats the SQL to any arbitrary level in a sequence of measurements.

In [74], a "predictive sense" of the SQL was proposed to suggest that the SQL was not beaten in that sense. This predictive sense can be described by the stipulation that prior to any measurement, α_1' and α_2' in (7.84) and (7.85) are unknown and random, thus α_1'' is also more random than n_1 and indeed obeys the SQL. But since we know we *will* have the reading value α' available at t which would be subtracted from (7.84), the reading α'' at t, we *can indeed predict* we will get $\langle \Delta \hat{X}^2(0) \rangle$, $\langle \Delta \hat{X}^2(t) \rangle$, and so on, arbitrarily small. Thus, the SQL is beaten by (7.80) in the predictive sense. Further elucidation of this point and discussion on the working of this scheme (7.80)–(7.85) was provided in [75].

Actually, this issue would not even arise if the measurement

$$|\mu\nu 0\omega\rangle\langle\mu'\nu'\alpha\omega| \tag{7.86}$$

is employed instead of (7.80), for which the state after measurement always has $\langle \hat{X} \rangle = \langle \hat{P} \rangle = 0$. This measurement is a special degenerate case of Gordon–Louisell measurement, and thus realizable by Ozawa's theorem. Indeed, an explicit Hamiltonian realization can be developed for (7.86) [76,77].

Since the positions of a free mass can be repeatedly measured accurately, it is not appropriate to say that $\hat{X}$ is not a QND observable. The term QND measurement is often used just to refer to a first-kind measurement, which is an acceptable terminology. What has never been demonstrated is that there is, in principle, any observable which is not a QND observable in the generic sense. In fact, it should be clear from the development in this section, and it can indeed be readily shown in principle, that any observable can be repeatedly measured arbitrarily accurately in the absence of particular constraints. The key point is that, as in (7.80), the state used to compute the measurement probability and the state after measurement need not be the same.

References

(All unpublished manuscripts by this author are available upon request.)

1. L. Mandel and E. Wolf, *Coherence and Quantum Optics*, (Cambridge University Press, Cambridge 1996)
2. H.P. Yuen, Phys. Rev. A **13**, 2226 (1976). The notation $|\beta\rangle_g$ in this reference is equivalent to $|\mu\nu\alpha\rangle$ in this paper with $\beta = \mu\alpha + \nu\alpha^*$
3. H.P. Yuen, Phys. Lett. A **56**, 105 (1976)
4. H.P. Yuen, "Generalized Coherent States and Optical Communications", in Proc. 1975 Conf. Information Sciences and System, Johns Hopkins Univ., pp. 171–177, 1975
5. J.H. Shapiro, H.P. Yuen and J.A. Machado Mata, IEEE Trans. Inform. Theory, IT–**25**, 179 (1979)
6. J.M. Wozencroft and I.M. Jacobs, *Principles of Communication Engineering*, (Wiley, New York 1965)
7. R.G. Gallager, *Information Theory and Reliable Communication*, (Wiley, New York 1968)
8. H.L. Van Trees, *Detection, Estimation, and Modulation Theory, Part I*, (Wiley, New York 1968)

9. T.M. Cover and J.A. Thomas, *Elements of Information Theory*, (Wiley, New York 1991)
10. H.P. Yuen, "Quantum Information Theory, the Entropy Bound, and Mathematical Rigor in Physics", in *Quantum Communication, Computing, and Measurement*, ed. by Hirota, etc., (Plenum, New York 1997) pp. 17–23
11. C.E. Shannon, Bell Sys. Tech. J. **27**, 423 (1948); **27**, 623 (1948)
12. C.E. Shannon, IRE National Convention Record, Part 4, pp. 142–163 (1959)
13. C.E. Shannon, Proc. IRE **37**, 10 (1949)
14. D. Slepian, Proc. IEEE **64**, 292 (1976)
15. H.P. Yuen, "A New Approach to Quantum Computation", in *Quantum Communications, Computations, and Measurements*, ed. by P. Kumar, H. Hirota, and M. d'Ariano, (Plenum, New York 1999)
16. M. Ozawa, "Realization of Measurement and the Standard Quantum Limit", in *Squeezed and Nonclassical Light*, ed. by P. Tombesi and E.R. Pike, (Plenum, New York 1989) pp. 263–286
17. G.D. Forney, Jr., "The Concept of State and Entropy in Quantum Mechanics", S. M. thesis, Dept. of Electrical Engineering, 1963
18. J.P. Gordon, "Noise at Optical Frequency and Information Theory", in *Quantum Electronics and Coherent Light*, Proceedings of the International School of Physics "Enrico Fermi", XXXI, ed. P.A. Miles, Academic Press, pp. 156 (1964)
19. C.W. Helstrom, *Quantum Detection and Estimation Theory*, (Academic Press, New York 1976)
20. A.S. Holevo, *Probabilistic and Statistical Aspects of Quantum Theory*, (North–Holland, Amsterdam, 1982)
21. H.P. Yuen and M. Lax, IEEE Trans. Inform. Thoery **19**, 740 (1973); H. P. Yuen, R. S. Kennedy, and M. Lax, IEEE Trans. Inform. Theory **21**, 125 (1975)
22. H.P. Yuen and J.H. Shapiro, IEEE Trans. Inform. Theory, IT–**26**, 78 (1980)
23. S.L. Braunstein, C.M. Caves, and G.J. Milburn, Am. Phys. (N.Y.) **247**, 135 (1996)
24. H. Takahasi, "Information Theory of Quantum Mechanical Channels", in *Advances in Communication Systems*, ed. by A. Balakrishnan, Vol. 1 (Academic Press, New York 1965), pp. 227–310
25. H.P. Yuen and J.H. Shapiro, IEEE Trans. Inform. Theory **24**, 657 (1978)
26. O. Hirota and S. Ikehara,Trans. IECE of Japan, *EGI*, 273 (1978)
27. J.H. Shapiro, Opt. Lett. **5**, 351 (1986)
28. J.I. Bowen, IEEE Trans. Inform. Theory **13**, 230 (1967)
29. A.S. Holevo, M. Sohma and O. Hirota, Phys. Rev. A **59**, 1820 (1999)
30. M. Sohma and O. Hirota, Phys. Rev. A **62**, 052312 (2000)
31. A.S. Holevo, M. Sohma and O. Hirota, Rept. Math. Phys. **46**, 343 (2000)
32. M. Hall, Phys. Rev. A **55**, 100 (1997)
33. G.M. D'Ariano, C. Macchiavello and M.F. Sacchi, Phys. Lett. A **248**, 103 (1998)
34. G.M. D'Ariano and P. Lo Presti, Phys. Rev. Lett. **86**, 4195 (2001)
35. H.P. Yuen, "High–rate Strong–signal Quantum Cryptography", in *Proceedings of the 1995 Conference on Squeezed States and Uncertainty Relations*, NASA Conference publication 3322, pp. 363–368, 1996
36. D. Gottesman and J. Preskill, Phys. Rev. A **63**, 022309 (2001)
37. C.M. Caves and P.D. Drummond, Rev. Mod. Phys. **88**, 481 (1994)

38. P. Yuen and M. Ozawa, Phys. Rev. Lett. **70**, 363 (1993)
39. J.P. Gordon, Proc. IRE **50**, 1898 (1962)
40. P.L. Zador, Bell Telephone Laboratories Technical Memorandum, MM65–1359–4, Murray Hill, NJ, 1965
41. A.S. Holevo, Probl. Inf. Transm. **9**, 177 (1973)
42. P. Hausladen, R. Josza, B. Schumacher, M. Westmoreland and W. Wooters, Phys. Rev. A **54**, 1869 (1996)
43. A.S. Holevo, "The Capacity of Quantum Channel with General Signal States", preprint
44. H.P. Yuen and J.H. Shapiro, IEEE Trans. Inform. Theory **24**, 657 (1978)
45. H.P. Yuen, "Quantum Communication, Quantum Measurement, TCS and QND", in *Quantum Optics, Experimental Gravity, and Measurement Theory*, eds. P. Meystre and M.O. Scully, (Plenum, New York 1983), pp. 249–268
46. H.P. Yuen, "Nonclassical Light", in *Photons and Quantum Fluctuations*, ed. E.R. Pike and H. Walther, Adam Hilger, pp. 1–9, 1988
47. H.P. Yuen, "Optical Communication with Novel Quantum Devices", in *Quantum Aspects of Optical Communications*, Lecture Notes in Physics 378, eds. C. Benjakallab, O. Hirota, and S. Reynard, (Springer, Berlin Helidelberg New York 1991), pp. 333–341
48. H.P. Yuen, Quantum Semiclass. Opt. **8**, 939 (1996)
49. G.M. D'Ariano, Int. J. Mod. Phys. B **6**, 1291 (1992)
50. G.M. D'Ariano, C. Macchiavello, N. Sterpi and H. P. Yuen, Phys. Rev. A **54**, 4712 (1996)
51. J.H. Shapiro and S.R. Shepard, Phys. Rev. A **43**, 3795 (1991)
52. H.P. Yuen and V. Chan, Opt. Lett. **8**, 177 (1983)
53. H.P. Yuen, Opt. Lett. **12**, 789 (1987)
54. H.P. Yuen, "Photon Number Duplication and Quantum Nondemolition Measurements", unpublished manuscript, 1991; presented at the Oct. 91 Optical Society of America meeting
55. H.P. Yuen, Opt. Lett. **17**, 73 (1992)
56. G.D. Bartolini, D.K. Serkland, P. Kumar and W. L. Kath, IEEE Photon. Technol. Lett. **9**, 1020 (1997)
57. A. Mecozzi, P. Kumar and H.P. Yuen, unpublished manuscript, 1998
58. H.P. Yuen, "Quantum versus Classical Noise Cryptography", in the book referred to in [15]
59. H.P. Yuen, "The Ultimate Quantum Limits on the Accuracy of Measurements" in *Proceedings of the Workshop on Squeezed States and Uncertainty Relations*, NASA Conference Publication 3135, pp. 13–22, 1991
60. C.M. Caves, Phys. Rev. D **23**, 1693 (1981)
61. R.S. Bondurant and J.H. Shapiro, Phys. Rev. D **30**, 2548 (1984)
62. H.P. Yuen, Phys. Rev. Lett. **56**, 2176 (1986)
63. H.P. Yuen, "Generalized Coherent States of the Radiation Field", unpublished manuscript, 1975
64. H.P. Yuen, Nuclear Phys. B **6**, 309 (1989)
65. H.P. Yuen and J. H. Shapiro, Opt. Lett. **4**, 334 (1979)
66. V.B. Braginskii and Yu.I. Vorontsov, Sov. Phys. Usp. **17**, 644 (1975)
67. C.M. Caves, etc., Rev. Mod. Phys. **52**, 341 (1980)
68. W. Pauli, Handbuch der Physik, vol. 5, (Springer, Berlin 1958)
69. H.P. Yuen, Phys. Rev. Lett. **51**, 719 (1983)

70. H.P. Yuen, Phys. Rev. Lett. **52**, 1730 (1984)
71. H.P. Yuen, "Violation of the Standard Quantum Limit by Realizable Quantum Measurements", unpublished manuscript, 1985
72. M. Ozawa, Phys. Rev. Lett. **60**, 385 (1988)
73. M. Ozawa, J. Math. Phys. **25**, 79 (1984)
74. R. Lynch, Phys. Rev. Lett. **54**, 1599 (1985)
75. H.P. Yuen, "General Quantum Measurements and the Standard Quantum Limit", unpublished manuscript, 1985
76. M. Ozawa, Phys. Rev. A **41**, 1735 (1990)
77. M. Ozawa, private communications, 1998

8 Novel Spectroscopy with Two-Level Atoms in Squeezed Fields

S. Swain

Squeezed light is an example of a nonclassical light field – that is, a field for which quantum mechanics is essential for its description. Since the quantum-mechanical nature of squeezed light is its distinguishing feature, it is clearly of interest to identify situations in which this field behaves in a radically different way from its nearest classical equivalent. The obvious area to examine is the interaction with atomic systems, and in this chapter, we describe situations in which the use of squeezed light leads to novel effects in atomic spectroscopy. It is demonstrated that the interaction of the squeezed vacuum with even the simplest atomic system, the two-level atom, produces an astonishingly rich range of phenomena. Squeezed light may interact with atomic systems in totally different ways to ordinary light. By isolating the unique characteristics of the squeezed vacuum in its interaction with atomic systems, we open the way to finding new applications. In this chapter, we concentrate on effects which occur in single two-level atoms or systems of two-level atoms, as three-level atoms are treated in a separate chapter of this book. The emphasis is on identifying phenomena which occur in a squeezed vacuum, but which do not occur for fields with a classical analogue. We shall not describe methods for generating squeezed light.

We begin with a brief survey of early investigations. The dynamical response of a two-level atom interacting with a single mode squeezed state was first considered by Milburn [1], who showed that, depending upon the direction of the squeezing, an increase or decrease of the collapse time occurred. The topic received a great stimulus when Gardiner [2] in 1986 showed that the two dipole quadratures of a two-level atom in a squeezed vacuum field may decay at markedly different rates. This modification of the basic radiative processes has significant consequences in spectroscopy, which are reviewed here. Savage and Walls [3] then considered absorptive optical bistability with a squeezed vacuum input and showed that tunneling times may be increased, and the intrinsic stability of the device substantially improved. Milburn [4] pointed out that atomic level shifts would be modified by the presence of the squeezed vacuum. Carmichael, Lane and Walls [5,6] examined the influence of squeezed light on resonance fluorescence, demonstrating that, under strong laser excitation, the heights and widths of the three spectral lines could be significantly increased or decreased depending upon the phase of the excitation. Janszky and Yushin [7] considered its influence on multi-photon processes. Subsequently, a large number of papers dealt with various aspects of the resonance fluorescence of a single two-level atom (including optical double resonance) [8–28], and with the related problem of the probe absorption spectrum [29–38]. Optical bistability [39,40] and the Jaynes–Cummings model [41–46] were investigated, along with a number of other features [47–54], and the effect of the squeezed vacuum on the dynamics of a two-level atom in the cavity environment has also received much attention [55–61]. The interaction of a squeezed vacuum with multilevel atoms [7,56,60,62–77] and systems of two-level atoms has been treated by many authors [48,52,78–85].

Although a large number of phenomena have been predicted in which the quantum nature of the squeezed vacuum plays an essential role in modifying atomic properties, only the linear dependence of the two-photon absorption rate of a three-level ladder atom has been confirmed experimentally by Georgiades et al. [67]. (Classical sources show a quadratic dependence.) An important series of experiments by Turchette et al. [86] has recently been reported which investigated the interaction of a beam of Cs atoms with squeezed light in the Fabry–Perot cavity environment. The principal phenomena examined were the modification of the transmission spectrum of a weak, tunable probe beam due to the presence of the squeezed light, and the variation of the transmission modulation with the phase of the squeezed vacuum relative to that of an applied, saturating, coherent field. A major objective was to find phenomena which could be attributed unequivocally to the quantum nature of the squeezed vacuum. However, this was impeded by experimental complications. Further verifications of nonclassical effects, in particular of hole-burning and dispersive profiles [21,24] where the squeezed field induces qualitatively different features, would be a powerful demonstration of the ability of squeezed light to modify atomic responses in a fundamental way.

On the other hand, there have been many proposed practical applications of squeezed light, and some experiments have already demonstrated its potential in precision measurements [87] and improvement of spectroscopic sensitivity [67,88–90]. Squeezed light may be used in optical communications to enhance signal-to-noise ratio [87,91], gravitational wave detection [92] and quantum non-demolition measurements [93]. It has been suggested that very low temperatures can be reached using a squeezed vacuum field to reduce the spontaneous emission on the optical transition [53,94–96]. Gea-Banacloche [97] and Marte et al. [98] have suggested that the laser linewidth can be significantly reduced by employing a squeezed vacuum input to the laser cavity. The latter authors have also calculated the squeezing spectrum of the output laser beam and have reported a reduction in the output fluctuations, which leads to sub-Poissonian photon statistics in the laser output. Recently, an experiment has demonstrated an improvement of the sensitivity in ultrahigh frequency measurements [99], using two-photon excitation of three-level atoms by the squeezed vacuum [99].

The wider applications of squeezed light are numerous. An idea of the range and variety of applications is provided by the following (random) selection of recent papers: the protective measurement of a single squeezed harmonic oscillator state [100], coherent nonlinear spectroscopy using number squeezed light, the improvement of quantum nondemolition measurements using a squeezed meter input [101], sub-shot-noise Doppler anemometry with amplitude squeezed light [102], noiseless transfer of nonclassical light through bistable systems [103], generation of few photon states from squeezed atoms [104], and spin squeezing in a collection of atoms illuminated with squeezed light [105].

There have been a number of reviews which deal with various aspects of squeezed light [106–125]. We have emphasized those aspects of the atom-squeezed light interaction which have not received much attention in these works. Since over a thousand papers have been published which deal with squeezing, we have been forced to be selective in the papers we have referenced: it is thus inevitable that significant references will have been omitted from this work.

Whilst two-level atoms provide the simplest system for studying the interaction of atoms with the squeezed vacuum, there are special effects which can only occur in three-level or multilevel systems, since two-photon transitions are possible in the latter. As two-photon correlations are important in the squeezed vacuum, we would expect there to be profound differences between using the squeezed vacuum and a classical light source. This is in fact the case, and was exploited in the first experiment to show the modification of atomic optical properties by the squeezed vacuum. The two-photon correlations in the squeezed vacuum permit a two-photon transition in multi-level systems to take place in a "single-step" process, which is directly proportional to the intensity of the squeezed field. By contrast, a classical field causes the transition to take place by a two-step process, giving rise to a quadratic dependence on the intensity. This fact has led to the first experiment in spectroscopy with squeezed light [67], which measured this difference on the intensity dependence. We do not deal further with three-level systems here as they are treated in detail in another chapter of this book.

The chapter is organized as follows. We begin with a description of the squeezed light produced by the parametric oscillator, and we take this as our model source of such light throughout the review. In order to present analytic expressions, we first treat the squeezed vacuum as a broadband reservoir. Now typical sources of squeezed light have bandwidths of the order of atomic linewidths, and so this approximation is not well satisfied with squeezed sources currently available. It does however provide a basis for a physical understanding of the various phenomena. (It should be noted however that some effects owe their existence to the finite bandwidth of the squeezed vacuum [54,126]). After this model has been used to describe and interpret the basic phenomena, we shall then consider how to treat the more realistic finite bandwidth situation. It has been found that many of the phenomena predicted for a broadband squeezed vacuum persist for even a relatively narrowband squeezed vacuum. We then describe the wide variety of novel effects which occur in the interaction of the squeezed vacuum with single, two level systems, treating both the free space and the cavity situations. Finally, the new features which can arise in systems of n two-level atoms, are dealt with.

8.1 Theoretical Description of the Interaction of Squeezed Light with a Two-Level Atom

There are a variety of theoretical methods for dealing with the interaction between the squeezed vacuum and atoms, and we briefly outline two here. For a full description of these, see the review [127]. In the broadband case, a relatively simple master equation describes the interaction between a two-level atom and a squeezed vacuum [2]. The two-level atom is defined in terms of the Pauli matrices

$$\sigma_1 = \sigma_+ + \sigma_- \ , \ \sigma_2 = -i\left(\sigma_+ - \sigma_-\right) \ , \ \sigma_3 = \left|e\right\rangle\left\langle e\right| - \left|g\right\rangle\left\langle g\right| \ , \tag{8.1}$$

where $\sigma_+ = \left|e\right\rangle\left\langle g\right|$ and $\sigma_- = \left|g\right\rangle\left\langle e\right|$, with $\left|e\right\rangle$ being the excited state and $\left|g\right\rangle$ the ground state of the two-level atom. In free space, the master equation for the reduced density matrix of the atomic system, ρ, in a frame rotating at the atomic resonance frequency, is given by

$$\begin{aligned}
\dot{\rho} = &-i\delta/2\left[\sigma_3,\rho\right] + i\Omega/2\left[\sigma_+\exp(-i\phi_L) + \sigma_-\exp(i\phi_L),\rho\right] \\
&+\gamma\left(N+1\right)/2\left(2\sigma_-\rho\sigma_+ - \sigma_+\sigma_-\rho - \rho\sigma_+\sigma_-\right) \\
&+\gamma N/2\left(2\sigma_+\rho\sigma_- - \sigma_-\sigma_+\rho - \rho\sigma_-\sigma_+\right) \\
&-\gamma M\exp[i\Phi - 2i\left(\omega_S - \omega_L\right)t]\sigma_+\rho\sigma_+ \\
&-\gamma M\exp[-i\Phi + 2i\left(\omega_S - \omega_L\right)t]\sigma_-\rho\sigma_- \ ,
\end{aligned} \tag{8.2}$$

where $\delta = \omega_A - \omega_L$ is the detuning between the atomic Bohr frequency ω_A and the laser driving frequency ω_L, Ω is the Rabi frequency, $\Phi = 2\phi_L - \phi_S$, ϕ_L being the laser phase, and γ is the Einstein A coefficient (spontaneous decay rate). The parameters N and M determine squeezing, and are given in (4.61). Clearly, the equation is much simpler if we assume $\omega_S = \omega_L$, that is, that the center frequency of the squeezed vacuum is equal to the laser frequency, and we henceforth assume this to be the case unless explicitly stated otherwise. For simplicity we neglect the Lamb shifts.

If we regard N and M not as quantities determined by the dynamics of the parametric oscillator, but as arbitrary parameters, it can be shown that they must satisfy the relation

$$M = \eta\sqrt{N\left(N+1\right)}, \qquad 0 \le \eta \le 1 \tag{8.3}$$

if all decay rates are to be positive. We term η the *degree of correlation*. In order to see the maximum effects of squeezing, we assume that the squeezed vacuum is *ideal* throughout: that is we take $\eta = 1$, or

$$M = \sqrt{N(N+1)} \,. \tag{8.4}$$

When this equality holds, we shall say that we have an *ideal squeezed vacuum*: we then have the interesting property that the magnitude of the correlation between two different photons, as evinced by (4.59), is greater than the magnitude of the correlation of a photon with itself, as evinced by (4.60). This is a nonclassical characteristic, because if we assume that the average in (4.60) may be represented as the integral over a positive-definite Glauber–Sudarshan P-function [128], which defines a *field with a classical analogue*, we obtain the inequality

$$M_{\text{class}} \leq N \,. \tag{8.5}$$

Thus for a field with a classical analogue, M can never exceed N. It has become the convention to call a field with $M = N$, a "classically squeezed field", or CSF for short. (The equivalent term "maximally-correlated classical field" is also sometimes used.) It should be borne in mind that a CSF is usually treated by quantum mechanics in this review.) A field that does not have a classical analogue is called a *nonclassical field*. If we find that a certain property exists for $M = \sqrt{N(N+1)}$, but disappears for $M \leq N$, we shall attribute it to the nonclassical nature of the squeezed vacuum. Thus we expect to see distinctive (i.e. nonclassical) behavior of the squeezed vacuum when M satisfies the constraints $N < M \leq \sqrt{N(N+1)}$. It is the excess two-photon correlations of the squeezed vacuum over the value N which are responsible for the unusual features in its interactions with atomic systems.

In several experiments it is better to use the nondegenerate parametric oscillator instead of the degenerate parametric oscillator. In this system, the functions N and M have two peaks, separated by the frequency 2α:

$$N(\omega) = \frac{\lambda_-^2 - \lambda_+^2}{8}\left[\frac{1}{(\omega-\alpha)^2+\lambda_+^2} - \frac{1}{(\omega-\alpha)^2+\lambda_-^2} + \frac{1}{(\omega+\alpha)^2+\lambda_+^2} - \frac{1}{(\omega+\alpha)^2+\lambda_-^2}\right], \tag{8.6}$$

$$M(\omega) = \frac{\lambda_-^2 - \lambda_+^2}{8}\left[\frac{1}{(\omega-\alpha)^2+\lambda_+^2} + \frac{1}{(\omega-\alpha)^2+\lambda_-^2} \right. \tag{8.7}$$

$$\left. + \frac{1}{(\omega+\alpha)^2+\lambda_+^2} + \frac{1}{(\omega+\alpha)^2+\lambda_-^2}\right]. \tag{8.8}$$

In the narrowband case, we may employ a suitable generalization of (8.2) [126,129], or we may use the master equation derived by Parkins and Gardiner on the basis of their coupled systems approach [130,131]. (See also [132].) In the latter case the density matrix ρ describes both the parametric oscillator field mode (a and $a^\dagger$) and the two-level atom (σ_+ and σ_-). From the master equations, using such techniques as the quantum regression theorem [133,134], all the quantities of interest may be calculated. For simplicity, we assume $\omega_A = \omega_L = \omega_S$. The master equation is then

$$\begin{aligned}\dot{\rho} = &\frac{1}{4}\epsilon\kappa\left[\left(a^\dagger\right)^2 - a^2, \rho\right] + \frac{1}{2}\kappa\left(2a\rho a^\dagger - a^\dagger a\rho - \rho a^\dagger a\right)\\ &-\sqrt{\zeta\kappa\gamma}\left\{\left[\sigma_+, a\rho\right] + \left[\rho a^\dagger, \sigma_-\right]\right\}\\ &+\frac{1}{2}i\left[\Omega\sigma_- + \Omega^*\sigma_+, \rho\right] + \frac{1}{2}\gamma\left(2\sigma_-\rho\sigma_+ - \sigma_+\sigma_-\rho - \rho\sigma_+\sigma_-\right) . \end{aligned} \tag{8.9}$$

The first two terms on the right hand side describe the parametric oscillator, the final two terms the two-level atom as before, and the third term the coupling between the two systems. This coupling is chosen to be one-way that is, the output from the parametric oscillator drives the atom, but the atom does not act back upon the parametric oscillator. The parameter ζ measures the coupling between the two systems.

8.2 The Two-Level Atom in Free Space

8.2.1 The Optical Bloch Equations

The simplest atomic system is the single two-level atom, and its interaction with the squeezed vacuum gives rise to many interesting features, which we shall discuss in this Section. We shall emphasize those phenomena in which excitation by squeezed light as opposed to classical light sources, leads to qualitatively different results. By these means, we hope to identify those effects which are most important for experimental testing, and, by isolating the unique characteristics of the squeezed vacuum in its interaction with atomic systems, to pave the way for possible new applications.

We shall first of all consider the broadband squeezed vacuum, which has the advantage of enabling analytic treatments to be undertaken, and thus facilitates physical understanding of the various phenomena. After this model has been used to describe and interpret the basic phenomena, we shall then consider how to treat the more realistic finite bandwidth situation. We begin by examining the free space situation, and then discuss cavity-based phenomena.

In this section we discuss in more detail some phenomena which are strongly modified by the presence of the squeezed vacuum, using the master equation (8.2), and assuming that the center frequency of the squeezed vacuum ω_S is equal to ω_L. The resulting Bloch equations are

$$\langle\dot{\sigma}_-\rangle = -\gamma\left(N + 1/2\right)\langle\sigma_-\rangle - i\delta\langle\sigma_-\rangle - \gamma M e^{i\Phi}\langle\sigma_+\rangle - \frac{i}{2}\Omega\langle\sigma_3\rangle ,$$

$$\begin{aligned}\langle\dot{\sigma}_+\rangle &= -\gamma\,(N+1/2)\,\langle\sigma_+\rangle + i\delta\,\langle\sigma_+\rangle - \gamma M e^{-i\Phi}\,\langle\sigma_-\rangle + \frac{i}{2}\Omega\,\langle\sigma_3\rangle\ ,\\ \langle\dot{\sigma}_3\rangle &= -\gamma\,(1+2N)\,\langle\sigma_3\rangle - \gamma + i\Omega\,(\langle\sigma_+\rangle - \langle\sigma_-\rangle)\ .\end{aligned} \tag{8.10}$$

The phase difference $\Phi = 2\phi_L - \phi_S$ between the driving laser and the applied squeezed vacuum is an important parameter in the problem. In terms of σ_1 and σ_2, the Bloch equations may be expressed:

$$\begin{aligned}\langle\dot{\sigma}_1\rangle &= -\gamma_+\,\langle\sigma_1\rangle + (\gamma M\sin\Phi - \delta)\,\langle\sigma_2\rangle\ ,\\ \langle\dot{\sigma}_2\rangle &= -\gamma_-\,\langle\sigma_2\rangle + (\gamma M\sin\Phi + \delta)\,\langle\sigma_1\rangle + \Omega\,\langle\sigma_3\rangle\ ,\\ \langle\dot{\sigma}_3\rangle &= -\gamma\,(2N+1)\,\langle\sigma_3\rangle - \gamma - \Omega\,\langle\sigma_2\rangle\ ,\end{aligned} \tag{8.11}$$

where $\gamma_\pm \equiv \gamma\,(N \pm M\cos\Phi + 1/2)$.

The latter have the steady-state solutions

$$\begin{aligned}\langle\sigma_1\rangle &= (\gamma M\sin\Phi - \delta)\,/D\ ,\\ \langle\sigma_2\rangle &= -\Omega\gamma\gamma_+/D\ ,\\ \langle\sigma_3\rangle &= -\gamma\{\gamma^2\left[(N+1/2)^2 - M^2\right] + \delta^2\}/D\ ,\end{aligned} \tag{8.12}$$

where $D \equiv \gamma\,(2N+1)\,\{\gamma^2[(N+1/2)^2 - M^2] + \delta^2\} + \Omega^2\gamma_+$.

Equations (8.10), (8.11) and (8.12) are the basic equations for discussing the properties of a two-level atom interacting with the squeezed vacuum. In the next few subsections we assume that the atom and laser are in resonance, $\delta = 0$.

8.2.2 The Dipole Decay Rates

The real and imaginary parts of the dipole moment of a two-level atom are proportional to the Pauli matrices σ_1 and σ_2. For such an atom immersed in a squeezed vacuum, Gardiner [2] showed, from the master equation (8.2) with $\Omega = 0$ (no driving field), that the two quadratures of the dipole moment have different decay rates, one faster and one slower than the natural decay rate $\gamma/2$:

$$\begin{aligned}\langle\dot{\sigma}_1\rangle &= -\gamma_L\,\langle\sigma_1\rangle\ ; \quad \gamma_L = \gamma\left(N + \frac{1}{2} + M\right) \geq \gamma/2\ ,\\ \langle\dot{\sigma}_2\rangle &= -\gamma_S\,\langle\sigma_2\rangle\ ; \quad \gamma_S = \gamma\left(N + \frac{1}{2} - M\right) \leq \gamma/2\ .\end{aligned} \tag{8.13}$$

These equations result from (8.11) on setting $\Omega = \Phi = \delta = 0$. In the absence of the squeezed vacuum, $N = M = 0$, the two decay rates are equal: $\gamma_L = \gamma_S = \gamma/2$, whereas for $M = \sqrt{N(N+1)}$ and $N \gg 1$, we have that $\gamma_L = 2N\gamma \gg \gamma/2$ is much enhanced, whilst $\gamma_S = \gamma/8N \ll \gamma/2$ is greatly decreased. Note that for a CSF, where $M = N$, we have $\gamma_S = \gamma/2$, so that a value for γ_S less than the natural linewidth is not possible for a field with a classical analogue: subnatural linewidths are a signature of a quantum field.

In the early days, attention was focused on large values of N, as the interest was in making γ_S arbitrarily small. This appeared to be the most favorable circumstance for high precision measurements. However, for many interesting effects, it is the strength of the two-photon correlations contained in (4.60) which are important. The relative difference between $M_{\text{class}} = N$ and $M_{\max} = \sqrt{N(N+1)}$ is greatest for small N, and many interesting effects can in fact arise for very small values of N [25,135].

It was emphasized by Gardiner that it was essential that *all* the modes with which the atom interacts be squeezed if the decay rates are to be given by (8.13). This poses difficulties if the atom is located in free space, for it means that the modes occupying all 4π solid angles of space about the atom must be squeezed. For this reason, the cavity situation is considered a more likely scenario for actually observing these effects, because only the modes in a small solid angle about the privileged cavity mode need be squeezed. The free space situation is however conceptually simpler; and for this reason we concentrate on the free space case initially and defer discussion of the cavity situation.

In Gardiner's paper, it was assumed that only a single channel of the radiation field was squeezed, and that the atom couples only to that channel. A generalization to the interaction of a two-level atom with a multi-channel squeezed vacuum, which includes squeezed plane waves, has been considered by Hegerfeldt, Sachse and Sondermann [136]. They report the existence of unusual spectra when the observation is such that interference is possible between the applied squeezed field and the scattered light. Some of the line-shapes they predict are very similar to those calculated earlier in anomalous resonance fluorescence, which is discussed in Sect. 8.2.4.

8.2.3 Resonance Fluorescence

In 1987, Carmichael, Lane and Walls [5] showed that the resonance fluorescence spectrum of two-level atom is much modified by the presence of a squeezed vacuum. The atom is driven by a resonant monochromatic field of frequency ω_L, characterized by its Rabi frequency Ω and phase ϕ_L. The phase difference $\Phi = 2\phi_L - \phi$ between the driving laser and the applied squeezed vacuum is an important parameter in the problem. It is assumed that the center frequency of the squeezed vacuum ω_S is equal to ω_L. The incoherent part of the resonance fluorescence spectrum is given by [137]

$$G(\omega) = \frac{1}{\pi}\text{Re}\int_0^{\infty} d\tau \, \langle \sigma_+(0), \sigma_-(\tau) \rangle \, e^{i\omega\tau} , \tag{8.14}$$

where $\langle x, y \rangle = \langle xy \rangle - \langle x \rangle \langle y \rangle$.

Assuming that all the modes with which the atom interacts are squeezed, the resonance fluorescence spectrum may be obtained from (8.14) and (8.10) by means of the quantum regression theorem [133,134]. For weak driving

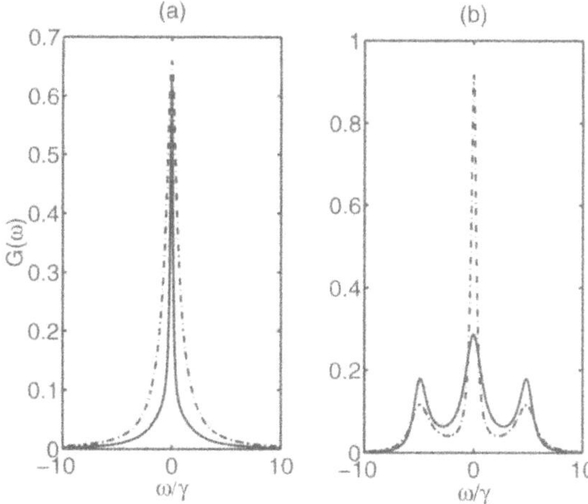

Fig. 8.1. The resonance fluorescence spectra in an ideal squeezed vacuum for **(a)** the weak field case: $N = 0.5$, $\Omega = 0.2\gamma$, and **(b)** the strong field case, $N = 0.1$ and $\Omega = 5\gamma$ for $\Phi = 0$ (*solid line*) and $\Phi = \pi$ (*dot-dashed line*). In frame **(a)**, the *dot-dashed* line is for the equivalent CSF

fields ($\Omega \ll \gamma$) and $N \geq 1$, the resonance fluorescence spectrum is found to consist of a subnatural linewidth peak (width $\simeq \gamma_S \ll \gamma/2$) superimposed on a broad background peak (width $\simeq \gamma_L \gg \gamma/2$), both centered on the resonant frequency ω_L. This is shown as the solid line in Frame (a) of Fig. 8.1. The spectra in the weak driving limit are essentially independent of Φ.

For strong driving fields ($\Omega \gg \gamma$), the resonance fluorescence spectrum consists of three peaks whose relative heights and widths depend strongly on the squeezing phase Φ. For $\Phi = 0$ the central peak is suppressed and broadened relative to the $N = 0$ case (where the ratio of the height of the central peak to that of the side-peaks is 3:1), whereas for $\Phi = \pi$, it is enhanced and narrowed. The side-peaks for $\Phi = 0$ are narrower than for $\Phi = \pi$, but the side-peaks are broadened for all values of Φ relative to the $N = 0$ case if $N > 1/8$. Quang et al. [138] and Banerjee [26] have pointed out that for $N < 1/8$, the side-peaks for the $\Phi = 0$ case are narrowed relative to the $N = 0$ case, acquiring the $N = 0$ linewidth for $N = 1/8$. As we shall see later, the side-peak widths behave in a somewhat different way under narrowband excitation. In Fig. 8.1b, we show the strong field resonance fluorescence spectra, choosing a small value for N in order not to make the spectra for $\Phi = 0$ and $\Phi = \pi$ too different to be plotted conveniently on the same graph.

The strong phase-dependence of the spectra is not itself a manifestation of the nonclassical nature of the squeezed vacuum. It has been shown that a very similar phase dependence can be obtained from a two-level atom driven by a monochromatic source to which an appropriate noise field has been added [139]. However, it is not possible to obtain subnatural linewidths with a classical field. The subnatural features apparent in the plots of Fig. 8.1

are nonclassical effects – that is, they disappear for a CSF[1]. This applies to both the central and side-peak narrowing. We illustrate this only in frame (a), where the dot-dashed line is for the equivalent CSF (with $M = N$). The CSF peak possess the natural linewidth.

The strong-excitation features we have just described may be easily understood in the semiclassical dressed state picture when the secular approximation is valid. In general, the two-level semiclassical dressed states are defined as

$$|\alpha\rangle = (s\,|g\rangle + c\,|e\rangle)\exp(-i\phi_L) \;\; ; \;\; |\beta\rangle = (c\,|g\rangle - s\,|e\rangle)\exp(-i\phi_L) \;\; , \tag{8.15}$$

where

$$s = \sqrt{\frac{\Omega' - \delta}{2\Omega'}} \, , \quad c = \sqrt{\frac{\Omega' + \delta}{2\Omega'}} \, , \quad \text{with} \quad \Omega' = \sqrt{\Omega^2 + \delta^2} \, . \tag{8.16}$$

For the secular approximation to be valid, the generalized Rabi frequency Ω' must be much larger than all the other frequencies that appear in the Bloch equations. When it holds, the equations for the diagonal density matrix equations in general take a rate equation form, and the off diagonal density matrix equations are uncoupled from the diagonal ones. That is, we have equations of the form

$$\begin{aligned} \dot{\rho}_{\alpha\alpha} &= -W_{\alpha\beta}\rho_{\alpha\alpha} + W_{\beta\alpha}\rho_{\beta\beta} \, , \\ \dot{\rho}_{\beta\beta} &= -W_{\beta\alpha}\rho_{\beta\beta} + W_{\alpha\beta}\rho_{\alpha\alpha} \, , \\ \dot{\rho}_{\alpha\beta} &= (i\Omega' - \Gamma)\,\rho_{\alpha\beta} \, , \end{aligned} \tag{8.17}$$

where, in the case of the secular approximation, the transition rates W_{ij} are equal to the spontaneous decay rates γ_{ij} between the dressed states $|i\rangle$ and $|j\rangle$ [25,135,141]. It may easily be shown from the quantum regression theorem that this system gives rise to the resonance fluorescence spectrum

$$G_{\text{coh}}(\omega) = c^2 s^2 (P_\alpha - P_\beta)^2\,\delta(\omega) \, , \tag{8.18}$$

$$\begin{aligned} G_{\text{inc}}(\omega) = \frac{\gamma}{\pi}\Bigg\{ &\frac{4c^2 s^2 P_\alpha P_\beta \Gamma_0}{\omega^2 + \Gamma_0^2} + \frac{c^4 P_\alpha \Gamma}{(\omega - \Omega')^2 + \Gamma^2} \\ &+ \frac{s^4 P_\beta \Gamma}{(\omega + \Omega')^2 + \Gamma^2}\Bigg\} \, , \end{aligned} \tag{8.19}$$

where $\Gamma_0 = W_{\alpha\beta} + W_{\beta\alpha}$, and

$$P_\alpha = \frac{W_{\beta\alpha}}{\Gamma_0} \text{ and } P_\beta = \frac{W_{\alpha\beta}}{\Gamma_0} \tag{8.20}$$

[1] It is not quite true that subnatural linewidths in resonance fluorescence occur only when the atom is driven by a squeezed vacuum. Mollow [137] showed that, for the $N = 0$ case, for certain values of Ω, the resonance fluorescence linewidth is *slightly* subnatural. Rice and Carmichael [140] showed that this subnatural linewidth is associated with the *generation* of squeezed light in the output field of this system.

are the steady-state occupation probabilities of the dressed states $|\alpha\rangle$ and $|\beta\rangle$ respectively. $G_{\mathrm{coh}}(\omega)$ is the coherent part of the spectrum and $G_{\mathrm{inc}}(\omega)$ the incoherent part – the part of most interest here. The frequency ω is measured from the laser frequency.

The incoherent spectrum consists of a central component whose halfwidth Γ_0 is the sum of the diagonal decay rates, with two side peaks displaced by the frequency Ω' from the central component, with halfwidths Γ given by the decay rate of the off-diagonal element. Using the explicit form for the dressed density matrix equations of motion given in [141] for example, we find that for resonant excitation of both the driving and squeezed fields, $c^2 = s^2 = 1/2$, and

$$\Gamma_0 = \gamma\left(N + \frac{1}{2} + M\cos\Phi\right) ,$$
$$\Gamma = \gamma\left(6N + \frac{3}{2} - M\cos\Phi\right) . \tag{8.21}$$

From expressions (8.19) and (8.21), the nature of the spectra in Fig. 8.1b may be understood. It is also easy to see that $\Gamma \le 3\gamma/2$, the value of the sideband halfwidths in the absence of the squeezed vacuum, when $N \le 1/8$ (and $M = \sqrt{N(N+1)}$).

8.2.4 Anomalous Resonance Fluorescence

For the parameter values employed in the previous subsection, the squeezed vacuum changes the relative heights and widths of the spectral peaks in the resonance fluorescence spectra, but does not lead to spectra which have *qualitatively* different features to those that occur in the absence of the squeezed vacuum. Recently, Smart and Swain have shown that under appropriate conditions, the presence of the squeezed vacuum can lead to resonance fluorescence lineshapes [21,24] which are quite distinct from those seen in its absence. The term '*anomalous resonance fluorescence*' has been used to describe these unusual spectral profiles. They occur at intermediate field strengths, $\Omega \simeq \gamma$, below the threshold at which the number of peaks in the spectrum changes from one to three, and for only a narrow range of Rabi frequencies about a certain value.

For resonant excitation of the atom and the laser frequency equal to the central frequency of the squeezed vacuum, the only case considered in this section, the anomalous profiles vary from a single broad peak with a subnatural hole bored into it at line center for $\Phi = 0$, to dispersive profiles for $\Phi = \pi/2$, and then to simply asymmetric peaks for larger values of Φ. Four examples are shown in Fig. 8.2, where we show the progression through hole burning and dispersive profiles to asymmetric spectra as Φ varies from $\Phi = 0$ to $\Phi = 4\pi/5$. We have chosen a wide range of N values. In Frame (a), where $\Phi = 0$, we illustrate that hole burning can occur for very small

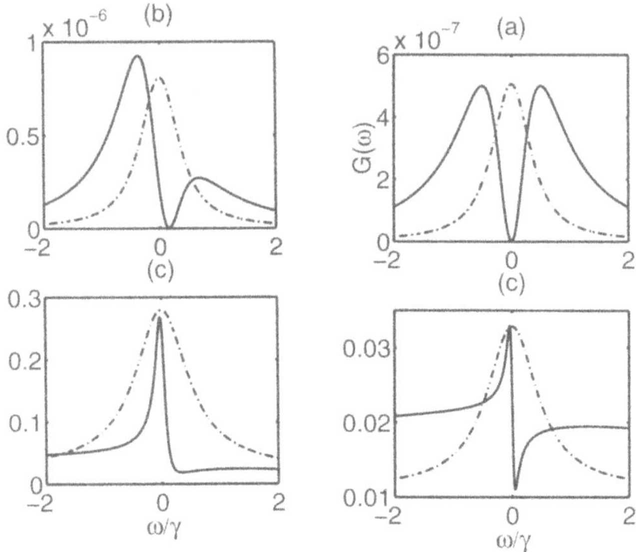

Fig. 8.2. The incoherent resonance fluorescence spectrum $G(\omega)$ for **(a)** $N = 1 \times 10^{-6}, \Phi = 0$ and $\Omega = 0.0225\gamma$; **(b)** as in Frame **(a)** except that now $\Phi = \pi/10$; **(c)** $N = 5, \Phi = \pi/2$ and $\Omega = 0.69\gamma$; and **(d)** $N = 2, \Phi = 0.8\pi$ and $\Omega = 1.58\gamma$. The *solid line* is for M maximal and the *dotted* for $M = N$. (The latter has been scaled to fit the plots and in **(c)** we have added 0.01 to the CSF plot)

values of N – in this case, $N = 1 \times 10^{-6}$. The same value of N is employed in frame (b), but now $\Phi = \pi/10$. The symmetry is destroyed. In order to see the dispersive profiles clearly, we need $\Phi = \pi/2$, and a flat background which requires a large value of N. This is demonstrated in Frame (c), where we take $N = 5$. Finally, Frame (d) provides an example of the spectra for large Φ, $\Phi = 0.8\pi$. The anomalous features are most pronounced for Φ in the region $0 \leq \Phi < \pi/2$, and they disappear as $\Phi \to \pi$.

We emphasize that these unusual forms of spectra only occur for values of Ω close to the values specified for each frame. If values of Ω significantly different from these are employed, the spectra revert to profiles similar to those shown in Frame (a) of Fig. 8.1. This strong dependence on Ω is itself an unusual feature. In the absence of the squeezed vacuum, the spectrum for these values of Ω is just a single peak of width approximately equal to the natural width γ.

The anomalous features are associated with nonclassical aspects of the squeezed vacuum. This is illustrated in Fig. 8.2, where the solid lines are for an ideal squeezed vacuum, and the broken for a CSF. It is apparent that the unusual features disappear completely for the equivalent classically squeezed field, so that observation of these anomalous features would be striking confirmation of the quantum properties of the squeezed vacuum.

Some understanding of the origin of the anomalous features may be obtained as follows [27,28]. We concentrate on the $\Phi = 0$ case. As in the absence of the squeezed vacuum, the incoherent spectrum below threshold may be expressed as the sum of three Lorentzians

$$G(\omega) = \sum_{i=1}^{3} \frac{C_i z_i}{\omega^2 + z_i^2} \tag{8.22}$$

with different weights C_i [137], where ω is measured from the laser frequency. (In addition, there is a coherent component of the form $C_0 \delta(\omega)$). One of the weights, say C_1, is always positive and one of the weights, say C_3, is always negative, and in the absence of an applied squeezed vacuum, the remaining weight C_2 is always positive. Mollow pointed out that the negative value of C_3 can give rise to a spectrum with subnatural linewidth, and Rice and Carmichael [140] showed that the negative value is due to the presence of squeezing in the output field produced by resonance fluorescence.

In the presence of the squeezed vacuum, there is a range of values of Ω for which C_2 is negative [27,28], and the width z_2 associated with this negative weight is less than the other two widths. Thus for a value of Ω close to the value which minimizes C_2, we obtain a hole burned in the spectrum. The range of values of Ω for which C_2 is negative defines the values of Ω for which anomalous features occur. This range gets smaller rapidly as N increases. Thus for $N = 0.1, C_2$ is negative over the range $0.19 < \Omega/\gamma < 0.42$ whereas for $N = 1$, the range of negative values is $0.36 < \Omega/\gamma < 0.48$ and for $N = 10$, the range is only $0.475 < \Omega/\gamma < 0.5$. For values outside this range, the spectra revert to the normal form typified by Fig. 8.1a. The situation for $N = 1$ is illustrated in Fig. 8.3.

The anomalous profiles as shown in the other frames of Fig. 8.2b also only persist for a small range of Ω values, but for values of $\Phi \neq 0$, a different explanation is needed [28].

Another physical explanation of the origin of the anomalous features is provided by the observation that the atomic system has evolved into a pure state when the anomalous features arise. This is discussed further below.

The anomalous features are associated with a number of other unusual characteristics of the two-level atom/squeezed vacuum system. We detail these below:

(1) In the squeezed vacuum, the total incoherent fluorescence intensity exhibits a minimum as a function of the Rabi frequency Ω. The value of Ω giving rise to this minimum is close to the value which produces the anomalous features.

With no squeezed vacuum, the total incoherent intensity increases monotonically with Ω. However, in the presence of the squeezed vacuum, for a fixed value of N and Φ, there is a large drop in the total intensity of the incoherently scattered radiation as the Rabi frequency passes through the value

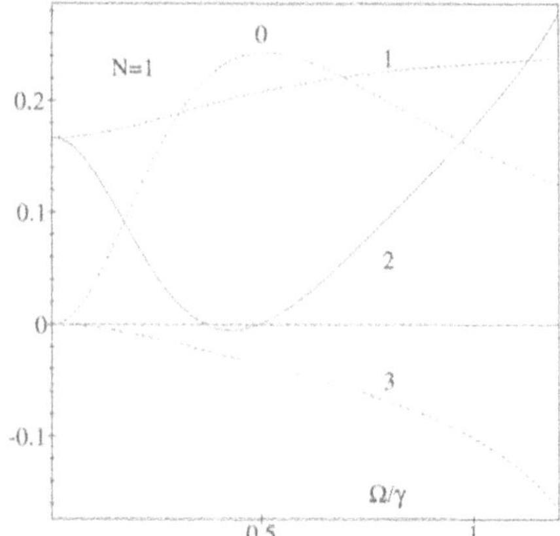

Fig. 8.3. The weights C_K in the resonance fluorescence spectrum as a function of the Rabi frequency Ω, for $N = 1$. (Only the value of K is labeled.) The anomalous features only arise for the narrow range of Ω values where C_2 is negative. In the absence of the squeezed vacuum, C_2 is always positive

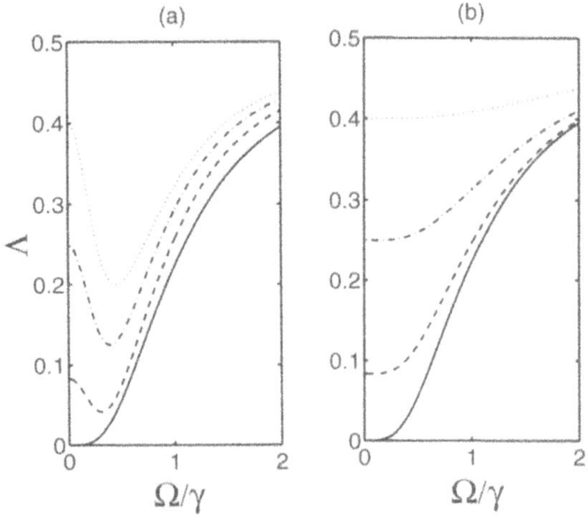

Fig. 8.4. The total incoherent intensity as a function of the driving field Rabi frequency Ω for **(a)** M maximal and **(b)** $M = N$ with $\Phi = 0$ and $N = 0$ (*solid line*), 0.1 (*dotted line*), 0.5 (*dot-dash line*) and 2 (*dotted line*)

of Ω which produces the anomalous features. The existence of this minimum is a non-classical effect, as we may see in Fig. 8.4, which compares the total incoherent intensity (i.e. the incoherent spectrum integrated over all frequencies) as a function of the driving field Rabi frequency for an ideal squeezed vacuum in Frame (a) and for the corresponding CSF's in Frame (b). A clear minimum is present for an ideal squeezed vacuum at the value of Ω very close to that for which the anomalous features arise.

We may exploit this property to obtain an expression for the values of Ω which produce anomalous features. The minimum in G_{tot} is found to occur at [28]

$$\frac{\Omega^2}{\gamma^2} = \xi \frac{4\hat{N}^2 - \hat{N} - \xi}{4\hat{N}^2 + \hat{N} - \xi}, \qquad \xi \equiv \frac{\hat{N}}{2\left(\hat{N} + M \cos \Phi\right)}, \tag{8.23}$$

where $\hat{N} = N + 1/2$. This expression predicts the values of Ω at which the anomalous features arise very well. In the case $M = \sqrt{N(N+1)}$ and $N \geq 1$, it simplifies to

$$\Omega = \frac{\gamma}{2\cos(\Phi/2)}, \tag{8.24}$$

which yields the values $\Omega = \gamma/2$ for $\Phi = 0$, $\Omega = \gamma/\sqrt{2}$ for $\Phi = \pi/2$ and $\Omega \to \infty$ as $\Phi \to \pi$.

We note that a much sharper minimum is found if we plot the incoherent fluorescence intensity at line center as a function of Ω instead of G_{tot}, but this minimum is not a purely quantum effect: a minimum occurs also for the CSF, which is albeit much shallower than the case of an ideal squeezed vacuum. The origin of the minimum in the CSF case is different however: it is associated with the spectrum splitting from a single-peaked to a triply-peaked form.

(2) The anomalous features are very sensitive to the degree of two photon correlation.

As we have seen in (8.3), if we treat N and M as parameters, we have $M = \eta\sqrt{N(N+1)}$ where η satisfies the inequality $0 \leq \eta \leq 1$. It is found that, the larger the value of N, the closer η must lie to 1 for the anomalous features to occur. This graphically demonstrates the importance of the two-photon correlations. For large values of N, the sensitivity of the anomalous features to the value of η is extreme [141]. For example, for $N = 10$, $\Omega = \gamma/\sqrt{2}$ and $\Phi = \pi/2$, a pronounced dispersive profile exists for $\eta = 1$, is significantly less pronounced for $\eta = 0.999999$, and is completely absent for $\eta = 0.999995$! This means that it is not sufficient for the field to be non-classical for the anomalous features to occur: the degree of two-photon correlation must be very close to its maximum permitted value. However, for small values of N, the sensitivity of the anomalous features to the degree of two-photon correlation is less critical.

The existence of anomalous features is also connected with other phenomena which are described below.

8.2.5 Pure Atomic States

It has been shown by Palma and Knight that a system of two two-level atoms interacting with the squeezed vacuum may decay into a pure state [48].

Agarwal and Puri considered a system of n two-level atoms interacting with the squeezed vacuum, with and without an external coherent driving field [52]. They obtained exact expressions for the system density matrix and showed that under appropriate conditions on the Rabi frequency of the driving field and the relative phase Φ between the driving field and the squeezed vacuum, this atomic system could also decay into a pure state. Tucci [142] discussed the thermodynamic properties of a single, driven two-level atom as it evolved toward a pure state under the influence of the squeezed vacuum.

Here we concentrate on the case of the single two-level atom, studying first the case of resonant excitation. Swain and Zhou [27,28] have shown that the existence of the anomalous features discussed in the previous subsection is also associated with the evolution of the system into this pure state under the influence of the squeezed vacuum. We may adapt Agarwal and Puri's condition to the case of a single atom, when we find that the atom may be approximately in a pure state when

$$\frac{\Omega}{\gamma} = \frac{\sqrt{M}}{\left(\sqrt{N+1}+\sqrt{N}\right)\cos\Phi/2} . \tag{8.25}$$

The atomic system may exactly achieve a pure state only when $\Phi = 0$. For $\Phi \simeq 0$, the above expression gives the value of Ω for which the atom best approximates a pure state.

The existence of the pure state is a quantum feature of the field, as we demonstrate below. The degree of purity of a two-level atom may be measured by the quantity

$$\Sigma \equiv 2\mathrm{Tr}\left(\rho^2\right) - 1 = \langle\sigma_1\rangle^2 + \langle\sigma_2\rangle^2 + \langle\sigma_3\rangle^2 , \tag{8.26}$$

where the σ_i are the Pauli matrices [142]. The value $\Sigma = 1$ corresponds to a pure state whilst $\Sigma = 0$ describes a completely mixed state. Note that in the absence of the squeezed vacuum, $N = 0$, Σ decreases monotonically with Ω. The behavior is quite different when the squeezed vacuum is introduced. The purity is shown in Fig. 8.5a as a function of Ω/γ and Φ for the case of an ideal squeezed vacuum with $N = 1$. The plot confirms that for $\Phi = 0$, the atom may exactly achieve a pure state for a particular value of the Rabi frequency ($\Omega \simeq \gamma/2$, in this case), but for other values of Φ the atomic system may, at best, only approximate a pure state. In general, the purity tends to decrease as $\Phi \to \pi$ and $\Omega \to \infty$. Frame (b) of the figure plots the purity for the equivalent CSF, and it is clear that whilst Σ has a maximum value, the atomic system is nowhere well approximated by a pure state. (The global maximum value of Σ in the CSF case is only about 0.2.) Thus the evolution of the atomic system into a pure state is a purely quantum feature of the squeezed vacuum.

We note that the total dipole fluctuation, $\mathrm{Var}(\sigma_1) + \mathrm{Var}(\sigma_2) + \mathrm{Var}(\sigma_3) = 3 - \Sigma$, which drives the incoherent resonance fluorescence, is a minimum when Σ takes its maximum value of unity, and so we can understand why

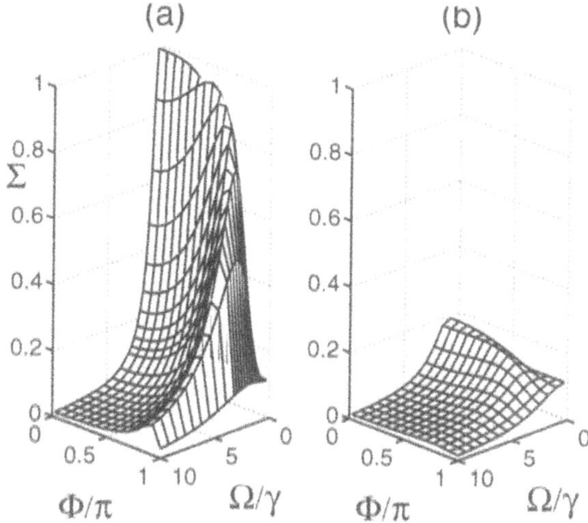

Fig. 8.5. Plot of Σ for $N = 1$ as a function of Φ and Ω/γ, **(a)** for the case of an ideal squeezed vacuum, and **(b)** for the equivalent CSF

the minimum in the total fluorescence (c.f. Fig. 8.4) occurs when the system is in a pure state.

For resonant excitation, the collapse of the atom into a pure state is only possible if $\Phi = 0$. If detunings $\delta \neq 0$ are permitted, it has been shown that such a collapse into a pure steady-state may also occur for the case $\Phi = \pi$ [50]. Zhou and Swain [143] have given a general discussion, and shown that the collapse of the atomic state into a pure state may occur for arbitrary Φ, the requirement being that given Φ, Ω and $\Delta \equiv \delta/\gamma$ are chosen to satisfy $\Sigma = 1$ [143]. The general pure state is found to have the form

$$|\psi\rangle = \sqrt{N+1-M}|g\rangle - \exp(i\theta_o)\sqrt{M-N}|e\rangle \ , \tag{8.27}$$

where θ_o is given by

$$\theta_o = \arctan\left(\frac{N + \frac{1}{2} + M\cos\Phi}{\Delta + M\sin\Phi}\right) \ . \tag{8.28}$$

The conditions for the pure state (8.27) to occur for a few specific cases are given below:

$$\begin{aligned} \Phi = 0, \quad \Delta = 0\ , \quad \Omega &= \frac{\gamma\sqrt{M}}{\sqrt{N+1}+\sqrt{N}}\ , \quad \theta_o = \frac{\pi}{2} \\ \Phi = \frac{\pi}{2}\ , \quad \Delta = N + \frac{1}{2} - M\ , \quad \Omega &= \frac{\gamma\sqrt{2M}}{\sqrt{N+1}+\sqrt{N}}\ , \quad \theta_o = \frac{\pi}{4} \\ \Phi = \pi\ , \quad \Delta \gg N + \frac{1}{2} - M\ , \quad \Omega &= \frac{2\Delta\gamma\sqrt{M}}{\sqrt{N+1}-\sqrt{N}}\ , \quad \theta_o = 0\ . \end{aligned} \tag{8.29}$$

Notice that for resonant excitation, an exact pure state is only possible if $\Phi = \Delta = 0$. In general, the pure state (8.27) describes a completely polarized atom with the Bloch vector **B** lying on the Bloch sphere with polar angle β and azimuthal angle θ_o,

$$\mathbf{B} = \cos\theta_o \, \sin\beta \, \mathbf{e}_x + \sin\theta_o \, \sin\beta \, \mathbf{e}_y + \cos\beta \, \mathbf{e}_z \ , \tag{8.30}$$

where

$$\beta = \arccos\left(-\frac{M-N}{M+N}\right) \ . \tag{8.31}$$

When $\Phi = 0$, then $\theta_o = \pi/2$, and the atomic Bloch vector (polarization) is in the Y–Z plane, whereas if $\Phi = \pi$, we have $\theta_o = 0$ and the atom polarizes in the X–Z plane.

8.2.6 Optimum Squeezing in the Output Field

Squeezing may be detected by directly homodyning the total radiation field and the local oscillator without first frequency filtering. This method gives the squeezing in terms of the total normally-ordered variances of the phase quadratures, and can be expressed in terms of the steady state solution of the Bloch equations (8.10) as [144]

$$S_\theta = \langle : E_\theta, E_\theta : \rangle = 1 + \langle \sigma_3 \rangle - \left(\langle \sigma_1 \rangle \cos\theta - \langle \sigma_2 \rangle \sin\theta \right)^2 \ , \tag{8.32}$$

where $E_\theta = \mathcal{E}^{(+)} \exp(-i\theta) + \mathcal{E}^{(-)} \exp(i\theta)$ is the θ-phase quadrature of the atomic fluorescence field, measured by homodyning with a local oscillator having a controllable phase θ relative to the driving laser. $E_{\theta=0}$ and $E_{\theta=\pi/2}$ are usually the in-phase (X) and out-of-phase (Y) quadratures of the fluorescent field, respectively. S_θ is the total normally-ordered variance of the θ-phase quadrature of the fluorescent field. The field is said to be squeezed when $S_\theta < 0$. The normalization we have chosen is such that maximum squeezing corresponds to $S_\theta = -0.25$. Equation (8.32) implies that the squeezing occurs at large values of the atomic coherences, $\langle \sigma_{1(2)} \rangle$.

It was shown by Walls and Zoller [144] that, in the absence of an applied squeezed field, the output field in the resonance fluorescence produced by a two-level atom is itself squeezed, as is the atomic dipole. It can be shown from their expressions that the largest squeezing is produced when $\Omega = \gamma/\sqrt{6}$. However, this is not the largest amount of squeezing which is theoretically possible from a two-level atom [145–148]. In the absence of an applied squeezed vacuum, it is not possible to achieve optimum squeezing in the output fluorescence, and the atom cannot evolve into a pure state. With a squeezed vacuum applied, Ficek and Swain [149] have shown that, under conditions of resonant excitation of the atom by the field, optimum squeezing occurs when the atom is in the particular pure state (8.27) with $N = 1/8$. It is then found that the emitted field exhibits 100% squeezing, as measured by (8.32). This is much greater squeezing than that present in the input squeezed vacuum. The squeezing occurs in the out-of-phase component, $\theta = \pi/2$. (If, however the squeezing is measured by the squeezing spectrum [150], it is found that the squeezing attains a minimum value that corresponds to 75% squeezing, again for $\theta = \pi/2$.) For resonant excitation, the conditions for optimum squeezing are found to be $\Phi = 0$, $N = 1/8$ and $\Omega = 0.433\gamma$. It is evident that in this region, the spectral intensity is much reduced, indicating a reduction of the spontaneous emission noise. This reduction of spontaneous emission leads to strong squeezing properties in the fluorescence field.

If detuned excitation is allowed, then it is possible to achieve 100% squeezing in the output [143] for different values of the quadrature phase. It is not difficult to show that for an arbitrary detuning $\Delta = \delta/\gamma$ and phase Φ, maximum squeezing occurs in the quadrature phase S_θ given by $\theta = \theta_o$, where θ_o is the same angle defined in (8.28):

$$\theta_o = \arctan\left(\frac{N + \frac{1}{2} + M\cos\Phi}{\Delta + M\sin\Phi}\right) . \tag{8.33}$$

However, it can be shown that maximal squeezing, $S_{\theta_o} = -0.25$, can only be achieved if $N = 1/8$, and then the atom is in the pure state (8.27). In terms of Φ, the appropriate values of Ω and Δ are then

$$\begin{aligned} \Delta &= \frac{1}{4}\tan(\Phi/2) , \qquad \Omega = \frac{\sqrt{3}}{4}\gamma\sec(\Phi/2) , \\ \tan\theta_0 &= 1/\tan(\Phi/2) , \qquad (N = 1/8) . \end{aligned} \tag{8.34}$$

Thus for $N = 1/8$, maximal squeezing is possible for all Φ such that $0 \leq \Phi < \pi$. For $\Phi = 0$, this occurs when $\Delta = 0$ and $\Omega = \sqrt{3}\gamma/4$ in the out-of-phase (Y) quadrature component, i.e. $\theta_o = \pi/2$, and for $\Phi = \pi/2$ it occurs in the $\pi/4$-phase quadrature, $\theta_o = \pi/4$, with $\Delta = 1/4$ and $\Omega = \sqrt{6}\gamma/4$ [151]. As $\Phi \to \pi, \Delta \to \infty$ and $\Omega \to \infty$.

For small values of N, large squeezing in the fluorescence is possible, even if it is not maximal. In particular, when $\Phi = \pi$ and $\Delta \gg N + \frac{1}{2} - M$, then $\theta_o = 0$, and optimal squeezing is always in the in-phase (X) quadrature [143,152].

We have also considered the squeezing spectrum, defined as the normally-ordered noise spectrum of the fluorescence field in the steady–state, given by [128]

$$S_{X(Y)}(\omega) = \int_{-\infty}^{\infty} d\tau \lim_{t\to\infty} \langle : E_{X(Y)}(\mathbf{r}, t+\tau), E_{X(Y)}(\mathbf{r}, t) :\rangle\, e^{i\omega\tau} , \tag{8.35}$$

where $E_{X(Y)}(\mathbf{r}, t)$ is the slowly varying in-phase (out-of-phase) quadrature operator of the fluorescent radiation field, under resonant and off-resonant excitation [152], and including the effects of a finite laser bandwidth due to phase diffusion. For resonant excitation at low intensities, the fluorescence field exhibits narrow bandwidth squeezing in the out-of-phase quadrature, centered at the laser frequency. Otherwise, squeezing occurs only in the in-phase quadrature. Specifically, for slightly off-resonant and weak excitation, the squeezing still centers at the laser frequency, with a finite bandwidth of the order of γ. More importantly, for far off-resonant and strong excitation, the resonance fluorescence exhibits two-mode squeezing around the Rabi sideband frequencies. Thus, we have a source of two-mode squeezing which is very easily frequency-tuned simply by adjusting the generalized Rabi frequency. The presence of a laser linewidth substantially reduces the degree of squeezing obtainable. The fluorescence shows no squeezing at all when the laser linewidth is greater than the atomic spontaneous emission linewidth.

8.2.7 Amplification of a Weak Probe Beam

Another way of examining the properties of the interacting atom/squeezed field system is to measure the absorption and refractive index of a weak, tunable probe beam which is passed through the system. The steady-state probe absorption spectrum is proportional to the quantity

$$A(\omega) = \frac{1}{\pi}\mathrm{Re}\int_0^\infty d\tau \, \langle [\sigma_-(\tau), \sigma_+(0)] \rangle \, e^{i\omega\tau} \, . \tag{8.36}$$

This has been evaluated for finite bandwidth squeezed light excitation by Ritsch and Zoller [29,153–155], and by An et al. [30,156] for the broadband case. An absorption peak of subnatural linewidth was found. This is a quantum feature, the spectrum in the case of the corresponding CSF having the natural linewidth.

Ficek and Dalton [31] considered the case of off-resonant squeezed vacuum excitation in a Fabry–Perot cavity ($\omega_S \neq \omega_0$) with no coherent driving field present, where they found that perfect probe transparency was possible for a particular probe frequency. They also reported the existence of a threshold value M_{thr} of the two-photon correlation strength M, which depends upon the detuning and cavity parameters. The absorption spectrum may be expressed as the sum of two terms. Above threshold, $M > M_{\mathrm{thr}}$, both terms are absorptive, but below threshold, one is absorptive and the other emissive, resulting in a hole in the absorption spectrum. For sufficiently large N $(N \geq 2)$, the absorption at the center of the hole approaches zero, resulting in transparency of the probe beam at this frequency. For a CSF, there is no hole in the spectrum.

Next we consider the consequences of adding a coherent driving field in addition to the squeezed vacuum. It was shown by Ficek et al. [32,34] that not only absorption, but amplification of the probe beam *at line center* is possible in the presence of the squeezed vacuum, and that this is an example of amplification without population inversion. There is no population inversion present in the bare state basis because we have only a two-level system, and

Table 8.1. Critical values for the Rabi frequency

	Ideal Squeezed Vacuum	**Equivalent CSF**
Absorption if	$\Omega < 0.5\gamma$	$\Omega < \left(N + \frac{1}{2}\right)\gamma$
Transparency if	$\Omega_{\text{crit}} = 0.5\gamma$	$\Omega_{\text{crit}} = \left(N + \frac{1}{2}\right)\gamma$
Amplification if	$\Omega > 0.5\gamma$	$\Omega > \left(N + \frac{1}{2}\right)\gamma$

the dressed states have equal populations, because we are dealing with the case of resonant excitation.

The Rabi frequency Ω needs to be sufficiently large for amplification to occur: we have the existence of a threshold, as mentioned earlier [36]. The critical value of Ω, i.e. the value of Ω at which the medium is perfectly transparent to the pulse, is given approximately by (8.23). Thus for $N \geq 1$ and $\Phi = 0$, the threshold occurs at $\Omega \simeq 0.5\gamma$. For values of Ω less than this, we have absorption, and for values of Ω greater than the threshold, we have amplification. For the corresponding CSF, the critical value is $\Omega_{\text{crit}} = \left(N + \frac{1}{2}\right)\gamma$. Thus for $N \geq 1$ and $\Phi = 0$, the position is summarized in Table 8.1. The critical value of Ω is larger for the CSF, and any amplification which does occur is much less than for an ideal squeezed vacuum.

The off resonant situation ($\omega_S \neq \omega_0$) was considered by Zhou and Swain with the optional coherent field resonant with the squeezed vacuum ($\omega_L = \omega_S$) both in free space [36] and in the bad cavity limit [37]. Similar features to those reported by Ficek and Dalton [31] were found, including a region of almost zero absorption and the existence of a threshold value of M in the presence of the coherent driving field.

It is also possible to find features analogous to the anomalous spectral profiles of resonance fluorescence in the probe absorption spectra. These have been discussed by Zhou, Ficek and Swain [35]. A variety of interesting and distinctive spectral profiles were reported, including narrow regions of pronounced amplification. A couple of examples are presented in Fig. 8.6. In Frame (a), where we take $N = 2$ and $\Phi = 0$, we show the progression from absorption for $\Omega = 0.35\gamma$, to almost flat transparency at $\Omega = 0.5\gamma$, to amplification for $\Omega = \gamma$. For values of $\Phi \neq 0$ or π, the spectra become asymmetric. In Frame (b) of Fig. 8.6, where $N = 5$, we show hole burning for $\Phi = 0$ and dispersive profiles for $\Phi = \pi/2$. The value of Ω for each plot was chosen so as to maximize the anomalous features. The plot for the equivalent CSF for Frame (b) is a structureless absorption peak which depends only very weakly on Φ. The properties of the anomalous spectra in probe absorption are very similar to those in resonance fluorescence, including extreme sensitivity to the values of Ω and η for the larger values of N.

Finally, we remark that the refractive index μ of the probe is proportional to the imaginary part of the quantity in (8.36):

$$\mu \propto B(\omega) = \frac{1}{\pi}\text{Im} \int_0^\infty d\tau \, \langle [\sigma_-(\tau), \sigma_+(0)] \rangle \, e^{i\omega\tau} \, . \tag{8.37}$$

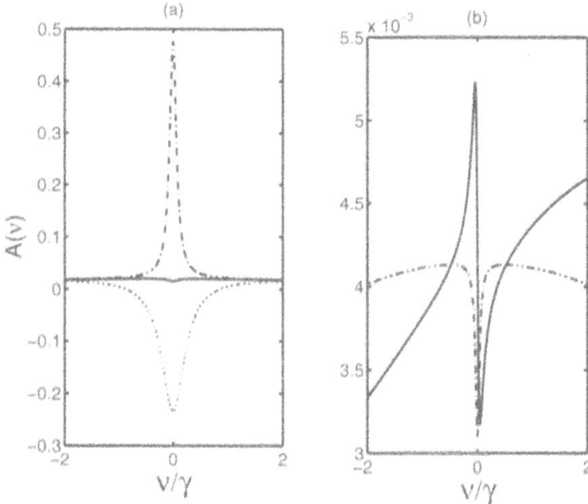

Fig. 8.6. Examples of probe absorption spectra. In **(a)** $N = 2$ and $\Phi = 0$ with $\Omega = 0.35\gamma$ (*dot-dash line*), $\Omega = 0.5\gamma$ (*solid line*) and $\Omega = \gamma$ (dotted line). In **(b)**, $N = 5$ and $\Omega = 0.5\gamma$ with $\Phi = 0$ (*dot-dash line*) and $\Omega = 0.707\gamma$ and $\Phi = \pi/2$ (*solid line*, $\times 5$)

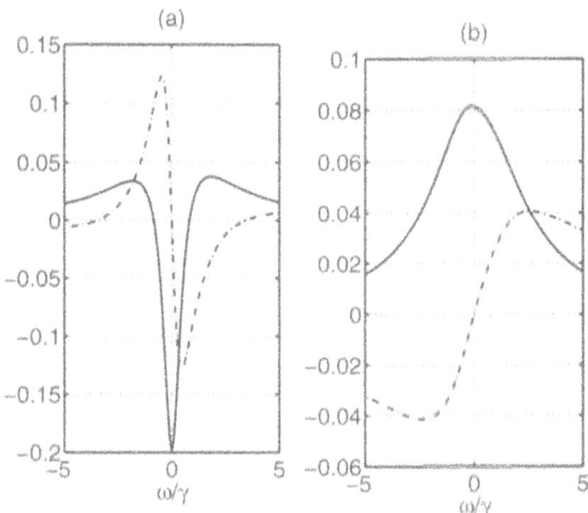

Fig. 8.7. The absorption (*solid line*) and refractive index (*dot-dash line*) in arbitrary units for $N = 1$ and $\Phi = 0$. Frame **(a)**, the ideal squeezed vacuum and **(b)**, the equivalent CSF

It is possible in the squeezed vacuum, even for a resonant driving field, to arrange for the refractive index to be near maximal at a frequency where the probe shows zero absorption. Alternatively, we can achieve maximum absorption or amplification at zero dispersion [157,158].

This is illustrated in Fig. 8.7. Frame (a) shows the absorption and refractive index for an ideal squeezed vacuum, and (b) for the equivalent CSF. In

Frame (a), the absorption is zero at $\omega \simeq \gamma$, and the refractive index is close to a maximum, whilst the amplification is a maximum at line center, where the refractive index is zero. For the CSF, there is absorption at line center instead of amplification.

8.2.8 Arbitrary Intensity Probe

In the treatments above, the amplitude of the probe beam was assumed to be very small compared to that of the pumping laser, so that linear response theory may be applied. The case of the absorption of an arbitrary intensity probe has been considered by Zhou and Swain [38]. The ratio of the Rabi frequency of the probe Ω_P to that of the driving field Ω is defined to be α, which may be arbitrary. The behavior as a function of the various parameters is rich and complicated, so we present only a brief summary of *some* of the features here. First we consider the situation for $N = 0$. For Ω large, there is much additional structure for $\alpha \simeq 1$ between the Rabi sideband peaks, which occur at $\nu = \pm\Omega$ for $\alpha \ll 1$, but move to larger values as α increases. For $\alpha \ll 1$, there is a small peak at line center. This grows and broadens as α increases, and for $\alpha \simeq 1$, it is dominant (with much superimposed structure).

We next introduce the squeezed vacuum, and consider first the situation where $\Phi = 0$. For Ω large and $N = 1$ say, the effect of the squeezed vacuum for α not too large is to wash out the structure in the central region, which represents a broad amplifying zone. As α approaches $\alpha \simeq 0.5$ however, the amplifying region takes the form of a narrow dip at line center, and as α is increased further ($\alpha \simeq 1$), we obtain a broad absorption peak with some structure. As N increases, say to $N = 2$, the structure becomes less apparent so that for $\alpha \simeq 1$, only a broad smooth peak is evident around line center. However, as N continues to increase, an extremely narrow peak at line center appears on top of the broad peak. For $N = 5$ for example, a peak with a width of a few percent of the natural width is clearly evident. What is even more remarkable, a fine resolution of this 'peak' reveals it to be in fact a doublet, the hole between the two peaks having the extremely small halfwidth of about 0.005γ. For $\Phi = \pi/2$ the narrow doublet becomes a narrow dispersive profile at line center, and for $\Phi = \pi$, the narrow doublet becomes a single peak. These very narrow features are quantum effects.

We illustrate these effects in Fig. (8.8) where we take $N = 1, \Phi = 0, \Omega = 10$ and (a) $\alpha = 1$, (b) $\alpha = 1.025$, (c) $\alpha = 1.05$ and (d) $\alpha = 1.1$. It is evident that the spectral profile changes very rapidly with α. The narrow features at line center are in fact doublets when viewed under a fine scale. The doublet structure becomes clearer for larger values of Ω.

Again, there is a threshold value of Ω, defined to be that value of Ω for which the medium is transparent to the probe at line center, in the weak probe limit. For Ω small but away from the values that give rise to the anomalous features, the absorption spectra are simple, and practically independent of the value of α apart from the scale. For Ω close to the anomalous values, for

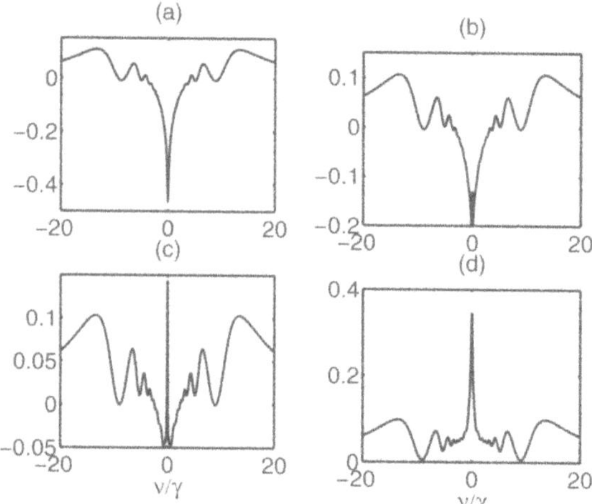

Fig. 8.8. The absorption spectrum for $N = 1$, $\Omega = 10\gamma$ and $\Phi = 0$ for **(a)** $\alpha = 1$, **(b)** $\alpha = 1.025$, **(c)** $\alpha = 1.05$ and **(d)** $\alpha = 1.1$

example $\Phi = 0$ and $\Omega = 0.5\gamma$, the spectra are again relatively simple, but show very sharp structures at line center. For large Ω, and $\alpha \simeq 1$, the spectra are very complicated, even for $N = 0$. For $\alpha < 0.5$, they show remarkably narrow structures at line center.

8.2.9 Dressed State Population Trapping

Here, we assume that the driving laser may be detuned from the atomic resonance frequency, but that the center frequency of the squeezed vacuum is still equal to the laser frequency, $\omega_S = \omega_L$. One of the most striking phenomena which can occur for $\delta = \omega_A - \omega_L \neq 0$ is dressed state population trapping, which we now describe. The dressed states $|\alpha\rangle$ and $|\beta\rangle$ are defined in (8.15).

The dressed states populations are profoundly affected by the squeezed vacuum. First of all, we consider exact resonance, $\delta = 0$. Then, in the absence of the squeezed vacuum, or in the presence of the squeezed vacuum for the particular phase values $\Phi = 0$ or $\Phi = \pi$, it is easy to see that the populations of the dressed states must be equal. It is clear from (8.38) that equal dressed state populations imply a symmetric resonance fluorescence spectrum – at least, within the secular approximation. In the presence of the squeezed vacuum with $\Phi \neq 0$ or π, the dressed states populations are different, and this results in asymmetric spectra even for resonant excitation [72]. For $\delta = 0$ and $\Phi = \pi/2$ for example, the dressed state populations exhibit the greatest difference for $\Omega \simeq \gamma/\sqrt{2}$ for large N. (These parameter values are those for which dispersive profiles arise in resonance fluorescence.) Then, the ratio of the dressed states populations can reach as high a value as 5.83. The difference between the dressed state populations is much reduced if the ideal squeezed vacuum is replaced by a CSF.

There is also a qualitative difference (for $\Phi \neq 0, \pi$) in the dressed state population behavior as a function of the excitation number N in the ideal squeezed vacuum as compared to the CSF: in the former case, the population increases (or decreases) monotonically with N, saturating to a value different from 0.5, whereas in the latter, the population at first increases (or decreases), then tends to the value 0.5 as $N \to \infty$.

The secular approximation, which is valid if the Rabi frequency is much greater than all other frequencies in the problem, predicts equal dressed state populations. Thus we find that the dressed state populations tend to the value 0.5 as $\Omega \to \infty$. Note, however, this happens only very slowly with increasing Ω, and thus results obtained on the basis of the secular approximation must be used with some caution.

A clear physical interpretation of the resonance fluorescence spectra under strong excitation is provided by the 'dressed atom' picture. In the usual dressed atom treatment [159,160], the secular approximation is made, which results in simple expressions for the weights of the well-separated lines in terms of the steady state occupation probabilities P_α and P_β of the dressed states. These may be obtained by integrating the term representing each line in (8.19) over all frequencies as:

$$\begin{aligned} I_0 &= 4\gamma c^2 s^2 P_\alpha P_\beta \ , \\ I_+ &= \gamma c^4 P_\alpha, \ I_- = \gamma s^4 P_\beta \ , \\ I_{\text{coh}} &= \gamma c^2 s^2 \left(P_\alpha - P_\beta\right)^2 \ , \end{aligned} \tag{8.38}$$

where I_0 and $I_\pm$ the weights of the central and sidepeaks respectively of the incoherent part of the spectrum, and I_{coh} is the weight of the coherent, or elastically scattered contribution, which we have added for completeness.

The assumption of the secular approximation limits the validity of the treatment to large applied fields Ω or large detunings δ. However, it is not essential to make the secular approximation, and dressed state treatments which do not assume it have been performed [25,72,135,141]. The last two papers showed that the resonance fluorescence spectrum could be drastically modified by the presence of even very weak squeezed fields, say $N \simeq 10^{-5}$.

Assuming $\delta \neq 0$ enables us to find conditions where *all* the population resides in just one dressed state – that is, we have dressed state population trapping. This phenomenon was predicted by Courty and Reynaud [50], who showed that, working within the confines of the secular approximation, total population trapping occurred when $\Phi = \pi$ and at a particular value of the detuning and Rabi frequency.[2] We may understand the conditions for population trapping as follows, using the approach of Swain and Dalton [135], in terms of the spontaneous decay rates $\gamma_{\alpha\beta}$ and $\gamma_{\beta\alpha}$ between the dressed states. In the secular approximation, there are only spontaneous transitions between

[2] They actually treated the bad cavity situation. However, as the equations in this case are formally equivalent to those in free space, we discuss their work here.

the dressed state populations (the stimulated transitions are negligible). We then have $W_{ij} = \gamma_{ij}$ in the notation of (8.17). Suppose the spontaneous decay rate $\gamma_{\alpha\beta}$ from state $|\alpha\rangle$ to state $|\beta\rangle$ vanishes. Then, from (8.20) we see that $P_\beta = 0$ – there will be dressed state population trapping. All the population must reside in the state $|\alpha\rangle$, as the rate $\gamma_{\beta\alpha}$ does not also vanish. From the explicit expression for $\gamma_{\alpha\beta}$ given in [135], it may be seen that this quantity is only zero if $\Phi = \pi, M = \sqrt{N(N+1)}$, and the detuning is given by

$$\delta = \pm \frac{\Omega\left(\sqrt{N+1} - \sqrt{N}\right)}{2\left[N(N+1)\right]^{1/4}} . \tag{8.39}$$

It can be shown [25] that the system is then to be found in the pure state (8.27). As a consequence of the trapping of all the population in just one dressed state, the resulting incoherent resonance fluorescence spectrum has only one peak (instead of the three peaks that occur in the absence of population trapping). This may be seen from (8.38) or (8.19). If we replace the squeezed vacuum with the equivalent CSF, $\gamma_{\alpha\beta}$ cannot be made to vanish. There is no population trapping, and the usual three peaked spectrum is recovered. Thus the introduction of the squeezed vacuum leads to qualitatively new features in the resonance fluorescence spectrum, and these are attributable to the quantum nature of the squeezed vacuum. However, we should emphasize that all these predictions are made under the secular approximation.

Using an exact treatment, we have shown that it is impossible to achieve complete population trapping and the associated peak suppression if Φ is restricted to the value π [135]. We did however demonstrate that complete suppression can occur for different values of Φ and δ. If we do not make the secular approximation, then there are both spontaneous and stimulated transitions between the dressed state populations. The condition for population trapping is then that the total transition rate $W_{\alpha\beta}$ should vanish. Unfortunately, it is not possible to obtain a simple analytic expression in the general case corresponding to (8.39) for the secular approximation.

The dressed state population for $\Omega = 5\gamma$ and $N = 0.25$ as a function of Φ and δ are shown in Fig. 8.9, for (a) an ideal squeezed vacuum, and (b) for the equivalent CSF. The two plots are broadly similar, but the ideal squeezed vacuum case has more structure. First, we note that for $\Phi = \pi$, the population in frame (a) tends almost to zero at $\delta = 2.2\gamma$, and almost to unity at $\delta = -2.2\gamma$, these values corresponding to almost complete population trapping of the dressed states. This is the situation described by Courty and Reynaud. The resonance fluorescence spectra evaluated at ($\Phi = \pi$, $\delta = \pm 2.2\gamma$) will be mirror images of each other.

However, the point $(\Phi, \delta) = (\pi, -2.2\gamma)$ does not correspond to a *global* maximum. Although it is not very pronounced, the global maximum is at $(\Phi_1, \delta_1) = (0.95\pi, -2.05\gamma)$, at which point the population has a value extremely close to unity. Thus at this point there is practically complete popu-

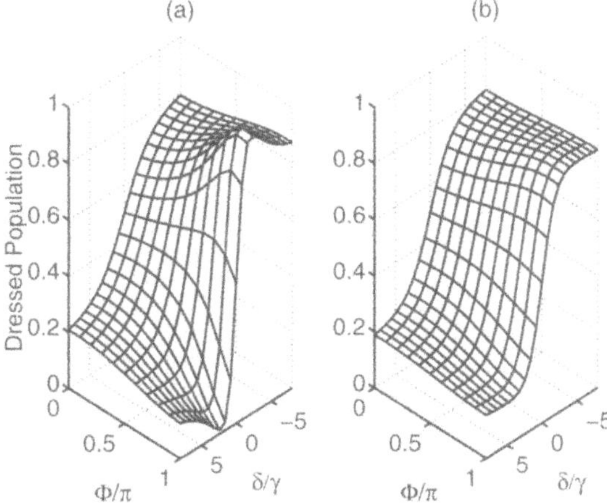

Fig. 8.9. The dressed state population $\rho_{\alpha\alpha}$ for $\Omega = 5\gamma$ and $N = 0.25$ as a function of the squeezing phase and the dimensionless detuning δ/γ for **(a)** an ideal squeezed vacuum, and **(b)** the equivalent MCCF

lation trapping and complete suppression of two of the three peaks of the resonance fluorescence spectrum occurs. Note that the symmetry is now broken: there is no global minimum at the reversed detuning point $(0.95\pi, +2.05\gamma)$. This means that the spectra evaluated at the points (Φ_1, δ_1) and $(\Phi_1, -\delta_1)$ will not be mirror images of each other.

The resonance fluorescence spectra at the optimum conditions for population trapping predicted by the secular approximation are presented in Fig. 8.10a. The solid lines give the *exact* spectra for $\Omega = 5\gamma$, $N = 0.25$ and $\Phi = \pi$, and $\delta = 2.2\gamma$. Small residual peaks are visible, showing that the population trapping is incomplete, reflecting a failure of the secular approximation. The dot-dashed line gives the spectrum for the equivalent CSF. There is a great contrast as compared with the ideal squeezed vacuum case, and there is no suggestion of peak suppression.

The resonance fluorescence spectra corresponding to the global maximum of Fig. 8.9a are illustrated in Fig. 8.10b. The solid line is the spectrum under conditions of complete population trapping $(\Phi, \delta) = (0.95\pi, -2.05\gamma)$. It can be seen that there is complete suppression of two peaks in this case. If we reverse the detuning, so that $(\Phi, \delta) = (0.95\pi, +2.05\gamma)$ (the dot-dashed line) there is now no peak suppression, although there is a reversal of asymmetry. In this case, there is obviously no symmetry between the $\delta = 2.05\gamma$ and $\delta = -2.05\gamma$ spectra. Observation of population trapping would provide powerful confirmation of the nonclassical character of the squeezed vacuum.

So far, we have only considered the case where the laser frequency is resonant with the carrier frequency of the squeezed vacuum. The situation where $\Delta_S \equiv \omega_L - \omega_S \neq 0$ has been considered by Ficek and Sanders [23], with $\delta = 0$. They showed that, with no driving field present, there is a squeezing

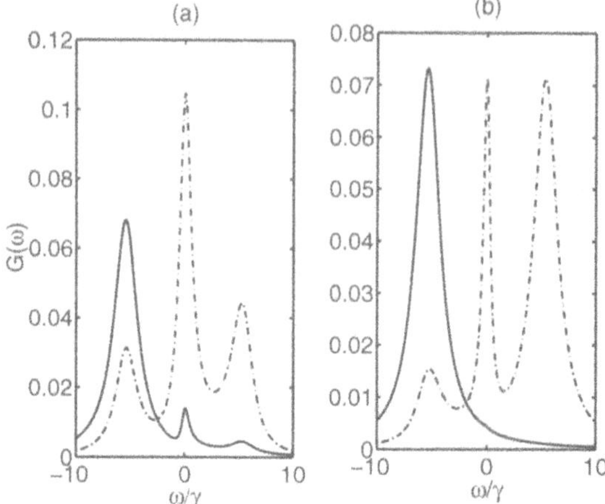

Fig. 8.10. **(a)** The resonance fluorescence spectrum $G(\omega)$ in arbitrary units for $\Omega = 5\gamma$, $\Phi = \pi$, $\delta = 2.2\gamma$ and $N = 0.25$, corresponding to the local maximum in the dressed state population, as a function of the dimensionless frequency ω/γ. The solid line is for an ideal squeezed vacuum and the dot-dashed line is for a CSF. (The latter is multiplied by the factor 0.5.) The spectrum for the detuning $\delta = -2.2\gamma$ would be the mirror image of the solid line (not shown). **(b)** The same as in Frame **(a)**, except that $\delta = 2.05\gamma$ and $\Phi = 0.95\pi$, corresponding to the global maximum in the dressed state population. The dot-dashed line in this case is for an ideal squeezed vacuum with reversed detuning, $\delta = -2.05\gamma$

threshold. Above the threshold, the polarization decay rate may be reduced, and below threshold, there is a squeezing induced two-peaked spectrum, and a frequency shift analogous to the Bloch–Siegert shift. When a driving field is introduced, the spectrum is asymmetric for $\Omega < \Delta_S$, but becomes symmetric as $\Omega/\Delta_S \to \infty$.

8.3 Two-Level Atoms in the Cavity Environment

As we have remarked previously, to observe the phenomena described so far for the free space situation, *all* the modes with which the atom interacts must be squeezed – that is, the squeezed modes must occupy all 4π solid angles of space [2]. This is not feasible. Instead the cavity environment, where only those modes centered around the privileged cavity mode need be squeezed, provides a much more realistic arrangement for the experimental investigation of these effects.

The two-level atom within the cavity, which may be single-sided, is driven by a coherent field of Rabi frequency Ω either through one mirror or through the open sides of the cavity. In addition, it is illuminated by the squeezed vacuum through one mirror. Photons are lost from the cavity at the total rate κ for both mirrors, and the atom may emit spontaneously through the

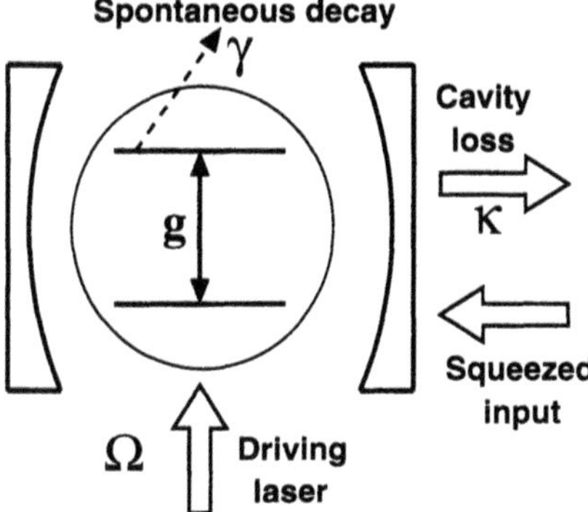

Fig. 8.11. The cavity environment. The two-level atom inside the cavity is driven by an applied laser field and the squeezed vacuum. It can spontaneously emit photons out of the side of the cavity

open sides of the cavity at the rate γ. The latter process provides the means of observing the resonance fluorescence spectrum (Fig. 8.11). The coupling constant for the two atomic levels is g.

The cavity experiment of Turchette et al. [86], the only one reported to date, concentrated on the line narrowing that is predicted in the transmission spectrum of a weak probe beam passing through the Cs atoms in the cavity, in the presence of squeezed light. The experimental difficulties were very great, and it was not possible to distinguish between the effects of nonclassical and classical excitation. The probe transmission modulation for a squeezed field rotating in phase with respect to an applied coherent field was also examined. The trends were much as expected, but the overall magnitude of the effects was smaller than anticipated. However, good qualitative agreement between theory and experiment was obtained. The effect of the experimental complications may be understood with reference to (8.3). Even if a perfectly correlated source of squeezed light, with $\eta = 1$, is employed, the experimental complications act to considerably reduce the effective value of η as far as atomic interactions are concerned. The effective η may be expressed as $\eta_{\text{eff}} = \eta_1\eta_2 \ldots$, where the η_k represent the effect of different experimental factors, such as fluctuations in the atomic beam properties. As each of these must be less than unity, the value of η_{eff} is rapidly reduced if there are several such factors. It may be so reduced that the two-photon correlation strength lies in the classical, not the quantum, domain.

8.3.1 The Mean Photon Number

One of the simplest observables of this system is the mean photon number. Cabrillo and Swain [161] have shown that when the system is driven by a coherent field, a remarkable sensitivity of the mean photon number $\langle n \rangle$ to the relative phase between the driving field and the squeezed vacuum appears.

The effect is strongly dependent on the degree of two-photon correlations: the change in $\langle n \rangle$ as Φ varies from 0 to π is most pronounced for an ideal squeezed vacuum, but very slight for the equivalent CSF.

8.3.2 The Bad Cavity Limit

The simplest cavity model is the *bad cavity limit*, which is defined by the conditions

$$\kappa \gg g \gg \gamma \,, \quad \text{with} \quad C = g^2/\kappa\gamma \quad \text{finite} \,, \tag{8.40}$$

where C is the single-atom cooperativity parameter. The cavity decay dominates, so that the atom always experiences the cavity mode in the state induced by the vacuum modes, enabling us to eliminate the cavity mode and obtain effective Bloch equations. The condition $g \gg \gamma$ ensures that coherent evolution dominates. Smyth and Swain have shown that, in this limit, the cavity field may be eliminated to obtain effective Bloch equations which are formally equivalent to those in free space, but with 'cavity renormalized' parameters [61]. Rice and Pedrotti [162] studied the incoherent resonance fluorescence spectrum and found that the cavity enhanced spontaneous emission could be suppressed, and a subnatural line width obtained, for an appropriate choice of the phase of the squeezed vacuum.

The Bloch equations in the bad cavity limit may be written in the form [61]

$$\begin{aligned} \langle \dot{\sigma}_- \rangle &= -\gamma_C (1 + 2N_C) \langle \sigma_- \rangle + \gamma_C M_C e^{-i\Phi} \langle \sigma_+ \rangle + \Omega \langle \sigma_3 \rangle \,, \\ \langle \dot{\sigma}_3 \rangle &= -\gamma_C - \gamma_C (1 + 2N_C) \langle \sigma_3 \rangle - \frac{1}{2} (\langle \sigma_- \rangle + \langle \sigma_+ \rangle) \,, \end{aligned} \tag{8.41}$$

where

$$\begin{aligned} &\gamma_C = \gamma (1 + C) \,, \\ &N_C = \beta N \,, \; M_C = \beta M \,, \; \beta = C/(1 + C) \,. \end{aligned} \tag{8.42}$$

The quantity γ_C is the cavity enhanced spontaneous decay rate, and β is often referred to as the *beta factor* of the cavity.

The equations (8.41) are formally identical to the Bloch equations (8.10) in free space: the only difference is the replacement of the free space parameters γ, N and M by cavity modified ones, γ_C, N_C and M_C. An important consequence of this renormalization is that if an ideal squeezed vacuum, with $M^2 = N(N + 1)$, is put into the cavity then the squeezed vacuum inside the cavity is not ideal: $M_C^2 < N_C (N_C + 1)$. We have in fact, $M_C = \eta_C \sqrt{N_C (N_C + 1)}$, where

$$\eta_C = \eta \sqrt{\frac{\beta (N + 1)}{(\beta N + 1)}} < 1 \,, \tag{8.43}$$

and η_C is the effective degree of two-photon correlation. Thus we lose the option of using an ideal squeezed vacuum in the cavity environment [61].

We are particularly interested in observing the anomalous spectra, as they represent the most striking aspects of resonance fluorescence in a squeezed vacuum. It has been shown that, in spite of the degradation of the squeezing implied by (8.43), the hole-burning profiles are readily observable in the bad cavity environment [61]. However, the dispersive profiles are more difficult to observe, since large values of N are needed to flatten out the background so that the dispersive profiles are clearly exhibited.

8.3.3 The Fabry–Perot Microcavity

Next we consider a more detailed model, that of a Fabry–Perot microcavity, in which two plane, parallel mirrors which lie parallel to the (x, y)-plane, one with reflectivity $R < 1$ and the other perfectly reflecting, are located a distance L apart. We assume the imperfect mirror to be illuminated by a cone of squeezed light of half-angle Θ. The axis of the cone defines the z axis. The coherent driving field may be admitted through the open sides of the cavity. The two-level atom is assumed to be located a distance z from the perfect mirror along the axis of the cavity. In the microcavity, the structure of the modes available to the atom for spontaneous emission is modified, thus changing the spontaneous decay rate [163].

This situation has been treated by Parkins and Gardiner [56], who found that the inhibition of phase decays discussed in Sect. 8.2.2 is still observable with an input beam of modest angular spread θ if the characteristics of the input beam are suitably matched to those of the cavity.

From consideration of the boundary conditions at the mirrors, we find the following expression for the electric field inside the cavity [60]:

$$\mathbf{E}^{(+)}(\mathbf{r}, t) = i \sum_{s} \int d^3\mathbf{k} \left(\frac{2ck}{\varepsilon_o \hbar (2\pi)^3} \right)^{1/2} Y(\theta_k, L) \times e^{i\mathbf{k}^{\|} \cdot \mathbf{r}} \hat{\mathbf{e}}_{ks}^{\|} \sin(k_z r) \hat{a}(\mathbf{k}, s) \ , \tag{8.44}$$

where $Y(\theta_k, L)$ is the cavity transfer function

$$Y(\theta_k, L) = \frac{i\left(1 - R^2\right)^{1/2}}{1 - R \exp\left[2ikL\cos\theta_k\right]} \ , \tag{8.45}$$

θ_k is the angle between the $\mathbf{k}$ vector and the z-axis (cavity axis), $\hat{a}(\mathbf{k}, s)$ is the annihilation operator of the cavity mode of wave vector $\mathbf{k}$ and polarization s, and $\hat{\mathbf{e}}_{ks}^{\|}$ and $\mathbf{k}^{\|}$ are projections onto the (x, y)-plane of the polarization vectors $\hat{\mathbf{e}}_{ks}$ and $\mathbf{k}$ respectively.

The function $|Y(\theta_k, L)|^2$ is the Airy function of the cavity [164]. If R is close to 1 and $L = \lambda/2$, where λ is the resonance wavelength, $|Y(\theta_k, L)|^2$ exhibits a sharp peak centered at $\cos\theta_k = 1$. This means that the field inside the cavity is affected only by those modes which are contained in a small solid angle around the cavity axis. By squeezing of these modes we can achieve perfect matching between the input squeezed field and the atom.

The modified electric field inside the cavity changes the spontaneous emission rates from those corresponding to a normal vacuum, γ, into $\gamma_C = a\gamma$, [163,165], where

$$a = \frac{3}{2}\frac{(1+R)}{(1-R)} \int_0^1 du \left(1+u^2\right) \frac{\sin^2(kzu)}{1+F\sin^2(kLu)} , \tag{8.46}$$

with $u = \cos\theta_k$, $\mathbf{r} = (x, 0, z)$ is the position of the atom, and

$$F = 4R/(1-R)^2 . \tag{8.47}$$

We studied anomalous spectra in a microcavity with one perfectly reflecting mirror, assuming both a perfectly mode matched input squeezed beam and a Gaussian polarization matched beam [166]. Mirror separations of $L = \lambda/2$ and $L = \lambda$ were both examined, but we describe only the former situation here. The arrangement is similar to ones considered previously involving the interaction between atoms and squeezed light [56,60]. The Bloch equations in the Fabry–Perot cavity are

$$\begin{aligned} \langle\dot{\sigma}^-\rangle &= -\frac{\gamma}{2}(a+2|b|N)\langle\sigma^-\rangle - i\Omega\langle\sigma_3\rangle \\ &\quad -\gamma|b|M\exp(i\Phi_G)\langle\sigma^+\rangle , \end{aligned} \tag{8.48}$$

$$\langle\dot{\sigma}_3\rangle = -\gamma(a+2|b|N)\langle\sigma^-\rangle + i\frac{\Omega}{2}\left(\langle\sigma^+\rangle - \langle\sigma^-\rangle\right) - \frac{1}{2}\gamma a , \tag{8.49}$$

where the parameter b is introduced by the squeezed vacuum, and depends upon the nature of the matching of the squeezed input to the cavity [166]. For perfect matching of the squeezed field to the cavity, we have

$$\begin{aligned} b &= \frac{3}{2}\frac{(1+R)}{(1-R)}\frac{1}{\mathcal{N}} \\ &\quad \times \left[\int_{\cos\Theta}^1 du \left(1+u^2\right) \frac{\sin^2(kzu)}{1+F\sin^2(kLu)} \mathcal{J}_0\left(kx\left(1-u^2\right)^{1/2}\right)\right]^2 , \end{aligned} \tag{8.50}$$

where $\mathcal{N}$ is the normalization constant

$$\mathcal{N} = \int_{\cos\Theta}^1 du \left(1+u^2\right) \frac{\sin^2(kzu)}{1+F\sin^2(kLu)} , \tag{8.51}$$

$\mathcal{J}_0$ is the zeroth-order Bessel function, and Θ is the solid angle over which the squeezed vacuum is propagated. If we set $a = b = 1$, the Bloch equations reduce to those in free space.

The parameters a and b depend upon the angle of squeezing Θ and the position $(x, 0, z)$ of the atom in the cavity. For example, Fig. 8.12a shows how b evolves with z and Θ, for a mirror reflectivity $R = 0.99$. Notice that b attains its maximum value quite rapidly with Θ. This is important because it

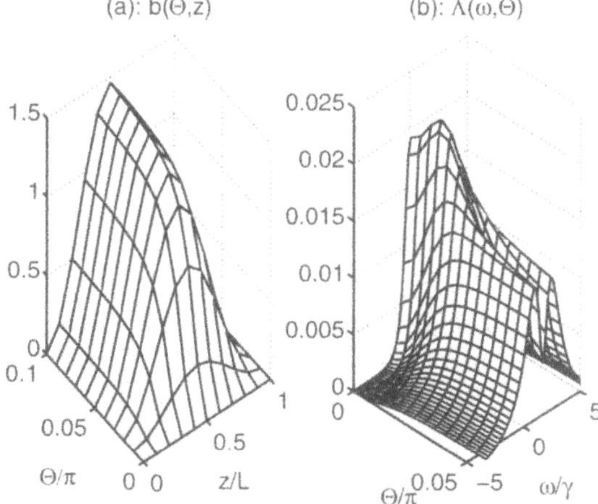

Fig. 8.12. **(a)** The parameter b as a function of the angle of squeezing Θ and the z position of the atom in the cavity. ($N = 0.05$, $R = 0.99$ and $L = \lambda/2$.) **(b)** The resonance fluorescence spectrum as a function of the angular divergence Θ of the squeezed source for perfect mode matching with the Rabi frequency chosen to exhibit the most pronounced anomalous features for the same parameters as in **(a)**

shows that only modest squeezing angles are required for the atom to interact predominantly with the squeezed light. Frame (b) shows how the hole burning develops as the angle Θ is increased. Note that only modest values of Θ are needed for the hole burning to be pronounced.

For $\Phi \neq 0, \pi$, marked dressed state population inversions are possible. For example, for $N = 0.2, \Phi = \pi/2, \Theta = \pi/10$ and $\Omega = 0.65\gamma$, the maximum population of the dressed state is about 0.75. (In the absence of the squeezed vacuum the dressed state population would have the value 0.5).

As in the case of the bad cavity model, it is found that the hole burning spectra persist strongly whilst the dispersive profiles are not very prominent for the low values of N that are experimentally feasible at present.

8.3.4 Bichromatic Excitation

With monochromatic excitation, observation of the anomalous spectral features in resonance fluorescence in a squeezed vacuum is hampered by the coherent scattering which may overwhelm the incoherent part of the spectrum. If we employ instead a bichromatic field [167,168], then there is no coherent component at line center. The anomalous spectral features may be observed under bichromatic excitation in either the free space or the bad cavity situation, but below we concentrate on the cavity case.

It is assumed that the two-level atom is coupled to a broadband squeezed vacuum via the lossy mirror of a single-ended cavity. It is driven directly by a bichromatic laser field with equal amplitude modes of frequencies $\omega_L \pm \delta_L$. The atomic Bohr frequency, the average frequency of the bichromatic laser, the cavity resonance frequency and the center frequency of the squeezed vacuum are all taken to be identical. After eliminating the cavity mode following the procedure in [27,35,37,61], the modified Bloch equations in the bad cavity limit are found to be

$$\begin{aligned} \langle\dot{\sigma}_-\rangle &= -\gamma_C\hat{N}_C\langle\sigma_-\rangle - \gamma_C M_C e^{-i\Phi}\langle\sigma_+\rangle - i\Omega\cos\delta_L t\langle\sigma_3\rangle \ , \\ \langle\dot{\sigma}_3\rangle &= -2\gamma_C\hat{N}_C\langle\sigma_3\rangle + 2i\Omega\cos\delta_L t(\langle\sigma_+\rangle - \langle\sigma_-\rangle) - \gamma_C \ , \end{aligned} \tag{8.52}$$

where Ω is the Rabi frequency, ϕ_L the phase, and δ_L half of the frequency difference of the bichromatic field, and as before, $N_C = \beta N, M_C = \beta M$, $\beta = C/(1+C)$, with $\gamma_C = \gamma(1+C)$ $\hat{N}_C = N_C + 1/2$, and $\Phi = 2\phi_L - \phi_s$. Equations (8.52) are formally identical to the optical Bloch equations in free space [167,168], but with modified parameters. The quantity γ_C is the cavity-enhanced spontaneous decay rate, and N_C and M_C are the effective values of the squeezing parameters seen by the atom in the cavity environment. As for monochromatic excitation, the cavity reduces the photon number and the degree of two–photon correlation of the injected squeezed vacuum by the same β-factor, so that the squeezing inside the cavity is no longer ideal: $M_C < \sqrt{N_C\,(N_C+1)}$.

We find that the anomalous spectral features at line center persist over a large range of δ_L and Ω values, which presents an advantage over monochromatic excitation [167,168]. For example, with the parameter values $C = 10$, $\Omega = 10\gamma$, $N = 0.005$, and $\Phi = 0$ the hole-burning profile with subnatural linewidth can occur over the range $33\gamma \le \delta_L \le 60\gamma$, and is most pronounced around $\delta_L \sim 45\gamma$. If $\Phi \ne 0$, the spectrum becomes asymmetric, and for small values of the phase, for instance $\Phi = \pi/10$, the spectrum becomes dispersive for $35\gamma \le \delta_L \le 50\gamma$. The dispersive profiles however are more fragile than the hole burning ones, and are much less pronounced in the cavity situation than in free space. The anomalous features disappear as Φ increases much more rapidly in the bichromatic case than in the monochromatic case.

As in the monochromatic case, the anomalous features are quantum mechanical in origin: they disappear if the ideal squeezed vacuum is replaced by the equivalent CSF.

Surprisingly, our numerical calculations show that the anomalous features reduce in magnitude if we increase the squeezed photon number. For $N = 0.05$, $\Phi = 0$, $\Omega = 10\gamma$, $\delta_L = 25\gamma$, the hole-burning profile in the case of $C = 3.2$ has almost disappeared, contrary to the monochromatic case where the larger the squeezed photon number, the more significant the anomalous features [27,35,37,61]. The small N situation is more readily achieved experimentally. For a degenerate optical parametric oscillator, $N = 0.005$ represents a comfortable working point of about 7% of threshold, while $N = 0.05$ is about 21%.

8.4 Finite Bandwidth Sources

We have previously assumed that the bandwidth of the squeezed vacuum is much larger than the other linewidths which arise in the model. However, the bandwidth obtained in experimentally realizable sources is typically of the order of a few atomic linewidths, and for a realistic description of the interaction between an atom and squeezed light, it is desirable to take account of the finite bandwidth of the squeezed source. Such studies have been conducted by Gardiner et al. [169], Parkins and Gardiner [170] and Ritsch and Zoller [155], using numerical stochastic methods, and applied to discussions of the narrowing of emission and absorption lines. It was verified that linewidth narrowing did occur, but was degraded by the effects of the finite bandwidth. It was later shown by Parkins [171,172] however, that for strong driving fields a finite bandwidth of squeezing can actually decrease the width of the Rabi sidebands of the Mollow triplet. Different effects were found depending upon whether the source of squeezing was a degenerate or nondegenerate parametric oscillator. In the former case, it is possible to narrow either the sidebands or the central peak, and in the latter case, it is possible to narrow all three peaks simultaneously.

The suppression of spontaneous emission by a squeezed vacuum in a cavity has been discussed by Cirac and Sanchez-Soto [57], who also reported that finite bandwidths could increase the suppression. Vyas and Singh [173] have considered resonance fluorescence with only a few modes squeezed, and have also examined the photon statistics. Parkins, Zoller and Carmichael [59] have discussed the finite bandwidth squeezed vacuum excitation of a vacuum Rabi resonance in a cavity, effectively forming a two-level system, and shown that the linewidth narrowing predicted by Gardiner occurs in the emission spectra.

Recently, Yeoman and Barnett [126] have derived an analytic master equation for the case of a two-level atom interacting with resonant squeezed and driving fields. Their analytic results agree with the numerical findings of Parkins. They also show that the width of the central peak of the spectrum depends solely on the squeezing present at the Rabi sideband frequencies. The Yeoman–Barnett approach has been used by Ficek et al. [129] in the cavity situation. They report that, within the secular approximation, the effect of squeezing is more evident in the linewidths of the Rabi sidebands than in the linewidth of the central component. With a two-mode cavity and a two-mode squeezed vacuum, they find that the signature of squeezing is present in all spectral linewidths, and that the narrowing is very sensitive to the detuning of the driving field from atomic resonance.

We continue with a brief investigation of the effects of finite bandwidths on the anomalous resonance fluorescence spectra, using the coupled systems approach of Gardiner and Parkins [131]. The lineshapes for $N(\omega)$ and $M(\omega)$ are given by (4.61). We restrict our attention to the bad cavity environment. If the linewidths are very narrow (say less than about 2γ, roughly), then the anomalous features disappear. However, the linewidths do not have to be very

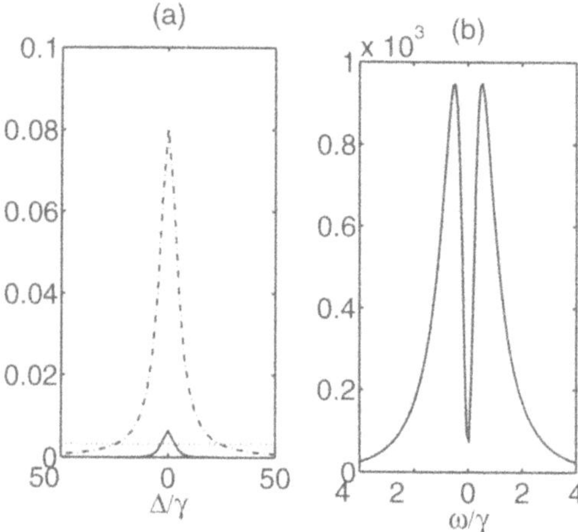

Fig. 8.13. (a) $N(\Delta)$ (*solid line*) and $M(\Delta)$ (*dashed line*) for $\kappa = 10\gamma$ and $\varepsilon = 0.2\gamma$. The horizontal dotted line is at one half the maximum height of the N curve. **(b)** The corresponding spectrum with $\Phi = 0$ and $\Omega = 0.17\gamma$

large for pronounced hole burning to occur in the $\Phi = 0$ case. Figure 8.13a shows the lineshapes for $\kappa = 10\gamma$ and $\varepsilon = 0.2\gamma$, which gives a halfwidth for the N curve of about 3γ. Notice that the value of N here is very small. The M curve is much higher, and broader. The hole burning in the corresponding spectrum (Fig. 8.13b) is very clear.

As expected, it is more difficult to observe the dispersive profiles. We need stronger squeezing, and a broader spectrum ($\sim 7\gamma$ for the N-width), as shown in Fig. 8.14a, where we take $\kappa = 40\gamma$ and $\varepsilon = 8\gamma$. Even then, the dispersive features are not very clear.

Finally, we consider the situation where the two-level atom is driven only by the squeezed vacuum, which can be of arbitrarily small bandwidth, again employing the Parkins–Gardiner approach. There is no monochromatic driving field here. We also examine the situation where, in place of the squeezed vacuum, the atom is driven by a field possessing two-photon correlations of arbitrary strength [174,175]. The single two-level atom is supposed to be driven by the output from a degenerate optical parametric oscillator, whose N and M fields are given by (4.61). The widths are determined by E, the pump amplitude (assumed real) for the optical parametric oscillator, and κ the total width of its cavity. Remarkable spectral profiles for the scattered light are obtained for squeezing bandwidths of the order of the natural atomic width and smaller, including the existence of a narrow hole at line-center. Under appropriate conditions, the spectrum may be both subnatural and narrower than the spectrum of the incident field, a phenomenon which does *not* depend upon strong coupling between the atom and the applied field. These features are illustrated in Fig. 8.15, where we take $E = \kappa/100$, and $\kappa = 1, 0.5, 0.15$ and 0.01 in frames (a)–(c) respectively. It is clear that complete quenching of

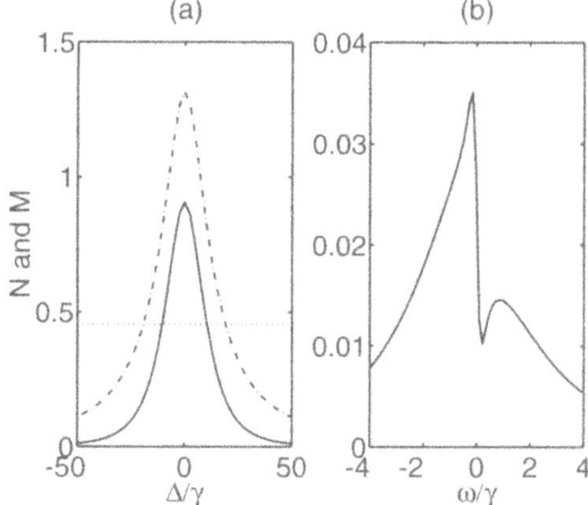

Fig. 8.14. **(a)** $N(\Delta)$ (*solid line*) and $M(\Delta)$ (*dashed line*) for $\kappa = 40\gamma$ and $\varepsilon = 8\gamma$. The horizontal *dotted line* is at one half the maximum height of the N curve. **(b)** The corresponding spectrum with $\Phi = \pi/2$ and $\Omega = 0.565\gamma$

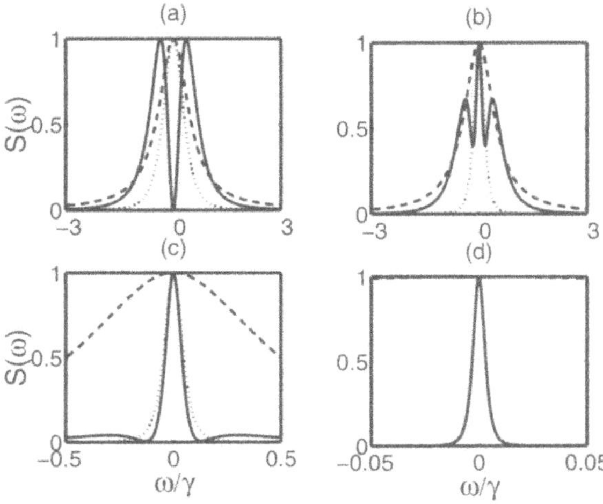

Fig. 8.15. Fluorescence spectrum emitted by a two-level atom driven by finite-bandwidth squeezed light from a single-ended ($\kappa_b = 0 \Rightarrow r = 1$) degenerate parametric oscillator. Parameters are $E = \kappa/100$ and $\lambda = 1$. The *solid curves* in **(a)**, **(b)**, **(c)** and **(d)** are for $\kappa/\gamma = 1, 0.5, 0.15, 0.01$, respectively. The *dashed curve* in each frame is a normalized Lorentzian of width γ, whilst the dotted curve is $N(\omega)/N(0)$ from (4.61)

the fluorescence can take place at line center for $\kappa = \gamma$. To explain the role the classical or non-classical nature of the driving field plays, they considered driving the atom with the alternative system which provides an arbitrary bandwidth light field with arbitrary two-photon correlations. In addition to numerical solutions, analytic solutions were obtained which were remarkably accurate for their domain of validity. These are discussed in detail in [175], and provide additional physical insight. Surprisingly, spectral narrowing does not require squeezing: narrowed spectra may appear for driving with a blackbody field. Then the atomic fluorescence can be expressed as the product of the atomic response with the incident spectrum. The resulting narrowing is maximized for incident bandwidths of the order of the natural width. However, for the appearance of the narrow hole at line-center, a non-classical driving field is essential. If the squeezing is ideal, the fluorescence at line-center exactly vanishes. The effects of experimentally pertinent factors such as the solid angle of coupling were discussed.

8.5 Systems of N_A Two-Level Atoms

The interaction of a collection of N_A two-level atoms with squeezed light is another topic which has received much attention. One of the striking features of such systems is that the atomic system may, under appropriate conditions, achieve a steady-state which is a pure state [52,78,82–85,176]. The discussion is much simpler when all the atoms are confined to a volume whose spatial dimensions are much less than the resonant wavelength of an atom – the so-called Dicke regime [177]. First we consider the simplest such system – one consisting of just two two-level atoms. We also restrict attention to the case where the center frequency of the squeezed vacuum is equal to the Bohr frequency of the two-level atom.

8.5.1 Two Two-Level Atoms in The Dicke Regime

In the limit of small interatomic separations [177], the two-atom Dicke model has close parallels with the three-level cascade system [48,81]. For example, the system exhibits a linear dependence on intensity of the two-photon transition rate, one-photon population inversions are possible, and the atomic system can achieve a pure steady-state. A distinction is that the system of two identical atoms cannot produce two-photon population inversions.

This system may be represented by the states [177–179]

$$\begin{aligned} |1\rangle &= |g_1\rangle\,|g_2\rangle\ , \\ |2\rangle &= \frac{1}{\sqrt{2}}\left(|e_1\rangle\,|g_2\rangle + |g_1\rangle\,|e_2\rangle\right)\ , \\ |3\rangle &= |e_1\rangle\,|e_2\rangle\ , \\ |4\rangle &= \frac{1}{\sqrt{2}}\left(|e_1\rangle\,|g_2\rangle - |g_1\rangle\,|e_2\rangle\right)\ . \end{aligned} \tag{8.53}$$

where $|g_i\rangle$ is the ground and $|e_i\rangle$ the excited state of the ith atom. It can be shown that if the state $|4\rangle$ is initially unpopulated, it always remains so. The system then has the same structure as the particular three-level cascade system in which $\alpha = \gamma_2/\gamma_1 = 1$, and one would expect it to show similar properties. Indeed, it has been shown [48,81] that the populations of the excited states $|2\rangle$ and $|3\rangle$ depend on the squeezing correlation strength M, and for an ideal squeezed vacuum, $M = \sqrt{N(N+1)}$, the state $|2\rangle$ has zero population in the steady-state. The atoms are then in a pure state, called the *pairwise atomic state* [52,78,82,83]. However, since the decay rates of the transitions $|3\rangle \to |2\rangle$ and $|2\rangle \to |1\rangle$ are necessarily equal in the two-atom Dicke model ($\alpha = 1$), complete inversion ($\rho_{33} = 1$) and two-photon population trapping are not possible in this system, in contrast to the three-level atom case.

The concept of pairwise atomic states applies not only to a pair of two-level atoms, but in general to a system with an even number of atoms. It was shown by Agarwal and Puri [52] that the N_A – atom Dicke system interacting with a squeezed vacuum decays to a state whose density operator is given by

$$\rho = \mathcal{N}_o\left(\mu S^- + \nu S^+\right)^{-1}\left(\mu S^+ + \nu S^-\right)^{-1} \quad \text{if} \quad N_A \quad \text{is even}\ , \tag{8.54}$$

and

$$\rho = |\psi_0\rangle\langle\psi_0| \quad \text{if} \quad N_A \quad \text{is odd}\ , \tag{8.55}$$

where the $S^{\pm} \equiv \Sigma_{i=1}^{N_A}\sigma_i^{\pm} = \Sigma_{i=1}^{N_A}|e_i\rangle\langle g_i|$ are the collective atomic operators, $\mathcal{N}_o$ is a normalization constant such that $\mathrm{Tr}(\rho) = 1$, $\mu = \sqrt{N+1}, \nu = \sqrt{N}$, and $|\psi_0\rangle$ is defined by

$$\left(\mu S^- + \nu S^+\right)|\psi_0\rangle = 0\ . \tag{8.56}$$

Thus, for even number of atoms the stationary state is the pairwise atomic state:

$$|\psi_0\rangle = \mathcal{N}\exp\left(\theta S_z\right)\exp\left(-i\pi S_y/2\right)|0\rangle\ , \tag{8.57}$$

where the (S_x, S_y, S_z) are the angular momentum operators corresponding to a spin of value $N_A/2$, $|0\rangle$ is the eigenvector of S_z corresponding to the eigenvalue $m = 0$, $\mathcal{N}$ is a normalization factor and $\theta = \ln\left(M/\sqrt{N+1/2}\right)$.

Agarwal and Puri have generalized these expressions to the situation where an external driving is also present [52].

8.5.2 Effect of Finite Separations

We return to the study of just two two-level atoms. The pairwise atomic states are characterized by zero stationary population of the state $|2\rangle$, which may be monitored by observing the intensity of the fluorescence from this state. However, it is difficult experimentally to fulfill the requirement that interatomic separations be much smaller than the resonant wavelength: in fact, present atom trapping and cooling techniques can trap two atoms only within distances of the order of a resonant wavelength [180]. It is therefore of interest to examine the effect of increasing the interatomic separation so that the Dicke model eventually ceases to be valid. With a finite interatomic separation, it is apparent that the simple three-state representation of two atoms cannot be applied, and the two-atom system must be represented by the full four-level system of (8.53).

The additional asymmetric intermediate state $|4\rangle$ is known as a *trapping state* and can show a large stationary population, independent of the value of M. This may destroy the purity of the stationary state [82,84].

To obtain the stationary populations, we need the master equation for a two-atom system driven by a squeezed vacuum field [81,179]. It has the form

$$\begin{aligned}\frac{d\rho}{dt} &= -\frac{i}{\hbar}\left[H_\sigma,\rho\right] - \zeta\sum_{i,j}\gamma_{ij}m_i\left(\left[\sigma_i^+\rho,\sigma_j^+\right] + \left[\sigma_j^+,\rho\sigma_i^+\right] + \text{H.c.}\right)\\ &\quad -\zeta\sum_{i,j}\gamma_{ij}N_i\left(\rho\sigma_j^-\sigma_i^+ + \sigma_j^-\sigma_i^+\rho - 2\sigma_i^+\rho\sigma_j^-\right)\\ &\quad -\sum_{i,j}\gamma_{ij}\left(\zeta N_i + 1\right)\left(\rho\sigma_i^+\sigma_j^- + \sigma_i^+\sigma_j^-\rho - 2\sigma_j^-\rho\sigma_i^+\right)\ ,\end{aligned} \tag{8.58}$$

where $m_i = M(\omega_i)\exp(-2i\omega_s t + i\phi_S)$ and $N_i = N(\omega_i)$ are squeezing parameters, which depend upon the lineshape of the squeezing, (4.61) and (8.7), $\sigma_i^+ = |e_i\rangle\langle g_i|$ is the atomic raising operator for the i-th atom, $\sigma_i^- = \left(\sigma_i^+\right)^\dagger$, $\gamma_{11} = \gamma_{22} = \gamma$ is the spontaneous decay rate for a single atom, and $\gamma_{12} = \gamma_{21}$ is the collective damping rate defined as [178,179]

$$\begin{aligned}\gamma_{12} &= \frac{3}{2}\gamma\left\{\left[1-(\hat{\boldsymbol{\mu}}\cdot\hat{\boldsymbol{r}})^2\right]\frac{\sin(kr)}{kr}\right.\\ &\quad\left.+\left[1-3(\hat{\boldsymbol{\mu}}\cdot\hat{\boldsymbol{r}})^2\right]\left[\frac{\cos(kr)}{(kr)^2}-\frac{\sin(kr)}{(kr)^3}\right]\right\}\ ,\end{aligned} \tag{8.59}$$

where $k = 2\pi/\lambda_0$, r is the interatomic separation, $\hat{\boldsymbol{\mu}}$ and $\hat{\boldsymbol{r}}$ are unit vectors along the transition dipole moment and the interatomic axis, respectively, and λ_0 is the resonant wavelength. The system Hamiltonian for two atoms is given by [178,179]

$$H_\sigma = \sum_{i=1}^{2}\omega_i\left[\sigma_i^+\sigma_i^-,\rho\right] + i\sum_{i\neq j}^{2}\Omega_{ij}\left[\sigma_i^+\sigma_j^-,\rho\right]\ , \tag{8.60}$$

where ω_i is the transition frequency of the ith atom, and

$$\begin{aligned}\Omega_{ij} &= \frac{3}{2}\gamma\left\{-\left[1-(\hat{\boldsymbol{\mu}}\cdot\hat{\boldsymbol{r}})^2\right]\frac{\cos(kr)}{kr}\right.\\ &\quad\left.+\left[1-3(\hat{\boldsymbol{\mu}}\cdot\hat{\boldsymbol{r}})^2\right]\left[\frac{\sin(kr)}{(kr)^2}+\frac{\cos(kr)}{(kr)^3}\right]\right\}\ ,\end{aligned} \tag{8.61}$$

is the collective shift of the atomic levels due to their dipole-dipole interaction.

The parameters (8.59) and (8.61), which both depend on the interatomic separation, determine the collective properties of the two-atom system. For $kr \ll 1$, we see that γ_{12} reduces to γ, and Ω_{ij} reduces to the static dipole-dipole interaction:

$$\begin{aligned} \gamma_{12} &\approx \gamma , \\ \Omega_{ij} &\approx \frac{3\gamma}{2\,(kr)^3}\left[1 - 3\,(\hat{\boldsymbol{\mu}}\cdot\hat{\boldsymbol{r}})^2\right] . \end{aligned} \tag{8.62}$$

From the master equation, we may obtain the steady state atomic populations, assuming perfect matching ($\zeta = 1$) of the squeezed modes to the vacuum modes coupled to the atoms. It is found that for small interatomic separations, the population of the antisymmetric state $|4\rangle$ does not change in time. Thus an initially unpopulated antisymmetric state remains unpopulated for all times, and the population is distributed only between the three Dicke states. For larger separations, $\gamma_{12} \neq \gamma$, and the antisymmetric state is coupled to the three Dicke states. In this case the steady-state populations of the excited collective states are:

$$\begin{aligned} \rho_{22} &= \frac{(n^2-1)}{4n^2} - \frac{aM^2\,(2n^2-a)}{D} , \\ \rho_{33} &= \frac{(n-1)^2}{4n^2} + \frac{a^2M^2\,(2n-1)}{D} , \\ \rho_{44} &= \frac{(n^2-1)}{4n^2} + \frac{aM^2\,(2n^2+a)}{D} , \end{aligned} \tag{8.63}$$

where

$$\begin{aligned} n &= 1 + 2N , \qquad a = \gamma_{12}/\gamma , \\ D &= n^2\left[n^4 + 4M^2\,(a^2 - n^2)\right] . \end{aligned} \tag{8.64}$$

In Fig. 8.16a, we plot the steady-state populations as a function of the interatomic separation for $M^2 = N\,(N+1)$ with $N = 0.1$ and $\hat{\boldsymbol{\mu}}$ perpendicular to $\hat{\boldsymbol{r}}$. It is clear that for small interatomic separations the antisymmetric state is the most populated state of the system, and the symmetric state remains unpopulated. This leads to the destruction of the purity of the stationary state of the system. The purity is measured by the quantity [82]

$$\Sigma = \mathrm{Tr}\left(\rho^2\right) = \rho_{11}^2 + \rho_{22}^2 + \rho_{33}^2 + \rho_{44}^2 + 2\,|\rho_u|^2 , \tag{8.65}$$

where

$$\begin{aligned} \rho_u &= \frac{2an^3M}{D} , \\ \rho_{11} &= 1 - \rho_{22} - \rho_{33} - \rho_{44} . \end{aligned} \tag{8.66}$$

In Fig. 8.16b, we plot Σ as a function of the interatomic separation for an ideal squeezed vacuum and different N. It is evident that the stationary state of the system is always a mixed state.

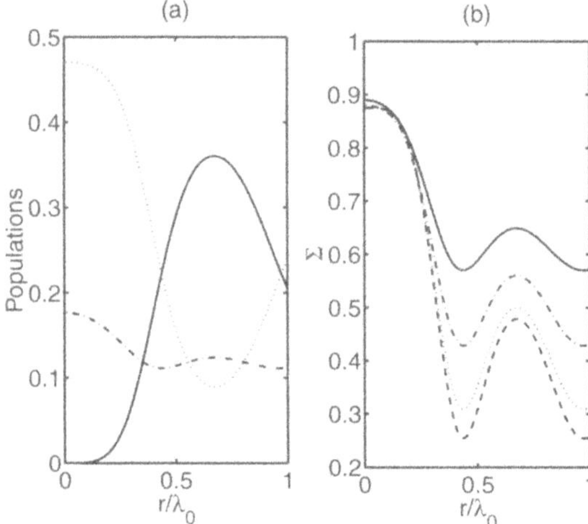

Fig. 8.16. **(a)** The populations of state $|2\rangle$ (*solid line*), state $|3\rangle$ (*dot-dash line*) and state $|4\rangle$ (*dotted line*) for $\hat{\boldsymbol{\mu}} \cdot \hat{\boldsymbol{r}} = 0$ and $N = 1$ against r/λ_0. **(b)** The purity Σ against r/λ_0 for $\hat{\boldsymbol{\mu}} \cdot \hat{\boldsymbol{r}} = 0$ and $N = 0.1$ (*solid line*), 0.2 (*dot-dash line*), 0.5 (*dotted line*) and 10 (*lower solid line*)

It is still possible, however, to create the pure pairwise atomic state for the two-atom system with the interatomic separation included, provided the observation time is order of γ^{-1}. The antisymmetric state $|4\rangle$ decays on a time scale $\sim (\gamma - \gamma_{12})^{-1}$, and for $\gamma_{12} \approx \gamma$ the decay rate of the antisymmetric state is much longer than γ^{-1}. By contrast, the state $|2\rangle$ decays on a time scale $\sim (\gamma + \gamma_{12})^{-1}$, which for $\gamma_{12} \approx \gamma$ is shorter than γ^{-1}. For observation times of the order γ^{-1}, the antisymmetric state does not participate in the interaction and only the states $|1\rangle$ and $|3\rangle$ are populated. The quasi steady-state is the pairwise atomic state.

8.5.3 Two Non-Identical Atoms

By employing two non-identical atoms of significantly different transition frequencies, driven by a two-mode squeezed vacuum field [85], it is possible to achieve the pure pairwise atomic state with the interatomic separation comparable to the resonant wavelength, and the antisymmetric state fully participating in the interaction. Assuming that $|\omega_1 - \omega_2| \gg \gamma$ and $\omega_S = (\omega_1 + \omega_2)/2$, the following steady-state populations are obtained:

$$\rho_{33} = \frac{1}{4}\left[\frac{(n-2)}{n} - \frac{1}{(n^2 - 4a^2M^2)}\right] ,$$

$$\rho_{22} = \rho_{44} = \frac{1}{4}\left[1 - \frac{1}{(n^2 - 4a^2M^2)}\right] . \tag{8.67}$$

For $a = 1$, we see that $\rho_{22} = \rho_{44} \approx 0$. This shows that for a system of two different atoms, the antisymmetric state is no longer a trapping state, and the system may attain the pairwise atomic state despite the presence of the antisymmetric state.

Acknowledgments

This work was supported by the EPSRC. We gratefully acknowledge helpful conversations with S.M. Barnett, C. Cabrillo, B.J. Dalton, Z. Ficek, T.A.B. Kennedy, P.L. Knight, W.S. Smyth and P. Zhou.

References

1. G.J. Milburn, Optica Acta **31**, 671 (1984)
2. C.W. Gardiner, Phys. Rev. Lett. **56**, 1917 (1986)
3. C.M. Savage and D.F. Walls, Phys. Rev. Lett. **57**, 2164 (1986)
4. G.J. Milburn, Phys. Rev. A **34**, 4882 (1986)
5. H.J. Carmichael, A.S. Lane and D.F. Walls, J. Mod. Opt. **34**, 821 (1987)
6. H.J. Carmichael, A.S. Lane and D.F. Walls, Phys. Rev. Lett. **58**, 2539 (1987)
7. J. Jansky and Y. Yushin, Phys. Rev. A **35**, 1288 (1987)
8. W.P. Zhang and W.H. Tan, Opt. Commun. **69**, 128 (1988)
9. A. Joshi and S.V. Lawande, Phys. Rev. A **41**, 2822 (1990)
10. J.I. Cirac and L.L. Sanchez-Soto, Opt. Commun. **77**, 26 (1990)
11. B.N. Jagatap, S.V. Lawande and Q.V. Lawande, Phys. Rev. A **43**, 6316 (1991)
12. A. Joshi and R.R. Puri, Phys. Rev. A **43**, 6428 (1991)
13. A.D. Gasasyan and B.G. Sherman, Opt. Spectrosc. **70**, 398 (1991)
14. B.N. Jagatap and S.V. Lawande, Phys. Rev. A **44**, 6030 (1991)
15. S. Smart and S. Swain, Phys. Rev. A **45**, 6863 (1992)
16. S. Smart and S. Swain, Phys. Rev. A **45**, 6857 (1992)
17. H. Moya-Cessa and A. Vidiella-Barranco, J. Mod. Opt. **39**, 2481 (1992)
18. A.D. Wilson-Gordon, V. Buzek and P.L. Knight, Czech. J. Phys. **42**, 975 (1992)
19. S. Smart and S. Swain, J. Mod. Opt. **40**, 1939 (1993)
20. N.H. Moin and M.R.B. Wahiddin, Opt. Commun. **100**, 105 (1993)
21. S. Smart and S. Swain, Phys. Rev. A **48**, R48 (1993)
22. I.E. Lyublinskay and R. Vyas, Phys. Rev. A **48**, 3966 (1993)
23. Z. Ficek and B.C. Sanders, J. Phys. B **27**, 809 (1994)
24. S. Swain, Phys. Rev. Lett. **73**, 1493 (1994)
25. C. Cabrillo, W.S. Smyth, S. Swain and P. Zhou, Opt. Commun. **114**, 344 (1995)
26. A. Banerjee, Phys. Rev. A **52**, 2472 (1995)
27. S. Swain and P. Zhou, Phys. Rev. A **52**, 4845 (1995)
28. S. Swain and P. Zhou, Opt. Commun. **123**, 310 (1996)
29. H. Ritsch and P. Zoller, Opt. Commun. **64**, 523 (1987)
30. S. An, M. Sargent and D.F. Walls, Opt. Commun. **67**, 373 (1988)
31. Z. Ficek and B.J. Dalton, Opt. Commun. **102**, 231 (1993)
32. Z. Ficek, W.S. Smyth and S. Swain, Opt. Commun. **110**, 555 (1994)
33. S.S. Hassan, O.M. Frege and N. Nayak, J. Opt. Soc. Am. B **12**, 1177 (1995)
34. Z. Ficek, W.S. Smyth and S. Swain, Phys. Rev. A **52**, 4126 (1995)

35. P. Zhou, Z. Ficek and S. Swain, J. Opt. Soc. Am. B **13**, 768 (1996)
36. P. Zhou and S. Swain, Quantum Semiclass. Opt. **8**, 959 (1996)
37. P. Zhou and S. Swain, Opt. Commun. **131**, 153 (1996)
38. P. Zhou and S Swain, Phys. Rev. A **55**, 772 (1997)
39. S.F. Haas and M. Sargent, Opt. Commun. **79**, 366 (1990)
40. S. Singh, J. Rai, C.M. Bowden and A. Postan, Phys. Rev. A **45**, 5160 (1992)
41. V. Buzek, Phys. Lett. A **139**, 231 (1989)
42. K.A. Rustamov, E.I. Aliskenderov, T.D. Ho and A.S. Shumovsky, Physica A **158**, 649 (1989)
43. T. Maqbool and M.S.K. Razmi, Phys. Rev. A **44**, 6147 (1991)
44. Y. Ben-Aryeh, C.A. Miller, H. Risken and W. Schleich, Opt. Commun. **90**, 259 (1992)
45. D. Cohen, Y. Ben-Aryeh and A. Mann, Phys. Rev. A **49**, 2040 (1994)
46. H.T. Dung, EI. Aliskenderov and L. Knoll, J. Mod. Opt. **42**, 1069 (1995)
47. M.V. Satyanarayana, P. Rice, R. Vyas and H.J. Carmichael, J. Opt. Soc. Am. B **6**, 228 (1989)
48. G.M. Palma and P.L. Knight, Phys. Rev. A **39**, 1962 (1989)
49. M.L. Zou and G.C. Guo, J. Phys. B. **22**, 2205 (1989)
50. J.M. Courty and S. Reynaud, Europhys. Lett. **10**, 237 (1989)
51. J.I. Cirac and L.L. Sanchez-Soto, Phys. Rev. A **40**, 3743 (1989)
52. G.S. Agarwal and R.R. Puri, Phys. Rev. A **41**, 3782 (1990)
53. A.S. Parkins and R. Muller, J. Mod. Opt. **43**, 2554 (1996)
54. A.S. Parkins, Phys. Rev. A **53**, 2893 (1996)
55. J. Gea-Banacloche, R.R. Schlicher and M.S. Zubairy, Phys. Rev. A **38**, 3514 (1988)
56. A.S. Parkins and C.W. Gardiner, Phys. Rev. A **40**, 3796 (1989)
57. J.I. Cirac and L.L. Sanchez-Soto, Phys. Rev. A **44**, 1948 (1991)
58. J.I. Cirac, Phys. Rev. A **46**, 4354 (1992)
59. A.S. Parkins, P. Zoller and H.J. Carmichael, Phys. Rev. A **48**, 758 (1993)
60. Z. Ficek and P.D. Drummond, Europhys. Lett. **24**, 455 (1993)
61. W.S. Smyth and S. Swain, Phys. Rev. A **53**, 2846 (1996)
62. J. Gea-Banacloche, Phys. Rev. Lett. **62**, 1603 (1989)
63. J. Javanainen and P.L. Gould, Phys. Rev. A **41**, 5088 (1990)
64. V. Buzek, P.L. Knight and I.K. Kudryavtsev, Phys. Rev. A **44**, 1931 (1991)
65. Z. Ficek and P.D. Drummond, Phys. Rev. A **43**, 6258 (1991)
66. Z. Ficek and P.D. Drummond, Phys. Rev. A **43**, 6247 (1991)
67. N.P. Georgiades, E.S. Polzik, K. Edamatsu, H.J. Kimble and A.S. Parkins, Phys. Rev. Lett. **75**, 3426 (1995)
68. M.R. Ferguson, Z. Ficek and B.J. Dalton, J. Mod. Opt. **42**, 679 (1995)
69. P. Zhou and S. Swain, Phys. Rev. A **54**, 2455 (1996)
70. Q.V. Lawande, B.N. Jagatap and S.V. Lawande, Phys. Rev. A **43**, 535 (1991)
71. A. Joshi and R.R. Puri, Phys. Rev. A **45**, 2025 (1992)
72. S. Smart and S. Swain, Quantum Opt. **5**, 75 (1993)
73. S. Smart and S. Swain, J. Mod. Opt. **41**, 1055 (1994)
74. M.R. Ferguson, Z. Ficek and B.J. Dalton, Phys. Rev. A **54**, 2379 (1996)
75. P. Zhou and S. Swain, Opt. Commun. **123**, 297 (1996)
76. M.R. Ferguson, Z. Ficek and B.J. Dalton, Phys. Rev. A **56**, 4125 (1997)
77. S. Clark and S. Parkins, J. Opt. B: Quantum and Semiclass. Opt. **5**, 145 (2003)

78. S.M. Barnett and M.A. Dupertuis, J. Opt. Soc. Am. B **4**, 505 (1987)
79. G.M. Palma and P.L. Knight, Opt. Commun. **73**, 131 (1989)
80. G.M. Palma and P.L. Knight, Phys. Rev. A **42**, 1829 (1990)
81. Z. Ficek, Phys. Rev. A **42**, 611 (1990)
82. Z. Ficek, Phys. Rev. A **44**, 7759 (1991)
83. M.R. Muhamad, S.K. Ng and M.R.B. Wahiddin, Phys. Rev. A **50**, 1939 (1997)
84. Z. Ficek, Opt. Commun. **82**, 130 (1991)
85. Z. Ficek and M.R.B. Wahiddin, Opt. Commun. **134**, 387 (1997)
86. C.J. Hood, H.J. Kimble, Q.A. Turchette, N.Ph. Georgiades and A.S. Parkins, Phys. Rev. A **58**, 4056 (1998)
87. L.A. Wu M. Xiao and H.J. Kimble, Phys. Rev. Lett. **59**, 278 (1987)
88. E.S. Polzik, J. Carri and H.J. Kimble, Phys. Rev. Lett. **68**, 3020 (1992)
89. E.S. Polzik, J. Carri and H.J. Kimble, App. Phys. B **55**, 279 (1992)
90. S. Lathi, S. Kasapi and Y. Yamamoto, Opt. Lett. **22**, 478 (1997)
91. B. Yurke, P. Grangier, E.R. Slusher and A. LaPorta, Phys. Rev. Lett. **59**, 2153 (1987)
92. C.M. Caves, Phys. Rev. D **23**, 1693 (1981)
93. M.D. Reid, M.D. Levenson, R.M. Shelby and D.F. Walls, Phys. Rev. Lett. **57**, 2473 (1986)
94. Y. Shevy, Phys. Rev. Lett. **64**, 2905 (1990)
95. R. Graham, D.F. Walls and W.P. Zhang, Phys. Rev. A **44**, 7777 (1991)
96. J.I. Cirac and P. Zoller, Phys. Rev. A **47**, 2191 (1993)
97. J. Gea-Banacloche, Phys. Rev. Lett. **59**, 543 (1987)
98. H. Ritsch, M.A.M. Marte and D.F. Walls, Phys. Rev. A **38**, 3577 (1988)
99. E.S. Polzik, N.Ph. Georgiades and H.J. Kimble, Phys. Rev. A **55**, R1605 (1997)
100. O. Alter and Y. Yamamoto, Phys. Rev. A **53**, 2911 (1996)
101. S. Schiller, R. Bruckmeier, K. Schneider and J. Mlynek, Phys. Rev. Lett. **78**, 1243 (1997)
102. M. Xiao, Y.Q. Li, P. Lynam and P.J. Edwards, Phys. Rev. Lett. **78**, 3105 (1997)
103. Y.M. Golobev and M.I. Kolobov, Phys. Rev. Lett. **79**, 399 (1997)
104. H. Saito and M. Ueda, Phys. Rev. Lett. **79**, 3869 (1997)
105. K. Molmer, A. Kuzmich and E.S. Polzik, Phys. Rev. Lett. **79**, 4782 (1997)
106. D.F. Walls, Nature **306**, 141 (1983)
107. L. Mandel, Phys. Script. T**12**, 42 (1986)
108. D.F. Walls, Nature **324**, 210 (1986)
109. R. Loudon and P.L. Knight, J. Mod. Opt. **34**, 709 (1987)
110. R.E. Slusher and B. Yurke, Scientific American **258**, 50 (1988)
111. R.E. Slusher and B. Yurke, J. Lightwave Tech. **8**, 466 (1990)
112. D.F. Walls, Science Progress **74**, 291 (1990)
113. K. Zaheer and M.S. Zubairy, Adv. At. Mol. Opt. Phys. **28**, 143 (1990)
114. V.P. Bykov, Sov. Phys. Usp. **34**, 910 (1991)
115. H.J. Kimble, Phys. Rep. **219**, (1992)
116. C. Fabre, Phys. Rep. **219**, (1992)
117. Y. Ben-Aryeh, Laser Phys. **4**, 733 (1994)
118. C.M. Caves and P.D. Drummond, Rev. Mod. Phys. **66**, 481 (1994)
119. H.J. Kimble, O. Carnal, Z. Hu, H. Mabuchi, E.S. Polzik, R.J. Thompson and Q.A. Turchette, Ann. New York Acad. Sci. **755**, 87 (1995)

120. U. Leonhardt and H. Paul, Prog. Quant. Elect. **19**, 89 (1995)
121. A. Gatti, A. Marzoli, G.L. Oppo, L.A. Lugiato, S.M. Barnett and H. Wiedemann, J. Nonlin. Opt. Phys. and Mat. **5**, 809 (1995)
122. S.E. Bialkowski, Crit. Rev. Anal. Chem. **26**, 101 (1996)
123. P. Hariharan and B.C. Sanders, Prog. Opt. **36**, 49 (1996)
124. D.N. Klyshko, Sov. Phys. Usp. **39**, 573 (1996)
125. A. Gatti, L.A. Lugiato and H. Wiedemann, in *Quantum fluctuations*, eds. E. Giacobino, S. Reynaud and J. Zinn-Justin, (Elsevier, Amsterdam 1997), p. 431
126. G. Yeoman and S.M. Barnett, J. Mod. Opt. **43**, 2037 (1996)
127. B.J. Dalton, Z. Ficek and S. Swain, J. Mod. Opt. **46**, 379 (1999)
128. D.F. Walls and G.J. Milburn, *Quantum Optics*, (Springer, Berlin Heidelberg New York 1994)
129. Z. Ficek, B.J. Dalton and M.R.B. Wahiddin, J. Mod. Opt. **44**, 1005 (1997)
130. C.W. Gardiner, Phys. Rev. Lett. **70**, 2269 (1993)
131. C.W. Gardiner and A.S. Parkins, Phys. Rev. A **50**, 1792 (1994)
132. H.J. Carmichael, *An Open Systems Approach to Quantum Optics*, (Springer, Berlin Heidelberg New York 1992)
133. M. Lax, Phys. Rev. **129**, 2342 (1963)
134. S. Swain, J. Phys. A **14**, 2577 (1981)
135. S. Swain and B.J. Dalton, Opt. Commun. **147**, 187 (1997)
136. T.I. Sachse, G.C. Hegerfeldt and D.G. Sondermann, Quantum Semiclass. Opt. **9**, 961 (1997)
137. B.R. Mollow, Phys. Rev. **188**, 1969 (1969)
138. T. Quang, M. Kozierowski and L.H. Lan, Phys. Rev. A **39**, 644 (1989)
139. P. Zhou and S. Swain, Phys. Rev. Lett. **89**, (1999)
140. P.R. Rice and H.J. Carmichael, J. Opt. Soc. Am. B **5**, 1661 (1988)
141. W.S. Smyth and S. Swain, Opt. Commun. **112**, 91 (1994)
142. R.R. Tucci, Int. J. Mod. Phys. **5**, 1457 (1991)
143. P. Zhou and S. Swain, Phys. Rev. A **59**, (1999)
144. D.F. Walls and P. Zoller, Phys. Rev. Lett. **47**, 709 (1981)
145. R. Tanas, Z. Ficek and S. Kielich, Phys. Rev. A **29**, 2004 (1984)
146. K. Wodkiewicz and J.H. Eberly, J. Opt. Soc. Am. B **2**, 458 (1985)
147. P.K. Aravind, J. Opt. Soc. Am. B **3**, 1712 (1986)
148. S.M. Barnett and P.L. Knight, Phys. Scr. **21**, 5 (1988)
149. Z. Ficek and S. Swain, J. Opt. Soc. Am. B **14**, 258 (1997)
150. M.J. Collett and D.F. Walls, Phys. Rev. A **32**, 2887 (1985)
151. S. Bali, Z.H. Lu and J.E. Thomas, Phys. Rev. Lett. **81**, 3635 (1998)
152. P. Zhou and S. Swain, Phys. Rev. A **51**, 841 (1999)
153. H. Ritsch and P. Zoller, Opt. Commun. **66**, 333 (1988)
154. H. Ritsch and P. Zoller, Phys. Rev. Lett. **61**, 1097 (1988)
155. H. Ritsch and P. Zoller, Phys. Rev. A **38**, 4657 (1988)
156. S. An and M. Sargent, Phys. Rev. A **39**, 3998 (1989)
157. U. Akram, K.T. Lim, M.R. Muhamad, M.R.B. Wahiddin and Z. Ficek, Phys. Rev. A **57**, 2072 (1998)
158. M.R.B. Wahiddin, U. Akram and Z. Ficek, Phys. Lett. A **238**, 117 (1998)
159. C. Cohen-Tannoudji and S. Reynaud, J. Phys. B **10**, 345 (1977)
160. C. Cohen-Tannoudji and S. Reynaud, in *Multiphoton Processes* edited by J.H. Eberly and P. Lambropoulos, (Wiley, New York 1978), p. 103

161. C. Cabrillo and S. Swain, Phys. Rev. Lett. **77**, 478 (1996)
162. P.R. Rice and L.M. Pedrotti, J. Opt. Soc. Am. B **9**, 2008 (1992)
163. P. Mataloni, L. Crescentini, F. De Martini, M. Marrocco and R. Loudon, Phys. Rev. A **42**, 2480 (1991)
164. M. Born and E. Wolf, *Principles of Optics*, (MacMillan, New York 1964)
165. P.W. Milonni and P.L. Knight, Opt. Commun. **9**, 119 (1973)
166. Z. Ficek, W.S. Smyth, S. Swain and M. Scott, Phys. Rev. A **57**, 585 (1998)
167. P. Zhou, S. Swain and Z. Ficek, Phys. Rev. A **55**, 2340 (1997)
168. Z. Ficek, P. Zhou and S. Swain, Opt. Commun. **148**, 159 (1997)
169. A.S. Parkins, C.W. Gardiner and M.J. Collett, J. Opt. Soc. Am. B **4**, 1683 (1987)
170. A.S. Parkins and C.W. Gardiner, Phys. Rev. A **37**, 3867 (1988)
171. A.S. Parkins, Phys. Rev. A **42**, 6873 (1990)
172. A.S. Parkins, Phys. Rev. A **42**, 4352 (1990)
173. R. Vyas and S. Singh, Phys. Rev. A **45**, (1992)
174. W.S. Smyth and S. Swain, Phys. Rev. A **60**, (1999)
175. W.S. Smyth and S. Swain, J. Mod. Opt. **46**, (1999)
176. Z. Ficek, Opt. Commun. **88**, (1992)
177. R.H. Dicke, Phys. Rev. **93**, 99 (1954)
178. R.H. Lehmberg, Phys. Rev. A **2**, 883 (1970)
179. Z. Ficek and R. Tanaś, Phys. Rep. **372**, 369 (2002)
180. R.G. DeVoe and R.G. Brewer, Phys. Rev. Lett. **76**, 2049 (1996)

9 Spectroscopy with Three-Level Atoms in a Squeezed Field

Z. Ficek

Highly stabilized sources of light generate electromagnetic fields which, even in a vacuum state, exhibit inherent and unavoidable small fluctuations. These fluctuations are often unobservable, however, they are always present and are the source of what we call quantum noise. The fluctuations arise from the quantum nature of electromagnetic field, and it has the long been thought that they were an insuperable barrier to accuracy. In the last two decades theoretical studies followed by experimental measurements have shown that this limitation could be circumvented. The search for light fields with reduced or even completely suppressed fluctuations has become a new subject for physicists to study [1,2]. The possibility of obtaining light fields whose fluctuations are less than those naively expected from the laws of quantum mechanics allows researchers to perform experiments with greater precision than possible with laser light.

There have been a few successful applications of light fields with reduced "squeezed" fluctuations in the optical measurements [3] and in atomic spectroscopy [4,5]. Further experimental applications of squeezed light include quantum teleportation of optical coherent states [6], generation of spin squeezed atomic states [7], and fourth-order bi-photon polarization interferometry [8]. Other applications of squeezed light have been suggested including ultra-sensitive measurements in physics and new phenomena in laser physics, or even technical uses in optical telecommunication and computing. One of the most encouraging applications of squeezed fields, on which the present theoretical and experimental research is focused, is the field of atomic spectroscopy [9,10]. This interest stems mostly from the fact that radiative properties of atoms are intimately related to fluctuations in the electromagnetic field. Moreover, the study of squeezing effects in atomic spectroscopy is of undoubted interest as it can lead to novel physical effects not obtainable with conventional radiation sources and also provides the further possibility of verifying directly the predictions of quantum electrodynamics.

Radiative properties of atoms depend on the state of the reservoir to which they are coupled [11]. The fluctuations of the reservoir field are determined by the two quadrature components of the electromagnetic field. If the fluctuations of a given component of the reservoir field are squeezed, the relaxation of the dipole moment of an atom may be modified. Gardiner [12] first pointed out that in a squeezed reservoir the relaxation of one of the two components of the atomic dipole moment can be reduced or even completely suppressed. This modification can lead to subnatural linewidths in the fluorescence and absorption spectra [13–28]. This effect could be used in laser spectroscopy to improve the resolution of spectral lines.

The narrowing of spectral lines is one of the large number of examples of unusual effects predicted in spectroscopy with squeezed light. Many of those effects are characteristic of multilevel systems and do not appear in a simple two-level system. Three-level atoms are such an example, and in this chapter we describe the major effects resulting from the interaction of a three-level cascade atom with a squeezed vacuum field. In Sect. 9.1, we present a master equation technique used for analyzing the interaction of squeezed light with atoms, which is applicable only for Markovian squeezed reservoirs. Next, in Sect. 9.2, we review squeezing induced phenomena with a consideration of the most fundamental three-level atom processes. In particular, we consider transient as well as stationary radiative properties of single three-level atoms such as quantum beats, modified atomic decay rates and an altered atomic population distribution. In the field of atomic population distribution, a squeezed vacuum field can drive atomic populations in ways not possible with classical fields. Examples include a violation of a Boltzmann distribution of the population as well as two-photon inversions.

In contrast to the significant theoretical advances in this area, experimental work has proven to be extremely difficult with only one experiment so far demonstrating a nonclassical effect in spectroscopy with squeezed light [4]. The experiment has demonstrated a departure from the quadratic intensity dependence, characteristic of classical light sources, of the population of the upper state of a three-level cascade atom [29–34]. The observed departure is an evidence of the quantum nature of the excitation process. However, additional experimental attempts to observe nonclassical effects in quantum interference [5], and spontaneous decay of the atomic polarization [35] have not convincingly demonstrated that these effects arise from the quantum nature of the squeezed field used in the experiments. The phase-dependent absorption spectra and quantum interference, observed in the experiments, can be produced by classically correlated fields [36,37].

The dearth of experimental observations of the nonclassical effects is attributed to two principal difficulties: (1) Most of the predicted nonclassical effects appears for small intensities of the squeezed field and in fact they become more pronounced as the intensity decreases. For a weak excitation field the signal to be observed is also very weak, which makes it difficult to collect an efficient amount of the scattered radiation from the interaction. (2) The predicted effects are very sensitive to the coupling efficiency η of the squeezed field with the atom. For example, the atomic decay rate γ can be altered in the squeezed vacuum as [12,13]

$$\gamma = \gamma_0 \left[1 + \eta \left(: \Delta\hat{x}^\theta :\right)^2\right] , \tag{9.1}$$

where $\left(: \Delta\hat{x}^\theta :\right)^2$ is the normally ordered variance of the squeezed field quadrature $\hat{x}^\theta = \hat{a}\exp(-i\theta) + \hat{a}^\dagger\exp(i\theta)$, and γ_0 is the atomic decay rate in the ordinary vacuum. Since $\left(: \Delta x^\theta :\right)^2 \geq -1$, the decay rate can be reduced below γ_0 only for sufficiently large η. In practice, coupling efficiencies are very small, e.g., for an atom in free space illuminated by a focused Gaussian beam of squeezed light, η is of the order of $10^{-1} - 10^{-2}$, or even smaller. With such small η, the squeezing effects would be so small as to be practically indistinguishable from that induced by the ordinary vacuum fluctuations in modes not occupied by the squeezed field.

Different schemes have been proposed to increase η, including an approximate one-dimensional coupling situation in which only few modes need be squeezed to have the perfect coupling efficiency [38,39]. In practice, this one-dimensional model could be realized inside optical microcavities, where an atom strongly interacts with the privileged cavity modes whose propagation vectors lie inside a small solid angle around the cavity axis [40–42]. These modes could be easily covered by the squeezed field giving the effective $\eta = 1$ coupling. Following this idea, Turchette et al. [35] have performed a series of experiments which investigated the interaction of a beam of Cs atoms with squeezed light in a high finesse optical cavity. The major objective of the experiments was to find nonclassical effects in the transmission spectrum of a weak probe beam monitoring a coherently driven two-level system. However, this was impeded by experimental complications to achieve the perfect $\eta = 1$ coupling inside the cavity, which appear very difficult to circumvent. Including the total experimental losses, the estimated coupling efficiency was very small ($\eta \approx 0.3$). At this level of η, the experiments have not been able to identify effects which could be regarded as nonclassical.

Therefore, squeezing induced nonclassical effects, which persist independently of η and are not limited to very weak squeezed fields are extremely interesting from the experimental point of view. Although the subject of spectroscopy with squeezed light has been extensively investigated in the past and a large number of phenomena have been predicted, it is remarkable that only few nonclassical effects have been found that are insensitive to η.

9.1 Three-Level Atom in a Squeezed Vacuum: Master Equation

The three-level system considered here is shown in Fig. 9.1. The system consists of three unequally spaced levels $|1\rangle, |2\rangle$ and $|3\rangle$, with energies E_i $(i = 1, 2, 3)$, in the cascade configuration. The levels $|i\rangle$ and $|j\rangle$ are separated

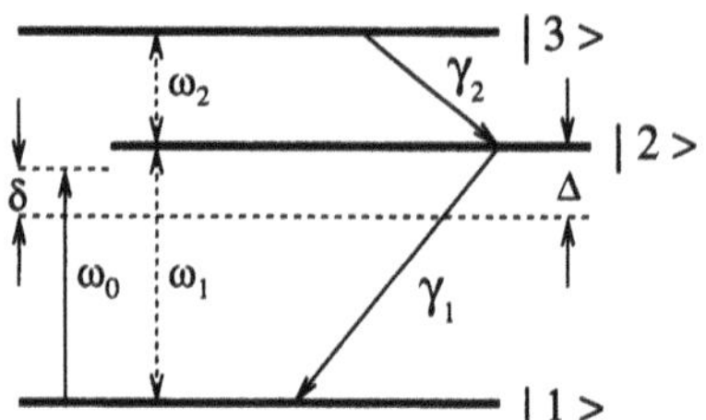

Fig. 9.1. Schematic diagram of a three-level atom in the cascade configuration

by the energy $\hbar\omega_{ij} = E_i - E_j$, where ω_{ij} is the transition frequency, and are connected by the electric-dipole moment μ_{ij}. We assume that non-zero electric-dipole moments are between the intermediate level $|2\rangle$ and the ground level $|1\rangle$, and between the upper level $|3\rangle$ and the level $|2\rangle$. The levels $|1\rangle$ and $|3\rangle$ are not connected by an electric dipole moment ($\mu_{13} = \mu_{31} = 0$), and the transition $|1\rangle \rightarrow |3\rangle$ is allowed in the higher order quadrupole approximation. Hence, the upper level $|3\rangle$ could be reached from the ground level $|1\rangle$ via two pathways; either the two dipole stepwise transition $|1\rangle \rightarrow |2\rangle \rightarrow |3\rangle$, or the quadrupole one step $|1\rangle \rightarrow |3\rangle$ transition.

The system is coupled to a three-dimensional multi-mode vacuum field. Moreover, the system is driven by two coherent fields, one coupled to the electric-dipole transitions, and the other coupled to the quadrupole transition. The Hamiltonian for the system interacting with the driving fields is given by

$$H = H_o + H_F + H_I \ , \tag{9.2}$$

where

$$H_o = \sum_{i=1}^{3} E_i \left|i\right\rangle \left\langle i\right| \tag{9.3}$$

is the Hamiltonian of the system,

$$H_F = \hbar c \sum_s \int d^3\boldsymbol{k} \left|\boldsymbol{k}\right| a^{\dagger}_{\boldsymbol{k}s} a_{\boldsymbol{k}s} \tag{9.4}$$

is the Hamiltonian of the multimode vacuum plus driving lasers field.

The interaction Hamiltonian H_I is composed of three terms

$$H_I = H_L^d + H_L^q + H_V \ , \tag{9.5}$$

where

$$\begin{aligned} H_L^d = &-\frac{1}{2} i\hbar\Omega_1 \left\{\sigma_{21}^{+} \exp[-i\left(\omega_L t + \phi\right)] - \sigma_{21}^{-} \exp[i\left(\omega_L t + \phi\right)]\right\} \\ &-\frac{1}{2} i\hbar\Omega_2 \left\{\sigma_{32}^{+} \exp[-i\left(\omega_L t + \phi\right)] - \sigma_{32}^{-} \exp[i\left(\omega_L t + \phi\right)]\right\} \end{aligned} \tag{9.6}$$

is the interaction between the system and a classical laser field of frequency ω_L and phase ϕ coupled to the electric-dipole moments,

$$H_L^q = -i\hbar |q| \left\{\sigma_{31}^+ \exp[-2i(\omega_q t + \phi_o)] - \sigma_{31}^- \exp[2i(\omega_q t + \phi_o)]\right\} \tag{9.7}$$

is the interaction between the system and a laser field of angular frequency ω_q, phase ϕ_o and the Rabi frequency q coupled to the electric-quadrupole moment of the two-photon $|1\rangle \to |3\rangle$ transition, and

$$\begin{aligned} H_V = {} & i\hbar \sum_s \int d^3\boldsymbol{k} \left[\boldsymbol{\mu}_{21} \cdot \boldsymbol{g}_{\boldsymbol{k}s}(\boldsymbol{r})\, a_{\boldsymbol{k}s} \sigma_{21}^+ - \text{H.c.}\right] \\ & + i\hbar \sum_s \int d^3\boldsymbol{k} \left[\boldsymbol{\mu}_{32} \cdot \boldsymbol{g}_{\boldsymbol{k}s}(\boldsymbol{r})\, a_{\boldsymbol{k}s} \sigma_{32}^+ - \text{H.c.}\right] \end{aligned} \tag{9.8}$$

is the interaction of the atomic dipole moments with the multimode vacuum field. In (9.4)–(9.8), $a_{\boldsymbol{k}s}^\dagger$ and $a_{\boldsymbol{k}s}$ are the creation and annihilation operators for the vacuum modes of the wave vector $\boldsymbol{k}$ and polarization s, $\sigma_{ij}^\pm$ $(i > j = 1, 2, 3)$ are the atomic operators, and $\Omega_j = \boldsymbol{\mu}_{ij} \cdot \boldsymbol{E}_L/\hbar \quad (i = 1, 2)$ are the Rabi frequencies of the laser field of amplitude $\boldsymbol{E}_L$ driving the dipole transitions $|1\rangle \to |2\rangle$ and $|2\rangle \to |3\rangle$, respectively. The coupling of the system and the second laser, driving the two-photon transition, is given by the quadrupole term q, which can be interpreted as a two-photon Rabi frequency [43], while the coupling $\boldsymbol{\mu}_{ij} \cdot \boldsymbol{g}_{\boldsymbol{k}s}(\boldsymbol{r})$ determines the interaction of the multimode vacuum field with the atom, located at a point $\boldsymbol{r}$. The later is in the electric-dipole approximation only.

The standard rotating wave approximation (RWA) has been used to eliminate the nonresonant terms within the interaction Hamiltonian (9.5). The evolution of the atom depends on the driving fields amplitudes and frequencies, as well as on the state of the multi-mode vacuum field. We consider the multi-mode field as a three-dimensional reservoir to the atom. In this case the evolution can be described by the reduced density operator ρ of the system obtained by taking a trace over the reservoir variables

$$\rho = \text{Tr}_R(\rho_{AF}) , \tag{9.9}$$

where ρ_{AF} is the density operator of the total atom + field system. The density operator ρ satisfies the Liouville equation

$$i\hbar\dot{\rho}(t) = \text{Tr}_R([H_V(t), \rho_{AF}(t)]) , \tag{9.10}$$

with the condition that the atom and reservoir are independent at the initial time t = 0, i.e. $\rho_{AF}(0) = \rho(0)\rho_F(0)$, where $\rho_F(0)$ is the density operator of the reservoir, and $H_V(t)$ is given in (9.8). We can solve (9.10) via iteration, which in the weak coupling or Born approximation gives

$$\dot{\rho}(t) = -\frac{1}{\hbar^2} \int_0^t dt' \text{Tr}_R \{[H_V(t), [H_V(t'), \rho(t')\rho_F(0)]]\} . \tag{9.11}$$

Substituting the interaction Hamiltonian (9.8), we find that the evolution of the density operator depends solely on the second order correlation functions of the reservoir operators. We assume that a part of the reservoir modes is in a squeezed vacuum state for which the correlation functions are given by [32,44]

$$\begin{aligned} \left\langle a^{\dagger}_{\boldsymbol{k}s} a_{\boldsymbol{k}'s'} \right\rangle &= N(\omega)\, U^{*}_{s}(\boldsymbol{k})\, U_{s'}(\boldsymbol{k})\, \delta(\omega - \omega')/k^2 , \\ \left\langle a^{\dagger}_{\boldsymbol{k}s} a^{\dagger}_{\boldsymbol{k}'s'} \right\rangle &= M(\omega)\, U_{s}(\boldsymbol{k})\, U_{s'}(\boldsymbol{k}')\, \delta(2\omega_o - \omega - \omega')/kk' , \end{aligned} \tag{9.12}$$

where the parameters $N(\omega)$ and $M(\omega) = |M(\omega)| \exp(i\phi_s)$ characterize squeezing intensity and multi-mode correlations, given in (4.59) and (4.60), respectively, ω_o is the carrier frequency of the squeezed field and ϕ_s is its phase. The parameter $U_s(\boldsymbol{k})$ determines the mode function of the output squeezed field and includes only directions $\boldsymbol{k}$ of the field modes confined to a solid angle $\theta_k \leq \theta_m$, where θ_m is a maximum value of the solid angle over which the squeezed field is propagated. The explicit form of the mode function depends on the method of propagation and focusing the squeezed field.

In order to optimize the squeezing effects on the atom, the mode function $U_s(\boldsymbol{k})$ should be perfectly matched to the mode structure $\boldsymbol{g}_{\boldsymbol{k}s}(\boldsymbol{r})$ of the three-dimensional vacuum field coupled to the atom. Such a requirement of perfect matching is practically impossible to achieve in present experiments [35]. Therefore we consider mode structures which correspond to an imperfect matching of the mode function $U_s(\boldsymbol{k})$ to the vacuum field mode structure $g_{\boldsymbol{k}s}(\boldsymbol{r})$. In this case, we can write the mode function $U_s(\boldsymbol{k})$ as

$$U_s(\boldsymbol{k}) = \begin{cases} [\mathcal{N}(k)]^{-1/2}\, \boldsymbol{\mu}_i \cdot \boldsymbol{g}^{*}_{\boldsymbol{k}s}(\boldsymbol{r})\, D(k) & \text{for} \quad \theta_k \leq \theta_m , \\ 0 & \text{for} \quad \theta_k > \theta_m , \end{cases} \tag{9.13}$$

where $\mathcal{N}(k)$ is the normalization constant such that $|U_s(\boldsymbol{k})|^2 = 1$, and the parameter $D(k)$ determines the coupling efficiency of the squeezed field mode function $U_s(\boldsymbol{k})$ to the vacuum field mode function $g_{\boldsymbol{k}s}(\boldsymbol{r})$. For perfect coupling efficiency $|D(k)| = 1$, whereas $|D(k)| < 1$ for an imperfect coupling. The parameter $D(k)$ contains both the amplitude and phase coupling, and its explicit form depends on the method of propagation and focusing of the squeezed field. For example, in the case of a Gaussian profile of a focused squeezed field, the parameter $D(k)$ is given by [39,40]

$$D(k) = \exp\left[-W_0 \sin^2\theta_k - ikz_f \cos\theta_k\right] , \tag{9.14}$$

where W_0 is the beam spot size at the focal point z_f.
In a cavity situation, for example, the parameter $D(k)$ is identified as the cavity transfer function, the absolute value square of which is the Airy function of the cavity [40,41].

Before returning to the derivation of the master equation, we should remark that in the squeezing propagation case in which the squeezed modes lie inside the cone of angle $\theta_m < \pi$, we assume that the modes outside the cone are in their ordinary vacuum state. In practice, the modes will be in a finite temperature black-body state, which means that inside the cone the modes are in mixed squeezed vacuum and black-body states. However, this is not a serious practical problem as experiments are usually performed at low temperatures where the black-body radiation is negligible. In principle, we can include the black-body radiation effect (thermal noise) to the problem by replacing $N(\omega)$ in (9.12) by $N(\omega) + \bar{N}$, where $\bar{N}$ is proportional to the photon number in the black-body radiation.

We now return to the derivation of the master equation of the atom in a squeezed vacuum field. Substituting (9.8),(9.12) and (9.13) into (9.11), we obtain

$$\begin{aligned}\dot{\rho}(t) = \sum_{i>j=1}^{3} \sum_{n>m=1}^{3} \{ & \left[\sigma_{ij}^{-} Y_{nm,ij}(t,\tau), \sigma_{nm}^{+}\right] + \left[\sigma_{ij}^{-}, Y_{nm,ij}(t,\tau)\sigma_{nm}^{+}\right] \\ & + \left[\sigma_{ij}^{+} N_{nm,ij}(t,\tau), \sigma_{nm}^{-}\right] + \left[\sigma_{ij}^{+}, N_{nm,ij}(t,\tau)\sigma_{nm}^{-}\right] \\ & + \left[\sigma_{ij}^{+} M_{nm,ij}(t,\tau), \sigma_{nm}^{+}\right] + \left[\sigma_{ij}^{+}, M_{nm,ij}(t,\tau)\sigma_{nm}^{+}\right] \\ & + \left[\sigma_{ij}^{-} M_{nm,ij}^{*}(t,\tau), \sigma_{nm}^{-}\right] + \left[\sigma_{ij}^{-}, M_{nm,ij}^{*}(t,\tau)\sigma_{nm}^{-}\right]\} , \end{aligned} \tag{9.15}$$

where the time dependent operators are

$$\begin{aligned} Y_{ij,mn}(t,\tau) &= \int dk k^2 \left[\chi_{ij,mn}^{(N)}(t) + \chi_{ij,mn}(t)\right] \\ &\quad \times \int_0^t d\tau \rho(t-\tau) \exp[i(\omega_{ij} - \omega_k)\tau] , \\ N_{ij,mn}(t,\tau) &= \int dk k^2 \chi_{ij,mn}^{(N)}(t) \\ &\quad \times \int_0^t d\tau \rho(t-\tau) \exp[-i(\omega_{ij} - \omega_k)\tau] , \\ M_{ij,mn}(t,\tau) &= \int dk k (2k_o - k) \chi_{ij,mn}^{(M)}(t) \\ &\quad \times \int_0^t d\tau \rho(t-\tau) \exp[i(\omega_{ij} - \omega_k)\tau] , \end{aligned} \tag{9.16}$$

with the mode functions

$$\begin{aligned} \chi_{ij,mn}(t) &= \int d\Omega_k \sum_s \left[\boldsymbol{\mu}_{ij}^{*} \cdot \boldsymbol{g}_{\boldsymbol{k}s}^{*}(\boldsymbol{r})\right] \left[\boldsymbol{\mu}_{mn} \cdot \boldsymbol{g}_{\boldsymbol{k}s}(\boldsymbol{r})\right] \\ &\quad \times \exp\left[i(\omega_{ij} - \omega_{mn})t\right] , \\ \chi_{ij,mn}^{(N)}(t) &= N(\omega) |D(k)|^2 \exp\left[-i(\omega_{ij} - \omega_{mn})t\right] \\ &\quad \times \sum_{s,s'} \left[\int\int_{\Omega_s} d\Omega_k d\Omega_{k'} \left|\boldsymbol{\mu}_{ij} \cdot \boldsymbol{g}_{\boldsymbol{k}s}(\boldsymbol{r})\right|^2 \left|\boldsymbol{\mu}_{mn} \cdot \boldsymbol{g}_{\boldsymbol{k}'s'}(\boldsymbol{r})\right|^2\right]^{\frac{1}{2}} , \end{aligned}$$

$$\chi_{ij,mn}^{(M)}(t) = M(\omega)|D(k)|^2 \exp[i(2\omega_o - \omega_{ij} - \omega_{mn})t]$$
$$\times \sum_{s,s'} \left[\int \int_{\Omega_s} d\Omega_k d\Omega_{k'} \left|\boldsymbol{\mu}_{ij} \cdot \boldsymbol{g}_{\boldsymbol{k}s}(\boldsymbol{r})\right|^2 \left|\boldsymbol{\mu}_{mn} \cdot \boldsymbol{g}_{\boldsymbol{k}'s'}(\boldsymbol{r})\right|^2 \right]^{\frac{1}{2}} , \tag{9.17}$$

and Ω_s is the solid angle over which the squeezed modes are propagated.

The master equation (9.15) with parameters (9.16) and (9.17) is quite general in terms of the matching of the squeezed modes to the vacuum modes and the bandwidth of the squeezed field relative to the atomic linewidths. The master equation is in the form of an integro-differential equation, and we can simplify the form by employing the Markov approximation. In this approximation the integral over the time delay τ contains functions which decay to zero over a short correlation time τ_c. This correlation time is of the order of the inverse bandwidth of the squeezed field, and the short correlation time approximation is formally equivalent to assume that squeezing bandwidths are much larger than the atomic linewidths. Over this short time-scale the density operator would hardly have changed from $\rho(t)$, thus we can replace $\rho(t-\tau)$ by $\rho(t)$ in (9.16) and extend the integral to infinity. Next, we can perform the integration and find

$$\lim_{t\to\infty} \int_0^t d\tau \rho(t-\tau) e^{ix\tau} \approx \rho(t) \left[\pi\delta(x) + i\frac{P}{x}\right] , \tag{9.18}$$

where P indicates the principal value of the integral. Finally, to carry out the polarization sums and integrals over $d\Omega_k$ in (9.17), we use the plane-wave description of the vacuum mode structure $\boldsymbol{g}_{\boldsymbol{k}s}(\boldsymbol{r})$, in which

$$\boldsymbol{g}_{\boldsymbol{k}s}(\boldsymbol{r}) = \left(\frac{ck}{2\varepsilon_o\hbar(2\pi)^3}\right)^{1/2} \hat{\boldsymbol{e}}_{\boldsymbol{k}s} e^{i\boldsymbol{k}\cdot\boldsymbol{r}} , \tag{9.19}$$

where $\hat{\boldsymbol{e}}_{\boldsymbol{k}s}$ is the unit polarization vector of a plane wave of the wave-vector $\boldsymbol{k}$ with polarization s.

With the choice of spherical representation for the polarization vectors as

$$\begin{aligned} \hat{\boldsymbol{e}}_{\boldsymbol{k}1} &= (-\cos\theta_k \cos\psi_k, -\cos\theta_k \sin\psi_k, \sin\theta_k) , \\ \hat{\boldsymbol{e}}_{\boldsymbol{k}2} &= (\sin\psi_k, -\cos\psi_k, 0) , \end{aligned} \tag{9.20}$$

and assuming that the dipole moments are parallel $(\boldsymbol{\mu}_1 \| \boldsymbol{\mu}_2)$, the sums over s and the integrals over $d\Omega_k$ in (9.17) lead to

$$\begin{aligned} Y_k(t,\tau) &= \sqrt{\gamma_i\gamma_j}\left[1 + N(\omega_i)|D(k_i)|^2 v(\theta_m)\right]\rho(t) e^{i(\omega_i-\omega_j)t} , \\ N_k(t,\tau) &= \sqrt{\gamma_i\gamma_j} N(\omega_i)|D(k_i)|^2 v(\theta_m)\rho(t) e^{-i(\omega_i-\omega_j)t} , \\ M_k(t,\tau) &= \sqrt{\gamma_i\gamma_j} M(\omega_i)|D(k_i)|^2 v(\theta_m)\rho(t) e^{i(2\omega_o-\omega_i-\omega_j)t} , \end{aligned} \tag{9.21}$$

where $\gamma_i = k_i^3 \mu_i^2 / 6\pi\varepsilon_o \hbar$ is the spontaneous emission rate of the ith transition, and

$$v(\theta_m) = \frac{1}{2}\left[1 - \frac{1}{4}\left(3 + \cos^2\theta_m\right)\cos\theta_m\right] . \tag{9.22}$$

In the derivation of (9.21), we have ignored the principal value parts which contribute to the energy shifts of the atomic levels. In fact, the shifts are very small for a broadband squeezed field and their contribution to the atomic dynamics are negligible [45]. We note here that the energy shifts are very sensitive to the bandwidth of the squeezed field and can be important in a narrow-bandwidth squeezed field [15]. Despite this, it is always true that the shifts vanish, independent of the squeezing bandwidth, when the ith transition is resonant with the squeezing carrier frequency [32].

With the parameters (9.21), the master equation of the three-level atom in a broadband squeezed vacuum reads

$$\begin{aligned}
\frac{\partial \rho}{\partial t} = &-\frac{1}{2}\sum_{i,j=1}^{2} \gamma_{ij}\left[1 + \tilde{N}(\omega_i)\right]\left(\rho\sigma_i^+\sigma_j^- + \sigma_i^+\sigma_j^-\rho - 2\sigma_j^-\rho\sigma_i^+\right)e^{i(\omega_i-\omega_j)t} \\
&-\frac{1}{2}\sum_{ij} \gamma_{ij}\tilde{N}(\omega_i)\left(\rho\sigma_i^-\sigma_j^+ + \sigma_i^-\sigma_j^+\rho - 2\sigma_j^+\rho\sigma_i^-\right)e^{-i(\omega_i-\omega_j)t} \\
&-\frac{1}{2}\sum_{ij} \gamma_{ij}\tilde{M}(\omega_i)\left(\left[\sigma_i^+\rho, \sigma_j^+\right] + \left[\sigma_j^+, \rho\sigma_i^+\right]\right)e^{i(2\omega_o-\omega_i-\omega_j)t} \\
&-\frac{1}{2}\sum_{ij} \gamma_{ij}\tilde{M}^*(\omega_i)\left(\left[\sigma_i^-\rho, \sigma_j^-\right] + \left[\sigma_j^-, \rho\sigma_i^-\right]\right)e^{-i(2\omega_o-\omega_i-\omega_j)t} \\
&-\frac{i}{\hbar}\left[H_L^d + H_L^q, \rho\right] ,
\end{aligned} \tag{9.23}$$

where $\sigma_1^\pm \equiv \sigma_{21}^\pm$, $\sigma_2^\pm \equiv \sigma_{32}^\pm$, $\gamma_{ij} = (\gamma_i\gamma_j)^{1/2}$, H_L^d and H_L^q are given in (9.6) and (9.7), respectively, and

$$\begin{aligned}
\tilde{N}(\omega_i) &= \eta N(\omega_i) , \\
\tilde{M}(\omega_i) &= \eta M(\omega_i) ,
\end{aligned} \tag{9.24}$$

with $\eta = |D(k_i)|^2 v(\theta_m)$.

The parameter η determines the matching of the squeezed modes to the modes surrounding the atom. This includes the coupling efficiency of the mode functions, given by the parameter $D(k)$, and the angular dimensions, given by the angle θ_m, of the squeezed modes. For an imperfect matching the master equation (9.23) is formally identical to that for perfect matching: the only difference is the replacement of the squeezing parameters N and M by the matching modified parameters $\tilde{N}$ and $\tilde{M}$.

Equation (9.23) is the final form of the master equation that gives us an elegant description of the physics involved in the dynamics of a three-level atom interacting with a squeezed vacuum field. In the following sections, we will illustrate specific examples of the applications of the master equation to evaluate dynamics and stationary properties of a three-level atom interacting with a squeezed vacuum field. Particular attention will be paid to the role of quantum correlations in the atomic dynamics and population distribution between the atomic energy levels.

9.2 Equations of Motion for the Density Matrix Elements

The master equation (9.23) allows us to calculate the transient and stationary dynamics of the cascade three-level atom in a squeezed vacuum field, which describe the behavior of the atom. Here, we will focus on the dynamics of the atomic excited states populations, which are given by the diagonal density matrix elements $\rho_{22} = \langle 2|\,\rho\,|2\rangle$ and $\rho_{33} = \langle 3|\,\rho\,|3\rangle$. From the master equation (9.23), we find that the density matrix elements satisfy the following system of eight coupled differential equations

$$\begin{aligned}
\dot{\rho}_{22} &= \tilde{N}\gamma_1 - \left[\tilde{N}\gamma_2 + \left(2\tilde{N}+1\right)\gamma_1\right]\rho_{22} + \left[\left(\tilde{N}+1\right)\gamma_2 - \tilde{N}\gamma_1\right]\rho_{33} \\
&\quad + \tilde{M}\gamma_{12}\tilde{\rho}_{13} + \tilde{M}^*\gamma_{12}\tilde{\rho}_{31} - \frac{1}{2}\Omega_1\left(\tilde{\rho}_{12}+\tilde{\rho}_{21}\right) + \frac{1}{2}\Omega_2\left(\tilde{\rho}_{23}+\tilde{\rho}_{32}\right) , \\
\dot{\rho}_{33} &= -\left(\tilde{N}+1\right)\gamma_2\rho_{33} + \tilde{N}\gamma_2\rho_{22} - \frac{1}{2}\tilde{M}\gamma_{12}\tilde{\rho}_{13} - \frac{1}{2}\tilde{M}^*\gamma_{12}\tilde{\rho}_{31} \\
&\quad - |q|\left(\tilde{\rho}_{13}+\tilde{\rho}_{31}\right) - \frac{1}{2}\Omega_2\left(\tilde{\rho}_{23}+\tilde{\rho}_{32}\right) , \\
\dot{\tilde{\rho}}_{13} &= \left(\dot{\tilde{\rho}}_{31}\right)^* = -\left\{\frac{1}{2}\left[\tilde{N}\gamma_1 + \left(\tilde{N}+1\right)\gamma_2\right] + 2i\delta\right\}\tilde{\rho}_{13} + \frac{3}{2}\gamma_{12}\tilde{M}^*\rho_{22} \\
&\quad + 2|q|\,\rho_{33} + \frac{1}{2}\Omega_1\rho_{23} - \frac{1}{2}\Omega_2\rho_{12} - \frac{1}{2}\left(\gamma_{12}\tilde{M}^* + 2|q|\right) , \\
\dot{\tilde{\rho}}_{12} &= \left(\dot{\tilde{\rho}}_{21}\right)^* = -\frac{1}{2}\Omega_1 - \left\{\frac{1}{2}\left[\left(2\tilde{N}+1\right)\gamma_1 + \tilde{N}\gamma_2\right] - i\left(\Delta-\delta\right)\right\}\tilde{\rho}_{12} \\
&\quad + \tilde{M}^*\gamma_1\tilde{\rho}_{21} + \gamma_{12}\left(\tilde{N}+1\right)\tilde{\rho}_{23} - \frac{1}{2}\left(\gamma_{12}\tilde{M}^* - 2|q|\right)\tilde{\rho}_{32} \\
&\quad + \frac{1}{2}\Omega_2\tilde{\rho}_{13} + \frac{1}{2}\Omega_1\left(2\rho_{22}+\rho_{33}\right) , \\
\dot{\tilde{\rho}}_{23} &= \left(\dot{\tilde{\rho}}_{32}\right)^* = -\left\{\frac{1}{2}\left[\left(2\tilde{N}+1\right)\gamma_2 + \left(\tilde{N}+1\right)\gamma_1\right] + i\left(\Delta+\delta\right)\right\}\tilde{\rho}_{23} \\
&\quad + \tilde{M}^*\gamma_2\tilde{\rho}_{32} + \gamma_{12}\tilde{N}\tilde{\rho}_{12} - \frac{1}{2}\left(\gamma_{12}\tilde{M}^* + 2|q|\right)\tilde{\rho}_{21} \\
&\quad - \frac{1}{2}\Omega_1\tilde{\rho}_{13} + \frac{1}{2}\Omega_2\left(\rho_{33}-\rho_{22}\right) ,
\end{aligned} \tag{9.25}$$

where $\gamma_i = \gamma_{ii}, \tilde{N} = \tilde{N}(\omega_1) = \tilde{N}(\omega_2)\,, \tilde{M} = \tilde{M}(\omega_1) = \tilde{M}(\omega_2)\,,$

$$\tilde{\rho}_{13} = \left(\tilde{\rho}_{31}\right)^* = \rho_{13}\exp\left[-i\left(2\omega_o - \omega_1 - \omega_2\right)t\right] ,$$

$$\begin{aligned} \tilde{\rho}_{12} &= (\tilde{\rho}_{21})^* = \rho_{12} \exp\left[-i\left(\omega_o - \omega_1\right)t\right] \ , \\ \tilde{\rho}_{23} &= (\tilde{\rho}_{32})^* = \rho_{23} \exp\left[-i\left(\omega_o - \omega_2\right)t\right] \ , \end{aligned} \tag{9.26}$$

are the slowly varying parts of the off-diagonal matrix elements, and

$$\Delta = \frac{1}{2}\left(\omega_1 - \omega_2\right) \ , \qquad \delta = \omega_o - \frac{1}{2}\left(\omega_1 + \omega_2\right) \ , \tag{9.27}$$

are half of the frequency difference between the two atomic transitions and the detuning of the driving fields from the average atomic frequency respectively. In the derivation of (9.25), we have assumed that the frequencies of the driving laser fields are equal to the carrier frequency of the squeezed vacuum field, i.e. $\omega_L = \omega_q = \omega_o$, and the laser fields phases are equal to zero $(\phi = \phi_o = 0)$.

The set of the equations (9.25) can be written in a matrix form as

$$\dot{\boldsymbol{X}}\left(t\right) = A\boldsymbol{X}\left(t\right) + \boldsymbol{I} \ , \tag{9.28}$$

where $\boldsymbol{X}(t)$ is a column vector composed of the density matrix elements, $\boldsymbol{I}$ is a column vector composed of the inhomogeneous terms, and A is an 8×8 matrix obtained from the coefficients appearing in the equations of motion of the density matrix elements. Since we are interested in the time evolution of the density matrix elements, we will need explicit expressions for the components X_i of the vector $\boldsymbol{X}(t)$ in terms of their initial values. This can be done by a direct integration of (9.28). Thus, if t_o denotes an arbitrary initial time, the integration of (9.28) leads to the following formal solution for $\boldsymbol{X}(t)$

$$\boldsymbol{X}\left(t\right) = \boldsymbol{X}\left(t_o\right) e^{At} - \left(1 - e^{At}\right) A^{-1}\boldsymbol{I} \ . \tag{9.29}$$

Because the determinant of the matrix A is different from zero, there exists a complex invertible matrix U which diagonalizes A, and $\lambda = U^{-1}AU$ is the diagonal matrix of complex eigenvalues. By introducing $\boldsymbol{Y} = U^{-1}\boldsymbol{X}$ and $\boldsymbol{L} = U^{-1}\boldsymbol{I}$, we can rewrite (9.29) as

$$\boldsymbol{Y}\left(t\right) = \boldsymbol{Y}\left(t_o\right) e^{\lambda t} - \left(1 - e^{\lambda t}\right) \lambda^{-1}\boldsymbol{L} \ , \tag{9.30}$$

or in component form

$$Y_i\left(t\right) = Y_i\left(t_o\right) e^{\lambda_i t} - \sum_{j=1}^{8} \left(\lambda^{-1}\right)_{ij} \left(1 - e^{\lambda_j t}\right) L_j \ . \tag{9.31}$$

To obtain solutions for $X_i(t)$ we must determine the eigenvalues λ_i and eigenvectors $Y_i(t)$, which are readily evaluated by a numerical diagonalization of the matrix A.

The steady-state values of the density matrix elements can be found from (9.31) by taking $t \to \infty$, or more directly by setting the left-hand side of (9.28) equal to zero, and then

$$\boldsymbol{X}\left(\infty\right) = -A^{-1}\boldsymbol{I} \ , \tag{9.32}$$

or in component form

$$X_i(\infty) = -\sum_{j=1}^{8} \left(A^{-1}\right)_{ij} I_j \,. \tag{9.33}$$

We will use the solutions (9.31) and (9.33) to calculate the transient and steady-state populations of the atomic excited states.

One of the most interesting questions which our calculations can answer concerns the population distribution and to identify intrinsically quantum optical field effects. We will compare the atomic population for the atom interacting with a classical squeezed field with the population when the classical squeezed field is replaced by a quantum squeezed field. Three different situations will be discussed: (a) the atom driven by a squeezed field alone $(\Omega_1 = \Omega_2 = q = 0)$, (b) the atom driven by a squeezed field and the coherent field coupled to the two one-photon transitions $(\Omega_1, \Omega_2 \neq 0, q = 0)$, (c) the atom driven by a squeezed field and the coherent field coupled to the two-photon transition $(\Omega_1 = \Omega_2 = 0, q \neq 0)$.

9.3 Spontaneous Emission in a Squeezed Vacuum

The solution (9.31) of the coupled differential equations for the density matrix elements fully describes the dynamics of the atomic system in a squeezed vacuum as well as driven by coherent laser fields. We will first consider a simple case of spontaneous emission where the atom is interacting with a squeezed vacuum in the absence of the coherent fields $(\Omega_1 = \Omega_2 = q = 0)$, and analyze the time evolution and stationary distributions of the atomic population. In Fig. 9.2, we show the time evolution of the populations $\rho_{22}(t)$ and $\rho_{33}(t)$ for $N = 0.1, \alpha = 1, \delta = 0$ and $|M|^2 = N(N+1)$. The atom was initially in its ground state. The population in the intermediate state increases in time from its zero value at $t = 0$, attains a maximum at $t \approx (\gamma_1/2)^{-1}$ and next decreases to zero for long times. On the other hand, the population in the upper state always increases in time and attains a maximum value for long times. Consequently, in the steady-state only the ground and the upper states are populated. In this case a Boltzmann distribution of the population of the atomic states is violated. There is an inversion of the population between the $|3\rangle$ and $|2\rangle$ states. Figure 9.3 shows the populations for the same parameters as in Fig. 9.2, but for a classical squeezed field with $|M| = N$. It is seen that there is no population inversion and for all times $\rho_{22}(t) > \rho_{33}(t)$ indicating that the population satisfies the Boltzmann distribution. It is apparent that there is no population inversion for a classical squeezed field, so that observation of the inversion would be striking confirmation of the quantum nature of the squeezed vacuum field.

The population inversion has been predicted assuming that the double carrier frequency of the squeezed field is resonant with the two-photon transition frequency ω_3, i.e. the two-photon detuning $\delta = 0$. This simple case

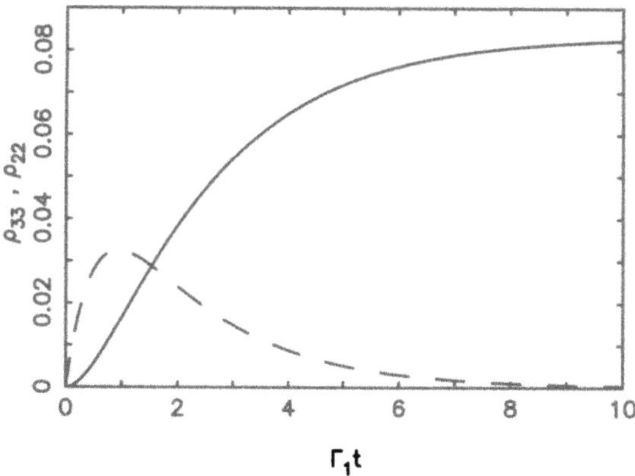

Fig. 9.2. Time evolution of the populations $\rho_{33}(t)$ (*solid line*) and $\rho_{22}(t)$ (*dashed line*) for $N = 0.1, \alpha = 1, \eta = 1, \delta = 0$, $\Gamma_1 = \gamma_1/2$ and $|M| = \sqrt{N(N+1)}$. The atom was initially in its ground state

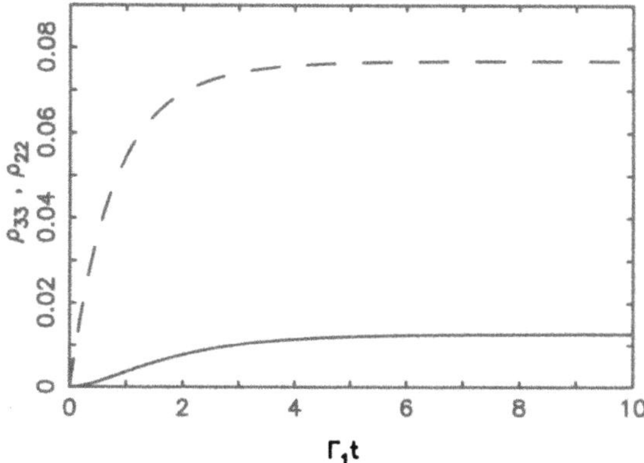

Fig. 9.3. Time evolution of the populations $\rho_{33}(t)$ (*solid line*) and $\rho_{22}(t)$ (*dashed line*) for $N = 0.1, \alpha = 1, \eta = 1, \delta = 0$, $\Gamma_1 = \gamma_1/2$ and a classical squeezed field with $|M| = N$

arises from the fact that a non-zero detuning can diminish the effect of the two-photon correlations $|M|$ on the atomic dynamics. It is easy to see from the master equation (9.23) that a detuning $\delta \neq 0$ introduces an oscillating term $\exp(\pm 2i\delta t)$ to the terms induced by M and M^*. For large detunings this term rapidly oscillates resulting in a reduction of the magnitude of the terms proportional to M and M^*.

A non-zero detuning δ generally reduces the role of the squeezing correlations M in the atomic dynamics but, on the other hand, can lead to other interesting features not observed for the resonant interaction. For example, it has been shown that in an off-resonant squeezed vacuum, the fluorescence spectrum of a two-level atom exhibits a strong asymmetry [46–48]. The ab-

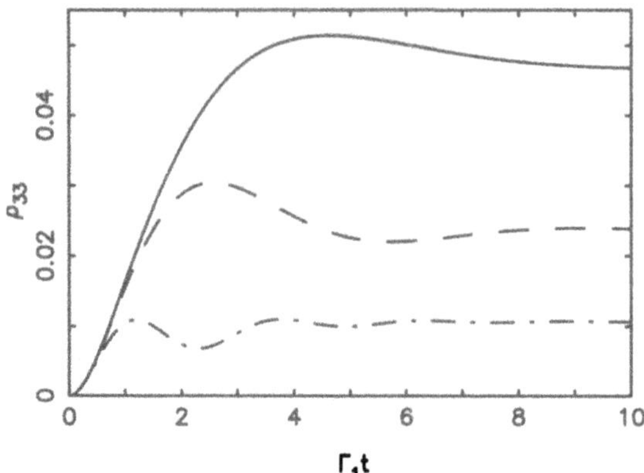

Fig. 9.4. Time evolution of the population $\rho_{33}(t)$ for $N = 0.1, \alpha = 1, \eta = 1$, $\Gamma_1 = \gamma_1/2$, $|M| = \sqrt{N(N+1)}$ and different δ : $\delta = 0.5\gamma_1$ (*solid line*), $\delta = \gamma_1$ (*dashed line*), $\delta = 2.5\gamma_1$ (*dashed-dotted line*)

sorption spectrum of a weak probe beam shows interesting phenomena such as squeezing-induced transparency [49,50] and a large index of refraction accompanied by vanishing absorption [51]. In a two-level atom the squeezed correlations and the detuning alter only the transition dipole moment [12,13,52]. The population is not sensitive to the correlation $|M|$ and the detuning.

The situation is different if we consider a three-level cascade atom. Figure 9.4 shows the time evolution of the population of the state $|3\rangle$ for $|M|^2 = N(N+1), N = 0.1, \alpha = 1, \eta = 1$ and different δ. We see that for small detunings the population monotonically increases in time. However, for moderate detunings a pronounced sinusoidal modulation (quantum beats) appears in the time evolution of the population. The frequency of the modulation and its amplitude depend on δ and increases with increasing δ.

In Fig. 9.5, we plot the time evolution of the population $\rho_{33}(t)$ for $N = 0.1, \delta = 2.5\gamma_1, \alpha = 1$ and the two values of $|M|$: $|M| = N$ corresponding to a classical squeezing and $|M| = \sqrt{N(N+1)}$ of the quantum squeezing. Surprisingly, the classical analog does not show quantum beats. The lack of the quantum beats for the classical squeezed field can be explained by the fact that the two-photon correlations $|M|$ for a classical squeezed field are much weaker than for the corresponding quantum squeezed field. In the case shown in Fig. 9.5, $N = 0.1$ and then $|M| = 0.1$ for the classical squeezed field, whereas $|M| = 0.33$ for the corresponding quantum field.

We now proceed to illustrate the stationary populations of the excited states and the effect of the imperfect matching, $\eta < 1$. From (9.25) and (9.33), we find that the stationary populations of the states $|2\rangle$ and $|3\rangle$ are [32,53,54]:

$$\rho_{22} = \frac{\eta N(\eta N + 1) - \eta^2 |M|^2}{3\eta^2 N^2 + 3\eta N + 1 - 3\eta^2 |M|^2}, \tag{9.34}$$

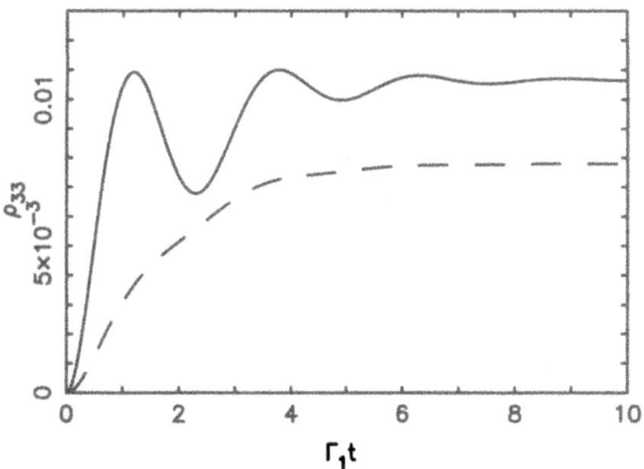

Fig. 9.5. Time evolution of the population $\rho_{33}(t)$ for $N = 0.1, \alpha = 1, \eta = 1, \delta = 2.5\gamma_1$, $\Gamma_1 = \gamma_1/2$ and different $|M|$: $|M| = \sqrt{N(N+1)}$ (*solid line*), $|M| = N$ (*dashed line*)

$$\rho_{33} = \frac{\eta^2 N^2 W - \eta^2 |M|^2 (W - 1 - \alpha)}{W\left(3\eta^2 N^2 + 3\eta N + 1 - 3\eta^2 |M|^2\right)} , \tag{9.35}$$

where $W = \alpha + \eta N (1 + \alpha)$, and $\alpha = \gamma_2/\gamma_1$.

For a classical squeezed field with $|M| = N$ the stationary populations of the excited states reduce to:

$$\rho_{22} = \frac{\eta N}{3\eta N + 1} , \tag{9.36}$$

$$\rho_{33} = \frac{\eta^2 N^2 (1 + \alpha)}{W(3\eta N + 1)} . \tag{9.37}$$

In this case, the ratios

$$\frac{\rho_{33}}{\rho_{22}} = \frac{\eta N (1 + \alpha)}{\alpha + \eta N (1 + \alpha)} , \tag{9.38}$$

and

$$\frac{\rho_{22}}{\rho_{11}} = \frac{\eta N W}{W(\eta N + 1) + \alpha \eta N} \tag{9.39}$$

are smaller than one, indicating that in the classical squeezed field the populations obey a Boltzmann distribution with $\rho_{11} > \rho_{22} > \rho_{33}$. Moreover, in the limit of low intensities ($N \ll 1$) of the squeezed field the population ρ_{33} is proportional to $(\eta N)^2$, showing that in classical fields the population exhibits a quadratic dependence on the intensity. This dependence reflects a two step excitation $|1\rangle \to |2\rangle$ and $|2\rangle \to |3\rangle$, each proportional to ηN. Similar results have been obtained by Thomsen and Wiseman [55], who considered the interaction of a cascade three-level atom with squashed light [56].

The situation is different for a quantum squeezed field with maximal correlations $|M|^2 = N(N+1)$. In this case the populations are given by

$$\rho_{22} = \frac{\eta(1-\eta)N}{3\eta(1-\eta)N+1} , \tag{9.40}$$

$$\rho_{33} = \frac{\eta^2 N^2 + \eta^2(1-\eta)(1+\alpha)N^2}{(\alpha+\alpha\eta N+\eta N)[3\eta(1-\eta)N+1]} . \tag{9.41}$$

The population ρ_{33} exhibits both the linear and quadratic dependence on N. The quadratic term vanishes for a good matching $(\eta \approx 1)$. For weak fields $(N \ll 1)$ the linear term dominates indicating that in a low intensity quantum squeezed field the population ρ_{33} depends linearly on the intensity [29–34,54]. This distinctive feature reflects a direct modification of the two-photon absorption in that the photon correlations $|M|$ enable the transition $|1\rangle \to |3\rangle$ to occurs in a "single step" proportional to N. Moreover, the ratio between the populations ρ_{33} and ρ_{22} is now given by

$$\frac{\rho_{33}}{\rho_{22}} = \frac{\eta[1+(1-\eta)(1+\alpha)N]}{(1-\eta)[\alpha+\eta(1+\alpha)N]} , \tag{9.42}$$

which can be larger than one. This happens for $\eta > \alpha/(1+\alpha)$, indicating that for a good matching the level populations no longer obey the Boltzmann distribution. Furthermore, for $\alpha \ll 1$ and $\eta \approx 1$ the population in the state $|3\rangle$ approaches one $(\rho_{33} \approx 1)$ indicating that a nearly-complete population inversion can be achieved in a quantum squeezed field.

The population inversions are very sensitive to the coupling efficiency of the atom to the squeezed field. This is a practical problem as the present sources of squeezed light generate a beam which can be focused on the atom with small solid angles $(\theta_m < 20°)$. In this case the values that the matching parameter η takes are also small, $\eta < 0.05$, making the observation of the population inversions very difficult experimentally. However, from the population inversion condition, $\eta > \alpha/(1+\alpha)$, it is clear that for α small enough an inversion can be obtained even with an imperfect matching. It follows that stationary population inversions can be created by decay processes in a quantum squeezed field. On the other hand, the linear dependence on intensity of the population ρ_{33} is not disturbed by an imperfect matching and can be observed even for a small efficiency η, independent of α. From (9.41), we find that the population ρ_{33}, for $N \ll 1$, can be written as

$$\rho_{33} \approx \frac{\eta^2}{\alpha}\left[N+(1-\eta)(1+\alpha)N^2\right] . \tag{9.43}$$

Equation (9.43) shows, that the population consists of not only a linear dependence term N, but also a quadratic dependence term N^2. The linear term is entirely due to the nonclassical excitation, while the quadratic term arises from the classical excitation. The magnitude of the quadratic term depends on η and vanishes for perfect coupling $\eta = 1$. In this case the population exhibits the purely nonclassical (linear) dependence on N. However, for $\eta \ll 1$ both terms contribute equally to the population, indicating that in the experimentally realistic situations the nonclassical excitation is always accompanied by the classical excitation process. Thus, the purely linear dependence cannot be observed, only a departure from the quadratic behavior. The linear term of (9.43) dominates for small intensities of the squeezed field and becomes more pronounced as the intensity decreases. For larger intensities, the quadratic term dominates and masks the linear term. Therefore, to observe a large departure from the quadratic intensity dependence, it is important to apply a very weak squeezed field. This imposes an additional limit on the experiment that the upper state population, which ultimately determines the size of the signal to be observed, is also very weak and hence more difficult to detect. Nevertheless, a departure from the N^2 dependence was observed experimentally [4], which is evidence for the nonclassical nature of the squeezed field. In the experiment the squeezed vacuum field was generated by an optical parametric amplifier whose output consists of two low-intensity but very strongly correlated beams of frequencies ω_1 and ω_2, symmetrically located about the carrier frequency $\omega_o = (\omega_1 + \omega_2)/2$. The output exhibits the two-mode correlations and properties, as characterized by (4.60). This squeezed vacuum field was then focused into a cloud of cesium atoms in a magneto-optic trap. The atomic cesium behaves as a three-level cascade atom with transition wavelengths $\lambda_{32} = 917$ nm and $\lambda_{21} = 852$ nm. The two beams of frequencies ω_1 and ω_2 were exactly tuned to those atomic transitions. By monitoring the fluorescence at 917 nm, which is proportional to the population ρ_{33}, the experimental team observed that the population ρ_{33} changes linearly with the squeezing intensity. The observed departure from the quadratic intensity dependence, observed with a classical field, gives compelling evidence for the nonclassical character of the squeezed vacuum field.

Another distinctive feature of the emitted fluorescence, not observed experimentally yet, which persists even for very small coupling efficiencies η are quantum beats. Figure 9.6 shows the time evolution of the population $\rho_{33}(t)$ for $N = 0.1$, $|M| = \sqrt{N(N+1)}, \alpha = 1, \delta = 2.5\gamma_1$, and two different η. We see that the quantum beats are not disturbed by an imperfect matching. Similar to the linear dependence on intensity of the population ρ_{33}, an imperfect matching only diminishes the intensity of the fluorescence field. This makes the quantum beats effect accessible for the present experiments [4,35].

We emphasize that in contrast to the linear dependence on intensity, the quantum beats are a signature of a nonclassical effect which is not accompanied by any classical effect and their appearance is not limited to very small intensities of the squeezed field. Thus, we have a very strong prediction that distinguishes between nonclassical versus classical effects in a much more

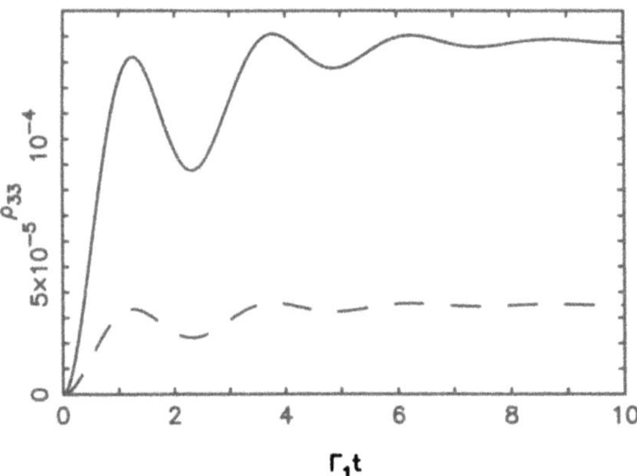

Fig. 9.6. Time evolution of the population $\rho_{33}(t)$ for $N = 0.1, \alpha = 1, \eta = 1, \delta = 2.5\gamma_1$, $\Gamma_1 = \gamma_1/2$, $|M| = \sqrt{N(N+1)}$ and different $\eta : \eta = 0.1$ (*solid line*), $\eta = 0.05$ (*dashed line*)

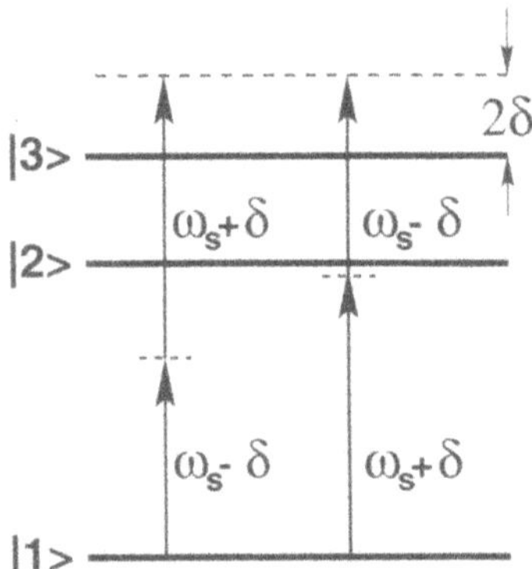

Fig. 9.7. Schematic diagram of a cascade three-level system driven by a detuned squeezed vacuum field

simple way that the difference between the linear and quadratic terms in the steady-state population.

We now identify the origin of the quantum beats. In Fig. 9.7, we present a diagram of the atomic system and the excitation squeezed field detuned from the two-photon resonance by 2δ. The atom can be excited to the upper state in a single step by a simultaneous absorption of two strongly correlated (entangled) photons from the squeezed field of frequencies $\omega_s + \delta$ and $\omega_s - \delta$, where ω_s is the carrier frequency of the squeezed field. Since the photons can be absorbed in two different orders, first the photon of the frequency $\omega_s + \delta$ and next the photon of the frequency $\omega_s - \delta$ and vice versa, this leads to two indistinguishable excitation pathways whose the amplitudes interfere coherently to result in quantum beats. In the case of a classically squeezed field the absorbed photons are not entangled, resulting in two distinguishable pathways of excitation.

9.4 Stationary Lineshape in a Squeezed Vacuum

The addition of a coherent driving field to the atom damped by a squeezed vacuum field leads to phase dependent dynamics and spectral properties of the resonance fluorescence. Carmichael et al. [13] have shown that the fluorescence spectrum of a two-level atom in a squeezed field is still a triplet, as in the case of the normal vacuum [57], but now the central peak of the spectrum can possess either a subnatural or supernatural linewidth depending on the relative phase between the driving field and the squeezed vacuum. The Rabi sidebands, however, exhibit a very weak dependence on the phase and can be narrowed, relatively to their normal vacuum bandwidth, only for low intensities of the squeezed field [58]. The sensitivity of the Rabi sidebands to the phase increases with decreasing bandwidth of the squeezed field and for a narrow bandwidth squeezed field all three peaks in the spectrum can be narrowed [12–16]. Thus, in a squeezed vacuum the spectrum can be modified quantitatively from the spectrum associated with the normal vacuum. Apart from the quantitative modifications several theoretical investigations have also predicted qualitative modifications to the spectrum. Smart and Swain [59,60] and Hegerfeldt et al. [61] have shown that, under appropriate conditions, a hole burning in the spectral line and dispersive profiles can appear in the spectrum. The term "anomalous" spectral features has been used to describe these unusual spectral profiles.

A number of interesting features have been reported in the investigations of the effect of squeezing on the fluorescence and absorption spectra of three-level atoms [62–66]. The results show that, in contrast to a two-level atom, the phase properties of the spectra of a three-level atom depend on the carrier frequency of the squeezed vacuum field and its bandwidth. In particular, for the carrier frequency resonant with one of the two atomic transitions the widths of only some spectral lines depend on the squeezing phase [65], and the Autler-Townes absorption spectrum depends on the phase only in a narrow-bandwidth squeezed vacuum [67].

Here, we consider the phase properties of the stationary lineshape of the fluorescence field emitted from a three-level cascade atom driven by a coherent laser field in a squeezed vacuum. The stationary lineshape is defined as the dependence of the population of the atomic excited states on detuning of the laser field from the atomic resonances [68–70]. The lineshape could be observed experimentally, in a similar way as the linear dependence on intensity of the population ρ_{33}, by monitoring the fluorescence from the state $|3\rangle$ as a function of detuning of the laser field from the two-photon resonance $\omega_o = (\omega_1 + \omega_2)/2$. It is of particular interest to observe the phase dependent dynamics, as they represent the most striking

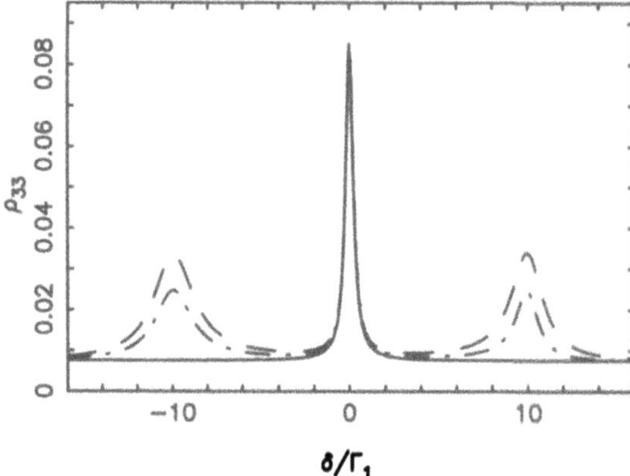

Fig. 9.8. The stationary population ρ_{33} as a function of δ/Γ_1 ($\Gamma_1 = \gamma_1/2$) for $N = 0.1, \eta = 1, \alpha = 1, \Delta = 10\gamma_1$, $|M| = \sqrt{N(N+1)}$ and $\Omega_1 = \Omega_2 = 0, \phi_s = 0$ (*solid line*); $\Omega_1 = \Omega_2 = \gamma_1, \phi_s = 0$ (*dashed line*); $\Omega_1 = \Omega_2 = \gamma_1, \phi_s = \pi$ (*dashed-dotted line*)

aspect of resonance fluorescence in a squeezed vacuum. In Fig. 9.8 we plot the stationary population ρ_{33} calculated from (9.33), as a function of δ for $N = 0.1, |M|^2 = N(N+1), \Delta = 10\gamma_1, \Omega_1 = \Omega_2 = \gamma_1$ and different phases ϕ_s. It is seen that in the absence of the coherent driving field the population shows a pronounced peak at $\delta = 0$. With the driving field included ($\Omega_1 = \Omega_2 \neq 0$), two sidebands emerge in addition to the central peak, at detunings $\delta = \pm\Delta$. The sidebands correspond to the driving field frequency resonant with the one-photon transitions $|1\rangle \to |2\rangle$ and $|2\rangle \to |3\rangle$. The amplitudes of the sidebands strongly depend on the squeezing phase, whereas the central peak at $\delta = 0$ exhibits a weak dependence on the phase.

Figure 9.9 shows that the central peak at $\delta = 0$ almost disappears when the quantum squeezed field is replaced by a classical squeezed field with $|M| = N$, whereas the phase properties of the sidebands remain unchanged. This implies that the central peak arises from the two-photon correlations $|M|$, and the phase property of the sidebands is not the strike signature of the quantum squeezed field. Nevertheless, the phase property of the stationary lineshape, discussed here, is an experimentally accessible approach to detect phase effects associated with the interaction of squeezed light with atoms.

9.5 Quantum Interference with Squeezed Light

To investigate further the role of the two-photon correlations in atomic spectroscopy with squeezed light, the experimental group at Caltech [5] have carried out another experiment beyond that described in Sect. 9.3. The objective of the experiment was to explore explicitly the issue of the phase-sensitivity of the excitation process in which the quantum squeezing effects

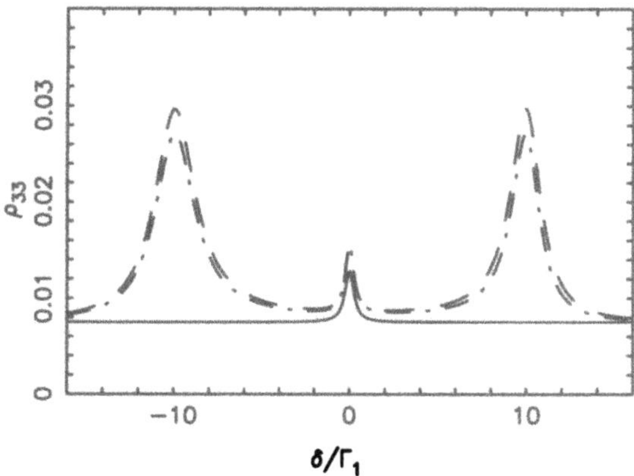

Fig. 9.9. The stationary population ρ_{33} as a function of δ/Γ_1 ($\Gamma_1 = \gamma_1/2$) for $N = 0.1, \eta = 1, \alpha = 1, \Delta = 10\gamma_1$, $|M| = N$ and $\Omega_1 = \Omega_2 = 0, \phi_s = 0$ (*solid line*); $\Omega_1 = \Omega_2 = \gamma_1, \phi_s = 0$ (*dashed line*); $\Omega_1 = \Omega_2 = \gamma_1, \phi_s = \pi$ (*dashed-dotted line*)

could be distinguished from that of a classical squeezing excitation. Moreover, the experiment offers a method of measuring quantum fluctuations of the electromagnetic field with ultra-high frequencies (10's – 100's THz). This frequency regime is well beyond the detection capabilities of conventional homodyne photo-detection schemes, and a new technique has been proposed for investigation of such high frequency correlations.

The principle of the experiment is that it is based on the combination of the two-photon nonclassical excitation of the state $|3\rangle$ in a cascade three-level atom, as studied in Sect. 9.3, together with a coherent "reference oscillator" excitation coupled to a quadrupole moment of the $|1\rangle \to |3\rangle$ transition. Hence, the state $|3\rangle$ could be reached via two pathways, i.e. either via two dipole stepwise absorption from the squeezed field, resulting in the $|1\rangle \to |2\rangle$ and $|2\rangle \to |3\rangle$ transitions, or via the quadrupole two-photon absorption from the reference coherent field, which leads to the direct $|1\rangle \to |3\rangle$ transition. The amplitudes for these two indistinguishable excitation pathways interfere coherently to result in a quantum interference which can manifest itself as a modulation of the excited state population ρ_{33} as the phase of the squeezed field or the reference coherent field is varied.

More specifically, in the experiment the pathway $|1\rangle \to |2\rangle \to |3\rangle$ corresponding to the $6S_{1/2}F = 4 \to 6P_{3/2}F' = 5 \to 6D_{5/2}F'' = 6$ transition in atomic cesium was excited by the output of a non-degenerate parametric oscillator whose frequencies $\omega_{s,i} = \omega_o \pm \Delta\omega$, resonant with the atomic transition frequencies ω_1 and ω_2, exhibit strong nonclassical two-photon correlations. Simultaneously, the atom was illuminated with a coherent laser field of frequency $\omega_o = (\omega_1 + \omega_2)/2$ corresponding to half that of the $6S_{1/2}F = 4 \to 6D_{5/2}F'' = 6$ transition frequency. Quantum interference between the amplitudes of these two excitation paths leads to a phase-sensitive dependence of ρ_{33} on the two-photon correlations M.

We illustrate the theory of the quantum interference effect using the equations of motion (9.25) for the density matrix elements of the atom interacting with a squeezed vacuum field and driven by a coherent laser field coupled to the quadrupole transition ($\Omega_1 = \Omega_2 = 0, q \neq 0$). In this case the stationary population of the upper state $|3\rangle$ is given by

$$\rho_{33} = \frac{\Lambda_1 + |q|\,\Lambda_2 + |q|^2\,\Lambda_3}{\Lambda_4 + |q|\,\Lambda_5 + |q|^2\,\Lambda_6}\ , \tag{9.44}$$

where

$$\begin{aligned}
\Lambda_1 &= \frac{1}{4}\alpha\left[\left(\tilde{N}+1\right)\alpha+\tilde{N}\right]\left\{\tilde{N}^2\left[\left(\tilde{N}+1\right)\alpha+\tilde{N}\right]\right. \\
&\quad \left. -\left|\tilde{M}\right|^2\left(\tilde{N}\alpha+\tilde{N}-1\right)\right\}\ , \\
\Lambda_2 &= \sqrt{\alpha}\left|\tilde{M}\right|\left[\left(\tilde{N}+1\right)\alpha+\tilde{N}\right]\cos\phi_s\ , \\
\Lambda_3 &= \left(\tilde{N}\alpha+\tilde{N}+1\right)\left[\left(\tilde{N}+1\right)\alpha+\tilde{N}\right]+4\alpha\left|\tilde{M}\right|^2\left(\cos^2\phi_s-1\right)\ , \\
\Lambda_4 &= \frac{1}{4}\alpha\left[\left(\tilde{N}+1\right)\alpha+\tilde{N}\right]^2\left(3\tilde{N}^2+3\tilde{N}+1-3\left|\tilde{M}\right|^2\right)\ , \\
\Lambda_5 &= \sqrt{\alpha}\left(1-\alpha\right)\left|\tilde{M}\right|\left[\left(\tilde{N}+1\right)\alpha+\tilde{N}\right]\cos\phi_s\ , \\
\Lambda_6 &= 2\left(1-\alpha\right)\left[\left(\tilde{N}+1\right)\alpha+\tilde{N}\right]+6\alpha\left|\tilde{M}\right|^2\left(\cos^2\phi_s-1\right)\ .
\end{aligned} \tag{9.45}$$

The steady-state solution (9.44) shows that in the presence of the two-photon pumping process the upper state population ρ_{33} depends on the squeezing phase [71]. This is much more clear when we simplify (9.44) assuming, as in the experiment, that the two-photon Rabi frequency q of the reference field is much stronger than the intensity of the squeezed field ($|q| \gg N$), and $\alpha \approx 1$. Then (9.44) simplifies to

$$\rho_{33} \approx \frac{4\,|q|^2}{\alpha^2}\left[1+\sqrt{\alpha}\frac{\left|\tilde{M}\right|}{|q|}\cos\phi_s\right]\ , \tag{9.46}$$

which explicitly shows the sinusoidal modulation of the population with the phase ϕ_s. The amplitude of the modulation depends on $|M|$ indicating that a squeezed vacuum is required in order to observe the modulation. However, the case $|M| \neq 0$ does not necessarily imply a nonclassical character of the oscillations. The oscillations are present for both a quantum squeezed field with $|M| = \sqrt{N(N+1)}$ and a classical squeezed field with $|M| = N$. In order to identify intrinsically quantum effects associated with the nonclassical nature of the quantum squeezed field, we concentrate on the region of small

$N\,(N \ll 1)$, where the distinction between classical and quantum squeezing is maximal [38,48]. For small N and $|M| = \sqrt{N\,(N+1)}$, we find that the phase sensitive modulations are proportional to $\sqrt{N}$, while for a classical squeezed field with $|M| = N$ the onset of the modulations is proportional to N. The distinction between $\sqrt{N}$ and N is apparently an unambiguous signature of a nonclassical effect.

Moreover, in the experiment the two modes of the output of the nondegenerate parametric oscillator were separated by 25 THz. Therefore, the quantum interference technique allows the phase-sensitive detection of quantum fluctuations of electromagnetic fields separated by tens of THz.

9.6 Conclusion

We have explored the role of squeezed light in spectroscopy with three-level atoms. Many interesting modifications from "classical" behavior of radiative properties of a three-level atom in the cascade configuration have been discussed including population inversions, quantum beats, linear dependence on intensity of the two-photon excitation process, phase dependent populations and phase dependent quantum interferences. The origin of these modifications are the two-photon correlations characteristic of squeezed light, which lead to a reduction of noise in one of the two quadrature components of the field components.

We have particularly explored those effects which have been observed experimentally: the linear dependence on intensity of the two-photon excitation process [4] and quantum interferences [5]. We have also studied other experimentally realistic effects in the hope of stimulating further experimental work.

Finally, we would like to point out that the novel effects discussed here are a part of a much broader class of effects predicted in spectroscopy with squeezed light [10]. This includes phase dependent dynamics and spectral properties of two-level atoms, collective multi-atom systems [72,73] and other (lambda and vee) configurations of three-level atoms [74,75].

Acknowledgments

This research was supported by the Australian Research Council. I am grateful to Peter Drummond, Bryan Dalton, Stuart Swain, Ridza Wahiddin, Barry Sanders, Ryszard Tanaś, Peng Zhou, William Smyth, Uzma Akram, Kiat Lim, Matt Ferguson and Martin Bosticky for their many contributions to this research.

References

1. Special issues of Journal Mod. Opt. **34**, nos. 6/7 (1987); and J. Opt. Soc. Am. B **4**, no. 10 (1987)

2. For references see a recent review on squeezed states V.V. Dodonov, J. Opt. B: Quantum Semiclass. Opt. **4**, R1 (2002)
3. E.S. Polzik, J. Carri and H.J. Kimble, Phys. Rev. Lett. **68**, 3020 (1992); Appl. Phys. B **55**, 279 (1992)
4. N.Ph. Georgiades, E.S. Polzik, K. Edamatsu, H.J. Kimble and A.S. Parkins, Phys. Rev. Lett. **75**, 3426 (1995)
5. N.Ph. Georgiades, E.S. Polzik and H.J. Kimble, Phys. Rev. A **55**, R1605 (1997)
6. A. Furusawa, J.L. Sorensen, S.L. Braunstein, C.A. Fuchs, H.J. Kimble and E.S. Polzik, Science **282**, 706 (1998)
7. J. Hald, J.L. Sorensen, C. Schori and E.S. Polzik, Phys. Rev. Lett. **83**, 1319 (1999)
8. M. Atature, A.V. Sergienko, B.M. Jost, B.E.A. Saleh and M.C. Teich, Phys. Rev. Lett. **83**, 1323 (1999)
9. A.S. Parkins, in *Modern Nonlinear Optics, Part II*, eds. M. Evans and S. Kielich (Wiley, New York 1993), p. 607
10. B.J. Dalton, Z. Ficek and S. Swain, J. Mod. Opt. **46**, 379 (1999)
11. W.H. Louisell, *Quantum Statistical Properties of Radiation* (Wiley, New York 1973)
12. C.W. Gardiner, Phys. Rev. Lett. **56**, 1917 (1986)
13. H.J. Carmichael, A.S. Lane and D.F. Walls, J. Mod. Opt. **34**, 821 (1987)
14. A.S. Parkins and C.W. Gardiner, Phys. Rev. A **40**, 3796 (1989); **42**, 5765(E) (1990)
15. A.S. Parkins, Phys. Rev. A **42**, 6873 (1990); **42**, 6873 (1990)
16. J.I. Cirac and L.L. Sanchez-Soto, Phys. Rev. A **44**, 1948 (1991)
17. G. Yeoman and S.M. Barnett, J. Mod. Opt. **43**, 2037 (1996)
18. R. Tanaś, Z. Ficek, A. Messikh and T. Fl-Shahat, J. Mod. Opt. **45**, 1234 (1998)
19. Z. Ficek, B.J. Dalton and M.R.B. Wahiddin, J. Mod. Opt. **44**, 1005 (1997)
20. A. Messikh, R. Tanaś and Z. Ficek, Phys. Rev. A **61**, 033811 (1999)
21. Z. Ficek, J. Seke, R. Kralicek and G. Adam, Opt. Commun. **147**, 289 (1998)
22. H. Ritsch and P. Zoller, Phys. Rev. A **38**, 4657 (1988)
23. S. An and M. Sargent, Phys. Rev. A **39**, 3998 (1989)
24. Z. Ficek and B.J. Dalton, Opt. Commun. **102**, 231 (1993)
25. Z. Ficek, W.S. Smyth and S. Swain, Opt. Commun. **110**, 555 (1994); Phys. Rev. A **52**, 4126 (1995)
26. R.R. Tucci, Opt. Commun. **118**, 241 (1995)
27. S.S. Hassan, H.A. Batarfi and R.K. Bullough, J. Opt. B: Quantum Semiclass. Opt. **2**, R35 (2000)
28. R. Tanaś, in *Coherence and Statistics of Photons and Atoms*, edited by J. Perina (Wiley, New York 2000), p. 289
29. J. Janszky and Y. Yushin, Phys. Rev. A **36**, 1288 (1987)
30. J. Gea-Banacloche, Phys. Rev. Lett. **62**, 1603 (1989)
31. J. Javanainen and P.L. Gould, Phys. Rev. A **41**, 5088 (1990)
32. Z. Ficek and P.D. Drummond, Phys. Rev. A **43**, 6247 (1991); **43**, 6258 (1991)
33. C.W. Gardiner and A.S. Parkins, Phys. Rev. A **50**, 1792 (1994)
34. P. Zhou and S. Swain, Phys. Rev. A **54**, 2455 (1996); Opt. Commun. **134**, 127 (1997)
35. Q.A. Turchette, N.Ph. Georgiades, C.J. Hood, H.J. Kimble and A.S. Parkins, Phys. Rev. A **58**, 4056 (1998)
36. N.Ph. Georgiades, E.S. Polzik and H.J. Kimble, Phys. Rev. A **59**, 123 (1999)

37. P. Zhou and S. Swain, Phys. Rev. Lett. **82**, 2500 (1999)
38. A.S. Parkins and C.W. Gardiner, Phys. Rev. A **38**, 111 (1988)
39. Z. Ficek and P.D. Drummond, Europhys. Lett. **24**, 455 (1993)
40. A. Yariv, *Quantum Electronics* (Wiley, New York 1989)
41. M. Born and E. Wolf, *Principles of Optics* (Macmillan, New York 1964), Chap.7
42. F. De Martini, M. Marrocco, P. Mataloni, L. Crescentini and R. Loudon, Phys. Rev. A **43**, 2480 (1991)
43. Z. Chen and H. Freedhoff, J. Phys. B **24**, 3373 (1991)
44. P.D. Drummond, Quantum Opt. **2**, 205 (1990)
45. G.J. Milburn, Phys. Rev. A **34**, 4882 (1986)
46. S. Smart and S. Swain, Quantum Opt. **5**, 75 (1993)
47. J.M. Courty and S. Reynaud, Europhys. Lett. **10**, 237 (1989)
48. A.S. Shumovsky and T. Quang, J. Phys. B **22**, 131 (1989)
49. P. Zhou and S. Swain, Quant. Semiclass. Opt. **8**, 959 (1996)
50. M.R.B. Wahiddin, Z. Ficek, U. Akram, K.T. Lim and M.R. Muhamad, Phys. Rev. A **57**, 2072 (1998)
51. U. Akram, M.R.B. Wahiddin and Z. Ficek, Phys. Lett. A **238**, 117 (1998)
52. G.M. Palma and P.L. Knight, Opt. Commun. **73**, 131 (1989)
53. V. Buzek, P.L. Knight and I.K. Kudryavtsev, Phys. Rev. A **44**, 1931 (1991)
54. Z. Ficek and P.D. Drummond, Physics Today, September 1997, p.34
55. L.K. Thomsen and H.M. Wiseman, Phys. Rev. A **64**, 043805 (2001)
56. H.M. Wiseman, Phys. Rev. Lett. **81**, 3840 (1998)
57. B.R. Mollow, Phys. Rev. **188**, 1969 (1969)
58. A. Banerjee, Phys. Rev. A **52**, 2472 (1995)
59. S. Smart and S. Swain, Phys. Rev. A **48**, R50 (1993)
60. S. Swain, Phys. Rev. Lett. **73**, 1493 (1994)
61. G.C. Hegerfeldt, T.I. Sachse and D.G. Sondermann, Quant. Semiclass. Opt. **9**, 961 (1997)
62. Q.V. Lawande, B.N. Jagatap and S.V. Lawande, Phys. Rev. A **43**, 535 (1991)
63. A. Joshi and R.R. Puri, Phys. Rev. A **45**, 2025 (1992)
64. S. Smart and S. Swain, J. Mod. Opt. **41**, 1055 (1994)
65. M.R. Ferguson, Z. Ficek and B.J. Dalton, Phys. Rev. A **54**, 2379 (1996); **56**, 4125 (1997)
66. M. Jakob and G.Yu. Kryuchkyan, Phys. Rev. A **57**, 1355 (1998)
67. M. Bosticky, Z. Ficek and B.J. Dalton, Phys. Rev. A **53**, 4439 (1996); Phys. Rev. A **57**, 3869 (1998); J. Opt. B: Quantum and Semiclass. Opt. **1**, 447 (1999)
68. H.J. Kimble and L. Mandel, Phys. Rev. A **13**, 2123 (1976)
69. Z. Ficek and M.R.B. Wahiddin, Quantum and Semiclass. Opt. **7**, 41 (1995)
70. D. Erenso and R. Vyas, Phys. Rev. A **65**, 063808 (2002)
71. L. Adhya, Phys. Rev. A **63**, 043807 (2001)
72. Z. Ficek and R. Tanaś, in *Modern Nonlinear Optics, Part 1, Second Edition, Advances in Chemical Physics, Volume 119*, ed. M. Evans (Wiley, New York 2001), p. 215
73. Z. Ficek and R. Tanaś, Phys. Rep. **372**, 369 (2002)
74. M.A. Anton, O.G. Calderon and F. Carreno, Phys. Lett. **A311**, 297 (2003)
75. F. Carreno, M.A. Anton and O.G. Calderon, Opt. Commun. **221**, 365 (2003)

10 Einstein–Podolsky–Rosen Correlations, Entanglement and Quantum Cryptography

M. D. Reid

The Einstein, Podolsky and Rosen [1] (EPR) argument was presented in 1935 as an attempt to show that quantum mechanics is incomplete. EPR defined as a starting point the premises of "reality" and "no action-at-a-distance" (often referred to as locality). The validity of these premises is taken for granted in all classical theories. The premise of realism implies that if one can predict with certainty the result of a measurement of a physical quantity at a location A, without disturbing the system at A, then the results of the measurement were predetermined. An "element of reality" is defined as a variable corresponding to this physical quantity, where the value of the "element of reality" is the value of the result of the measurement should it be performed. The element of reality is then a variable assigned to describe the system at A, that assumes one of the set of values that are the predicted results of the measurement. The no action-at-a-distance assumption postulates that the action of performing a measurement at a spatially separated location B cannot induce any immediate change to the subsystem at A. In this paper, and elsewhere, "local realism" is defined as the dual premise, where both realism and locality are assumed.

EPR next considered certain quantum systems composed of two, correlated, spatially-separated particles at locations A and B. They argued that, assuming the quantum prediction for the correlation to be correct and assuming the validity of "local realism", there exist two "elements of reality" that simultaneously describe with absolute definiteness the position and momentum of the single particle at a location A. Such an "element of reality" description for the particle at A itself is not consistent with a local quantum description, since both position and momentum are defined to a precision beyond the bounds given by the uncertainty principle. EPR took the view that local realism must be valid. They therefore argued that quantum mechanics must be incomplete. One would have to assume the existence of "hidden variables", that are not part of quantum theory, in order to describe the localized subsystems consistently with the quantum predictions. The argument is perhaps best viewed as a demonstration of the inconsistency between quantum mechanics as we know it (that is without "completion") and local realism.

Schrödinger's reply [2] to EPR in 1935 is now also well-known. In this reply Schrödinger introduced the concept of entangled quantum states, pointing to entanglement as a signature of states not compatible with classical notions, such as local realism. Schrödinger's reply is best known for his discussion of the paradox of the macroscopically entangled state, a "cat" in a quantum superposition of "alive" and "dead" states.

In 1986 a quantum nondemolition experiment confirming the quantum correlation of the quadrature phase amplitudes of two spatially separated fields was reported by Levenson et al. [3]. Two-mode squeezing with two spatially separated detectors was experimentally demonstrated using parametric oscillation in 1987 by Heidmann et al. [4]. With the experimental realization of such two-mode squeezing and quantum nondemolition measurements the possibility of testing for continuous variable EPR-type correlations became apparent [5]. In these proposals the conjugate "position" and "momentum" observables would be the two orthogonal quadrature phase amplitudes of the field. The high efficiency of detectors used in the quadrature phase amplitude ("squeezing") experiments was an important issue. Experimental tests of EPR correlations at the time were based on Bohm's [6] version of the EPR paradox, and in fact focussed on Bell's [7,8] extension of this work. These Bohm/Bell tests involved measurements performed on individual photon pairs, and used very inefficient detectors. Such inefficiencies meant that the Bell tests were inconclusive.

A specific proposal and a criterion to demonstrate the continuous variable EPR correlations [9] for real experiments by way of a realization of the 1935 EPR paradox was put forward in 1989. This proposal employed the two-mode squeezed state as the EPR source. Further theoretical work [10] presented a prediction of EPR-correlated fields for the two output fields of the parametric oscillator. The first experimental achievement, using the parametric oscillator, of the 1989 EPR criterion, for efficient measurements with continuous variable outcomes, was reported by Ou et al. [11] in 1992.

There have been further experimental observations [12,13] of continuous variable EPR correlations as evidenced by the 1989 EPR criterion. Zhang et al. [12] detected EPR correlations between the the intense output fields of the parametric oscillator above threshold. Silberhorn et al. [13] detected such correlations for pulsed fields. Further theoretical studies of EPR systems using continuous variables have been presented by Tara et al. [14] and Giovannettii et al. [15]. More recently there has been further experimental generation of even stronger EPR correlations in the parametric oscillator by Bowen et al. [16]. In addition recent experiments by Bowen et al. [16], Korolkova et al. [17] and Schori et al. [18] take advantage of new criteria developed by Simon [19] and Duan et al. [19] that allow deduction of entanglement based on continuous variable measurements, without the need to prove EPR correlations directly. These experiments give experimental proof of entanglement using continuous variable or quadrature phase amplitude measurements on the parametric and fiber systems.

The EPR fields have proven significant in enabling the experimetal realization of continuous variable quantum teleportation [20], and may have application also to quantum cryptography [21]. Again because of the efficient nature of the measurement of field quadrature phase amplitudes, and the intrinsic macroscopic nature of the correlated fields, continuous variable teleportation and cryptography may provide advantages over protocols using a finite number state basis, where signals of individual photons are sent [22,23].

The continuous variable EPR fields however are also significant in demonstrating, by way of the 1935 EPR paradox, the inconsistency of quantum mechanics with local realism, and this is the main focus of the paper here. In Sect. 10.1, I review the generalization of the EPR argument to provide a criterion for the demonstration of EPR correlations for real experimental fields. In Sect. 10.2, the theory of the generation and detection of EPR fields using a two-mode squeezed light source is reviewed. In Sect. 10.3, criteria for more general (weaker) EPR correlations is discussed, and in Sect. 10.4 the relationship of EPR correlations to quantum entanglement is treated. The application of the EPR fields to quantum cryptography is outlined in Sect. 10.5. This represents a potentially important practical application to communicating information in a way that cannot be tapped without giving away the presence of the eavesdropper.

10.1 Generalization of the EPR Argument to Give Criteria for EPR Correlations

Crucial to demonstrating EPR correlations is the definition of local realism [1], as defined above in the introduction. We summarize the EPR argument again as follows. EPR considered two spatially separated subsystems at A and B. Two observables $\hat{x}^A$ and $\hat{y}^A$ are defined for subsystem A, where $\hat{x}^A$ and $\hat{y}^A$ do not commute. We consider from this that they satisfy an uncertainty relation $\Delta\hat{x}^A \Delta\hat{y}^A \geq 1$, where we assume appropriate scaling has made the obervables dimensionless and the quantum bound 1. It is pointed out however that EPR did not introduce the uncertainty relation in their original paradox: this was done later [9] as a means to define EPR correlations for real experiments.

To continue the EPR argument, suppose one may predict with certainty the result of measurement $\hat{x}^A$ by a measurement performed at B. Also, for a different choice of measurement at B, suppose one may predict the result of measurement $\hat{y}^A$ at A. Such correlated systems are predicted by quantum theory. Assuming "local realism", EPR deduce the existence of an element of reality, $\tilde{x}^A$, for the physical quantity $\hat{x}^A$; and also an element of reality, $\tilde{y}^A$, for $\hat{y}^A$. Local realism implies the existence of two hidden variables $\tilde{x}^A$ and $\tilde{y}^A$ that simultaneously determine, with no uncertainty, the values for the result of an $\hat{x}^A$ or $\hat{y}^A$ measurement on subsystem A, should it be performed. This hidden variable state for the subsystem A is not describable within quantum mechanics, because of the uncertainty relation. Hence, EPR argued, if quantum mechanics is to be compatible with local realism, we must regard quantum mechanics to be incomplete.

Of course as stated by EPR in their 1935 argument, there is also the assumption that the predictions of quantum mechanics are correct, in that such correlations, called EPR correlations, may actually be observed. Such experimental tests and observations were not undertaken for some years after the EPR argument was put forward.

The EPR argument was translated to the case of spin measurements by Bohm [6]. In 1966 Bell proved theoretically that for such spin measurements that the actual predictions of quantum mechanics could differ from all predictions of local realism [7]. If the predictions of quantum mechanics were correct then, in this instance local realism itself would be predicted to fail experimentally. As a consequence the "completion" of quantum mechanics, as one of the alternatives following from the EPR argument (to allow for hidden variables which were compatible with local realism) would be ruled out. All local hidden variable theories would fail to predict results for certain actual experiments.

The first conclusive experimental tests of Bell's inequality were performed with photons [8], and gave data to support quantum mechanics, and were indicative of Bohm's EPR correlations for spin measurements. However the detectors used were so inefficient that irrefutable conclusions could not be drawn. More recently, experiments performed with trapped ions [24,25] have generated entangled Bell and Bohm-type spin states, without detector inefficiency loopholes, though only for very limited spatial separations of subsystems.

The first conclusive demonstration of EPR correlations, without detector efficiency difficulties, were obtained in 1992 for the continuous variable case. Later demonstrations followed in 2000, for macroscopic fields and pulsed fields. The comment must be made however that the separation of subsystems has not been sufficient to base the locality hypothesis on Einstein's causality. This requirement has been emphasized in Bell's theoretical work, and in several experimental tests of Bell's inequality. The basic theory behind the EPR continuous variable demonstrations will be presented here, summarized from the original manuscript [9] and from a recent article [26].

10.1.1 Generalized EPR Argument

For most practical situations occurring the correlations considered by EPR between results for measurements at A and B, and predicted for certain idealized quantum states, are not perfect. In order to demonstrate the existence of EPR correlations for real experiments where the correlation between spatially separated systems need not be maximum, we need to extend the EPR argument to situations where the result of measurement $\hat{x}^A$ at A cannot be predicted with absolute certainty [9,26]. This issue has some fundamental significance, since it is always assumed as part of the argument that the quantum predictions will be correct. Any inconsistency between quantum mechanics and the premise of local realism should be tested. As discussed later in this article in Sect. 10.4, the demonstration of EPR correlations is also a demonstration of entanglement, or quantum inseparability. Entanglement is now recognized as the key feature of quantum states useful to the field of quantum information and its experimental detection is crucial to this field.

To generalize the EPR argument then to a situation of limited correlation, we note that the assumption of local realism still allows us to deduce the existence of an "element of reality" ($\tilde{x}^A$) for $\hat{x}^A$ at A, in the sense that we can make a prediction of the result at A, without disturbing the subsystem at A, under the locality assumption. This prediction is subject to the result x_i^B of a measurement, $\hat{x}^B$ say, performed at B. The possible values for the "element of reality" are no longer a set of definite numbers with zero uncertainty, but are a set of distributions $P_i(x^A) = P(x^A|x_i^B)$, where $P(x^A|x_i^B)$ is the conditional probability for a result x^A at A given a result x_i^B for $\hat{x}^B$ at B.

In a similar manner one may infer the result of measurement $\hat{y}^A$ at A, based on a (different) measurement $\hat{y}^B$ at B. The results of the measurement $\hat{y}^B$ at B are denoted by y_j^B. The set $P_j(y^A) = P(y^A|y_j^B)$, where $P(y^A|y_j^B)$ is the conditional probability of a result y^A for $\hat{y}^A$ given the result y_j^B for the measurement $\hat{y}^B$ at B, form the predicted results of the measurement for $\hat{y}^A$ at A based on the measurement $\hat{y}^B$ at B. This set defines a fuzzy "element of reality" $\tilde{y}^A$ to describe the predetermined nature of the $\hat{y}^A$ at A.

It is possible then in principle to measure the individual variances $\Delta_i^2 x$ of the conditional distributions $P(x^A|x_i^B)$ (and also $\Delta_j^2 y^A$ for the $P(y^A|y_j^B)$). If each of the variances satisfy $\Delta_i x^A \Delta_j y^A < 1$ (for all i, j) it becomes transparent that this would imply the demonstration of the EPR paradox. This is because every outcome for the measurements ($\hat{x}^B$ or $\hat{y}^B$) performed at B indicates the result at A to a precision better than given by the uncertainty bound 1. "Elements of reality" exist for the physical quantities $\hat{x}^A$ and $\hat{y}^A$, to predetermine the results of measurements in a way to exclude a local quantum description for the subsystem A. This situation most truly reflects the original EPR gedanken experiment.

However more generally in an experiment, the conditional variance $\Delta_i x^A$ for some results x_i^B could be greater than 1. The exact boundary at each we can claim an EPR paradox needs defining. Before discussing the most general criteria we first present the most obvious criterion sufficient to demonstrate EPR correlations.

10.1.2 1989 Inferred Heisenberg Uncertainty EPR Criterion

The 1989 criterion sufficient to demonstrate the existence of EPR correlations is based on the average fuzziness $\Delta^2_{\text{inf}}\hat{x}^A = \sum_i P(x^B_i)\Delta^2_i x$ and $\Delta^2_{\text{inf}}\hat{y}^A = \sum_j P(y^B_j)\Delta^2_j p$ attributed to the predictions for $\hat{x}^A$ and $\hat{y}^A$ respectively. This criterion is particularly useful in applications to quantum cryptography as discussed in Sect. 10.5. Here μ_i and Δ_i are the mean and standard deviation, respectively, of the conditional distribution $P(x^A|x^B_i)$. We use the symbol "inf" to remind us that we make an inference about the result at A based on the outcome at B. The variance is the rms error in this inference, where we take the predicted value at A to be the mean of the conditional distribution. This variance reflects the minimum rms error possible, as another choice of estimate for the result at A will only increase the rms error. The best estimate [27] of the outcome of $\hat{x}^A$ at A, based on a result x^B_i for the measurement at B, is given by μ_i. We define $\Delta_i = \sqrt{\langle (x^A_i - \mu_i)^2 \rangle}$ as the root mean square of the error or deviation $\delta_i = x^A_i - \mu_i$ where the predicted value for the measurement at A is μ_i. (Here x/x^B_i is the value obtained for x^A given the result x^B_i at B). The EPR paradox is achieved when $\Delta_{\text{inf}}\hat{x}^A\Delta_{\text{inf}}\hat{y}^A < 1$. This is the criterion used experimentally to date to justify the claim of the EPR paradox.

We consider a measured error $\Delta_{\text{inf,est}}\hat{x}^A$ in the prediction for the outcome of measurement $\hat{x}^A$ at A, where the estimate for the prediction of the result at A is based on the result at B; and a similar measured error $\Delta_{\text{inf,est}}\hat{y}^A$ for the prediction of $\hat{y}^A$ at A. These errors will always be greater than or equal to that above defined through the conditional distributions. The 1989 criterion for demonstration of EPR correlations is then to find a violation of an inferred uncertainty principle,

$$\Delta_{\text{inf}}\hat{x}^A\Delta_{\text{inf}}\hat{y}^A < 1 , \tag{10.1}$$

(or $\Delta_{\text{inf,est}}\hat{x}^A\Delta_{\text{inf,est}}\hat{y}^A < 1$) since here "elements of reality" attributed (by local realism) to the subsystem A would give predicted results for measurements $\hat{x}^A$ and $\hat{y}^A$ that are determined more precisely than the limit given by the quantum uncertainty bound. For the interested reader, this is justified in somewhat more detail in the papers [9].

10.1.3 Estimate of Average Conditional Variances Through Linear Regression

To summarize for the ideal EPR experiment, one measures the physical quantity $\hat{x}^A$ at A, and simultaneously a physical quantity $\hat{x}^B$ at B, numerous times to establish the conditional probability distributions $P(x^A|x^B_i)$, the probability of obtaining a result x^A at A given a result x^B_i for the measurement at B. The result x^B_i at B allows us to infer the result of the measurement at A should it be performed, the estimate being the mean of the distribution x^A given x^B_i and the rms error in our estimate then given by the $\Delta_i x$ where $\Delta^2_i x$

is the variance of this distribution. The probability of the system A being in any one of states designated by these values is then the probability $P(x_i^B)$ of obtaining the result x_i^B at B. We then calculate the weighted variance for the element of reality

$$\tilde{x}^A = \sum_{x_i^B} P(x_i^B)\Delta_i^2 x^A , \tag{10.2}$$

to determine an average inference variance $\Delta^2_{\mathrm{inf}}\hat{x}^A$. A similar approach for the conjugate measurements enables determination of the inference error $\Delta^2_{\mathrm{inf}}\hat{y}^A$. The measured violation of the inequality $\Delta^2_{\mathrm{inf}}\hat{x}^A\Delta^2_{\mathrm{inf}}\hat{y}^A \geq 1$ equates to observation of EPR correlations, or, as we later show, entanglement.

The evaluation of the conditional probabilities for each outcome of the continuous variable x_i^B at B is not always practical. It is however possible to use other sets of measurements which are sufficient to indicate that the system is EPR correlated. These measurements are in fact closely related to measurements of squeezing, and the same measurement techniques may be employed. Such techniques are the basis for actual continuous variable EPR experiments to date, and were presented in 1989 [9]. This method will be applied in Sect. 10.2 to a particular EPR-correlated system, the two-mode squeezed state.

To explain, we propose, upon a result x_i^B for the measurement at B, that the predicted value for the result x^A at A is given linearly by the estimate $x_{\mathrm{est}} = gx_i^B + d$. The size of the deviation $\delta_i = x^A - (gx_i^B + d)$ in this linear estimate can then be measured (or calculated). We simultaneously measure $\hat{x}^A$ at A and $\hat{x}^B$ at B, to determine x^A and x_i^B and then to calculate initially for given x_i^B $\langle\delta^2\rangle_i = \sum_x (P(x^A, x_i^B)(x^A - (gx_i^B + d)^2)/\sum P(x_i^B)$. Averaging over the different values of x_i^B we obtain as a measure of error in our inference, based on the linear estimate:

$$\begin{aligned}\Delta_L^2\hat{x}^A &= \sum_{x_i^B} P(x_i^B)\langle\delta^2\rangle_i \\ &= \sum_{x,x_i^B} P(x^A, x_i^B)\{x - (gx_i^B + d)\}^2 \\ &= \langle\{\hat{x}^A - (g\hat{x}^B + d)\}^2\rangle .\end{aligned} \tag{10.3}$$

The best linear estimate x_{est} is the one that will minimize $\Delta_L^2\hat{x}^A$. This corresponds to the choice [27] $d = -\langle(\hat{x}^A - g\hat{x}^B)\rangle$. Denoting $\delta_0 = \hat{x}^A - g\hat{x}^B$, our choice of estimate optimized with respect to d gives a minimum error

$$\Delta_L^2\hat{x}^A = \langle\delta_0^2 - \langle\delta_0\rangle^2\rangle . \tag{10.4}$$

The best choice for g is discussed in [9] and in Sect. 10.2 below.

If the estimate x_{est} corresponds to the mean of the conditional distribution $P(x^A|x_j^B)$ then the variance $\Delta_L^2\hat{x}^A$ will correspond to the average conditional variance $\sum_{x_i^B} P(x_i^B)\Delta_i^2 x$ specified above. This is the case, with a certain choice of g, for the two-mode squeezed state used to model continuous variable EPR states generated to date. In general the variances of type $\Delta_L^2\hat{x}^A$ based on estimates will be greater than or equal to the optimal evaluated from the conditionals. The satisfaction then of

$$\Delta_L^2\hat{x}^A\Delta_L^2\hat{y}^A < 1 \tag{10.5}$$

will ensure EPR correlations.

10.2 EPR Correlations from Two-Mode Squeezed Light

EPR correlations for quadrature phase amplitudes were defined in [9], and are intrinsic to two-mode squeezed light. Two-mode squeezed light was considered by Caves and Schumaker [28]. Two-mode squeezed (or EPR) fields may be generated by the parametric interaction

$$H_I = i\hbar\kappa(\hat{a}^\dagger\hat{b}^\dagger - \hat{a}\hat{b}) \ . \tag{10.6}$$

The state generated after a time t, the two-mode squeezed quantum state, may be expressed as

$$|\psi\rangle = \sum_{n=0}^{\infty} c_n|n\rangle_{a_1}|n\rangle_{b_1} \ , \tag{10.7}$$

where $c_n = \tanh^n r/\cosh r$, with $r = \kappa t$. We define the quadrature phase amplitudes

$$\begin{aligned}
\hat{x}^A &= (\hat{a} + \hat{a}^\dagger) \ , \\
\hat{y}^A &= (\hat{a} - \hat{a}^\dagger)/i \ , \\
\hat{x}^B &= (\hat{b} + \hat{b}^\dagger) \ , \\
\hat{y}^B &= (\hat{b} - \hat{b}^\dagger)/i \ .
\end{aligned} \tag{10.8}$$

The Heisenberg uncertainty relation for the orthogonal amplitudes of mode $\hat{a}$ is

$$\Delta^2 x^A \Delta^2 y^A \geq 1 \ . \tag{10.9}$$

The output quadrature amplitudes are (recall $\hat{x}^A = \hat{x}^A(t), \hat{y}^A = \hat{y}^A(t)$)

$$\begin{aligned}
\hat{x}^A(t) &= \hat{x}^A(0)\cosh(\kappa t) + \hat{x}^B(0)\sinh(\kappa t) \ , \\
\hat{x}^B(t) &= \hat{x}^B(0)\cosh(\kappa t) + \hat{x}^A(0)\sinh(\kappa t) \ , \\
\hat{y}^A(t) &= \hat{y}^A(0)\cosh(\kappa t) - \hat{y}^B(0)\sinh(\kappa t) \ , \\
\hat{y}^B(t) &= \hat{y}^B(0)\cosh(\kappa t) - \hat{y}^A(0)\sinh(\kappa t) \ ,
\end{aligned} \tag{10.10}$$

where κ is proportional to the strength of parametric interaction and the $t = 0$ operators represent inputs. As $r = \kappa t$ increases, $\hat{x}^A(t)$ becomes increasingly correlated with $\hat{x}^B(t)$, and $\hat{y}^A(t)$ becomes increasingly correlated with

$-\hat{y}^B(t)$, the correlation becoming perfect in the limit $\kappa t \to \infty$. With output fields $\hat{a}$ and $\hat{b}$ spatially separated, this is the situation [9] of the 1935 EPR paradox, and thus illustrates EPR correlations.

For imperfect correlation, the degree of correlation may still be sufficient to ensure EPR correlations [9]. The results for measurements $\hat{x}^A(t)$ and $\hat{x}^B(t)$ (or $\hat{y}^A(t)$ and $\hat{y}^B(t)$) can be compared, yielding an estimate of the error in inferring the result of measurement $\hat{x}^A(t)$ on mode $\hat{a}$, based on a measurement $\hat{x}^B(t)$ on mode $\hat{b}$. We calculate a $\delta_{0x} = \hat{x}^A - g\hat{x}^B$ for the inference of $\hat{x}^A$ as defined in the last section, and define similarly a δ_{0y} for the inference of $\hat{y}^A$. In this example we have $\delta_{0x} = \hat{x}^A(t) - g\hat{x}^B(t)$ and $\delta_{0y} = \hat{y}^A(t) + g\hat{y}^B(t)$, where the factor g may be modified to give the minimum error. Following the details as explained in the previous section, one can calculate the variances associated with the linear inference of $\hat{x}^A$ from $\gamma\hat{x}^B$, and $\hat{y}^A$ from $\gamma\hat{y}^B$:

$$\begin{aligned} \Delta_L^2 \hat{x}^A &= \langle \delta_x^2 \rangle - \langle \delta_x \rangle^2 , \\ \Delta_L^2 \hat{y}^A &= \langle \delta_y^2 \rangle - \langle \delta_y \rangle^2 . \end{aligned} \tag{10.11}$$

The minimum variance $\Delta^2_{L,\text{min}}\hat{x}^A$ (and $\Delta^2_{L,\text{min}}\hat{y}^A$) occurs for a particular value of g. Finding the turning point with g yields

$$\Delta^2_{L,\text{min}}\hat{x}^A = \frac{\Delta^2\hat{x}^A(T)\Delta^2\hat{x}^B(T) - [\langle \hat{x}^A(T), \hat{x}^B(T)\rangle]^2}{\Delta^2\hat{x}^B(T)} \tag{10.12}$$

with

$$g = \langle \hat{x}^A(T), \hat{x}^B(T)\rangle / \Delta^2\hat{x}^B(T) , \tag{10.13}$$

where $\langle x, y\rangle = \langle xy\rangle - \langle x\rangle\langle y\rangle$ and one deduces a $\Delta^2_{L,\text{min}}\hat{y}^A$ in similar fashion.

EPR correlations are demonstrated when the product $\Delta_L^2\hat{x}^A\Delta_L^2\hat{y}^A$ drops below the quantum limit given by $\Delta^2 x^A \Delta^2 y^A \geq 1$ [9]: that is we have the EPR criterion (10.1) as given in the previous section by (10.5)

$$\Delta_L^2\hat{x}^A\Delta_L^2\hat{y}^A < 1 . \tag{10.14}$$

For arbitrary coherent input states, we predict from [9] ($g = \tanh 2\kappa t$)

$$\Delta^2_{L,\text{min}}\hat{x}^A = \Delta^2_{L,\text{min}}\hat{y}^A = 1/\cosh(2\kappa t) . \tag{10.15}$$

The EPR correlations are predicted possible for all nonzero values of the two-mode squeeze parameter $r = \kappa t$.

The first experimental observations by Ou et al. [11] of two-mode squeezed light and EPR correlations used intracavity parametric interactions (parametric oscillation) to generate the correlated fields. The fields were then detected external to the cavity, being transmitted from the cavity through "leaky" cavity end-mirrors. To determine the correlation between the transmitted external fields various input/output relations are used, as determined by the boundary conditions at the mirrors. The theory predicting the existence of EPR correlated fields satisfying the EPR criterion for the nondegenerate parametric oscillator was presented in [10].

Fig. 10.1. The local realist will describe the correlated statistics through classical variables λ (that can be shared by the two subsystems at A and B). These classical variables depict possible states of the subsystems. The $\Delta_\lambda x^A$ and $\Delta_\lambda y^A$ are the standard deviations of the distributions $P^A_{\theta=0}(x^A|\lambda)$ and $P^A_{\theta=\pi/2}(y^A|\lambda)$, respectively, that denote the probability of a result x^A upon measurement of $\hat{x}^A$ ($\theta = 0$) or $\hat{y}^A$ ($\theta = \pi/2$), given that the system is in a particular state λ. Similar definitions exist for the subsystem at B. The system is EPR-correlated in a general sense when it can be demonstrated that in order to predict the observed statistics, $\Delta_\lambda x^A \Delta_\lambda y^A < 1$ (where $\Delta\hat{x}^A \Delta\hat{y}^A \geq 1$ is the uncertainty relation) for at least one of the possible states λ

10.3 Generalized EPR Criteria

While in Sect. 10.1.2 we have considered one useful criterion that is sufficient to prove the existence of EPR correlations, we wish in this section to construct the most general (weakest) criteria that could in any way imply EPR correlations.

To define generalized criteria sufficient to prove the situation of EPR correlations we need to reexamine the general philosophy of the argument, in order to define what it means to claim "EPR correlations". Our approach here is to define EPR correlations in the more general way as correlations that cannot be predicted by "hidden variables" for each of the localized subsystems unless the hidden variables define the predicted results for measurements so precisely so as to exclude the possibility of a localized quantum description. The hidden variables must be consistent with EPR's no action-at-a-distance (locality) premise.

We propose, in order to provide a general mathematical formulation for EPR constraints *that the mathematical formulation taken by a local realistic theory must be of the type considered by Bell* [7]. With this postulate, the definition of EPR criteria is then closely related to that of quantum inseparability (entanglement) which will be discussed in Sect. 10.4.

To expand our argument then we begin with the initial assumption of EPR, that local realism is valid. We must extend the concept of local realism to situations of less than ideal correlation. We do this by first considering the "local realist", someone who believes in the validity of local realism. For the situation of less than perfectly correlated subsystems, as discussed in Sect. 10.1, we propose that the local realist, in order to describe the predetermined fuzzy "elements of reality", would still insist to use (hidden?) classical variables to describe the possible states of each subsystem (see Fig. 10.1). The classical variables are used in way that are compatible with EPR's premise of locality (no action-at-a-distance). The (hidden) variables however no longer specify the predicted result for $\hat{x}^A$ or $\hat{y}^A$ measurements with absolute certainty, but have an arbitrary indefiniteness in their predictions.

The (hidden) variables are symbolized by a set λ_a and specify the actual state of the subsystem A. There is also a set λ_b for B, but such (hidden) variables may be shared between subsystems to describe correlated statistics, so the total set $\{\lambda_a, \lambda_b\}$, which describes the hidden variables at A and also at B, we denote in accordance with tradition by $\{\lambda\}$. There will be a probability distribution $\rho(\lambda)$ to give the likelihood of possible values of λ. Each such possible actual (hidden) variable state, denoted by a set of specific values for $\{\lambda_a\}$, will define with some yet unspecified indeterminacy, the result of a measurement of position, *and* the result of a measurement of momentum, should one of these measurements be performed. These more general hidden variable theories that are still consistent with EPR's locality (no action-at-a-distance) were considered in 1966 by Bell [7] and these issues are well discussed in the literature.

To make the connection with the original paradox, where measurements at A and B are perfectly correlated in the fashion of the EPR argument, the local realistic viewpoint would always demand that the underlying (hidden) variables define the result of measurement position or momentum with perfect accuracy. In the case of weakly correlated subsystems, the *(hidden) variable description for the local subsystem cannot be deduced to predetermine the results with such accuracy*, and this is why the local realist must allow for an indefiniteness in the prediction for the result x^A and a result y^A, given a set of values for $\{\lambda\}$. This indeterminacy we represent as a value $\Delta_\lambda x^A$ and $\Delta_\lambda y^A$, to give a standard deviation in the predicted result of a measurement $\hat{x}^A$ or $\hat{y}^A$ respectively, given that the subsystem is described by the (hidden) variable set of values $\{\lambda\}$.

Perhaps the key point in understanding why it is possible to have more general criteria than that given by the 1989 EPR criterion (10.1) is to understand the following point. The values $\Delta_\lambda x^A$, $\Delta_\lambda y^A$ to describe the indeterminacy associated with the (hidden) variables are not generally related in a direct way to the values $\Delta_i x, \Delta_j y$ of the conditional distributions: for example the $\Delta_\lambda x$, $\Delta_\lambda y$ can be zero, while the $\Delta_i x, \Delta_j y$ can be significant as a result of a reduced correlation between the hidden variables for the subsystems. It is for this reason that the local realist must allow for general degrees of indefiniteness for his/her hidden variable predictions. It is true however that *certain degrees of correlation between the measurement results at A and B can only be explained through the use of underlying (hidden) variables with a sufficiently definite prediction in their prediction for results of momentum and position measurements.* Such tests then provide the mechanism to prove more general EPR correlations.

A further important point is to understand the meaning and use of the term "hidden variables". Importantly the local realist's theory, which uses local (hidden) variable descriptions, is not necessarily incompatible, depending on the statistics, with a local quantum description (even a quantum superposition) of the particle A. For this reason the term "hidden" has been bracketed throughout the above explanation. We would reserve the term "hidden variable" for the situation where it can be shown that the variables are not representing local quantum states for each subsystem, and therefore are in fact "hidden", to complete quantum mechanics. Indeed *by the very philosophy of the EPR argument, this is the situation defining EPR correlations.*

It is pointed out however that it was (presumably) the original intention of EPR to justify the incompleteness of quantum mechanics, based on the validity of local realism, using physical arguments only, without making any assumptions regarding the mathematical form a local realistic theory must take. In this respect it would be regarded that the violation of the 1989 EPR criteria are necessary for the demonstration of the original EPR paradox itself. This is an important philosophical point, and for this reason we distinguish between the two types of EPR criteria: the first, a stronger set demand constraints on variances of conditional distributions; the second set are based on the more general approach described in this section and we call these generalised EPR criteria, being criteria for EPR-type correlations. It is also pointed out that the 1989 EPR criteria are mathematically stronger, and these points are explained in the reference [9].

To summarize we propose that the most general application of the EPR premise of local realism to correlated systems is to postulate the existence of a local (hidden) variable theory (we may also call this a local realistic theory) to describe the correlations. The local realist will describe statistics of spatially separated subsystems through a general local (hidden) variable theory. In order to demonstrate EPR correlations, or the EPR paradox in the most general way, the object then is *to demonstrate the failure of any local realistic theory satisfying*

$$\Delta^x_\lambda A\Delta_\lambda y^A \geq 1 \tag{10.16}$$

to describe the statistics of the correlated, spatially-separated fields. This would indicate that at least part of the time, if local realism as manifested through any actual local realistic theory, is to correctly describe the statistics, the local subsystem at A must be in a *hidden* variable state satisfying $\Delta^x_\lambda A\Delta_\lambda y^A < 1$, a description that cannot be given by a local quantum wavefunction for the subsystem alone. The "incompleteness of quantum mechanics" from the point of view of the local realist then follows, and this is the EPR paradox in a most generalized sense.

Various methods or criteria to test for the failure of the premise that the local (hidden) variable theory, where all the (hidden) variable states satisfy $\Delta_\lambda x^A\Delta_\lambda y^A \geq 1$, can describe certain quantum correlations, and therefore to demonstrate EPR correlations, are presented in Sect. 10.4. The 1989 EPR criterion is such a criterion, as it must be. However it is also shown in Sect. 10.4 that the demonstration of a two-mode squeezing in both "position" and "momentum" amplitudes will imply a demonstration of a weaker generalised EPR correlation. Such observations do not always imply a satisfaction of the 1989 strong EPR criterion.

10.4 Generalized EPR Correlations and Entanglement

Entanglement was introduced by Schrödinger [2] in his reply to EPR in 1935. An entangled state is one such as

$$|\gamma\rangle_A|\delta\rangle_B + |\eta\rangle_A|\epsilon\rangle_B \ , \tag{10.17}$$

which cannot be expressed in the factorized form $|\alpha\rangle|\beta\rangle$. Here $|\gamma\rangle_A$ and $|\eta\rangle_A\rangle$ are orthonormal vectors for system A, $|\delta\rangle_B$ and $|\epsilon\rangle_B\rangle$ are orthonormal vectors for system B. Schrödinger pointed to entanglement between spatially separated systems A and B being the crucial property of quantum states exhibiting nonclassical behaviour. Entanglement is a necessary criterion for states which violate a Bell inequality, and therefore show a violation of the EPR premise of local realism. This extreme property of quantum mechanics is fundamental to the field of quantum information.

In this section we make the formal connection between entanglement and the original EPR paradox (or correlations), by showing that the demonstration of EPR correlations, through an EPR criterion such as that defined in Sect. 10.1.2, will also imply entanglement; and that the demonstration of entanglement of spatially separated subsystems will imply, if not a violation of a Bell-type inequality, then a proof of EPR correlations in the most general sense. In order to formally link entanglement with the observation of EPR correlations we need to make use of the generalized EPR argument, since it is certainly true that not all entangled states will predict the ideal correlations of the original EPR paradox. The intimate connection between EPR criteria and quantum inseparbility has been discussed recently by P. Grangier and F. Grosshans [20] (see also Braunstein et al. [20]) in relation to providing a reliable signature for quantum teleportation.

The demonstration of entanglement may be defined as the measured experimental violation of any one of the set of criteria following necessarily from the assumption of separability, where a separable quantum state is defined as being expressible by a density matrix of the form

$$\rho = \sum_r P_r \rho_r^A \rho_r^B \ , \tag{10.18}$$

where $\sum_r P_r = 1$. Necessary conditions for separability for finite-dimensional systems were explored by Peres and Horodecki et al. [29]. Such necessary conditions using continuous variable measurements, based on the use of the uncertainty bound, have been derived recently by Duan et al. [19] and Simon [19].

10.4.1 1989 EPR Criterion as a Signature of Entanglement

We consider two observables $\hat{x}^A$ and $\hat{y}^A$ for subsystem A, where $\hat{x}^A$ and $\hat{y}^A$ assume the uncertainty relation $\Delta\hat{x}^A\Delta\hat{y}^A \geq 1$. We consider the measured error $\Delta_{\mathrm{inf}}\hat{x}^A$ in the prediction for the outcome of measurement $\hat{x}^A$ at A, based on a result at B; and a measured error $\Delta_{\mathrm{inf}}\hat{y}^A$ for the prediction of $\hat{y}^A$ at A.

We will first show that separability will always imply $\Delta_{\mathrm{inf}}\hat{x}^A\Delta_{\mathrm{inf}}\hat{y}^A \geq 1$, meaning that a satisfaction of the 1989 criterion as given by (10.1), which is sufficient to demonstrate EPR correlations, will always imply quantum entanglement. We summarize the approach taken in a recent article [26].

The conditional probability of result x^A for measurement $\hat{x}^A$ at A given a simultaneous measurement of $\hat{x}^B$ at B with result x_i^B is $P(x^A|x_i^B) = P(x^A, x_i^B)/P(x_i^B)$ where, assuming separability,

$$P(x^A, x_i^B) = \sum_r P_r P_r(x_i^B) P_r(x^A) \ . \tag{10.19}$$

Here $|x^A\rangle, |x^B\rangle$ are the eigenstates of $\hat{x}^A$,$\hat{x}^B$ respectively, and $P_r(x^A) = \langle x^A|\rho_r^A|x^A\rangle$, $P_r(x_i^B) = \langle x_i^B|\rho_r^B|x_i^B\rangle$. The mean μ_i of this conditional distribution is

$$\begin{aligned}\mu_i &= \sum_{x^A} x^A P(x^A|x_i^B) \\ &= \{\sum_r P_r P_r(x_i^B)\langle x^A\rangle_r\}/P(x_i^B) \ ,\end{aligned} \tag{10.20}$$

where $\langle x^A\rangle_r = \sum_{x^A} x^A P_r(x^A)$.

The variance $\Delta_i^2 x$ of the distribution $P(x^A|x_i^B)$ is

$$\Delta_i^2 x = \{\sum_r P_r P_r(x_i^B) \sum_{x^A} P_r(x^A)(x^A - \mu_i)^2\}/P(x_i^B) \ . \tag{10.21}$$

For each state r, the mean square deviation $\sum_{x^A} P_r(x^A)(x^A - d)^2$ is minimized with the choice $d = \langle x^A\rangle_r$ [27]. Therefore for the choice $d = \mu_i$,

$$\begin{aligned}\Delta_i^2 x &\geq \{\sum_r P_r P_r(x_i^B) \sum_{x^A} P_r(x^A)(x^A - \langle x^A\rangle_r)^2\}/P(x_i^B) \\ &= \{\sum_r P_r P_r(x_i^B)\sigma_r^2(x^A)\}/P(x_i^B) \ ,\end{aligned} \tag{10.22}$$

where $\sigma_r^2(x^A)$ is the variance of $P_r(x^A)$. Taking the average variance over the x_i^B we get (recall from Sect. 10.1.2 that we may also define an inference error

$\Delta^2_{\text{inf,est}}\hat{x}^A$ where the prediction or estimate for the result at A, based on the measurement at B, might not be optimal ie giving the minimum rms error)

$$\begin{aligned}\Delta^2_{\text{inf,est}}\hat{x}^A &\geq \Delta^2_{\text{inf}}\hat{x}^A \\ &\geq \sum_{x_i^B} P(x_i^B)\{\sum_r P_r P_r(x_i^B)\sigma_r^2(x^A)\}/P(x_i^B) \\ &= \sum_r P_r\sigma_r^2(x^A)\sum_{x_i^B} P_r(x_i^B) \\ &= \sum_r P_r\sigma_r^2(x^A)\ . \end{aligned} \tag{10.23}$$

Also $\Delta^2_{\text{inf}}\hat{y}^A \geq \sum_r P_r\sigma_r^2(y^A)$, where $\sigma_r^2(y^A)$ is the variance of $P_r(y^A) = \langle y^A|\rho_r^A|y^A\rangle$, with $|y^A\rangle$ being the eigenstate of $\hat{y}^A$. This implies (from the Cauchy–Schwarz inequality)

$$\begin{aligned}\Delta^2_{\text{inf}}\hat{x}^A\Delta^2_{\text{inf}}\hat{y}^A &\geq \{\sum_r P_r\sigma_r^2(x^A)\}\{\sum_r P_r\sigma_r^2(y^A)\} \\ &\geq |\sum_r P_r\sigma_r(x^A)\sigma_r(y^A)|^2\ . \end{aligned} \tag{10.24}$$

For any ρ_r^A it is constrained, by the uncertainty relation, that $\sigma_r(x^A)\sigma_r(y^A) \geq 1$. We therefore conclude that for a separable quantum state

$$\Delta_{\text{inf}}\hat{x}^A\Delta_{\text{inf}}\hat{y}^A \geq 1\ , \tag{10.25}$$

(and also $\Delta_{\text{inf,est}}\hat{x}^A\Delta_{\text{inf,est}}\hat{y}^A \geq 1$). The experimental observation then of the EPR criterion $\Delta_{\text{inf}}\hat{x}^A\Delta_{\text{inf}}\hat{y}^A < 1$, as given by (10.1), will imply inseparability, that is entanglement.

10.4.2 A Signature of Entanglement Defined Through Observation of Two-Mode Squeezing

In general the variances of type $\Delta^2_L\hat{x}^A$, defined in Sect. 10.1.3, will be greater than or equal to the minimum estimate $\Delta^2_{\text{inf}}\hat{x}^A$ (defined in Sect. 10.1.2) of the result at A which is evaluated from the conditionals. The 1989 EPR criterion, as described in Sects. 10.1 and 10.2, measured using the linear estimates then implies inseparability, for any g and d defined in Sect. 10.1.3. This implies that experiments [11–13] claiming the observation of EPR correlations, using the linear estimate in association with the 1989 EPR criterion, are automatically demonstrations of entanglement.

Given though that this linear inference method is that actually used in the two-mode squeezing EPR experiments, it is worthwhile to demonstrate this explicitly. Also through the algebra we arrive at a weaker criterion still sufficient to demonstrate entanglement. This criterion shows that the measurement of two-mode squeezing for two pairs of appropriate quadratures is also a demonstration of entanglement. We proceed as follows.

Separability of the quantum state (defined in (10.18)) will imply the following upon optimizing d but keeping g general. Here, g and d are defined in Sect. 10.1.3 and we use [27]

$$\begin{aligned}\Delta_L^2\hat{x}^A &\geq \langle\{\hat{x}^A - \langle\hat{x}^A\rangle - g(\hat{x}^B - \langle\hat{x}^B\rangle)\}^2\rangle \\ &= \sum_{x,y}\sum_r P_r\langle x|\langle y|\rho_r^A\rho_r^B\{\hat{x}^A - \langle\hat{x}^A\rangle - g(\hat{x}^B - \langle\hat{x}^B\rangle)\}^2|x^A\rangle|x^B\rangle \\ &= \sum_r P_r\langle(\hat{\delta}_0 - \langle\hat{\delta}_0\rangle)^2\rangle_r \geq \sum_r P_r\langle(\hat{\delta}_0 - \langle\hat{\delta}_0\rangle_r)^2\rangle_r \,. \end{aligned} \tag{10.26}$$

Here $\hat{\delta}_0 = \hat{x}^A - g\hat{x}^B$ and $\langle \hat{y^B}\rangle_r$ denotes the average for state r given by density operator $\rho_r = \rho_r^A\rho_r^B$. Since ρ_r factorizes, $\langle\hat{x}^A\hat{x}^B\rangle_r = \langle\hat{x}^A\rangle_r\langle\hat{x}^B\rangle_r$. We have

$$\begin{aligned}\Delta_L^2\hat{x}^A &\geq \sum_r P_r(\langle\hat{\delta}_0^2\rangle_r - \langle\hat{\delta}_0\rangle_r^2) \\ &= \sum_r P_r(\Delta_r^2\hat{x}^A + g^2\Delta_r^2\hat{x}^B)\,, \end{aligned} \tag{10.27}$$

where $\Delta_r^2\hat{x}^A = \sigma_r^2(x^A)$ and $\Delta_r^2\hat{x}^B = \langle(\hat{x}^B)^2\rangle_r - \langle\hat{x}^B\rangle_r^2$. Also

$$\Delta_L^2\hat{y}^A \geq \sum_r P_r(\Delta_r^2\hat{y}^A + h^2\Delta_r^2\hat{y}^B)\,, \tag{10.28}$$

where $\Delta_r^2\hat{y}^A = \sigma_r^2(y^A)$ and $\hat{y}^B$ is the measurement at B used to infer the result for $\hat{y}^A$ at A. It follows (take $\Delta\hat{x}^B\Delta\hat{y}^B \geq 1$)

$$\begin{aligned}\Delta_L^2\hat{x}^A\Delta_L^2\hat{y}^A \geq &\sum_r P_r\{\sigma_r^2(x^A) + g^2\Delta_r^2\hat{x}^B\} \\ &\times\sum_r P_r\{\sigma_r^2(y^A) + h^2\Delta_r^2\hat{y}^B\}\,. \end{aligned} \tag{10.29}$$

Separability implies

$$\begin{aligned}\Delta_L^2\hat{x}^A\Delta_L^2\hat{y}^A &\geq \langle\{\sigma_r^2(x^A) + g^2\Delta_r^2\hat{x}^B\}\rangle\langle\{\sigma_r^2(y^A) + h^2\Delta_r^2\hat{y}^B\}\rangle \\ &\geq \langle\left[\{\sigma_r^2(x^A) + g^2\Delta_r^2\hat{x}^B\}\{\sigma_r^2(y^A) + h^2\Delta_r^2\hat{y}^B\}\right]^{1/2}\rangle^2 \\ &\geq \langle\{1 + g^2h^2 + g^2\sigma_r^2(y^A)\Delta_r^2\hat{x}^B \\ &\quad + h^2\sigma_r^2(x^A)\Delta_r^2\hat{y}^B\}^{1/2}\rangle^2\,. \end{aligned} \tag{10.30}$$

We notice immediately that

$$\Delta_L^2\hat{x}^A\Delta_L^2\hat{y}^A \geq 1 + g^2h^2 \tag{10.31}$$

meaning that the experimental observation of the EPR criterion

$$\Delta_L\hat{x}^A\Delta_L\hat{y}^A < 1 \tag{10.32}$$

implies not only EPR correlations but entanglement (inseparability).

In fact inseparability may be deduced through a weaker criterion based on the observation of a two-mode squeezing. Upon taking $g = 1$, and, in the final line of (10.30), we see from the above that another criterion necessary for separability (and therefore sufficient to prove entanglement) follows. We have

$$\begin{aligned}\Delta_L^2\hat{x}^A\Delta_L^2\hat{y}^A &= \langle\{\hat{x}^A - \langle\hat{x}^A\rangle - (\hat{x}^B - \langle\hat{x}^B\rangle)\}^2\rangle\langle\\ &\quad\times\{\hat{y}^A - \langle\hat{y}^A\rangle + (\hat{y}^B - \langle\hat{y}^B\rangle)\}^2\rangle\\ &\geq \langle\{2 + \sigma_r^2(y^A)\Delta_r^2\hat{y}^B + 1/\sigma_r^2(y^A)\Delta_r^2\hat{y}^B\}^{1/2}\rangle^2 \geq 4\ ,\end{aligned} \tag{10.33}$$

where we have used the inequality $x + 1/x \geq 2$ and where $\hat{x}^B$, $\hat{y}^B$ are the observables for the measurements made for system B, to allow inference of the result $\hat{x}^A$ and $\hat{y}^A$ respectively at A. The observation of

$$\begin{aligned}&\Delta_L^2\hat{x}^A\Delta_L^2\hat{y}^A\\ &= \langle\{\hat{x}^A - \langle\hat{x}^A\rangle - (\hat{x}^B - \langle\hat{x}^B\rangle)\}^2\rangle\\ &\times\langle\{\hat{y}^A - \langle\hat{y}^A\rangle + (\hat{y}^B - \langle\hat{y}^B\rangle)\}^2\rangle < 4\end{aligned} \tag{10.34}$$

will then be sufficient to demonstrate inseparability, that is entanglement. This criterion sufficient for entanglement. The criterion may be rewritten in terms of the quadrature amplitudes as

$$\Delta^2(x^A - x^B)\Delta^2(y^A + y^B) < 4\ . \tag{10.35}$$

This criterion might be identified as a "two-mode squeezing" criterion sufficient to prove entanglement, since the individual criterion

$$\langle\{\hat{x}^A - \langle\hat{x}^A\rangle - (\hat{x}^B - \langle\hat{x}^B\rangle)\}^2\rangle < 2\ , \tag{10.36}$$

($\Delta^2(x^A - x^B) < 2$) will demonstrate two-mode squeezing. In this way we see that fields that are two-mode squeezed with respect to both $x^A - x^B$ and $y^A + y^B$, so as to satisfy (10.34), are necessarily entangled. This criteria sufficient to prove entanglement is weaker (more easily achieved) than those given by (10.1) and (10.32) which are based on conditional probabilities.

10.4.3 Generalized EPR Correlations Deduced Through Demonstrations of Entanglement

In order to provide a general link between the demonstration of EPR correlations and the demonstration of entanglement, I first consider the generalized EPR argument [9,26] as discussed in Sect. 10.1.

As discussed in the Sect. 10.3, in order to demonstrate EPR correlations in the most general sense, *we require to rule out all possible local (hidden) variable distributions that satisfy an auxiliary assumption* stating that the variance of the predicted result for $\hat{x}^A$ and the variance for a predicted result for $\hat{y}^A$ for a given local (hidden) variable state $\{\lambda\}$ will satisfy the quantum uncertainty relation. We will show in what sense this is equivalent to demonstrating entanglement. We argue as follows:

At A there is the choice to measure either $\hat{x}^A$ or $\hat{y}^A$, a choice denoted by different values, 0 and $\pi/2$ respectively, of the parameter θ. Similarly at B there is the choice, denoted by ϕ, to measure $\hat{x}^B$ or $\hat{y}^B$. The (hidden) variable values $\{\lambda\}$) determine the results, or probabilities for results, of measurements if performed. There will be a probability $P_\theta(x^A|\lambda)$ for the result x^A of a measurement $\theta = 0$ at A, given the hidden variable state $\{\lambda\}$; similarly a $P_\phi(y^B|\lambda)$ is defined. In accordance with EPR's locality (no action-at-a-distance) assumption, this probability distribution is independent of the experimenter's choice ϕ of simultaneous measurement at B. Assuming a general local hidden variable theory the joint probability $P(x^A, x^B)$ of obtaining an outcome x^A at A and x^B at B is of the form

$$P(x^A, x^B) = \int_\lambda \rho(\lambda)\ P_\theta(x^A|\lambda) P_\phi(x^B|\lambda)\ d\lambda\ . \tag{10.37}$$

Such general local hidden variable theories were considered by Bell [7].

Before proceeding, we point out that from the local hidden variable decomposition given by (10.37) various constraints may be derived. These constraints would be classified as Bell-type inequalities. Not all entangled quantum states, that is quantum states violating the separability condition (10.18), will violate a particular Bell inequality based on certain types of measurements. It is certainly true that the two-mode squeezed state as given in Sect. 10.2 will not violate a Bell inequality based on results of $\hat{x}^A$ and $\hat{y}^A$ quadrature (or "position"/ "momentum" measurements). That this is true is seen on realizing that the Wigner function is positive for the two-mode squeezed state [7], and that the probabilities for results of measurements can be expressed in the form (10.37) using the Wigner function as the probability distribution and the Wigner function c-numbers x^A and y^A as hidden variables. Yet the two-mode squeezed state is clearly EPR correlated and as a result demonstrates the EPR inconsistency of quantum mechanics with local realism that is discussed in the introduction. Namely, that in order for local realism to hold, the hidden variables $\tilde{x}^A, \tilde{y}^A$ that simultaneously describe the localized subsystem A are specified so precisely that they defy the uncertainty principle bound, and therefore cannot be represented as a quantum state.

For our purposes, we wish to derive a general set of EPR criteria, expressed (as are Bell inequalities and the original 1989 EPR criterion (10.1)) in terms of the measurable probabilities and expectation values which give a clear experimental procedure. To do this we assume the local hidden variable form (10.37) that is based on local realism, but also make auxiliary assumptions in addition to (10.37), and consider the whole set of criteria following from necessarily from these assumptions. Our auxiliary assumption is to assume that for each (hidden) variable state $\{\lambda\}$, the variances $(\Delta_\lambda x^A)^2$, $(\Delta_\lambda y^A)^2$ of $P^A_{\theta=0}(x^A|\lambda)$, $P^A_{\theta=\pi/2}(y^A|\lambda)$ respectively are restricted by the quantum "uncertainty principle"

$$\Delta_\lambda x^A \Delta_\lambda y^A \geq 1\ . \tag{10.38}$$

A similar restriction is placed on the variances of the $P_\phi^B(x^B|\lambda)$.

To derive criteria sufficient to demonstrate EPR correlations, we assume the general local realistic description (10.37) with the proviso (10.38) and consider the whole set of criteria (inequalities or other constraints) following necessarily from these assumptions. The *violation of any one of these constraints will be a demonstration of the EPR paradox in the most general sense*, as explained in Sect. 10.3, since then we conclude that in order for a local realistic theory (the generalized consequence of local realism) to predict the quantum correlations given by this violation, we would need to specify the (hidden) variables with sufficient definiteness in their prediction for results of measurements (failure of the auxiliary assumption (10.38)). This definiteness excludes the possibility of the local (hidden) variables that depict the subsystem A (or B) being alternatively represented by a quantum state, in view of the uncertainty relation. The classical variables λ become "hidden" variables.

We will now show that the generalized criteria sufficient to demonstrate EPR correlations are precisely the set of criteria sufficient to demonstrate quantum inseparability (that is entanglement), where one has, in the derivation of the inseparability criterion, made use of the bounds given by the quantum uncertainty principle. As an example because the expression (10.37) is separable in its form and assuming the auxiliary assumption (10.38), then following the logic of (10.19) to (10.25), we are able to derive $\Delta_{\text{inf}}\hat{x}^A \Delta_{\text{inf}}\hat{x}^A \geq 1$. In other words the 1989 EPR criterion is a proof of a failure of the assumptions (10.37) and (10.38), which is justification of the criterion being sufficient to demonstrate the EPR paradox.

Other criteria sufficient to demonstrate inseparability, derived assuming the quantum separability (10.18) and using the bound given by the uncertainty principle, have been derived by Duan et al. and Simon [19]. In this paper we have also derived, using the quantum uncertainty principle bound, such criteria sufficient to demonstrate inseparability (entanglement), as given by $\Delta_{\text{inf}}\hat{x}^A \Delta_{\text{inf}}\hat{y}^A < 1$ and equation (10.34) relating to the boundary for a two-mode squeezing. These constraints are also examples of generalized EPR criteria.

To summarize, we have shown that a measured demonstration of entanglement will, if not equivalent to a violation of a Bell-type inequality, at least be equivalent to a demonstration of the EPR gedanken experiment. This is however only provided measurements and spatial separations between subsystems A and B allow justification of the locality assumption. We would claim that the inconsistency of quantum mechanics with local realism, a demonstration of the EPR paradox, is objectively demonstrable through entanglement criteria.

10.4.4 Relationship to Stronger Entanglement Criteria Based on Bell-Type Inequalities

It is mentioned that where one shows a violation of a constraint following from the separable from (10.37) *without making an auxiliary assumption* (10.38), then we have a violation of a Bell inequality. One approach, taken for continuous variables by Gilchrist et al. [30], is to define a binary outcome domain, for example to define a result +1 if $x \geq 0$, and a result -1 for $x < 1$. Defining $P_{\theta,\phi}(+,-)$ as the joint probability for outcome +1 at A and -1 at B, etc., the derivation of various Bell inequalities follows in the usual fashion from (10.37) as is well discussed in the literature [7] and will not be re-examined here. For example if we define the expectation value $E(\theta,\phi) = P_{\theta,\phi}(+,+)+P_{\theta,\phi}(-,-)-P_{\theta,\phi}(+,-)-P_{\theta,\phi}(-,+)$ then the following Bell inequality follows from the local hidden variable assumption (10.37)

$$E(\theta,\phi) - E(\theta,\phi') + E(\theta',\phi) + E(\theta',\phi') \leq 2 \,. \tag{10.39}$$

Now the violation of this Bell-type inequality will imply entanglement since the assumption of quantum separability (10.18) will also through the same algebra as follows from (10.37) imply the Bell inequality. Thus the violation of this Bell inequality will imply entanglement. However because no auxiliary assumption based on bounds (such as the uncertainty relation) given by a quantum state was used in the derivation of this Bell constraint, the violation of this type of inseparability criterion will allow a stronger conclusion than demonstration of EPR paradox. The stronger conclusion is that local realism itself is invalid, and that these quantum states cannot be represented by any more "complete" theory still satisfying local realism.

As discussed previously, it is possible for the two-mode squeezed state to predict EPR correlations satisfying the EPR condition $\Delta_{\mathrm{inf}}\hat{x}^A \Delta_{\mathrm{inf}}\hat{x}^A < C$ from a local hidden variable theory, derived from the quantum Wigner function, that gives agreement with the quantum predictions for the direct x^A, y^A measurements. This implies that certain Bell inequalities for $\hat{x}^A, \hat{y}^A$ measurements will not be violated in this case. The joint probabilities for observed measurements can be expressed using the Wigner function, which is positive for the two-mode squeezed state, in the form of a local hidden variable theory (10.37), where the Wigner function c-numbers x^A and y^A take on the role of the position and momentum hidden variables. Because the Wigner function is positive, this function gives the probability distribution for the hidden variables x^A, y^A, x^B, y^B.

The failure of such an experiment to show a violation of a Bell-type inequality might lead to the conclusion that the experiment could not show an incompatibility of quantum mechanics with local realism (by this we really mean local hidden variables), and that there is no paradox. This is not the case. The local realistic hidden variable theory used to give the quantum predictions is, necessarily, not actually quantum theory, since it must incorporate a description $\{\lambda_a\}$ for a state of the system at A or B in which the x^A and

y^A are prespecified to a variance better than the uncertainty principle. The separable local hidden variable theory based on the Wigner function is not a separable (local) theory in quantum mechanics since these simultaneously well-defined x^A and y^A are not quantum states.

10.4.5 Entanglement is Implied Through Demonstrations of EPR Correlations

It follows through the very meaning of the EPR paradox that we must use an entangled source if we are to demonstrate EPR, and that therefore demonstration of EPR correlations must imply entanglement. A separable (nonentangled source) as given by (10.18) has the interpretation that it is always in one of the factorizable states $\rho_r^A \rho_r^B$ with probability P_r; in each case the subsystem at A being describable by the quantum state ρ_r^A and the subsystem B being described by the quantum state ρ_r^B. These states ρ_r^A and ρ_r^B represent local descriptions for the subsystems at A and B respectively and may be symbolized by the classical variable λ to define a state for which the position and momentum measurements are sufficiently indefinite so that the uncertainty bound (10.16) is satisfied. The predicted statistics must be compatible with this local fuzzy description, and in being so cannot satisfy the general EPR criterion discussed in relation to equation (10.16) in Sect. 10.1.

Also crucial to Schrödinger's 1935 essay is the concept of a a Schrödinger cat state, generally defined as a quantum superposition of states macroscopically distinct. Of special interest is the macroscopically entangled state, a state of type $|\gamma\rangle_A|\delta\rangle_B + |\eta\rangle_A|\epsilon\rangle_B$ where $|\gamma\rangle_A|\delta\rangle_B$ and $|\eta\rangle_A|\epsilon\rangle_B$ are macroscopically distinct.

Macroscopic local realism has been defined as follows [32]: macroscopic local realism is based on a "macroscopic locality", which states that measurements at a location B cannot instantaneously induce macroscopic changes (for example the "dead" to "alive" state of a cat, or a change between macroscopically different photon numbers) in a second system A spatially separated from B. Macroscopic local realism also incorporates a "macroscopic realism", since it implies elements of reality with (up to) a macroscopic indeterminacy. The demonstration of macroscopic entanglement is then equivalent to an EPR argument based on macroscopic local realism [32]: the conclusions in line with the Einstein–Podolsky–Rosen argument are that macroscopic local realism can only be retained through certain theories alternative to quantum mechanics. In this way, the inconsistency of the "completeness of quantum mechanics" with "macroscopic local realism" is demonstrated. The interested reader is referred to further works [32].

Bell-type tests may then be used to rule out the possibility of these further alternatives, that would enable quantum mechanics to be consistent with macroscopic local realism, by "completing" quantum mechanics with hidden variables. With the demonstrations of macroscopic entanglement (Schrödinger's cat) using the continuous variable EPR criteria, it is not ruled out that one could adopt a local hidden variable theory which would enable quantum mechanics to be entirely compatible with the assumption of macroscopic local realism. Indeed where the source is the two-mode squeezed state, this is certainly possible.

In conclusion to this section, we point out the existence of special quantum states where it is possible to predict a violation of a Bell inequality using direct quadrature phase amplitude (continuous variable) measurements only [30]. Such states then demonstrate an outright incompatibility of quantum mechanics with macroscopic local realism, and the experimental verification would prove a failure of macroscopic local realism. In this way the existence of a Schrödinger's cat defying a macroscopic reality is proved. The interested reader is again referred to further works [33].

10.5 Application of EPR Two-Mode Squeezed States to Quantum Cryptography

Einstein–Podolsky–Rosen or entangled nature of the outputs of the two-mode squeezed state leads automatically to applications to the field of quantum information. There is now a significant interest in this field of research. The specific use of quantum effects to the field of cryptography was pioneered by Bennett and Brassard [23]. Later a proposal was put forward by Ekert to use the entangled EPR state considered by Bell and Bohm, predicted to violate a Bell inequality.

In the proposal of Ekert [23], two entangled fields or particles emitted by a common source propagate in two different directions towards two spatially separated observers (commonly called Alice and Bob). At location A Alice has the choice to make a certain measurement, which traditionally corresponds to the choice of angle of Stern–Gerlach apparatus, but for practical experiments which use photons will be a choice of angle θ for a polariser measurement. At location B Bob has a similar choice of measurement, the choice being denoted by ϕ. Where $\theta = \phi$ say, the outcome of the measurements made by Alice and Bob will always be the same, that is, the results are correlated. For more general angle choices, the measurements made at location A by Alice and at location B by Bob can test for the violation of a Bell inequality as predicted for the particular quantum state used. The idea is that Alice and Bob make measurements with the choice of measurement angle changing randomly between certain choices θ for Alice and ϕ for Bob, which enable the Bell inequality test. Where one uses the Bell–Bohm state there are only two outcomes $+1$ and -1 (this represents a bit value) for each measurement, corresponding to a photon being transmitted or not through the polariser (the "up" or "down" position of the Stern-Gerlach spin measurement apparatus). Alice and Bob record their measurement angle and result for each measurement.

Later Alice randomly selects a certain subset of her measurements and communicate through a public channel (meaning a communication line with no extra security) to Bob the measurement angle and bit value obtained for the selected subset. Bob makes a comparison with his own measurement angle and bit data, and is able to check for the predicted violation of a Bell inequality based on both his and Alice's results. If the violation of the Bell inequality is maintained at the level predicted by quantum mechanics, it is possible for him to conclude that there was no intervention (measurements) made on either communication channel, that transmitted to Alice or that transmitted to Bob. In this way he concludes that no eavesdropping occurred, and that the communication was secure. Alice may then communicate her chosen measurement angles to Bob, without communicating the associated bit values. Where the angle was chosen the same by both Bob and Alice they share a common bit value, and in this way they form a common sequence of secure bit values, called a key. The details of the proof of security for the proposal of Ekert will not be given here.

The use of quantum entanglement to provide keys for secure messages known only to Alice and Bob is intriguing and after the initial proposal by Bennett and Brassard interest in this field escalated. Initial interest, both theoretical and experimental, was restricted to situations involving measurements on single photons, along the lines of the traditional Bell inequality experiments, which use polariser measurements with one photon incident. A difficulty with this type of situation is that detector inefficiencies are relatively high, meaning that a fair fraction of the output photon pairs simply cannot be detected.

Quantum cryptography using squeezed EPR states where quadrature phase amplitude measurements have continuous variable outcomes was explicitly put forward by Ralph [21], and Hillery [21], who proposed different schemes for quantum cryptography, using squeezed states. These schemes were similar in many respects to the original proposal of Bennett and Brassard. A scheme more similar to that of Ekert but using EPR correlations instead of Bell inequalities was presented by this author [21], and there has been further theoretical proposals and progress by Navez et al. [21] and others, and very recent experimental interest by Pereira et al. [21], and Silberhorn et al. [21]. The readers are also referred to related work on quantum teleportation [20] which will not be discussed in this paper. Here I give a brief discussion of the essential elements of these schemes which use continuous variable entanglement for cryptography.

The EPR nature of the correlations of the two-mode squeezed state enables a proposal similar to that of Ekert, and we summarize the essence of this proposal here. The two-mode squeezed state emits two fields, one which propagates towards Alice at location A, and the other to Bob at location B (see Fig. 10.2). Alice selects to measure randomly either quadrature phase amplitude $\hat{x}^A = x^A$, corresponding to angle $\theta = 0$, or quadrature phase am-

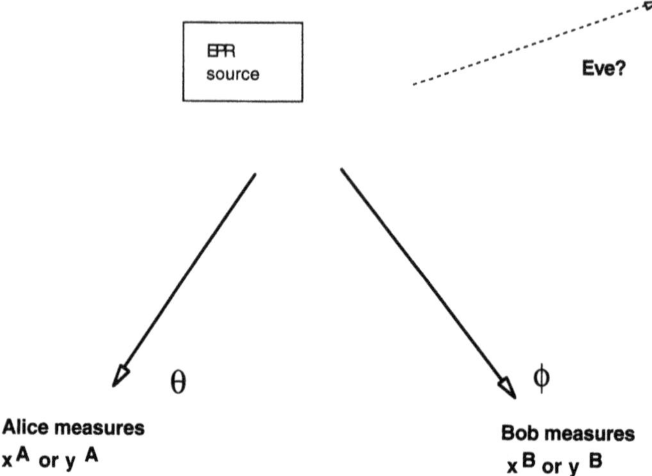

Fig. 10.2. Correlated EPR fields may be generated using for example the two-mode squeezed state. One field is sent to Alice; the other is sent to Bob. The correlated results of measurement ($\hat{x}^A, \hat{x}^B$ and $\hat{y}^A, \hat{y}^B$ are correlated) may form the shared bit values for a key. If Alice and Bob can prove that their fields are EPR-correlated according to the 1989 criterion (10.1), then it is impossible for any potential eavesdropper Eve at a third location to have an exact accurate replica of the results of the correlated measurements shared by Alice and Bob

plitude $\hat{y}^A = y^A$, corresponding to angle choice $\theta = \pi/2$, say. Similarly Bob will measure randomly either quadrature phase amplitude x^B, corresponding to angle $\phi = 0$, or quadrature phase amplitude y^B, corresponding to angle choice $\phi = \pi/2$. As discussed in Sect. 10.2, for the choices $\theta = \phi = 0$ as the two-mode squeeze parameter r becomes large, the results of measurement x^A for Alice and x^B for Bob will be identical. For the choices $\theta = \phi = \pi/2$ the results are also correlated: we have $y^A = -y^B$. In the style of the original quantum cryptographic proposals, Alice communicates after the measurements through a public channel her choice of measurement angle θ and the result x^A for the measurement, for a subensemble, randomly selected after the detections. Again Bob may check against his measurements and compare his value of x^B for measurements which should be correlated with Alice's values x^A. The observation through measurement by Alice and Bob of the EPR correlation is used to check for security, that is, that there could have been no intervention on the communication channel. Once security is established, the measurement angle for the remaining subset is shared, and where the choice of angle is to predict correlation between results, the sequence of common shared values becomes the bit values for the key.

We now examine the extent and proof of security for a finite r. We would operate in a regime of high r where there is good correlation between the results for Alice and Bob, for appropriately chosen angles. Alice and Bob use their shared subensemble to calculate the conditional probabilities $P(x^A|x_i^B)$ and $P(y^A|y_j^B)$ discussed in Sect. 10.1, used to demonstrate the existence of EPR correlations. Here x^A is Alice's result for $\hat{x}^A$ and x_i^B is the result for Bob's measurement correlated with $\hat{x}^A$. Similarly y^A is Alice's result for $\hat{y}^A$ and y_j^B is the result for Bob's measurement correlated with $\hat{y}^A$. If Bob measures x_i^B for his measurement correlated with Alice's $\hat{x}^A$ there is an error $\Delta_i\hat{x}^A$ (the standard deviation of $P(x^A|x_i^B)$), for his estimate of Alice's result x^A. Similarly there is an error $\Delta_j\hat{y}^A$ (the standard deviation of $P(y^A|y_j^B)$, in his estimate of Alice's result y^A for $\hat{y}^A$, based on his result y_j^B for the correlated measurement. For our particular two-mode squeezed state

$$\Delta_i\hat{x}^A = \Delta_j\hat{y}^A = 1/\sqrt{\cosh(2\kappa t)} \tag{10.40}$$

for all i, j.

It is understood that the choice of measurement angles θ, ϕ is not made until after the fields have been detected. Therefore an eavesdropper (Eve) cannot anticipate the angle choice and must make her own random selection of measurement angle. The uncertainty principle produces bounds on the extent to which she can measure $\hat{x}^A$ and/or $\hat{y}^A$ and leave the measured or regenerated state (transmitted on to Bob or Alice) unchanged. It is this change or feedback in the field transmitted on to Bob, or Alice, which induces a lack of correlation in the future selected subensemble, which can be measured by Alice and Bob, and exposes Eve.

Here we give a simple demonstration of the impossibility of Eve accessing with perfect accuracy Alice's results x, y, while still leaving the EPR nature of the correlation between Bob's and Alice's results unaltered. It is assumed as always that the choice of angles (whether to measure $\hat{x}^A$ or $\hat{y}^A$) of Bob's and Alice's measurements are randomly and independently chosen after the transmission of the fields to Alice and Bob. Let us assume the hypothesis which we wish to disprove, that Eve can perform a measurement (or otherwise has information available to enable her) to deduce with no uncertainty the value of Alice's result, whether it be an x^A or y^A measurement that Alice makes. If so, then Bob, if the EPR correlations with Alice's results are maintained, could measure $\hat{x}^A$ say, thus inferring Alice's result for $\hat{x}^A$ to absolute precision, and Eve could measure $\hat{y}^A$ to infer Alice's result for $\hat{y}^A$ to absolute precision. This would constitute a simultaneous measurement of Alice's x^A and y^A to a combined uncertainty in defiance of the uncertainty relation and therefore the hypothetical situation does not exist. (It is important to distinguish (see Sect. 10.1) an actual predicted violation of the uncertainty relation, and the "inferred violation" which occurs through the EPR argument for the elements of reality which exist only if one assumes local realism and which does not translate to a contradiction with the predictions of quantum mechanics.) Full details of this argument are presented in a recent preprint [31].

To summarize the essential procedure of the proof, Bob, if the perfect EPR correlations with Alice's results are maintained, could select to measure $\hat{x}^B$ say, thus inferring Alice's result for $\hat{x}^A$ to absolute precision. Eve, if also EPR-correlated with Alice's measurements so that she can gain the shared information, has the ability to infer Alice's result for $\hat{y}^A$ to absolute precision. This would constitute a simultaneous measurement (state reduction) of Alice's $\hat{x}^A$ and $\hat{y}^A$ to a combined uncertainty in defiance of the uncertainty relation and therefore the situation is never predicted within quantum mechanics. Therefore the measurement by Alice and Bob of perfect EPR correlations would indicate the impossibility of the existence of an Eve possessing a third perfect copy of Alice's and Bob's shared data.

The determination by Alice and Bob of an optimal EPR correlation

$$\Delta_i x = 0 \, , \qquad \Delta_j y = 0 \tag{10.41}$$

for all i, j, in their detected fields implies security that any hypothetical Eve cannot obtain the key sequence. One may view this result as a statement that if Eve has intercepted the channels in any way to obtain the information with any degree of accuracy (to give a noninfinite variance in her estimate of the values), the EPR correlation would necessarily have been reduced to give a nonzero result for at least one of the $\Delta_i x$, $\Delta_j y$. It has also been proved that there is no alternative set of quantum fields (source) available to Eve to allow her to obtain the information with any degree of accuracy and still retain the optimal EPR correlation between Bob and Alice.

The practical situation where a reduced EPR correlation is observed between Alice and Bob is more subtle. An extension of the above discussion of security for weaker EPR correlations has been presented recently in [31]. To summarize, Alice and Bob would expect a certain degree of EPR correlation (set of results for their $\Delta_i x, \Delta_j y$) based on measurements performed on their EPR source (as would be predicted from the value of r and perhaps an expected degree of loss on transmission). If their measured EPR correlation is noticeably reduced on one particular transmission then this could indicate Eve's presence (through extra loss) on Bob's channel and certainly it would be wise to retransmit. However *even in such a situation where eavesdropping may have occurred,* (or in the situation where the EPR correlation is apparently as expected but where it cannot be excluded that Eve has substituted a different quantum source), it is shown in [31] that the use of the uncertainty relation *allows a deduction of the maximum information on the key available to Eve.* The strategy would be to derive limits on Eve's knowledge of the key sequence, based only on the reliable measurement of a certain amount of EPR correlation between Alice and Bob.

References

1. A. Einstein, B. Podolsky and N. Rosen, Phys. Rev. **47**, 777 (1935)

2. E. Schrödinger, Naturwissenschaften **23**, 807 (1935)
3. M.D. Levenson, R.M. Shelby, M.D. Reid and D.F. Walls, Phys. Rev. Lett. **57**, 2473 (1986)
4. A. Heidmann, R.J. Horowicz, S. Reynaud, E. Giacabino, C. Fabre and G. Camy, Phys. Rev. Lett. **59**, 2555 (1987)
5. M.D. Reid and P.D. Drummond, Phys. Rev. Lett. **60**, 2731 (1988); P. Grangier, M.J. Potasek and B. Yurke, Phys. Rev. A **38**, 3132 (1988); B.J. Oliver and C.R. Stroud, Phys. Lett. A **135**, 407 (1989)
6. D. Bohm, *Quantum Theory* (Prentice-Hall, Englewood Cliffs, New Jork 1951)
7. J.S. Bell, Physics, **1**, 195 (1965); J.F. Clauser and A. Shimony, Rep. Prog. Phys. **41**, 1881 (1978); D.M. Greenberger, M. Horne and A. Zeilinger, in *Bell's Theorem, Quantum Theory and Conceptions of the Universe*, M. Kafatos, ed., (Kluwer, Dordrecht, The Netherlands 1989), p. 69; N.D. Mermin, Physics Today **43**(6), 9 (1990)
8. A. Aspect, P. Grangier and G. Roger, Phys. Rev. Lett. **49**, 91 (1982); A. Aspect, J. Dalibard and G. Roger, Phys. Rev. Lett. **49**, 1804 (1982); Y.H. Shih and C.O. Alley, Phys. Rev. Lett. **61**, 2921 (1988); Z.Y. Ou and L. Mandel, Phys. Rev. Lett. **61**, 50 (1988); J.G. Rarity and P.R. Tapster, Phys. Rev. Lett. **64**, 2495 (1990); J. Brendel, E. Mohler and W. Martienssen, Europhys. Lett. **20**, 575 (1992); P.G. Kwiat, A.M. Steinberg and R.Y. Chiao, Phys. Rev. A **47**, 2472 (1993); T.E. Kiess, Y.H. Shih, A.V. Sergienko and C.O. Alley, Phys. Rev. Lett. **71**, 3893 (1993); P.G. Kwiat, K. Mattle, H. Weinfurter, A. Zeilinger, A.V. Sergienko and Y. Shih, Phys. Rev. Lett. **75**, 4337 (1995); D.V. Strekalov, T.B. Pittman, A.V. Sergienko, Y.H. Shih and P.G. Kwiat, Phys. Rev. A **54**, 1 (1996); W. Gregor, T. Jennewein and A. Zeilinger, Phys. Rev. Lett. **81**, 5039 (1998); A. Zeilinger, Rev. Mod. Phys. **71**, 5288 (1998)
9. M.D. Reid, Phys. Rev. A **40**, 913 (1989); M.D. Reid, quant-ph/0112038
10. M.D. Reid and P.D. Drummond, Phys. Rev. A **40**, 4493 (1989); P.D. Drummond and M.D. Reid, Phys. Rev. A **41**, 3930 (1990)
11. Z.Y. Ou, S.F. Pereira, H.J. Kimble and K.C. Peng, Phys. Rev. Lett. **68**, 3663 (1992)
12. Y. Zhang, H. Wang, X. Li, J. Jing, C. Xie and K. Peng, Phys. Rev. A **62**, 023813 (2000)
13. Ch. Silberhorn, P.K. Lam, O. Weiss, F. Koenig, N. Korolkova and G. Leuchs, Phys. Rev. Lett. **86**, 4267 (2001)
14. K. Tara and G.S. Agarwal, Phys. Rev. A **50**, 2870 (1994)
15. V. Giovannetti, S. Mancini and P. Tombesi, Euro. Phys. Lett. **54**, 559 (2001)
16. W.P. Bowen, R. Schnabel, P.K. Lam and T.C. Ralph, Phys. Rev. Lett. **90**, 043601 (2003)
17. N. Korolkova, C. Silberhorn, O. Glockl, S. Lorenz, C. Marquardt and G. Leuchs, Eur. Phys. J. D **18**, 229 (2002)
18. C. Schori, J.L. Sorensen and E.S. Polzik, Phys. Rev. A **66**, 033802 (2002)
19. L. Duan, G. Giedke, J.I. Cirac and P. Zoller, Phys. Rev. Lett. **84**, 2722 (2000); R. Simon, Phys. Rev. Lett. **84**, 2726 (2000). See also S.L. Braunstein, C.A. Fuchs, H.J. Kimble and P. van Loock, Phys. Rev. A **64**, 022321 (2001)
20. A. Furasawa, J. Sorensen, S. Braunstein, C. Fuchs, H.J. Kimble and E. Polzik, Science **282**, 706 (1998); W.P. Bowen, N. Treps, B.C. Buchler, R. Schnabel, T.C. Ralph, H.A. Bachor, T. Symul and P.K. Lam, Phys. Rev. A **67**, 032302 (2003); L. Vaidmann, Phys. Rev. A **49**, 1473 (1994); S. Braunstein and H.J.

Kimble, Phys. Rev. Lett. **80**, 869 (1998); A. Kuzmich and E.S. Polzik, Phys. Rev. Lett. **85**, 5639 (2000); I.V. Sokolov, M.I. Kolobov, A. Gatti, L.A. Lugiato, Optics Commun. **193**, 175 (2001); F. Grosshans and P. Grangier, Phys. Rev. A **64**, 1 (2001); S.L. Braunstein, C.A. Fuchs, H.J. Kimble and P. van Loock, Phys. Rev. A **64**, 022321 (2001)

21. T.C. Ralph, Phys. Rev. A **61**, 303 (1999); Phys. Rev. A **62**, 062306 (2000); M. Hillery, Phys. Rev. A **61**, 2309 (1999); M.D. Reid, Phys. Rev. A **62**, 062308 (2000); L.M. Duan, J.I. Cirac and P. Zoller, Phys. Rev. Lett. **85**, 5643 (2000); N.J. Cerf, M. Levy and G. Van Assche, Phys. Rev. A **63**, 052311 (2001); S.F. Pereira, Z.Y. Ou and H.J. Kimble, Phys. Rev. A **62**, 042311 (2000); P. Navez, E. Brambilla, A. Gatti and L.A. Lugiato, Phys. Rev. A **65**, 013813 (2002); C. Silberhorn, T.C. Ralph, N. Lutkenhaus and G. Leuchs, Phys. Rev. Lett. **89**, 167901 (2002); C. Silberhorn, N. Korolkova and G. Leuchs, Phys. Rev. Lett. **88**, 167902 (2002); F. Grosshans and P. Grangier, Phys. Rev. Lett. **88**, 057902 (2002)
22. C.H. Bennett, G. Brassard, C. Crepeau, R. Jozsa, A. Peres and W.K. Wootters, Phys. Rev. Lett. **70**, 1895 (1993); D. Bouwmeester, J.W. Pan, K. Mattle, M. Eibl, H. Weinfurter and A. Zeilinger, Nature **390**, 575 (1997); D. Boshi, S. Branca, F. De Martini, L. Hardy and S. Popescu, Phys. Rev. Lett. **80**, 1121 (1998)
23. C.H. Bennett and G. Brassard, in Proceedings of IEEE International Conference on Computers, Systems and Signal Processing, Bangalore, India (IEEE, New York 1984), p. 175; A.K. Ekert, Phys. Rev. Lett, **67**, 661 (1991); C.H. Bennett, G. Brassard and N.D. Mermin, Phys. Rev. Lett. **68**, 557 (1992); A.K. Ekert, J.G. Rarity, P.R. Tapster and G.M. Palma, Phys. Rev. Lett **69**, 1293 (1992); A.K. Ekert, B. Huttner, G.M. Palma and A. Peres, Phys. Rev. A **50**, 1047 (1994); C.H. Bennett, F. Bessette, G. Brassard, L. Savail and J. Smolin, J. Cryptology **5**, 3 (1992); A. Muller, J. Breguet and N. Gisin, Europhys. Lett. **23**, 383 (1993); P.D. Townsend, Electron Lett. **30**, 809 (1994); J.D. Franson and H. Ilves, Appl. Opt. **33**, 2949 (1994); W.T. Buttler, R.J. Hughes and C.M. Simmons, Phys. Rev. Lett. **81**, 3283 (1998)
24. Q.A. Turchette, C.S. Wood, B.E. King, C.J. Myatt, D. Liebfried, W.M. Itano, C. Monroe and D.J. Wineland, Phys. Rev. Lett. **81**, 3631 (1998)
25. M.A. Rowe, D. Kielpinski, V. Meyer, C.A. Sackett, W.M. Itano, C. Monroe and D.J. Wineland, Nature **409**, 791 (2001)
26. M.D. Reid, quant-ph/0103142
27. This choice $d = \mu_i$ will minimize the rms error $\sqrt{\langle (x/y_i - d)^2 \rangle}$
28. C.M. Caves and B.L. Schumaker, Phys. Rev. A **31**, 3068 (1985); B.L. Schumaker, Phys. Rev. A **31**, 3093 (1985)
29. A. Peres, Phys. Rev. Lett. **77**, 1413 (1996); M. Horodecki, P. Horodecki and R. Horodecki, Phys. Lett. A **223**, 1 (1996)
30. A. Gilchrist, P. Deuar and M.D. Reid, Phys. Rev. Lett. **80**, 3169 (1998); Phys. Rev. A **60**, 4259 (1999); B. Yurke, M. Hillery and D. Stoler, Phys. Rev. A **60**, 3444 (1999); M. Hillery, B. Yurke and D. Stoler, Phys. Rev. A **63**, 062111 (2001); W.J. Munro and G.J. Milburn, Phys. Rev. Lett. **81**, 4285 (1998)
31. M.D. Reid, quant-ph/0112039
32. M.D. Reid, Europhys. Lett. **36**, 1 (1996); quant-ph/0101050; Phys. Rev. Lett. **84**, 2765 (2000); Phys. Rev. A **62**, 022110 (2000)
33. M.D. Reid, quant-ph/0101052

Bibliography

Useful texbooks on theory, experiments, generation and applications of quantum squeezing.

H.-A. Bachor, *A Guide to Experiments in Quantum Optics* (Wiley, New York 1998)

C.W. Gardiner, *Handbook of Stochastic Methods* (Springer, Berlin Heidelberg New York 1990)

C.W. Gardiner and P. Zoller, *Quantum Noise* (Springer, Berlin Heidelberg New York 2000)

L. Mandel and E. Wolf, *Optical Coherence and Quantum Optics* (Cambridge University Press Cambridge 1995)

W. Schleich, *Quantum Optics in Phase Space* (Wiley, New York 2000)

M.O. Scully and M.S. Zubairy, *Quantum Optics* (Cambridge University Press Cambridge 1997)

W.Vogel, D.-G. Welsch and S. Wallentowitz, *Quantum Optics: An Introduction* (Wiley, New York 2001)

D.F. Walls and G.J. Milburn, *Quantum Optics* (Springer, Berlin Heidelberg New York 1994)

Index

Springer Series on

ATOMS + PLASMAS

Editors: G. Ecker P. Lambropoulos I.I. Sobel'man H. Walter
Founding Editor: H.K.V. Lotsch

1 **Polarized Electrons**
2nd Edition
By J. Kessler

2 **Multiphoton Processes**
Editors: P. Lambropoulos and S.J. Smith

3 **Atomic Many-Body Theory**
2nd Edition
By I. Lindgren and J. Morrison

4 **Elementary Processes in Hydrogen-Helium Plasmas**
Cross Sections and Reaction Rate Coefficients
By R.K. Janev, W.D. Langer, K. Evans Jr., and D.E. Post Jr.

5 **Pulsed Electrical Discharge in Vacuum**
By G.A. Mesyats and D.I. Proskurovsky

6 **Atomic and Molecular Spectroscopy**
Basic Aspects and Practical Applications
3rd Edition
By S. Svanberg

7 **Interference of Atomic States**
By E.B. Alexandrov, M.P. Chaika, and G.I. Khvostenko

8 **Plasma Physics**
Basic Theory with Fusion Applications
3rd Edition
By K. Nishikawa and M. Wakatani

9 **Plasma Spectroscopy**
The Influence of Microwave and Laser Fields
By E. Oks

10 **Film Deposition by Plasma Techniques**
By M. Konuma

11 **Resonance Phenomena in Electron–Atom Collisions**
By V.I. Lengyel, V.T. Navrotsky, and E.P. Sabad

12 **Atomic Spectra and Radiative Transitions**
2nd Edition
By I.I. Sobel'man

13 **Multiphoton Processes in Atoms**
2nd Edition
By N.B. Delone and V.P. Krainov

14 **Atoms in Plasmas**
By V.S. Lisitsa

15 **Excitation of Atoms and Broadening of Spectral Lines**
2nd Edition, By I.I. Sobel'man, L. Vainshtein, and E. Yukov

16 **Reference Data on Multicharged Ions**
By V.G. Pal'chikov and V.P. Shevelko

17 **Lectures on Non-linear Plasma Kinetics**
By V.N. Tsytovich

18 **Atoms and Their Spectroscopic Properties**
By V.P. Shevelko

19 **X-Ray Radiation of Highly Charged Ions**
By H.F. Beyer, H.-J. Kluge, and V.P. Shevelko

20 **Electron Emission in Heavy Ion–Atom Collision**
By N. Stolterfoht, R.D. DuBois, and R.D. Rivarola

21 **Molecules and Their Spectroscopic Properties**
By S.V. Khristenko, A.I. Maslov, and V.P. Shevelko

22 **Physics of Highly Excited Atoms and Ions**
By V.S. Lebedev and I.L. Beigman

23 **Atomic Multielectron Processes**
By V.P. Shevelko and H. Tawara

24 **Guided-Wave-Produced Plasmas**
By Yu.M. Aliev, H. Schlüter, and A. Shivarova

25 **Quantum Statistics of Strongly Coupled Plasmas**
By D. Kremp, W. Kraeft, and M. Schlanges

26 **Atomic Physics with Heavy Ions**
By H.F. Beyer and V.P. Shevelko

Springer Series on

ATOMIC, OPTICAL, AND PLASMA PHYSICS

GPSR Compliance
The European Union's (EU) General Product Safety Regulation (GPSR) is a set of rules that requires consumer products to be safe and our obligations to ensure this.

If you have any concerns about our products, you can contact us on

ProductSafety@springernature.com

In case Publisher is established outside the EU, the EU authorized representative is:

Springer Nature Customer Service Center GmbH
Europaplatz 3
69115 Heidelberg, Germany

www.ingramcontent.com/pod-product-compliance
Ingram Content Group UK Ltd.
Pitfield, Milton Keynes, MK11 3LW, UK
UKHW021857190726
13853UKWH00003B/1315

* 9 7 8 3 6 6 2 0 9 6 4 6 8 *